Handbook of Experimental Pharmacology

Volume 187

Kian Fan Chung • John Widdicombe
Editors

Pharmacology and Therapeutics of Cough

Contributors

M.G. Belvisi, F. Bertram, S.S. Birring, D.C. Bolser, A.C. Bonham, J. Bric, B.J. Canning, M.J. Carr, C.-Y. Chen, Y.-L. Chou, K.F. Chung, P.W. Davenport, P.V. Dicpinigaitis, R. Eccles, E. Ernst, R. Gatti, P. Geppetti, Q. Gu, D.J. Hele, J.P. Joad, L.-Y. Lee, S. Materazzi, S.B. Mazzone, M.A. McAlexander, I.M. McFadzean, L. McGarvey, A.H. Morice, R. Nassini, C.P. Page, D.I. Pavord, A.A. Raj, T. Shirasaki, F. Soeda, D. Spina, K. Takahama, M. Trevisani, B.J. Undem, J.G. Widdicombe

Springer

Prof. Kian Fan Chung
National Heart & Lung Institute
Imperial College
Dovehouse Street
London SW3 6LY
UK
f.chung@imperial.ac.uk

Prof. Dr. John Widdicombe
University of London
116 Pepys Road
London SW20 8NY
UK
JohnWiddicombeJ@aol.com

ISBN 978-3-540-79841-5 e-ISBN 978-3-540-79842-2

Handbook of Experimental Pharmacology ISSN 0171-2004

Library of Congress Control Number: 2008929590

Cover Design: WMXDesign GmbH, Heidelberg

Printed on acid-free paper

9 8 7 6 5 4 3 2 1

springer.com

Preface

The last decade or so has seen remarkable advances in our knowledge of cough. This applies especially to its basic mechanisms: the types of airway sensors, the pharmacological receptors on their membranes, the brainstem organization of the 'cough centre', and the involvement of the cerebral cortex in the sensations and the voluntary control of cough. With the exception of the last of these, nearly all the studies have been on experimental animals rather than humans, for obvious reasons. One group of experimental studies has particular relevance to human patients, and that is the demonstration of the sensitization of cough pathways both in the periphery and in the brainstem. Similar sensitizations have been shown for patients with chronic cough or who have been exposed to pollutants, and it is reasonable to suppose that this is the basis of their cough and that the underlying mechanisms are generally similar in humans and other species.

Important advances are also being made in clinical cough research. For the three main causes of clinical cough, asthma, post-nasal drip syndrome, and gastro-oesophageal reflux disease, we are beginning to understand the pathological processes involved. There remains a diagnostically obdurate group of idiopathic chronic coughers, but even for them approaches are being devised to clarify underlying mechanisms and to establish diagnoses.

Perhaps surprisingly, the field in which there has been the least spectacular advance is the therapy of cough. This is not because current therapies work; indeed most seem to work little better than a placebo. This applies not only to the many remedies bought over the counter at the pharmacist and to those administered as part of complementary and alternative medicine, but also to those available on prescription (only codeine, pholcodine, and dextromethorphan in the UK). Basic studies are pointing to many potentially valuable approaches to the treatment of cough, based on understanding the basic peripheral receptor mechanisms, the brainstem pathways in the control of cough, and the sensitization processes that may apply in disease. The pharmacological industry is following up these leads, and clinicians are waiting hopefully for the fruits of their research.

An indication of the growth of interest in cough is the recent surge in publications dedicated to the subject. Before 1996, the editors can only think of two or three. Since then there have been two multiauthor books, at least ten international symposia, with the proceedings of nearly all of them being published as journal supplements, and at least five task-force reports set up by national and international organizations such as the American College of Chest Physicians and the European Respiratory Society. These publications will be frequently referred to in the chapters in the present volume. If asked 'Does this justify more description and analysis?', the answer is an emphatic yes! The field is being explored very fast; and new and emerging results are very important for understanding and alleviating one of the commonest disease symptoms of mankind. In this volume, we hope to show that basic mechanisms are helping us to understand clinical cough and also the other way round.

The editors are grateful to all the contributors, including co-authors, who, as is well known, often do most of the hard work; and to the diligent but tolerant publishers, especially Susanne Dathe, for their help and encouragement.

London, UK

Kian Fan Chung
John G. Widdicombe

Contents

Contributors

M.G. Belvisi
Respiratory Pharmacology, Airway Diseases, National Heart & Lung Institute, Faculty of Medicine, Imperial College, Guy Scadding Building, Dovehouse Street, London SW3 6LY, UK
m.belvisi@imperial.ac.uk

F. Bertram
Sackler Institute of Pulmonary Pharmacology, Division of Pharmaceutical Sciences, School of Biomedical and Health Sciences, King's College, London SE1 1UL, UK

S.S. Birring
Department of Respiratory Medicine, King's College Hospital, London SE5 9RS, UK
surinder.birring@kch.nhs.uk

D.C. Bolser
Department of Physiological Sciences, College of Veterinary Medicine, University of Florida, Gainesville, FL 32610-0144, USA
bolser@ufl.edu

A.C. Bonham
Department of Pharmacology, University of California, Davis School of Medicine, 4150 V Street, 1104 PSSB Sacramento, CA 95817, USA
ann.bonham@ucdmc.ucdavis.edu

J. Bric
Department of Pharmacology, University of California, Davis School of Medicine, 4150 V Street, 1104 PSSB Sacramento, CA 95817, USA

B.J. Canning
Johns Hopkins Asthma and Allergy Center, 5501 Hopkins Bayview Circle, Rm. 3A 24, Baltimore, MD 21224, USA
bjc@jhmi.edu

M.J. Carr
GlaxoSmithKline, 709 Swedeland Rd, King of Prussia, PA 19406, USA
michael.j.carr@gsk.com

C.-Y. Chen
Department of Pharmacology, Davis School of Medicine, University of California, 4150 V Street, 1104 PSSB Sacramento, CA 95817, USA

Y.-L. Chou
Johns Hopkins Asthma and Allergy Center, 5501 Hopkins Bayview Circle, Rm. 3A.24, Baltimore, MD 21224, USA

K.F. Chung
National Heart and Lung Institute, Imperial College, Dovehouse Street, London SW3 6LY, UK
f.chung@imperial.ac.uk

P.W. Davenport
Department of Physiological Sciences, Box 100144, HSC, University of Florida, Gainesville, FL 32610, USA
davenportp@mail.vetmed.ufl.edu

P.V. Dicpinigaitis
Einstein Division/Montefiore Medical Center, 1825 Eastchester Road, New York, NY 10461, USA
pdicpinigaitis@pol.net

R. Eccles
Common Cold Centre, Cardiff School of Biosciences, Cardiff University, Cardiff CF10 3US, UK
eccles@cardiff.ac.uk

E. Ernst
Complementary Medicine, Peninsula Medical School, Exeter EX2 4NT
eernst@pms.ac.uk⟨mailto:eernst@pms.ac.uk⟩

R. Gatti
Department of Critical Care Medicine and Surgery, University of Florence, Viale Pieraccini, 6, 50139, Florence, Italy

P. Geppetti
Department of Critical Care Medicine and Surgery, University of Florence, Viale Pieraccini, 6, 50139, Florence, Italy
pierangelo.geppetti@unifi.it

Q. Gu
Department of Physiology, University of Kentucky, Lexington, KY 40536-0298, USA

D.J. Hele
Respiratory Pharmacology, Airway Diseases, National Heart & Lung Institute, Faculty of Medicine, Imperial College, Guy Scadding Building, Dovehouse Street, London SW3 6LY, UK

J.P. Joad
Department of Pharmacology, University of California, Davis School of Medicine, 4150 V Street, 1104 PSSB Sacramento, CA 95817, USA

L.-Y. Lee
Department of Physiology, University of Kentucky, Lexington, KY 40536-0298, USA
lylee@uky.edu

S. Materazzi
Department of Critical Care Medicine and Surgery, University of Florence, Viale Pieraccini, 6, 50139, Florence, Italy

S.B. Mazzone
School of Biomedical Sciences, The University of Queensland, St. Lucia QLD 4072, Australia
s.mazzone@uq.edu.au

M.A. McAlexander
GlaxoSmithKline, 709 Swedeland Rd, King of Prussia, PA 19406, USA

I.M. McFadzean
Sackler Institute of Pulmonary Pharmacology, Division of Pharmaceutical Sciences, School of Biomedical and Health Sciences, King's College, London SE1 1UL, UK

L. McGarvey
Department of Medicine, Institute of Clinical Science, The Queen's University of Belfast, Grosvenor Road, Belfast, BT12 6BJ, UK
l.mcgarvey@qub.ac.uk

A.H. Morice
Cardiovascular and Respiratory Studies, Castle Hill Hospital, University of Hull Castle Road, Cottingham, East Yorkshire, HU16 5JQ, UK
a.h.morice@hull.ac.uk

R. Nassini
Department of Critical Care Medicine and Surgery, University of Florence, Viale Pieraccini, 6, 50139, Florence, Italy

C.P. Page
Sackler Institute of Pulmonary Pharmacology, Division of Pharmaceutical Sciences, School of Biomedical and Health Sciences, King's College, London SE1 1UL, UK
clive.page@kcl.ac.uk

A.A. Raj, D.I. Pavord
Department of Respiratory Medicine, King's College Hospital,
London SE5 9RS, UK
surinder.birring@kch.nhs.uk

T. Shirasaki
Department of Environmental and Molecular Health Sciences, Graduate School of Pharmaceutical Sciences, Kumamoto University, 5–1 Oe-honmachi, Kumamoto 862–0973, Japan

F. Soeda
Department of Environmental and Molecular Health Sciences, Graduate School of Pharmaceutical Sciences, Kumamoto University, 5–1 Oe-honmachi, Kumamoto 862–0973, Japan

D. Spina
Sackler Institute of Pulmonary Pharmacology, Division of Pharmaceutical Sciences, School of Biomedical and Health Sciences, King's College, London SE1 1UL, UK

K. Takahama
Department of Environmental and Molecular Health Sciences, Graduate School of Pharmaceutical Sciences, Kumamoto University, 5–1 Oe-honmachi, Kumamoto 862–0973, Japan
takahama@gpo.kumamoto-u.ac.jp

M. Trevisani
Department of Critical Care Medicine and Surgery, University of Florence
Viale Pieraccini, 6, 50139, Florence, Italy

B.J. Undem
Howard Florey Institute, University of Melbourne, VIC 3010, Australia

J.G. Widdicombe
University of London, 116 Pepys Road, London SW20 8NY, UK
JohnWiddicombeJ@aol.com

Cough: Setting the Scene

K.F. Chung(✉) and J.G. Widdicombe

Contents

1 Introduction

This introductory chapter is intended to bring a wide variety of physiological, clinical, and therapeutic aspects of cough together, but with a minimum of overlap. In general, it is true to say that the last decade or so has seen dramatic advances in our knowledge of the physiological mechanisms of acute cough in experimental animals, and that these are now moving in the direction of understanding increased sensitivity of cough in chronic conditions. This latter aspect has clear implications for patients in whom acute cough may be an irritation but is seldom a major cause of concern, while chronic cough can destroy the quality of life and arouse serious concerns in both patient and carer.

K.F. Chung
National Heart and Lung Institute, Imperial College, Dovehouse Street, London SW3 6LY, UK
f.chung@imperial.ac.uk

K.F. Chung, J.G. Widdicombe (eds.), *Pharmacology and Therapeutics of Cough*,
Handbook of Experimental Pharmacology 187,

2 Definition and Semantics

Physiology textbooks describe cough as consisting of a three- or four-phase action: (1) the inspiratory phase, consisting of a deep inspiration; (2) the compressive phase, with closure of the larynx and a forced expiratory effort; (3) the expulsive phase, when the larynx opens and rapid expiration occurs with characteristic first cough sound; and (4) the restorative phase, when a final deep breath is taken (Fig. 1). All phases are characteristic of a voluntary cough, but a reflex cough such as that evoked by inhalation of an irritant substance, or one occurring spontaneously in disease may be quite different. There may be a second (or third or even fourth) closure of the larynx during the expulsive phase, producing a second (or third or fourth) cough sound, although this is often absent. Little is understood about conditions that lead to these extra cough sounds or that are associated with their absence. The initial inspiration may be absent; this is seen with the expiration reflex from the larynx or tracheobronchial tree. This starts with closure of the larynx and a forced

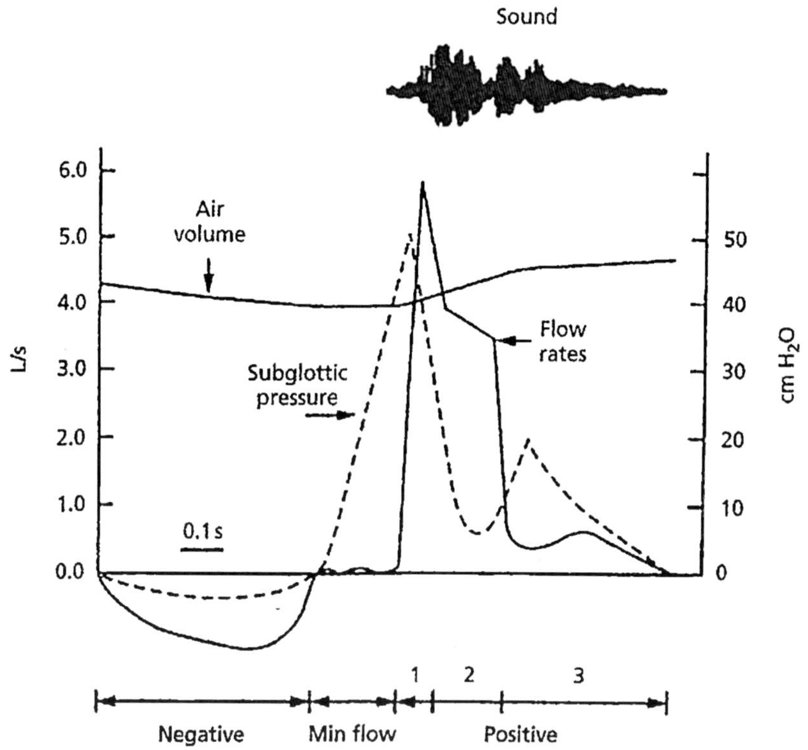

Fig. 1 The changes of the following variables during a representative cough: sound level, lung volume, flow rate, subglottic pressure. During inspiration the flow rate is negative; at the glottic closure the flow rate is zero; and during the expiratory phase the flow rate is positive. The last phase can be divided into three parts: growing, constant, and decreasing. (From Bianco and Robuschi 1989)

expiratory effort (the compressive phase) followed by an expulsive phase (Korpas and Tomori 1979; Widdicombe and Fontana 2006; Fontana and Widdicombe 2007; Tatar et al. 2008). Whether the expiration reflex should be called a cough is debated, although it possesses all the features of a cough except for the lack of the initial inspiration. It may have a different function from that of a cough, in that it should prevent aspiration of pharyngeal material into the lungs, whereas a cough expels material already in the lungs and requires a reserve of air deep in the lungs to make it efficient.

We have described the reflex cough and the expiration reflex as if they were isolated three- or four-phase activities, but in disease and on cough provocation they often consist of a sequence of expiratory efforts, usually interspersed with inspirations. These episodes are usually called cough epochs, but there is no clear distinction between a repeated cough or expiration reflex and an epoch. An arbitrary definition suggests that a 2-s gap between expulsive efforts is needed to call them separate events, and with less than a 2-s gap they become an epoch (Kelsall et al. 2008). There have been recent and valuable detailed analyses of the reflex cough and expiration reflex events during cough epochs (Smith Hammond 2008; Vovk et al. 2007; Kim et al. 2008). With auditory records of cough, it is difficult to analyze events during an epoch; however, it is possible with flow, pressure, or electromyographic records, although this may not be practical in the clinic. Therefore, for clinical purposes, it may be convenient to describe cough as "a forced expulsive manoeuvre, usually against a closed glottis and which is associated with a characteristic sound" (Morice et al. 2007). The presence of a forced expulsive maneuver and a characteristic sound of cough can be used to define cough clinically, and is the basis for many instruments used to measure coughs, either as single discrete events or as cough epochs.

The semantics of cough is confusing. You can have wet, dry, and moist coughs, depending on the ear of the listener. The first cough sound is usually called expulsive or explosive. The second cough sound has been called glottal or voiced. An epoch has been given a variety of names: bout, attack, peal, even peel. A new word "dystussia" has been suggested (Paul Davenport, personal communication) to describe a disordered cough airflow pattern (Smith Hammond 2008). This is a general term referring to a cough that is "abnormal" on the basis of altered cough patterns. It seems an attractive word for a cough that has reduced expiratory airflow rates and/or altered compression phase. In addition, perhaps, we should also consider eutussia (normal cough), atussia (absence of cough), hypotussia (reduced cough), and hypertussia (increased cough). Our view is that, while uniformity is desirable if it can be agreed, it is more important to define precisely what is being described, and to try to understand its mechanism.

3 Cough Triggers and Sensors

What transduces the cough response? A lot of research is still being undertaken to understand the "receptors" that can transduce the cough response since the first description of the irritant rapidly adapting receptor as being a cough "receptor" (Widdicombe 1954). We now prefer to call these sensors rather than receptors, since the latter term is now almost always used for membrane pharmacological receptors (Yu 2005). We also believe that there are an embarrassing number of cough sensors in the airways, i.e., those that can sense and transduce the cough response (Canning 2002, 2007; Canning and Chou 2008; Canning et al. 2006). Embarrassing because there is an embarras de richesses of both sensors and membrane receptors (Table 1). The sensors must all have slightly different reflex actions, and we do not know what the differences are. It is unlikely that they all cause identical respiratory reflex patterns of cough, and their nonrespiratory (e.g., cardiovascular, bronchomotor, mucosecretor, sensation) actions may also be different. Presumably, the primary first-order neurones in the vagi link up with different patterns of medullary second-order neurones. The different sensors may "sense" different irritant stimuli that cause cough (see below). The sensors include, for the tracheobronchial tree, bronchial C-fiber, Aδ-nociceptors, "cough receptors," and "rapidly adapting receptors" (Canning et al. 2006) and, for the laryngopharyngeal region, "irritant receptors" and C-fiber sensors (Widdicombe et al. 1988; Sant'Ambrogio and Sant'Ambrogio 1996). Full details of these sensors, their response to various stimuli, and the reflexes they induce are given elsewhere in this volume (Canning and Chou 2008).

Other airway and lung sensors may also influence cough, although they may not cause it. Pulmonary C-fiber sensors have been claimed to cause cough, but there is also evidence that they inhibit it (Tatar et al. 1988); their action may be determined or modulated by other influences such as inputs from other bronchopulmonary sensors and brainstem conditions. Slowly adapting pulmonary stretch receptors strongly enhance the expiration reflex, and probably also strengthen the reflex cough (Korpas and Tomori 1979; Tatar et al. 2008), although they do not themselves cause cough. Other bronchopulmonary sensors, e.g., neuroepithelial bodies (Adriaensen et al. 2003) and visceral pleural sensors (Pintelon et al. 2007), and other laryngeal sensors, such as "drive" and "temperature" sensors (Widdicombe et al. 1988), have

Table 1 Some characteristics of bronchopulmonary sensors probably responsible for cough

Receptor	Agonist	Nodose Aδ	Nodose C	Jugular C	Jugular Aδ	Nodose RARs
TRPV-1	Capsaicin, acid, heat, AA	No	Yes	Yes	Yes	No
5-HT_3	5-HT	No	Yes	No	No	No
P2X	Purines	No	Yes	No	No	Yes
ASIC	Acid	No	Yes	Yes	Yes	No
Nicotinic	Nicotine	No	Yes?	Yes?	Unknown	Unknown
BKB_2	Bradykinin	No	Yes	Yes	Yes	No
Adenosine	Adenosine	No	Yes	No	No	No

Modified from Kollarik and Undem (2006)

RARs rapidly adapting receptors, *TRPV*-1 transient receptor potential vanilloid-1, *AA* arachidonic acid, *5-HT* 5-hydroxytryptamine, *ASIC* acid-sensing ion channel, *BKB_2* bradykinin B_2

not been shown to influence cough, although no-one may have looked for this effect. How these sensors may interact and modulate the ultimate cough output remain to be explored.

There have been many studies on the morphology of sensors in the airways and lungs. Some, such as slowly adapting pulmonary stretch receptors and neuroepithelial bodies, have had their structures well delineated (Krauhs 1984; Adriaensen et al. 2003). But for the majority of sensors thought to be involved in cough, although they ramify within and below the airway epithelium, identification of the histological structure of the sensor is tenuous. Similarly, it is difficult to ascribe any one type of reflex associated with cough with a particular sensor and afferent pathway. A partial exception may be the expiration reflex from the larynx, which has a latency of 15–25 ms from mucosa to muscle in cats and humans (Tatar et al. 2008), and therefore must be conducted by myelinated afferent nerves, but this still leaves several possibilities open.

The membrane receptors on the sensors show as much diversity as the sensors themselves (Table 1). They include at least eight "specific" receptors, which when activated open ion-selective channels leading to production of action potentials. Other receptors, e.g., cannabinoid receptors, may close excitatory channels and thus inhibit cough, or they may open "inhibitory" channels (e.g., potassium channels). Extensive reviews of cough membrane receptors are included in this volume (Belvisi and Hele 2008; Lee and Gu 2008; Materazzi et al. 2008; Mazzone and Undem 2008). We will give one illustration. Acid stimulates at least three cough sensors (Aδ-nociceptors, C-fiber sensors, and probably rapidly-adapting receptors), but with different patterns and timings of neuronal activity (Kollarik and Undem 2006; Kollarik et al. 2007). The membrane receptors involved belong mainly to two families, the transient receptor potential and the acid-sensing ion channel families (note the word "family"!). The actual number of membrane receptor types may add up to dozens. Their concentration and distribution are not known in detail even for the guinea pig, the species most studied. Yet, there are great species differences in cough reflexes (Belvisi and Bolser 2002), so even these partial results cannot be applied accurately to humans.

The problem can be illustrated in a different way by three other examples. Firstly, alkalis such as ammonia are powerful stimulants of cough (Widdicombe 1954; Boushey et al. 1972; Van Hirtum and Berckmans 2004; Li et al. 2006; Rahman et al. 2007). Yet as far as we can discover, no-one has identified the sensors and membrane receptors that respond to alkali. It is unlikely that, when they cause cough, they have an "antiacid effect" or else they would inhibit cough; that is precisely what weak concentrations of ammonia have been shown to do to citric acid induced cough (Mercaux et al. 2000). Secondly, both hyperosmolar and hypoosmolar solutions of sodium chloride provoke cough (Koskela et al. 2008; Lavorini et al. 2007) (one might expect the two to have opposite actions), yet the mediating sensors and membrane receptors have not been identified. Thirdly, cold air is an established cause of cough (Cho et al. 2003), and may be one of the factors causing high-altitude cough. Yet, exercising polar explorers do not complain of cough (Mason and Barry 2007) and, as far as we can determine, no-one has identified airway sensors that respond to cold and cause cough. Possibly the laryngeal "cold" sensors are a candidate (Widdicombe et al. 1988); however, they also respond to airflow, which has not been shown to cause cough. There must be hundreds of different agents that can cause cough, but there are only basic detailed studies on acid and capsaicin, and to a lesser extent on nicotine, adenosine compounds, bradykinin, and 5-hydroxytryptamine.

The whole subject of cough sensors and membrane receptors is a tangled web; an appropriate term to apply to cough sensory neurones and their tortuous terminals at both ends. However, the web continues to be unraveled by persistent research.

4 Cough Pathways

As far as we know, all afferent pathways for cough, either from the larynx or from the tracheobronchial tree, travel in the vagus nerves. Vagotomy or vagal block with local anesthesia abolishes all reflex cough in humans and other animals (Korpas and Tomori 1979; Guz et al. 1970). There are some afferent pathways from the lungs that travel in the sympathetic nerves, for example, coming from neuroepithelial bodies and visceral pleural sensors, but these have not been shown to mediate cough. The vagal nerves have their cell bodies in the nodose and jugular ganglia; Undem and his colleagues have differentiated between cough sensors with cell bodies in each type of ganglion, and have shown that the ganglia have different embryonic origins, from placodes and neural crest, respectively (Kollarik et al. 2007). Whether they correspond to different types of cough is not known. The vagal afferent nerves (first-order neurones) travel to the nucleus of the solitary tract (NTS), especially the caudal part of the tract (Mutolo et al. 2008), where they synapse with second-order neurones. From thereon, the picture becomes very complicated. Second- (or later-) order neurones travel not only to cause cough, but also to influence modalities such as breathing, the cardiovascular system, skeletal muscle tone, airway mucosecretion, and sensation (inter alia). With the exception of cough, none of these connections has been worked out in any detail. When cough is initiated, the central respiratory rhythm generator is "switched off" and "gates" are thought to open to allow activation of the cough generator (Bolser and Davenport 2004; Bolser et al. 2006). Presumably these gates are closed in the absence of a stimulus that leads to a cough.

Detailed maps of the brainstem pathways that mediate cough have been determined (Shannon et al. 2000, 2004), and their relationship to the brainstem neuronal control of breathing described (Bolser and Davenport 2004). They are too complicated to summarize here, but their importance can be illustrated in five ways. Bolser et al. (2006) have proposed a "holarchical" system for cough in the brainstem whereby different functional control elements regulate the different behaviors related to cough, including cough itself. This system includes "gates" which, by opening and closing, can determine whether or not a particular activity, e.g., cough, is permitted. Secondly, they are a site of sensitization (and possibly desensitization) of the cough reflex (Bonham et al. 2006; Chen et al. 2008), and are therefore very relevant to what happens in diseases associated with cough. Thirdly, they are the site of action of many antitussive drugs (Bolser 2008; Takahama et al. 2008). The understanding of these medullary pathways could lead to important advances in antitussive therapy (Chung 2007, 2008). Fourthly, the circuitries for the tracheobronchial reflex cough and for the laryngeal expiration reflex have been established as different (Baekey et al. 2004). Since the two reflexes have different physiological and pharmacological controls (Tatar et al. 2008), this observation is of potential significance in the development of future antitussive therapeutic strategies. And fifthly, there seem to be different gating systems for cough from the larynx compared with cough from the lower airways (Bolser and Davenport 2004); this could correspond to the different circuitries for the expiration reflex compared with the reflex cough, and these have similar implications for therapy.

But the influence of cough inputs extends far beyond the brainstem. They cause the sensations of "irritation" and "urge-to-cough" (see later), and activate many parts of the cerebral cortex and upper brain. This has been well illustrated by the functional magnetic resonance brain imaging studies in humans that associate the urge-to-cough sensation with cortical neuronal activation pathways (Mazzone et al. 2007). We do not know whether these supramedullary pathways are enhanced in diseases associated with a chronic cough. A comparison with pain has been drawn; the latter elicits many reflexes, sensations, and emotive changes, and similar processes are now being studied in relation to cough (Gracely et al. 2007; Widdicombe 2008).

5 Peripheral and Central Cough Sensitization

There have been many recent studies that show that peripheral cough sensors can be "sensitized" in animals (Carr 2004, 2007; Carr and Lee 2006; McAlexander and Carr 2008); these studies show that exposure to an appropriate peripheral stimulus, for example, by development of an allergic sensitivity or irritation by pollutants or their constituents, can lower the threshold and increase the cough response to tussigenic agents such as citric acid or capsaicin, and increase the action potential response in fibers thought to originate from cough sensors. Histological examination of the sensors shows that their structure may change, in particular to contain more inclusions such as those of neuropeptides (Chuachoo et al. 2006). It seems almost certain that the same process of sensitization can occur in humans. For example, atmospheric exposure to pollutants or experimental exposure to ozone lowers the cough threshold to agents such as citric acid and capsaicin (Joad et al. 2007).

To what extent the same process applies to patients with cough is more difficult to decide. The disease process could cause a greater stimulus to cough sensors otherwise of "normal" sensitivity; for example, the presence of excess mucus, edema in the mucosa, and greater release of tussigenic agents such as bradykinin or neuropeptides could move the cough sensor response up the stimulus/response curve and give the impression of sensitization, while in reality it is the stimulus that is increased (Widdicombe 1996). It is not known whether mediator release in the airways' mucosa sensitizes the nerves there. One might even speculate that the increased acidity of the airway surface liquid in asthmatics (Koutsokera et al. 2008) is a sensitizing or cough-promoting agent. Inhalation of weak ammonia concentrations may inhibit cough (Moreaux et al. 2000). For the clinician, this should not be a semantic quibble; if there is an added cough stimulus (such as mucus) then it may be preferable to decrease the stimulus rather than depress the cough, whereas if the cough is sensitized, a symptomatic (antitussive) approach may be better.

Neural mechanisms of reflex cough are regulated by the inspiratory and expiratory networks of the brainstem, pons, and cerebellum, particularly in brainstem nuclei in the NTS where there are connections to respiratory related neurones in the central respiratory generator (Shannon et al. 2000, 2004). Changes in the central processing at the level of the ganglia or brainstem ("central sensitization") encompass changes in sensory pathways with the release of neurotransmitters or neuromodulators, or in excitability of postsynaptic neurones, or in a change in the structure of the nerve (Bonham et al. 2006). Central nervous system sensitization of the cough reflex has also been shown in animals, including primates In particular, the role of substance P released from first-order neurones and acting on second-order neurones in the NTS has been established, and the membrane receptor mechanisms on the second-order neurones have been analyzed in detail (Chen et al. 2008). With an upregulated cough reflex, due, for example, to inhaled pollutants, the substance P levels in the NTS are increased, just as they are in the first-order neurones (the sensory fibers) (Chen et al. 2008). Injections of substance P into the NTS enhance coughing due to a peripheral stimulus, and neurokinin 1 receptor antagonists depress cough in animals (Advenier and Emonds-Alts 1996; Bonham et al. 2006). For obvious reasons it is impossible to repeat these studies in humans, and in patients who have a sensitized cough reflex it is difficult to partition the response between periphery and brainstem. On theoretical grounds, it seems likely that both sites are involved. Disease processes in the airways will sensitize the sensors there, which in turn will cause sensitization at the first- and second-order neurones in the brainstem, as seems to happen in experimental animals. From the therapeutic point of view, effective neurokinin 1 receptor antagonists might act at both levels (Advenier and Emonds-Alts 1996).

6 Sensory Correlates of Cough

Reflex coughing, as distinct from voluntary or habit coughing, is often associated with unpleasant sensation in the chest or throat; however, this is not always present, especially with conditions in the lower airways involving, for example, excessive mucus. The terms used to describe the sensations are various, and include "irritation," "rawness," and even "pain" (Widdicombe 2008).

Urge-to-cough is a distinct sensation that, with increasing levels of cough stimulation, has a lower threshold and occurs before the cough itself (Davenport 2008;

Vovt et al. 2007), Other respiratory sensations, such as tightness, air-hunger, sense of effort and sense of lung volume are not usually associated with cough.

Patients with chronic cough often complain of a persistent tickling or irritating sensation in the throat (feeling of an itch) or a choking sensation, and it is sometimes felt in the chest, that often leads to paroxysms of coughing. Triggers such as changes in ambient temperature, taking a deep breath, laughing, talking over the phone for more than a few minutes, cigarette smoke, aerosol sprays, perfumes or eating crumbly dry food are common. Unpleasant sensation related to cough may be localized vertically, in the throat or in the chest, but not usually more precisely or laterally. Vagotomy or vagal anesthesia prevents the sensation (Petit 1970; Winning et al. 1988), and that from the chest is absent in patients with bilateral lung transplant (Butler et al. 2001)

Urge-to-cough has been extensively studied in the last few years, especially by Davenport and colleagues. It is described in detail elsewhere in this volume by Davenport (Davenport 2008). The parts of the cerebral cortex and upper brain that are activated by these sensations have also been mapped out (Mazzone et al. 2007). Urge-to-cough can occur with stimuli, such as aerosols of capsaicin, citric acid, and distilled water, and intravenous injections of lobeline and capsaicin, which are too weak to cause cough, and in the presence or absence of unpleasant sensation (Widdicombe 2008). While urge-to-cough has no particular location in the body, unpleasant sensation related to cough may be felt in the chest or the throat (Butler et al. 2001). In the latter case it must be referred from another site, since intravenously administered lobeline is thought to act on bronchopulmonary sensors but arouses a raw sensation in the larynx. A similar referred unpleasant sensation is seen with some patients with unilateral lung disease, when the sensation is identified as coming from the ipsilateral side of the face (Sarlani et al. 2003). But cough is not usually associated with this condition.

We cannot say which sensor or sensors in the lungs are responsible for the respiratory sensations, but it seems likely that there are different combinations of activity for cough, urge-to-cough, and rawness. For example, distilled water aerosol produces cough and urge-to-cough but no rawness (Lavorini et al. 2007), while anecdotally many lung conditions produce cough and rawness but no urge-to-cough, or cough with neither rawness nor urge-to-cough. The complexity of the airway sensory system mediating cough, as already described, makes it unlikely that identical pathways are responsible to all three reactions.

7 Epidemiology of Clinical Cough

Chronic cough is not uncommon and its prevalence varies from 9 to 33% of the population, and there is an association with cigarette smoking (Cullinan 1992; Ford et al. 2006; Zemp et al. 1999), in that chronic smokers have a threefold increase in prevalence of chronic cough compared with never smokers and ex-smokers (Zemp et al. 1999). Other associations are reported too with asthma or respiratory wheeze, or with symptoms of gastroesophageal reflux disease (GORD) (Janson et al. 2001; Ford et al. 2006). Exposure to environmental pollutants, particularly PM_{10} particulates, is also associated in adults and schoolchildren with productive cough or chronic nocturnal dry cough (Braun-Fahrlander et al. 1997; Pierse et al. 2006). Increases in levels of PM_{10} and of nitrogen dioxide have been correlated to reductions in peak expiratory flows and to increasing reporting of cough, sputum production, and sore throat in children. Clearly more research is needed to firm and explain the link with environmental pollution. For the respiratory physician, patients with a chronic cough probably account for 10–38% of his/her outpatient practice. Only a minority of the population identified in epidemiological surveys seek medical help or advice about their symptom. It is important to find out whether there are any medical associations with chronic cough in the community and of the natural history of this symptom.

8 Clinical Associations of Cough

Cough has been divided into an acute self-limiting cough lasting less than 3 weeks or a chronic persistent cough, usually defined as lasting for more than 8 weeks. Acute cough is usually the result of an upper respiratory tract virus infection that usually clears within 2 weeks in two thirds of people. Nonviral causes of acute cough include exacerbation of existing asthma or potential exposure to environmental pollutants. Other types of cough last for a limited period of 3–8 weeks, which is referred to as subacute cough, reported to be postinfective (Kwon et al. 2006). Eleven to 25% of patients with chronic cough report a postinfectious cough (Poe et al. 1989). Persistent cough following *Mycoplasma* or *Bordetella pertussis* infections have been highlighted (Davis et al. 1995), but no doubt other infections may be involved and further research in this area is needed.

In North America and Europe, the most common conditions associated with causing chronic cough, with normal findings on a chest radiograph, include the corticosteroid-responsive eosinophic airway diseases (asthma, cough-variant asthma, and eosinophilic bronchitis), and a range of conditions typically associated with an inhaled corticosteroid-resistant cough, including GORD and the postnasal drip syndrome or rhinosinusitis. The frequency of these causes has varied in different series depending on the location of the clinic and its particular interest, on the age of the patient, and on local definition of the disease entities (Chung and Pavord 2008). For example, with regard to the latter, in Japan, atopic cough and sinobronchial disease are more commonly diagnosed, while GORD is much less so (Niimi 2007; Kohno et al. 2006). The associations of various diseases with chronic cough still need to be worked out carefully, and the mechanisms of cough in disease are in need of clarification.

Asthma may present predominantly with cough, often nocturnal, and the diagnosis is supported by the presence of bronchial hyperresponsiveness. Three other conditions, cough-variant asthma, atopic cough, and eosinophilic bronchitis, are related to classic asthma, and are all associated with an eosinophilic airway inflammation and the cough responds well to inhaled corticosteroid therapy. This raises the possibility that eosinophils may directly contribute to increasing cough sensitivity.

GORD encompasses symptoms or complications such as heart burn, chest pain, sour taste, or regurgitation, and also a chronic persistent cough. Direct aspiration of gastric contents into the larynx and upper airways that could directly stimulate cough sensors and increases in tracheal acidity have been recorded during episodes of reflux (Jack et al. 1995). On the other hand, direct infusion of acid into the distal esophagus of patients with chronic cough due to GORD induces cough (Ing et al. 1994), through vagal cholinergic pathways. However, the majority of coughs in GORD do not coincide with an acid reflux episode (Ours et al. 1999; Irwin et al. 1989). Nonacid components such as pepsin, bile, and other gastric enzymes may induce cough. In addition, associated dysmotility of the esophagus is implicated but with not much evidence.

Postnasal drip ("nasal catarrh") is characterized by a sensation of nasal secretions or of a "drip" at the back of the throat, accompanied very often by frequent need to clear the throat ("throat-clearing") associated with nasal discharge or nasal stuffiness. The term "upper airway cough syndrome" is proposed as an alternative to stress the association of upper airways disease with cough (Pratter 2006). The pathogenesis of cough in the postnasal drip syndrome may be related to the direct pharyngeal, laryngeal, or sublaryngeal stimulation by the mucoid secretions from the rhinosinuses, which contain inflammatory mediators to induce cough.

9 Idiopathic Cough

Earlier series of chronic cough patients rarely identified patients in whom no identifiable cause was found or failure of treatment of identifiable causes occurred. More recent series have identified a significant proportion of patients labeled as "idiopathic" cough, ranging from 7 to 46%, despite thorough diagnostic workup (Irwin et al. 1981, 1990, 2006; Poe et al. 1989; O'Connell et al. 1994; Pratter et al. 1993; Smyrnios et al. 1995; Mello et al. 1996; French et al. 1998; McGarvey et al. 1998; Brightling et al. 1999; Birring et al. 2004b; Niimi et al. 2005; Kastelik et al. 2005; Fujimura et al. 2005; Shirahata et al. 2005; Palombini et al. 1999; Carney et al. 1997). It may be interesting to determine whether this represents a genuine change or whether different methods were being used regarding diagnostic approaches. The initiating cause of the cough may have disappeared, but its effect on enhancing the cough reflex may be more prolonged. An example could be the transient appearance of an upper respiratory tract virus infection or an exposure to toxic fumes that results in prolonged damage of the airways' mucosa. The repetitive mechanical and physical effects of coughing bouts on airway cells could activate the release of various chemical mediators that could enhance chronic cough through inflammatory mechanisms (Heino et al. 1990), providing a positive feed-forward system for cough persistence. It is quite possible that there is an induction of changes in the upper airways of inflammation and tissue remodeling induced by various causes associated with cough or by the act of coughing itself that could lead to an enhanced cough reflex, which in turn is responsible for maintaining cough. The cough becomes "idiopathic" when the primary inciting cause has resolved while cough is persistent. It is clear that more needs to be learned about idiopathic cough, and whether it is all "idiopathic" is the big question; in the meantime, it is reasonable to study this group as a separate entity.

10 Enhanced Cough

Patients with chronic cough often complain of a persistent tickling or irritating sensation in the throat (feeling of an itch), or a choking sensation and sometimes felt in the chest, that often leads to paroxysms of coughing. Triggers such as changes in ambient temperature, taking a deep breath, laughing, talking over the phone for more than a few minutes, cigarette smoke, aerosol sprays, perfumes, or eating crumbly dry food are common.

The mechanisms of idiopathic cough are unclear, but we assume that the initiating cause of the cough has disappeared, leaving an enhancement of the cough reflex which can be measured by the tussive response to inhalation of citric acid or capsaicin, as compared with noncoughers (Choudry and Fuller, 1992). The increase in cough sensitivity to capsaicin is related to the presence of a tickling or irritating sensation localized to the throat or lower-chest area that often leads to a paroxysm of coughing which patients with chronic cough find most distressing because it cannot be controlled. The paroxysm can be triggered in some patients by inhaling cold air, by a deep breath, by the act of laughing, and by breathing irritants such as cigarette smoke, aerosol sprays, or perfumes. The *urge-to-cough* is a sensory measure of this sensation of tickling or irritation that is induced at concentrations of inhaled capsaicin that are lower than those necessary to elicit a cough reflex, which is a motor cough behavior (Davenport et al. 2007), but may also be present in patients with chronic cough. This sensation may be a "referred" sensation since very often there are no visible abnormalities of the pharynx and larynx that are associated with it.

This enhanced cough reflex may result from an increased sensitivity of cough receptors with plasticity of the afferent innervation such as changes in nerve densities or in ion channels (peripheral sensitization) (Lee and Undem 2004; Carr and Lee 2006). The presence of increased expression of the transient receptor potential vanniloid-1 (TRPV-1) receptor in epithelial nerves of patients with nonasthmatic chronic cough indicates a potential mechanism of peripheral sensitization (Groneberg et al. 2004). Inflammation and remodeling of the airway submucosa with an increase in submucosal mast cells and airway wall remodeling with goblet cell hyperplasia, subepithelial fibrosis, and increased vascularity is reported in chronic cough patients (Niimi et al. 2005). Increased mast cells have also been observed in bronchoalveolar lavage fluid (McGarvey et al. 1999), with increased neutrophils (Jatakanon et al. 1999), and higher histamine, prostaglandins D_2 and

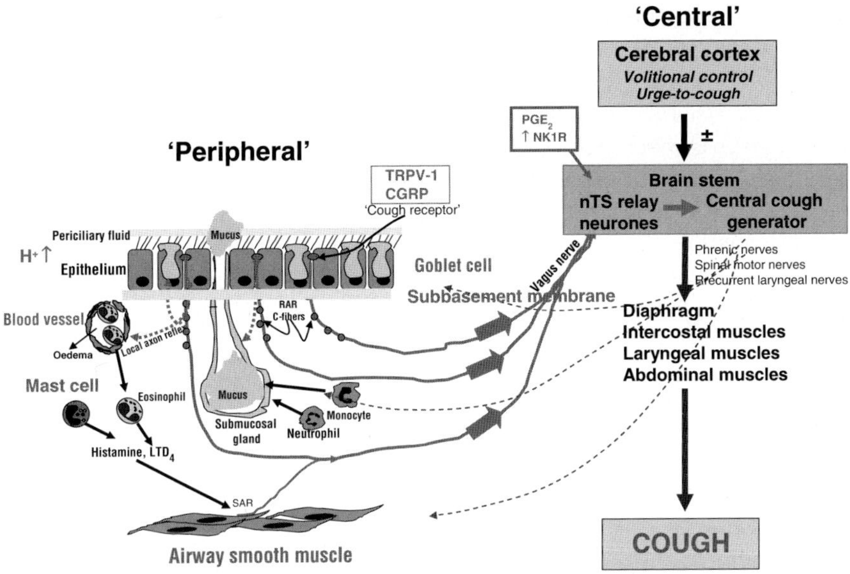

Fig. 2 Afferent pathways and central control of the cough reflex with peripheral and central sensitization of the reflex by a variety of mechanisms. *CGRP* calcitonin gene-related peptide, *nTS* nucleus of the solitary tract, LTD_4 leukotriene D_4, *NK* neurokinin, *PGE2* prostaglandin E2, *RAR* rapidly adapting receptors, *SAR* slowly adapting receptors, *TRPV-1* transient receptor potential vanilloid-1

E_2, tumor necrosis factor α, and interleukin-8 concentrations in induced sputum (Birring et al. 2004a). These inflammatory changes could certainly contribute to peripheral sensitization of the cough reflex. However, while the changes observed in the airways could also result from physical damage from the coughing act, they could nevertheless contribute to the chronicity of the cough, a possibility worth exploring. Some of the mechanisms underlying the enhanced cough response in chronic cough are illustrated in Fig. 2.

11 Measuring Cough

There has been a great deal of progress made in the field of cough measurement over the last 10 years (Chung 2006). Cough can be measured subjectively using symptom scores and specific quality-of-life measures, and objectively by measuring cough numbers and intensity, and by assessing the cough response to capsaicin or citric acid. Most previous reported clinical series of chronic cough do not state how the clinical response of the chronic cough patients was measured, and yet provide success of intervention as "yes/no." This could be the reason why there is a diversity of success in treating chronic cough in the literature. In the small studies of the antitussive effects of various agents, a variety of instruments have been used, including a cough scoring system or visual analogue scale completed by the patient, or tussive response to capsaicin or citric acid. Cough visual analogue scores are used most commonly in clinical trials. Typically these assess cough according to the patient's own experience: for example, the patient will be asked to rate his/her cough on a 10-cm scale fixed at both ends by "no cough" and "the worst cough ever." Assessments are responsive and repeatable but they are of no value in comparing cough severity between individuals or populations.

Cough-specific quality-of-life questionnaires have been used to identify the many different components of impaired health status seen in patients with chronic cough (French et al. 1998; Birring et al. 2003). The Leicester Cough Questionnaire comprises of 19 items and three domains made of physical, psychological, and social attributes, with a seven-point Likert response scale, and responsiveness to treatment has been shown in a group of patients with cough that were successfully treated (Birring et al. 2006).

The ultimate objective assessment of cough is to measure its frequency and intensity (Chung 2006) and there are now reliable ambulatory systems to measure cough, although measurement of cough intensity may not be easy (Birring and Yousaf 2008).

Measurement of the cough reflex has been studied using inhalation of citric acid or of capsaicin; both techniques have been well validated and the methods are well standardized. Capsaicin cough sensitivity is probably the most widely used test, as it induces cough reliably and assessments of the cough reflex with inhaled capsaicin are reproducible (Dicpinigaitis 2003; Dicpinigaitis and Alva 2005). An increase in cough sensitivity has been reported in most conditions associated with a chronic cough and improvements in cough sensitivity are seen in patients whose chronic cough has been successfully treated (O'Connell et al. 1994).

The need for objective measures in clinical trials is demonstrated by the more recent studies that have shown the limitation of available antitussives in the treatment of chronic cough. Codeine is probably the most commonly prescribed opioid-derived antitussive and recent studies using objective counts have shown that it is ineffective against the acute cough of the common cold (Freestone and Eccles 1997), or against cough in patients with COPD (Smith et al. 2006), despite the findings reported in previous publications on its antitussive effects. The correct use of these instruments to measure cough in the clinic as well as in the clinical trials assessing antitussive therapies will certainly be defined in the years to come (Pavord and Chung 2008).

12 Need for More Effective Antitussive Therapies

In patients with idiopathic cough or in those in whom treatments directed against associated cause (termed "specific antitussives") are not successful, there is a need for symptomatic antitussive therapies. The efficacy of codeine or dextrometorphan in chronic cough is limited at the recommended doses, and higher doses causes unacceptable side effects. There continues to be great interest from the pharmaceutical industry to develop new antitussives on the basis of our understanding of the mechanisms of the enhanced cough reflex (Chung 2005). New-generation opioids or inhibitors that target afferent nerves involved in sensitization of the cough reflex such as TRPV-1 antagonists, tachykinin receptor antagonists, or chloride channel blockers have been identified as potential antitussives. There has been little progress in translating this research to effective antitussives and few studies have been done in chronic cough patients (Chung 2005). This may relate to the fact that most targets for antitussive therapies are derived from animal models that differ from humans and that the human cough reflex pathway is difficult to study.

On the other hand, potential antitussives have arisen from clinical reports of existing drugs in the treatment of cough, particularly the use of centrally acting drugs such as the antiepileptics gabapentin and carbamazepine, and the antidepressants amitriptyline and paroxetine (Chung 2007). There is a pressing need to demonstrate their antitussive effects in controlled trials using appropriate cough-assessment tools and to investigate their mechanism of action. They may have an effect on the sensitization process, akin to the effect of these agents in controlling neuropathic pain, through inhibition of various neural inflammatory pathways, or through effects on supramedullary pathways (Widdicombe et al. 2006). For example, there have been reports of positive findings for the use of amitriptyline in 12 cough patients (Bastian et al. 2006), and in an open controlled study of postviral persistent cough (Jeyakumar et al. 2006). The applicants had anecdotal experience of the beneficial antitussive actions of this drug in chronic idiopathic cough patients. However, confirmation of the antitussive effects in double-blind controlled trials using validated measures of cough is urgently needed and the potential antitussive mechanisms need to be investigated.

13 Conclusions: The Future of Cough

In the next 10 years, we expect to continue to increase our understanding of the physiological mechanisms of acute and chronic cough, including from the point of view of peripheral and central sensitizations of the cough reflex. From the clinical aspect, we may expect to understand better the relationship of diseases that are linked to chronic cough to the pathogenesis of cough. Better tools available to measure cough should allow us to pick up efficacious treatments for chronic cough. As a result we should start to assess the real impact of potential symptomatic antitussives in chronic cough. If successful antitussives become available, they will change dramatically our approach to the management of chronic cough.

References

Adriaensen D, Broons I, Van Genechten J, Timmermans JP (2003) Functional morphology of pulmonary neuroepithelial bodies: Extremely complex airway receptors. Anat Rec 270A:25–40

Advenier C, Emonds-Alts X (1996) Tachykinin antagonists and cough. Pulm Pharmacol 9: 329–324

Baekey DM, Morris KF, Nuding SG, Segers LS, Lindsay BG, Shannon R (2004) Ventrolateral medullary respiratory network participation in the expiration reflex. J Appl Physiol 96:2057–2072

Bastian RW, Vaidya AM, Delsupehe KG (2006) Sensory neuropathic cough: A common and treatable cause of chronic cough. Otolaryngol Head Neck Surg 135:17–21

Belvisi MG, Bolser DC (2002) Animal models for cough. Pulm Pharmacol Ther 15:249–250

Belvisi MG, Hele DJ (2008) Cough sensors II: Opioid and cannabinoid receptors on vagal sensory nerves. In: Chung KF, Widdicombe JG (eds) Pharmacology and therapeutics of cough, Handbook of experimental pharmacology, vol 187. Springer, Berlin

Bianco S, Robuschi M (1989) Mechanics of cough. In: Braga PC, AllegraL (eds) Cough. Raven, New York, pp 29–36

Birring SS, Yousaf N (2008) Clinical Cough III: Measuring cough in the patient. In: Chung KF, Widdicombe JG (eds) Pharmacology and therapeutics of cough, Handbook of experimental pharmacology, vol 187. Springer, Berlin

Birring SS, Prudon B, Carr AJ, Singh SJ, Morgan MD, Pavord ID (2003) Development of a symptom specific health status measure for patients with chronic cough: Leicester Cough Questionnaire (LCQ). Thorax 58:339–343

Birring SS, Parker D, Brightling CE, Bradding P, Wardlaw AJ, Pavord ID (2004a) Induced sputum inflammatory mediator concentrations in chronic cough. Am J Respir Crit Care Med 169:15–19

Birring SS, Passant C, Patel RB, Prudon B, Murty GE, Pavord ID (2004b) Chronic tonsillar enlargement and cough: Preliminary evidence of a novel and treatable cause of chronic cough. Eur Respir J 23:199–201

Birring SS, Matos S, Patel RB, Prudon B, Evans DH, Pavord ID (2006) Cough frequency, cough sensitivity and health status in patients with chronic cough. Respir Med 100:1105–1109

Bolser DC (2008) Central mechanisms II: Pharmacology of brainstem pathways. In: Chung KF, Widdicombe JG (eds) Pharmacology and therapeutics of cough, Handbook of experimental pharmacology, vol 187. Springer, Berlin

Bolser DC, Davenport PW (2004) Functional organization of the central cough generation mechanism. Pulm Pharmacol Ther 15:221–226

Bolser DC, Poliacek I, Jakus J, Fuller DD, Davenport PW (2006) Neurogenesis of cough, other airway defensive behaviors and breathing: A holarchical system? Respir Physiol Neurobiol 152:255–265

Bonham AC, Sekizawa C-I, Chen C-Y, Joad JP (2006) Plasticity of brainstem mechanisms of cough. Respir Physiol Neurobiol 152:312–319

Boushey HA, Richardson PS, Widdicombe JG (1972) Reflex effects of laryngeal irritation on the pattern of breathing and total lung resistance. J Physiol 224:501–513

Braga PC, Allegra L (eds) (1989) Cough. Raven Press, New York, pp 29–36

Braun-Fahrlander C, Vuille JC, Sennhauser FH, Neu U, Kunzle T, Grize L, Gassner M, Minder C, Schindler C, Varonier HS, Wuthrich B (1997) Respiratory health and long-term exposure to air pollutants in Swiss schoolchildren. SCARPOL Team. Swiss Study on Childhood Allergy and Respiratory Symptoms with Respect to Air Pollution, Climate and Pollen. Am J Respir Crit Care Med 155:1042–1049

Brightling CE, Ward R, Goh KL, Wardlaw AJ, Pavord ID (1999) Eosinophilic bronchitis is an important cause of chronic cough. Am J Respir Crit Care Med 160:406–410

Butler JE, Anand A, Crawford MR, Glanville AR, McKenzie DK, Paintal AS, Taylor JL, Gandevia SC (2001) Changes in respiratory sensations induced by lobeline after human bilateral lung transplantation. J Physiol 534:583–593

Canning BJ (2002) Interactions between vagal afferent nerve subtypes mediating cough. Pulm Pharmacol Ther 15:187–192

Canning BJ (2007) Encoding of the cough reflex. Pulm Pharmacol Ther 20:396–401

Canning BJ, Chou Y-L (2008) Physiological and pharmacological properties of the afferent nerves regulating cough. In: Chung KF, Widdicombe JG (eds) Pharmacology and therapeutics of cough, Handbook of experimental pharmacology, vol 187. Springer, Berlin (in press)
Canning BJ, Mori N, Mazzone SB (2006) Vagal afferent nerves regulating the cough reflex. Respir Physiol Neurobiol 152:223–242
Carney IK, Gibson PG, Murree-Allen K, Saltos N, Olson LG, Hensley MJ (1997) A systematic evaluation of mechanisms in chronic cough. Am J Respir Crit Care Med 156:211–216
Carr MJ (2004) Plasticity of vagal afferent fibres mediating cough. Pulm Pharmacol Ther 17: 447–452
Carr MJ (2007) Plasticity of the afferent innervation of the airways: The role of ion channels. Pulm Pharmacol Ther 20:412–416
Carr MJ, Lee L-Y (2006) Plasticity of peripheral mechanisms of cough. Respir Physiol Neurobiol 152:298–311
Chen C-Y, Joad JP, Bric J, Bonham AC. (2008) Central mechanisms I: Plasticity of central pathways. In: Chung KF, Widdicombe JG (eds) Pharmacology and therapeutics of cough, Handbook of experimental pharmacology vol 187. Springer, Berlin (in press)
Cho YS, Park SY, Lee CK, Lee EY, Shin JH, Yoo B, Moon HB (2003) Enhanced cough response to hyperpnea with cold air challenge in chronic cough patients showing increased cough sensitivity to inhaled capsaicin. Allergy 58:486–491
Choudry NB, Fuller RW (1992) Sensitivity of the cough reflex in patients with chronic cough. Eur Respir J 5:296–300
Chuachoo B, Hunter DD, Myers AC, Kollarik M, Undem BJ (2006) Allergen-induced substance P synthesis in large-diameter sensory neurons innervating the lungs. J Allergy Clin Immunol 116:325–331
Chung KF (2005) Drugs to suppress cough. Exp Opin Investig Drugs 14:19–27
Chung KF (2006) Measurement of cough. Respir Physiol Neurobiol 152:329–339
Chung KF (2007) Effective antitussives for the cough patient: An unmet need. Pulm Pharmacol Ther 20:438–445
Chung KF (2008) Clinical cough V: The need for new therapies for cough: Disease-specific and symptom-related antitussives. In: Chung KF, Widdicombe JG (eds) Pharmacology and therapeutics of cough, Handbook of experimental Pharmacology, vol 187. Springer, Berlin (in press)
Chung KF, Pavord ID (2008) Prevalence, pathogenesis, and causes of chronic cough. Lancet 371:1364–1374
Cullinan P (1992) Persistent cough and sputum: Prevalence and clinical characteristics in south east England. Respir Med 86:143–149
Davenport PW (2008) The urge-to-cough: A respiratory sensation. In: Chung KF, Widdicombe JG (eds) Pharmacology and therapeutics of cough, Handbook of experimental pharmacology, vol 187. Springer, Berlin (in press)
Davenport PW, Bolser DC, Vickroy T, Berry RB, Martin AD, Hey JA, Danzig M (2007) The effect of codeine on the Urge-to-Cough response to inhaled capsaicin. Pulm Pharmacol Ther 20:338–346
Davis SF, Sutter RW, Strebel PM, Orton C, Alexander V, Sanden GN, Cassell GH, Thacker WL, Cochi SL (1995) Concurrent outbreaks of pertussis and Mycoplasma pneumoniae infection: Clinical and epidemiological characteristics of illnesses manifested by cough. Clin Infect Dis 20:621–628
Dicpinigaitis PV (2003) Short- and long-term reproducibility of capsaicin cough challenge testing. Pulm Pharmacol Ther 16:61–65
Dicpinigaitis PV, Alva RV (2005) Safety of capsaicin cough challenge testing. Chest 128:196–202
Fontana GA, Widdicombe JG (2007) What is cough and what should be measured? Pulm Pharmacol Ther 20:307–312
Ford AC, Forman D, Moayyedi P, Morice AH (2006) Cough in the community: A cross sectional survey and the relationship to gastrointestinal symptoms. Thorax 61:975–979
Freestone C, Eccles R (1997) Assessment of the antitussive efficacy of codeine in cough associated with common cold. J Pharm Pharmacol 49:1045–1049

French CL, Irwin RS, Curley FJ, Krikorian CJ (1998) Impact of chronic cough on quality of life. Arch Intern Med 158:1657–1661

Fujimura M, Abo M, Ogawa H, Nishi K, Kibe Y, Hirose T, Nakatsumi Y, Iwasa K (2005) Importance of atopic cough, cough variant asthma and sinobronchial syndrome as causes of chronic cough in the Hokuriku area of Japan. Respirology 10:201–207

Gracely RH, Undem BJ, Banzett RB (2007) Cough, pain and dyspnoea: Similarities and differences. Pulm Pharmacol Ther 20:433–437

Groneberg DA, Niimi A, Dinh QT, Cosio B, Hew M, Fischer A, Chung KF (2004) Increased expression of transient receptor potential vanilloid-1 in airway nerves of chronic cough. Am J Respir Crit Care Med 170:1276–1280

Guz A, Noble MIM, Eisele JH, Trenchard D (1970) Experimental results of vagal block in cardiopulmonary disease. In: Porter R (ed) Breathing: Hering-Breuer centenary symposium. Churchill, London, pp 315–328

Heino M, Juntunen-Backman K, Leijala M, Rapola J, Laitinen LA (1990) Bronchial epithelial inflammation in children with chronic cough after early lower respiratory tract illness. Am Rev Respir Dis 141:428–432

Ing AJ, Ngu MC, Breslin AB (1994) Pathogenesis of chronic persistent cough associated with gastroesophageal reflux. Am J Respir Crit Care Med 149:160–167

Irwin RS, Carrao WM, Pratter MR (1981) Chronic persistent cough in the adult: The spectrum and frequency of causes and successful outcome of specific therapy. Am Rev Respir Dis 123: 413–417

Irwin RS, Zawacki JK, Curley FJ, French CL, Hoffman PJ (1989) Chronic cough as the sole presenting manifestation of gastroesophageal reflux. Am Rev Respir Dis 140:1294–1300

Irwin RS, Curley FJ, French CL (1990) Chronic cough: The spectrum and frequency of causes, key components of the diagnostic evaluation, and outcome of specific therapy. Am Rev Respir Dis 141:640–647

Irwin RS, Ownbey R, Cagle PT, Baker S, Fraire AE (2006) Interpreting the histopathology of chronic cough: A prospective, controlled, comparative study. Chest 130:362–370

Jack CIA, Calverley PMA, Donnelly RJ, Tran J, Russell G, Hind CRK, Evans CC (1995) Simultaneous tracheal and oesophageal pH measurements in asthmatic patients with gastroesophageal reflux. Thorax 50:201–204

Janson C, Chinn S, Jarvis D, Burney P (2001) Determinants of cough in young adults participating in the European Community Respiratory Health Survey. Eur Respir J 18:647–654

Jatakanon A, Lalloo UG, Lim S, Chung KF, Barnes PJ (1999) Increased neutrophils and cytokines, TNF-alpha and IL-8, in induced sputum of non-asthmatic patients with chronic dry cough. Thorax 54:234–237

Jeyakumar A, Brickman TM, Haben M (2006) Effectiveness of amitriptyline versus cough suppressants in the treatment of chronic cough resulting from postviral vagal neuropathy. Laryngoscope 116:2108–2112

Joad JP, Sekizawa S-I, Chen C-Y, Bonham AC (2007) Air pollutants and cough. Pulm Pharmacol Ther 20:347–354

Kastelik JA, Aziz I, Ojoo JC, Thompson RH, Redington AE, Morice AH (2005) Investigation and management of chronic cough using a probability-based algorithm. Eur Respir J 25:235–243

Kohno S, Ishida T, Uchida Y, Kishimoto H, Sasaki H, Shioya T, Tokuyama K, Niimi A, Nishi K, Fujimura M, Matsuse H, Suzaki H (2006) The Japanese Respiratory Society guidelines for management of cough. Respirology 11(Suppl 4):S135–S186

Kelsall A, Decalmer S, Webster D, Brown N, McGuinness K, Woodcock A, Smith J (2008) How to quantify coughing? Correlations with quality of life in chronic cough. Eur Respir J in press

Kim J, Davenport P, Sapienza C (2008) Effect of expiratory muscle strength training on elderly cough function. Arch Geront Geriatr in press

Kollarik M, Undem BJ (2006) Sensory transduction in cough-associated nerves. Respir Physiol Neurobiol 152:243–254

Kollarik M, Ru F, Undem BJ (2007) Acid-sensitive vagal sensory pathways and cough. Pulm Pharmacol Ther 20:402–411

Korpas J, Tomori Z (1979) Cough and other respiratory reflexes. Karger, Basel
Koskela HO, Purokivi MK, Kontra KM, Taivainen AH, Tukiainen HO (2008) Hypertonic saline cough provocation test with salbutamol pre-treatment: Evidence for sensorineural dysfunction in asthma. Clin Exp Allergy
Koutsokera A, Loukides S, Gourgoulianis KI, Kostikas K (2008) Biomarkers in the exhaled breath condensate of healthy adults: Mapping the path towards reference values. Curr Med Chem 15:620–630
Krauhs JM (1984) Morphology of presumptive slowly adapting receptors in the dog trachea. Anat Res 210:73–84
Kwon NH, Oh MJ, Min TH, Lee BJ, Choi DC (2006) Causes and clinical features of subacute cough. Chest 129:1142–1147
Lavorini F, Fantana GA, Pantaleo T, Geri P, Piumelli R, Pistolesi M, Widdicombe J (2007) Fog-induced cough with impaired respiratory sensation in congenital central hypoventilation syndrome. Amer J Respir Crit Care Med 176: 825–832
Lee L-Y, Gu Q (2008) Cough sensors III: Nicotinic membrane receptors on cough sensors. In: Chung KF, Widdicombe JG (eds) Pharmacology and therapeutics of cough, Handbook of experimental pharmacology, vol 187. Springer, Berlin
Lee LY, Undem BJ (2004) Mechanisms of chronic cough. Pulm Pharmacol Ther 17:463–464
Lee M-G, Kollarik M, Chuaychoo B, Undem BJ (2004) Ionotropic and metabotropic receptor medicated airway sensory nerve activation. Pulm Pharmacol Ther 17:355–360
Li PB, Mu Y, Wang YG, Su WW (2006) Experimental studies on antitussive, expectorant and anti-asthmatic extracts from Citrus grandis var. tomentosa. Zhongguo Zhong Yao Za Zhi 31:1350–1352
Mason NP, Barry PW (2007) Altitude-related cough. Pulm Pharmacol Ther 20:388–395
Materazzi S, Nassini R, Gatti R, Trevisani M, Geppetti P. (2008) Cough sensors I: TRP membrane receptors on cough sensors. In: Chung KF, Widdicombe JG (eds) Pharmacology and therapeutics of cough, Handbook of experimental pharmacology, vol 187. Springer, Berlin
Mazzone SB, Undem BJ (2008) Cough sensors IV: Pharmacological modulation of cough sensors. In: Chung KF, Widdicombe JG (eds) Pharmacology and therapeutics of cough, Handbook of experimental pharmacology, vol 187. Springer, Berlin (in press)
Mazzone SB, McLennan L, McGovern AE, Egan GF, Farrell MJ (2007) Representation of capsaicin-evoked urge-to-cough in the human brain using functional magnetic resonance imaging. Am J Respir Crit care Med 176:327–332
McAlexander MA, Carr MJ (2008) Peripheral mechanisms I: Plasticity of peripheral pathways. In: Chung KF, Widdicombe JG (eds) Pharmacology and therapeutics of cough, Handbook of experimental pharmacology, vol 187. Springer, Berlin
McGarvey LP, Heaney LG, Lawson JT, Johnston BT, Scally CM, Ennis M, Shepherd DR, MacMahon J (1998) Evaluation and outcome of patients with chronic non-productive cough using a comprehensive diagnostic protocol. Thorax 53:738–743
McGarvey LP, Forsythe P, Heaney LG, MacMahon J, Ennis M (1999) Bronchoalveolar lavage findings in patients with chronic nonproductive cough. Eur Respir J 13:59–65
Mello CJ, Irwin RS, Curley FJ (1996) Predictive values of the character, timing, and complications of chronic cough in diagnosing its cause. Arch Intern Med 156:997–1003
Moreaux B, Nemmar A, Beerens D, Gustin P (2000) Inhibiting effect of ammonia on citric acid-induced cough in pigs: A possible involvement of substance P. Pharmacol Toxicol 87:279–285
Morice AH, Fontana GA, Belvisi MG, Birring SS, Chung KF, Dicpinigaitis PV, Kastelik JA, McGarvey LP, Smith JA, Tatar M, Widdicombe JG (2007) ERS guidelines on the assessment of cough. Eur Respir J 29:1256–1275
Mutulo D, Bongianni F, Cinelli E, Fontana G, Pantaleo T (2008) Modulation of the cough reflex by antitussive agents within the caudal aspect of the nucleus tractus solitarius in the rabbit. Am J Physiol Regul Integr Comp Physiol, in press
Niimi A (2007) Geography and cough aetiology. Pulm Pharmacol Ther 20:383–387
Niimi A, Torrego A, Nicholson AG, Cosio BG, Oates TB, Chung KF (2005) Nature of airway inflammation and remodeling in chronic cough. J Allergy Clin Immunol 116:565–570

O'Connell F, Thomas VE, Pride NB, Fuller RW (1994) Capsaicin cough sensitivity decreases with successful treatment of chronic cough. Am J Respir Crit Care Med 150:374–380

Ours TM, Kavuru MS, Schilz RJ, Richter JE (1999) A prospective evaluation of esophageal testing and a double-blind, randomized study of omeprazole in a diagnostic and therapeutic algorithm for chronic cough. Am J Gastroenterol 94:3131–3138

Palombini BC, Villanova CA, Araujo E, Gastal OL, Alt DC, Stolz DP, Palombini CO (1999) A pathogenic triad in chronic cough: Asthma, postnasal drip syndrome, and gastroesophageal reflux disease. Chest 116:279–284

Pavord ID, Chung KF (2008) Management of chronic cough. Lancet 371(9):1375–1384

Petit J. (1970) Sensory innervation of the airways. In: Porter R (ed) Breathing: Hering-Breuer centenary symposium. Churchill, London, pp 111–114

Pierse N, Rushton L, Harris RS, Kuehni CE, Silverman M, Grigg J (2006) Locally generated particulate pollution and respiratory symptoms in young children. Thorax 61:216–220

Pintelon L, Brouns I, De Proost I, Van Meir F, Timmermanns JP, Adriaensen D (2007) Sensory receptors in the visceral pleura: Neurochemical coding and live staining in whole mounts. Am J Respir Cell Mol Biol 36:542–551

Poe RH, Harder RV, Israel RH, Kallay MC (1989) Chronic persistent cough. Experience in diagnosis and outcome using an anatomic diagnostic protocol. Chest 95:723–728

Pratter MR (2006) Chronic upper airway cough syndrome secondary to rhinosinus diseases (previously referred to as postnasal drip syndrome): ACCP evidence-based clinical practice guidelines. Chest 129:63S–71S

Pratter MR, Bartter T, Akers S, DuBois J (1993) An algorithmic approach to chronic cough. Ann Intern Med 119:977–983

Rahman MH, Bratveit M, Moen BE (2007) Exposure to ammonia and acute respiratory effects in a urea factory. Int J Occup Environ Health 13:153–159

Sarlani E, Schwartz AH, Greenspan JD, Grace EG (2003) Facial pain as first manifestation of lung cancer: A case of lung cancer-related cluster headache and a review of the literature. J Orofac Pain 17:262–267

Sant'Ambrogio G, Sant'Ambrogio FB (1996) Sensory mechanisms in cough: Role of laryngeal afferens in cough. Pulm Pharmacol 9:309–314

Shannon R, Baekey DM, Morris KF, Li Z, Lindsey BG (2000) Functional connectivity among ventrolateral medullary neurones during fictive cough in the cat. J Physiol 525:207–224

Shannon R, Baekey DM, Morris KF, Nuding SC, Segers LS, Lindsey BG (2004) Production of reflex cough by brainstem respiratory networks. Pulm Pharmacol Ther 17:369–276

Shirahata K, Fujimoto K, Arioka H, Shouda R, Kudo K, Ikeda S (2005) Prevalence and clinical features of cough variant asthma in a general internal medicine outpatient clinic in Japan. Respirology 10:354–358

Smith J, Owen E, Earis J, Woodcock A (2006) Effect of codeine on objective measurement of cough in chronic obstructive pulmonary disease. J Allergy Clin Immunol 117:831–835

Smith Hammond C (2008) Cough and aspiration of food and liquids due to oral pharyngeal dysphagia. Lung 186(Suppl 1):S35–S40

Smyrnios NA, Irwin RS, Curley FJ (1995) Chronic cough with a history of excessive sputum production. The spectrum and frequency of causes, key components of the diagnostic evaluation, and outcome of specific therapy. Chest 108:991–997

Takahama K, Shirasaki T, Soeda F (2008) Central mechanisms III: Neuronal mechanisms of action of centrally-acting antitussives using electrophysiological and neurochemical study approaches. In: Chung KF, Widdicombe JG (eds) Pharmacology and therapeutics of cough, Handbook of experimental pharmacology, vol 187. Springer, Berlin (in press)

Tatar M, Webber SE, Widdicombe JG (1988) Lung C-fibre receptor activation and defensive reflexes in anaesthetized cats. J Physiol 402:411–420

Tatar M, Hanacek J, Widdicombe JG (2008) The expiration reflex from the trachea and bronchi. Eur Respir J 31:385–390

Van Hirtum A, Berckmans D (2004) Objective recognition of cough sound as biomarker for aerial pollutants. Indoor Air 14:10–15

Vovk A, Bolser DC, Hey JA, Danzig M, Vickroy T, Berry R, Martin ADS, Davenport PW (2007) Capsaicin exposure elicits complex airway defensive motor patterns in normal humans in a concentration-dependent manner. Pulm Pharmacol Ther 20:423–432

Widdicombe JG (1954) Receptors in the trachea and bronchi of the cat. J Physiol 123:71–103

Widdicombe JG (1996) Sensory neurophysiology of the cough reflex 98:584–9

Widdicombe JG (2008) Lung afferent activity: Implications for respiratory sensation. Respir Physiol Neurobiol

Widdicombe JG, Fontana G (2006) Cough: What's in a name. Eur Respir J 28:10–15

Widdicombe JG, Sant'Ambrogio G, Mathew OP (1988) Nerve receptors in the upper airway. In: Mathew OP, Sant'Ambrogio G (eds) Respiratory function of the upper airway. Marcel Dekker, New York pp. 193–232

Widdicombe JG, Eccles R, Fontana G (2006) Supramedullary influences on cough. Respir Physiol Neurobiol 152:320–328

Winning AJ, Hamilton RD, Shea SA, Guz A (1988) The effect of airway anaesthesia on the control of breathing and the sensation of breathlessness in man. Clin Sci (Lond) 68:215–225

Yu J (2005) Airway mechanosensors. Respir Physiol Neurobiol 148:217–243

Zemp E, Elsasser S, Schindler C, Kunzli N, Perruchoud AP, Domenighetti G, Medici T, Ackermann-Liebrich U, Leuenberger P, Monn C, Bolognini G, Bongard JP, Brandli O, Karrer W, Keller R, Schoni MH, Tschopp JM, Villiger B, Zellweger JP (1999) Long-term ambient air pollution and respiratory symptoms in adults (SAPALDIA study). The SAPALDIA Team. Am J Respir Crit Care Med 159:1257–1266

Cough Sensors. I. Physiological and Pharmacological Properties of the Afferent Nerves Regulating Cough

B.J. Canning(✉) and Y.-L. Chou

Contents

Abstract The afferent nerves regulating cough have been reasonably well defined. The selective effects of general anesthesia on C-fiber-dependent cough and the opposing effects of C-fiber subtypes in cough have led to some uncertainty about their regulation of this defensive reflex. But a role for C-fibers in cough seems almost certain, given the unique pharmacological properties of these unmyelinated vagal afferent nerves and the ability of many C-fiber-selective stimulants to evoke cough. The role of myelinated laryngeal, tracheal, and bronchial afferent nerve subtypes that can be activated by punctate mechanical stimuli, inhaled particulates, accumulated secretions, and acid has also been demonstrated. These "cough receptors" are distinct from the slowly and rapidly adapting intrapulmonary stretch receptors responding to lung inflation. Indeed, intrapulmonary rapidly and slowly adapting receptors and pulmonary C-fibers may play no role or a nonessential role in cough, or might even actively inhibit cough upon activation. A critical review of the studies of the afferent nerve subtypes most often implicated in cough is provided.

B.J. Canning
Johns Hopkins Asthma and Allergy Center, 5501 Hopkins Bayview Circle, Rm. 3A.24, Baltimore, MD 21224, USA
bjc@jhmi.edu

K.F. Chung, J.G. Widdicombe (eds.), *Pharmacology and Therapeutics of Cough*, Handbook of Experimental Pharmacology 187,

1 Introduction

Cough is a defensive reflex initiated primarily from the larynx, trachea, and large bronchi. Stimuli initiating cough include punctate mechanical stimuli, accumulated secretions, aspirate, particulate (e.g., powder, dust), capsaicin, bradykinin, and interventions that alter the pH or tonicity of airway surface liquid. Although afferents throughout the upper and lower airways and sensory nerves innervating the mediastinum, respiratory muscles, and chest wall all likely contribute to the encoding of cough thresholds and intensity, vagal afferent nerves innervating the large extrapulmonary and intrapulmonary airways are the primary regulators of cough. The physiological properties of airway vagal afferent nerves have been described in detail elsewhere (Canning et al. 2006). In this review, a description of the known physiological, morphological, and pharmacological properties of the vagal afferent nerve subtypes primarily implicated in cough is provided, as well as a summary of the important contributions of Widdicombe.

2 Widdicombe's Studies of Cough and Description of the "Cough Receptors"

The landmark studies by Widdicombe published in 1954 and cited in subsequent papers nearly 1,000 times since remain the best characterization of the afferent nerves regulating cough (Widdicombe 1954a,b,c). Three attributes of those studies account for their importance and lasting impact on the field. First, the methods for maintaining and monitoring respiration and respiratory reflexes while isolating the trachea and bronchi for selective afferent stimulation were highly novel and have served as the model for many subsequent studies of airway neural control. Second, the combination of respiratory reflex measurements with parallel single and/or multifiber afferent nerve recordings as well as phrenic nerve recordings in some preparations provided unmatched insight into the cause and effect of airway neural control. Finally, the rigor with which the studies were carried out – comparing different anesthetics (pentobarbital and chlorolase) with decerebrate preparations, the care with which the afferents were described (see Table 1), the identification of afferent nerve termination sites, the first-ever comparison of pulmonary stretch receptors with tracheal/bronchial stretch receptors, and the differentiation of nearly 300 units into four subtypes – have greatly influenced subsequent studies of airway neural control. Notably, these studies formed the basis of Widdicombe's graduate thesis (Widdicombe 2001). Because this work has been so influential and affirmed repeatedly in the years following, a summary of the key findings serves well as an introduction to this review.

Focusing initially on afferents stimulated by lung inflation in cats, Widdicombe described two mechanically sensitive afferent nerve subtypes. The majority of the afferents identified by lung inflation were slowly adapting, with adaptation indices

Table 1 Tracheal, bronchial, and lung stretch receptor subtypes in cats identified by Widdicombe

	Slowly adapting receptors	Rapidly adapting receptors[a]	Intermediate receptors
Stretch receptors responsive to whole lung inflation			
Adaptation index	<70	≥70	
Response to lung deflation	Not activated	Activated	
Pressure/volume threshold	Low	Moderate	
Tracheal/bronchial inflation	Not activated	Activated	
Tracheal/bronchial deflation	Not activated	Activated	
Termination sites	Peripheral airways/lung	Trachea/carina/ mainstem bronchi	
Stretch receptors responsive to tracheal/bronchial inflation and deflation[a]			
Pressure/volume threshold	Low	Not assessed	High
Active at eupnea	Yes	No	No
Termination site	Bronchi	Trachea/carina	Trachea/bronchi
Tracheal/bronchial mucosal probing	Unresponsive	Activated	Activated
Sulfur dioxide inhalation	11% sensitized	15% sensitized	80% sensitized
Powder inhalation	Not activated	Activated	Not activated
Response to topical anesthesia	Insensitive	Sensitive	Moderately sensitive
Response to repetitive stimulation	Sustained	Sustained	Decrementing

[a]Rapidly adapting receptors as defined by whole lung inflation correspond to the tracheal/bronchial stretch receptors described in the lower portion of the table. Many of the attributes listed (e.g., activated, not activated, low, high) are generalized, not uniformly expressed in each subtype. See the text for further details

of 40 or less. Amongst all fibers classified as *slowly adapting receptors* (SARs; adaptation index of less than 70), just 19% responded to lung deflation. The majority of SARs could be activated by mechanically probing the lung but not by tracheal/bronchial distension, suggesting a peripheral lung termination. About 80% of *rapidly adapting receptors* (RARs; defined by an adaptation index 70 or more) responded to lung deflation. Vagal cooling poorly differentiated the subtypes, while distension/volume thresholds were positively correlated with adaptation index.

Using catheters that allowed partition of the airways into pulmonary and tracheal/mainstem bronchial segments, each region subject to selective distension, and the trachea and bronchi accessible to mechanical probing, Widdicombe subsequently localized the terminals of RARs (as defined by whole lung inflation). In contrast to SARs, RARs terminated in the trachea and bronchi, not in the lung. When only the trachea and bronchi were distended, a highly heterogeneous group of afferent nerves was identified. The physiological properties of 166 such stretch receptors innervating the trachea and bronchi (thus differentiating this study from those of Adrian and of Knowlton and Larrabee (Adrian 1933; Knowlton and Larrabee 1946) were then characterized. Three subtypes were identified. Half of the tracheal and bronchial stretch receptors (82/166) gave a regular discharge in response to modest lung inflations and deflations. Adaptation indices amongst this group of stretch receptors varied widely, with about half of those fully characterized having adaptation indices to tracheal/bronchial inflation of 0–50%. Receptor discharge amongst these tracheal/bronchial SARs increased or decreased incrementally as the tracheal pressure was altered. Most (approximately 90%) SARs in Widdicombe's analysis of the tracheal/bronchial stretch receptors terminated in the mainstem bronchi. Tracheal/bronchial SARs were unaffected by topically applied procaine, just partially inhibited by ether vapor, and only modestly responsive to mechanical stimulation of the airway mucosa.

The two additional subtypes of tracheal and bronchial stretch receptors were described as RARs (46 of 166 fibers) and intermediate receptors (38 of 166). The tracheal and bronchial RARs produced a burst of activity only during the dynamic phases of either inflation or deflation of the trachea/bronchi, all having an adaptation index of 100%. The RARs were more responsive to airway deflation than inflation. The majority (approximately 90%) of tracheal/bronchial RARs terminated in the trachea or carina. Topically applied procaine and ether vapor prevented RAR discharge but had little effect on the other tracheal/bronchial stretch receptors. RARs were also more responsive to dust/powder inhalation and to punctate mechanical stimulation of the airway mucosa than the other airway stretch receptor subtypes. As the name suggests, the intermediate receptors had characteristics of both RARs and SARs, but with one significant difference being a progressive diminution of responsiveness to repeated airway inflations/deflations. The ability of sulfur dioxide to *sensitize* the majority (80%) of intermediate receptors to airway distension or collapse relative to its inability to sensitize tracheal/bronchial RARs (15%) or SARs (11%) to airway pressure changes also differentiated this subtype from the other tracheal/bronchial stretch receptors. Intermediate receptors were distributed throughout the trachea, carina, and bronchi and had inflation pressure thresholds for activation nearly 10 times that of the tracheal/bronchial or pulmonary SARs.

Because the tracheal/bronchial RARs identified by Widdicombe were most responsive to negative luminal pressures, their inflation pressure thresholds were not evaluated. It is also unclear whether tracheal/bronchial RARs and intermediate receptors would have been activated by whole lung inflation unless very high intratracheal pressures were sustained (see later).

Taken together and using responsiveness to whole lung and then tracheal/bronchial inflation and deflation for differentiation, Widdicombe described at least four subtypes of airway and lung stretch receptors in his initial studies: pulmonary SARs, tracheal/bronchial SARs (which adapt rapidly to whole lung inflation), tracheal/bronchial RARs, and intermediate receptors, with characteristics of both the tracheal/bronchial RARs and SARs (Table 1). In subsequent studies, Widdicombe and colleagues (Mills et al. 1969, 1970; Sellick and Widdicombe 1969, 1971; Widdicombe et al. 1962) described another subtype of stretch receptor, the lung irritant receptors, also known as the intrapulmonary RARs, which are described in a subsequent section. Only the pulmonary and tracheal/bronchial SARs and the intrapulmonary RARs (lung irritant receptors) are thought to have any activity at eupnea (owing to their low pressure thresholds). On the basis of the characteristics of the subtypes described above and parallel studies of the stimuli initiating cough,

Widdicombe concluded that tracheal/bronchial RARs and intermediate receptors regulate the coughing initiated by mechanical and chemical (sulfur dioxide) stimulation, respectively. These conclusions have not been refuted in the more than 50 years that have passed since their original publication.

In several places throughout the text of these papers, Widdicombe used the term "cough receptor" to describe the vagal afferent nerves that were activated by stimuli that initiated cough (Widdicombe 1954a,c). Used sparingly by Widdicombe and other physiologists since, the term has nevertheless achieved some degree of acceptance in the clinical literature but with little regard for what specific afferent nerves are being described (Barry et al. 1997; Chang et al. 1996; Fujimura et al. 1992, 1993; Malik et al. 1978). There are several arguments against using this term to describe an airway afferent nerve subtype. For starters, the term is nonsensical, and read literally may conjure images of one's hands, or perhaps handkerchief, amongst the more refined. Second, it is nonspecific, and might prompt grouping of afferent nerve subtypes that can all initiate coughing upon activation but otherwise share no other physiological attributes. The term is also limiting, implying that these afferents may subserve no other reflex functions, and also implying a somewhat circular approach to characterization. Yet for better or worse, it seems, cough researchers are stuck with the term "cough receptor." If so, then *cough receptor* should refer only to afferent nerves with attributes similar to those of the tracheal/bronchial RARs and intermediate receptors identified by Widdicombe, and should not be used to describe any other afferent nerve subtypes implicated in cough (e.g., C-fibers, intrapulmonary RARs) but possessing their own unique physiological characteristics.

3 Identification of the Afferent Nerves Regulating Cough in Anesthetized Guinea Pigs

Guinea pigs have become the most frequently used species in studies of cough despite the fact that up until the past decade or so very little information about the physiological properties of their airway and lung afferent nerves had been published. Most cough studies in guinea pigs have been carried out in conscious animals, arguably the most clinically relevant approach for pharmacological analysis but providing limited physiological insight (Belvisi and Bolser 2002; Canning 2008; Karlsson and Fuller 1999; Lewis et al. 2007). Still, there are many advantages to the guinea pig for cough research. Guinea pigs can be used in both conscious and anesthetized models of cough, largely infeasible in cats and dogs (because it is difficult to study conscious cough in these species), or in human subjects and nonhuman primates (because the desirable, invasive aspects of cough studies done in anesthetized animals are generally not possible in humans or nonhuman primates). Guinea pigs are also small enough (as are rabbits, cats, and the rodents rats and mice; guinea pigs may not be rodents; D'Erchia et al. 1996) such that stereotaxic methods for studying central processes relevant to cough can be employed, as well as tracing and in vitro electrophysiological studies. Guinea pigs also cough to the same stimuli (e.g., capsaicin, bradykinin, acid, punctate mechanical stimuli) that initiate coughing in human subjects, while there is some controversy as to whether mice or rats cough (Belvisi and Bolser 2002; Kamei et al. 1993; Ohi et al. 2004; Tatar et al. 1996, 1997). Studies from the laboratories of Advenier, Belvisi, Bolser, Chung, Fujimura, Kamei, Karlsson, McLeod, Morice, Sekizawa, Tatar and colleagues, and others have defined the pharmacology and pathophysiology of cough in guinea pigs (Bolser et al. 1991, 1994, 1997; Daoui et al. 1998; El-Hashim and Amine 2005; Forsberg et al. 1988; Fox et al. 1996; Gatti et al. 2006; Girard et al. 1995; Hara et al. 2008; Jia et al. 2002; Kamei and Takahashi 2006; Kamei et al. 2005; Karlsson et al. 1991a,b; Lalloo et al. 1995; Laude et al. 1993; Leung et al. 2007; Lewis et al. 2007; Liu et al. 2001; McLeod et al. 2001; O'Connell et al. 1994; Pinto et al. 1995; Plevkova et al. 2004; Tatar et al. 1996, 1997; Trevisani et al. 2004; Xiang et al. 2002). These advantages and the careful electrophysiological analyses by Bergren (Bergren 1997, 2001; Bergren et al. 1984; Bergren and Kincaid 1984; Bergren and Myers 1984; Bergren and Sampson 1982), Fox (Fox et al. 1993, 1995), Joad (Bonham et al. 1995, 1996; Joad et al. 1997, 2004; Mutoh et al. 1999), Tsubone (Sano et al. 1992; Tsubone et al. 1991), Undem (Canning et al. 2004; Chuaychoo et al. 2005, 2006; Kollarik and Undem 2002; Lee et al. 2005; McAlexander et al. 1999; McAlexander and Undem 2000; Riccio et al. 1996; Ricco et al. 1996; Undem et al. 2004), and colleagues have made it feasible to identify the afferent nerves regulating cough in guinea pigs.

Working from the pre-existing knowledge base about laryngeal, tracheal, and bronchial afferent nerves in guinea pigs, Canning et al. (2004) established a model in anesthetized guinea pigs whereby cough could be evoked electrically, mechanically, or by acid applied topically to the laryngeal and tracheal mucosa. Capsaicin or bradykinin applied topically to the tracheal mucosa of these anesthetized guinea pigs did not evoke coughing. It was also observed that cutting the recurrent laryngeal nerves prevented cough evoked from the rostral trachea and larynx, while cutting the superior laryngeal nerves was without effect. A subsequent analysis of the responsiveness and projections of the various afferent nerve subtypes innervating the trachea and larynx was compared with the results of these cough studies. Responsiveness to mechanical and acid stimulation did not reliably differentiate the afferents that regulate cough, nor did recurrent laryngeal nerve transections. Thus, all three known tracheal/laryngeal afferent nerve subtypes project axons to the trachea and larynx via the recurrent laryngeal nerves, and all subtypes, albeit with varying sensitivities, are responsive to acidic and mechanical stimuli (Kollarik and Undem 2002; Ricco et al. 1996; Undem et al. 2004). But the inability of capsaicin or bradykinin to acutely initiate coughing when applied topically to the tracheal and laryngeal mucosa of anesthetized guinea pigs strongly implicated the capsaicin-insensitive afferent nerves arising from the nodose ganglia (Myers et al. 2002; Ricco et al. 1996). Conversely, the inability of the superior laryngeal nerves to sustain a cough reflex following recurrent laryngeal nerve transection argued against afferent nerves arising from the jugular ganglia, given that very few nodose ganglia neurons project to the airways via the superior laryngeal nerves, while about half of the jugular ganglia neurons innervating the larynx and rostral trachea project to these airways via the superior laryngeal nerves. The evidence described above combined with the inability of capsaicin desensitization (which renders jugular ganglia neurons unresponsive to any stimuli) to prevent electrical-, mechanical-, or acid-induced coughing led to the conclusion that the capsaicin-insensitive nodose ganglia neurons innervating the trachea and larynx were both sufficient and necessary for initiating the cough reflex in anesthetized guinea pigs (Canning et al. 2004).

The nodose ganglia neurons innervating the larynx, trachea, and mainstem bronchi of guinea pigs had in previous studies by Undem and colleagues been called RARs on the basis of their response to punctate mechanical stimuli (McAlexander et al. 1999; Myers et al. 2002). To better define these afferent nerves regulating cough, we compared their physiological properties wit those of intrapulmonary receptors innervating the guinea pig airways and lungs. Using a whole lung preparation, we identified intrapulmonary afferent nerve subtypes that adapted either rapidly or slowly to distending pressures applied to the trachea. Like the tracheal afferent nerves regulating cough, both stretch receptor subtypes innervating the intrapulmonary airways and lung had cell bodies in the nodose ganglia. But in contrast to the tracheal/bronchial cough receptors, which have axonal conduction velocities of approximately $5\,\mathrm{m\,s^{-1}}$, the intrapulmonary stretch receptors have axon conduction velocities of approximately $16\,\mathrm{m\,s^{-1}}$. Other differences, including responsiveness to distending pressures, airway smooth muscle contraction, and ATP receptor activation, led to the conclusion that the afferent nerves regulating cough evoked from the trachea and larynx are distinct from the well-defined RARs, SARs, and C-fibers of the airways and lungs (Canning et al. 2004).

The afferent nerves regulating cough in guinea pigs are similar to those described by Widdicombe in the cat (Widdicombe 1954a). Differences may, however, exist. Widdicombe's studies leave it unclear whether all of the fibers responding to tracheal/bronchial pressure changes and comprising three tracheal/bronchial subtypes would have been identified as RARs in whole lung inflation studies (Widdicombe 1954a). In the guinea pig, cough receptors are unresponsive to changes in luminal pressure (Canning et al. 2004). In fact, the cough receptors described in cats were activated by selective tracheal/bronchial distension or collapse, stimuli that reportedly caused cough in this species, whereas the guinea pig cough receptors were insensitive to airway pressure changes, even highly unphysiological distending and collapsing pressures, and these pressure changes also failed to cause cough in guinea pigs. This may simply reflect the physical/structural differences in cat and guinea pig airways. It is also possible, however, that the apparent species difference is nonexistent. Thus, of the eight RARs (as identified by whole lung inflation) characterized by Widdicombe using both whole lung and tracheal/bronchial lung inflation and prompting his conclusion that lung RARs were in fact tracheal/bronchial stretch receptors, all eight were described as SAR-type tracheal/bronchial stretch receptors (Widdicombe 1954a). The tracheal/bronchial SARs are essentially identical in sites of termination (bronchi) and function to the intrapulmonary RARs described by Widdicombe in subsequent studies (Mills et al. 1969, 1970; Sellick and Widdicombe 1969, 1971). They are also the only tracheal/bronchial stretch receptor subtype *not* implicated in cough in cats by this investigator, and they represent only about half of the stretch receptors activated by tracheal bronchial luminal pressure changes. Also, while Widdicombe stated in the text that changes in tracheal/bronchial luminal pressures evoked coughing, the data presented in support of this assertion are not convincing (Widdicombe 1954c). As the reader can test, forceful expiratory and inspiratory efforts against a closed glottis can generate very large positive and negative airway pressures and sensations, and yet does not evoke coughing (Davies et al. 1984; Green and Kaufman 1990; Rao et al. 1981). In fact, Widdicombe argued against the notion that luminal pressure changes were a relevant stimulus for initiating cough, given the high pressure thresholds for the tracheal/bronchial RARs and intermediate receptors (Widdicombe 1954a, 1954c).

Widdicombe's studies in anesthetized cats (Widdicombe 1954a, 1954c) and our studies in anesthetized guinea pigs (Canning et al. 2004) concluded that the afferent nerves that regulate cough were myelinated (on the basis of axonal conduction velocities or sensitivity to vagal cooling). It has not been established whether the axons of cat cough receptors are also relatively slowly conducting (approximately $5\,\mathrm{m\,s^{-1}}$), as in guinea pigs. It is also unclear whether the cough receptors in cats are insensitive to smooth-muscle contraction, as are guinea pig cough receptors, in contrast to the intrapulmonary RARs in all species (see later).

Through the combination of electrophysiological studies with intravital labeling methods, retrograde neuronal tracing, organotypic cultures, and immunohistochemistry, the peripheral terminals of cough receptors in the guinea pig trachea and bronchus have been identified (Canning et al. 2006b). Terminating between the epithelium and smooth-muscle layers of the airways mucosa, the cough receptors assume a circumferential position in the extracellular matrix. Branching is extensive at the terminals, with axons projecting from longitudinal nerve bundles through the smooth-muscle layer. Similar structures have been described in the airway mucosa of other species but their identity as cough receptors is unclear (De Proost et al. 2007; Gaylor 1934; Larsell 1921, 1922; Yamamoto et al. 1995; Yu 2005). Immunohistochemistry confirms the selective expression of subtypes of Na^+–K^+–ATPase and Na^+–K^+–$2Cl^-$ transporter in guinea pig cough receptors (Canning et al. 2006b; Mazzone and McGovern 2006). More recently, Tetrodotoxin-insensitive Na^+ channels have been localized to these cough receptors (Kwong et al. 2008). Pharmacological analyses suggest that these regulators of ion flux and gradients, as well as Cl^- channels and voltage-sensitive K^+ channels may be critical to the regulation of cough receptor responsiveness to chemical (acid) and punctate mechanical stimuli (Canning 2007; Canning et al. 2006a; Fox et al. 1995; Mazzone and McGovern 2006; McAlexander and Undem 2000). No other stimuli thus far studied, including a variety of autacoids and neurotransmitters and ion channel modulators, alter cough receptor excitability or the ability of acid or mechanical stimuli to initiate coughing in guinea pigs.

4 Intrapulmonary Rapidly Adapting Receptors

The term "rapidly adapting receptor" (RAR) was originally used to describe a subtype of airway and lung stretch receptor that is activated during the dynamic phase of lung inflation, but that unlike SARs becomes quiescent during static lung inflation (Knowlton and Larrabee 1946; Widdicombe 1954a). While elegant in clarity and as a means of differentiating the stereotypical responses of these subtypes of airway sensory nerves to this specific stimulus, the term "rapidly adapting" has also been the source of considerable confusion. Often, describing an afferent nerve as *rapidly adapting* or *slowly adapting* based on responsiveness to one stimulus fails to adequately describe the properties of the spectrum of afferent nerves captured by this term (Canning et al. 2004; Widdicombe 1954a; Yu 2000, 2005). Indeed, there are reports of rapidly adapting SARs (Bergren and Peterson 1993; Widdicombe 1954a; Yu 2000, 2005). Other studies have shown that RARs, as defined by their response to static lung inflation, may adapt very slowly to lung deflation, airway smooth muscle contraction, pulmonary embolism, and carbon dust inhalation (Armstrong and Luck 1974; Bergren 1997; Canning et al. 2004; Ho et al. 2001; Mills et al. 1969, 1970; Sellick and Widdicombe 1969, 1971; Widdicombe, 1954a). Still another problem is use of the term "rapidly adapting receptors" to describe subtypes of airway sensory nerves (extrapulmonary *irritant* receptors (cough receptors), intrapulmonary irritant receptors or RARs) that share few physiological or morphological properties (Canning et al. 2004; Sellick and Widdicombe 1969, 1971; Widdicombe 1954a). Thus, in this review article, the term "rapidly adapting receptor" refers only to those intrapulmonary stretch receptors that rapidly adapt to sustained lung inflation.

Intrapulmonary RARs in most species display some activity during the dynamic phase of inspiration. Basal activity in RARs varies widely within and amongst species (Fig. 1), which may reflect heterogeneity of this subtype but also in how these receptors are defined. In general, RARs are considerably less active than SARs but more active than C-fibers during tidal breathing. Given their response to lung stretch/inflation, it is perhaps not surprising that RARs are activated (either directly or indirectly) by a variety of mechanical stimuli in the lung, including airway smooth muscle contraction, pulmonary edema, decreased lung compliance, lung collapse, and negative airway luminal pressures. The responsiveness of RARs to airway smooth muscle contraction implies that RARs may be associated with airway smooth muscle. Direct evidence for such an association in the intrapulmonary airways is, however, lacking, as there are no morphological studies adequately identifying the peripheral terminals of RARs.

Intrapulmonary RAR activation initiates bronchospasm and mucus secretion via parasympathetic reflexes and tachypnea, characterized by decreased postexpiratory/preinspiratory pause (Canning et al. 2001; Haxhiu et al. 1997, 2000; Mills et al. 1969; Raj et al. 1995; Widdicombe et al. 1962; Yu et al. 1989) (Fig. 2). But intrapulmonary RAR activation does not initiate coughing. Intrapulmonary RAR stimulants, including lung deflation/collapse, pulmonary embolism, decreased lung compliance, and airway smooth muscle contraction, are highly ineffective at initiating cough in either conscious or anesthetized animals or humans (Canning et al.

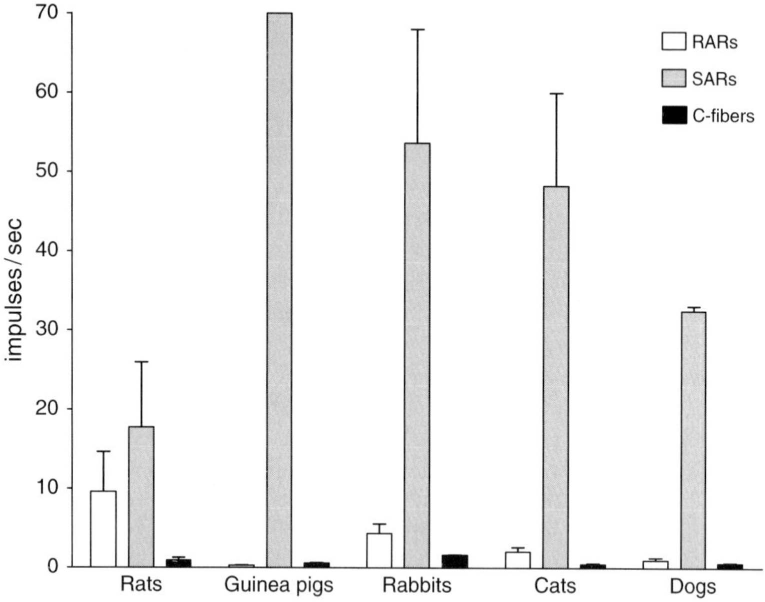

Fig. 1 Basal activity of rapidly adapting receptors (RARs), slowly adapting receptors (SARs), and C-fibers during tidal breathing in various species (Adcock et al. 2003; Armstrong and Luck 1974; Bergren 1997, 2001; Bergren and Peterson 1993; Bonham et al. 1995; Coleridge and Coleridge 1977, 1984; Coleridge et al. 1965, 1983; Davies et al. 1996; Delpierre et al. 1981; Green et al. 1986; Gu et al. 2003; Gu and Lee 2002; Hargreaves et al. 1993; Ho et al. 2001; Jonzon et al. 1986; Karlsson et al. 1993; Kaufman et al. 1980; Kunz et al. 1976; Lee and Morton 1993; Lin and Lee 2002; Matsumoto 1997; Matsumoto et al. 1990; Mills et al. 1970; Mohammed et al. 1993; Paintal 1969; Sellick and Widdicombe 1971; Vidruk et al. 1977; Yu et al. 1987, 1987, 1991)

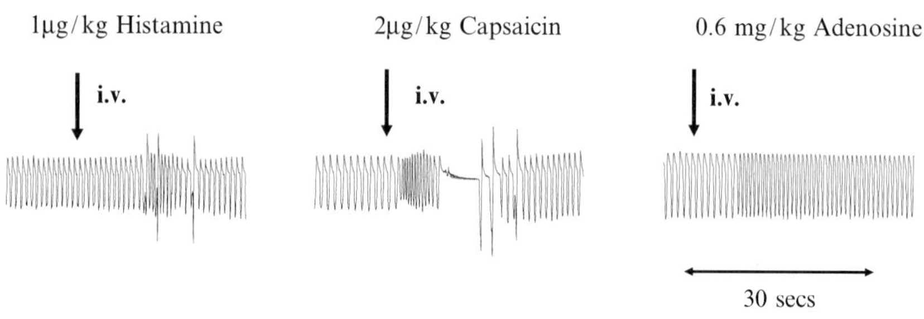

Fig. 2 Representative traces illustrating the respiratory reflex effects initiated in anesthetized guinea pigs by activation of RARs by intravenous histamine and C-fibers by intravenously administered capsaicin and adenosine. Adenosine acts selectively on C-fibers arising from the nodose ganglia in guinea pigs, whereas capsaicin activates all bronchopulmonary C-fibers in this species. Note that none of these challenges initiate coughing and the distinct differences in reflex effects initiated by the nonselective C-fiber stimulant capsaicin, which evokes both apnea and tachypnea, and the nodose C-fiber-selective stimulant adenosine, which evokes only tachypnea. In these same animals, mechanically probing the laryngeal, tracheal, or bronchial mucosa, or acid applied topically to the mucosa readily evokes coughing. These studies illustrate the distinct reflex effects initiated by RAR and C-fiber subtypes and facilitates differentiation of the afferent nerves regulating cough from other afferent nerve subtypes in the airways

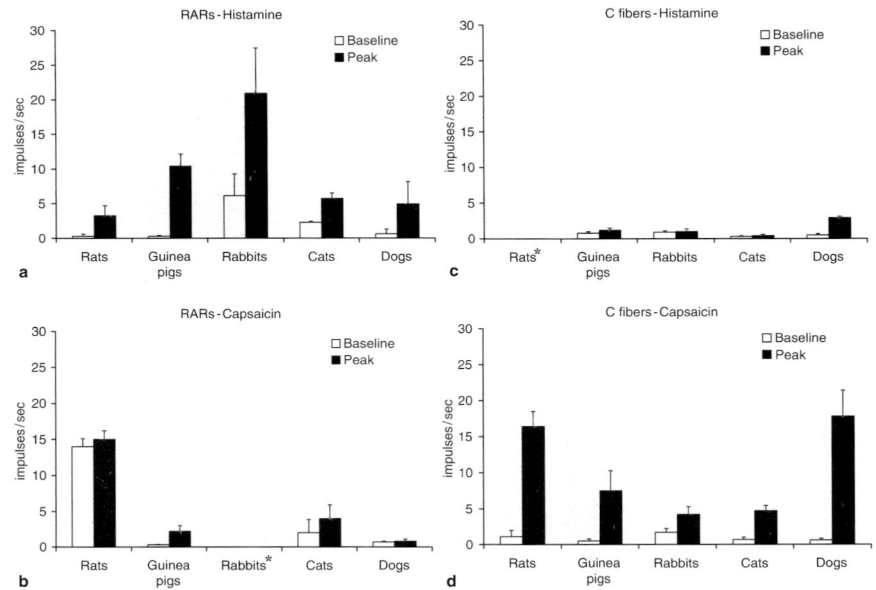

Fig. 3 a–d Mean data illustrating the responsiveness of airway and lung RARs and C-fibers in various species to histamine and capsaicin challenges. Open bars depict mean ± standard error of the mean (SEM) basal activity during tidal breathing. Filled bars depict mean ± SEM peak activity following challenge with either capsaicin or histamine. Where multiple studies have evaluated the actions of histamine or capsaicin on a specific afferent nerve subtype in a single species, the average of the results in these studies is presented (Adcock et al. 2003; Armstrong and Luck 1974; Bergren 1997, 2001; Bergren and Peterson 1993; Coleridge and Coleridge 1977, 1984; Coleridge et al. 1965; Delpierre et al. 1981; Ho et al. 2001; Lee and Morton 1993; Lin and Lee 2002; Mills et al. 1970; Mohammed et al. 1993; Paintal 1969; Sellick and Widdicombe 1971; Vidruk et al. 1977). Both stimuli were administered either intravascularly (bronchial artery or right atrial injection) or by inhalation at various doses and in animals that were either freely breathing or paralyzed and ventilated, with or without open chest. Despite these different experimental conditions, a clear pattern of selectivity for capsaicin (C-fibers) and histamine (RARs) is apparent. Comparable selectivity for C-fibers has been reported for bradykinin and 5-HT3 receptor selective agonists, while stimuli that like histamine evoke bronchospasm, such as substance P, methacholine, and efferent vagus nerve stimulation, have been found to be relatively selective for RARs *not previously studied

2006b). Conversely, stimuli that do initiate coughing (e.g., capsaicin, bradykinin, acid) are either modestly effective or ineffective at activating RARs (Armstrong and Luck 1974; Bergren 1997; Coleridge and Coleridge 1984; Ho et al. 2001; Mohammed et al. 1993) (Fig. 3 a–d). It is unclear whether intrapulmonary RARs can even modulate the cough reflex. Pulmonary edema is a potent RAR (and pulmonary C-fiber) stimulant but fails to induce cough and may even inhibit cough acutely (Korpas et al. 1993; Polacek et al. 1986; Sellick and Widdicombe 1969). In humans, there is a poor correlation between lung function and responsiveness to tussive stimuli. Asthmatics, for example, may have a cough response to experimental challenge that is indistinguishable from that of normal subjects (Dicpinigaitis 2007). Similarly, acute pretreatment with bronchodilators or bronchoconstrictors has no effect on cough responsiveness despite profoundly and oppositely altering airway smooth muscle tone and likely RAR activity (Fujimura et al. 1992, 1993). The data in animals are less clear. There is a very poor correlation between the ability of stimuli to initiate coughing and bronchospasm. The bronchoconstrictor substance P is ineffective at inducing cough and without effect on cough evoked subsequently by citric acid (El-Hashim and Amine 2005). We have found similar results in anesthetized guinea pigs using histamine as a means of inducing bronchospasm (Fig. 4). By contrast, in dogs, histamine and allergen (both of which induce bronchospasm) fail to evoke coughing but greatly enhance coughing evoked subsequently by mechanical stimulation of the airways (House et al. 2004).

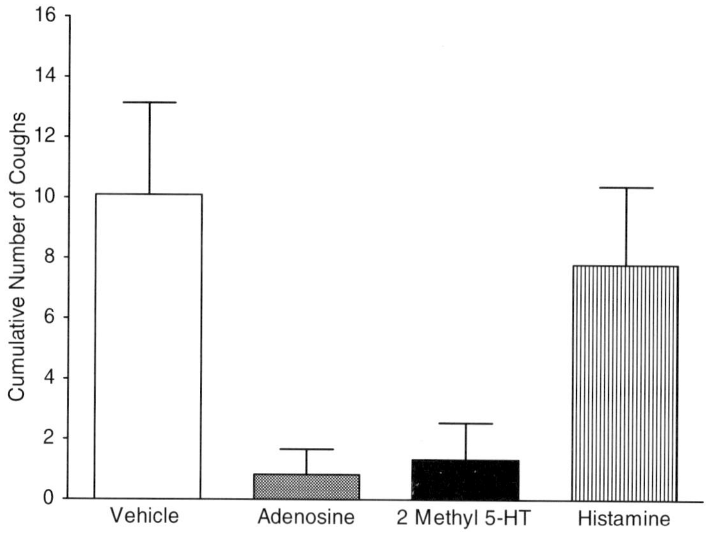

Fig. 4 Selective activation of pulmonary C-fibers arising from the nodose ganglia with either adenosine or 2-methyl-5-hydroxytryptamine (a 5-HT3 receptor selective agonist) markedly inhibits citric acid induced coughing in anesthetized guinea pigs, while intrapulmonary RAR activation by histamine is without effect. Citric acid (0.001–2 M) was applied topically to the tracheal mucosa in $100-\mu$L aliquots at 1-min intervals and in ascending concentrations. Each bar represents the mean ± SEM cumulative number of coughs evoked by citric acid in five to six experiments

Problematic in interpreting these results, however, are the indirect consequences of bronchospasm on the airways that might influence subsequently evoked reflexes, including changes in respiratory pattern and blood gases and perhaps a coincidently evoked reflex mucus secretion (Cohn et al. 1978; Coleridge et al. 1982; Green et al. 1986; Lin et al. 2005; Olgiati et al. 1981). Histamine may also modulate excitability or activate some airway C-fibers (Coleridge and Coleridge 1984; Lee and Morton 1993). Taken together, however, the data suggest more of a modulatory, nonessential role of intrapulmonary RARs in cough.

5 C-Fibers

Given their defining physiological attribute of an axonal conduction velocity of $2\,\mathrm{m\,s^{-1}}$ or less, bronchopulmonary C-fibers are the most readily identifiable vagal afferent nerve subtype innervating the airways. C-fibers can be activated by several chemical and mechanical stimuli, with responses depending upon the stimulus and the C-fiber subtype studied (Coleridge and Coleridge 1984; Lee and Pisarri 2001; Ricco et al. 1996; Undem et al. 2004). The majority of C-fibers innervating the airways and lungs of all species are activated by the TRPV1 receptor agonist capsaicin, a predictable observation, given the known expression patterns of TRPV1 in afferent C-fibers throughout the body of most species (Caterina et al. 1997). But it is inappropriate to conclude from these data that responsiveness to capsaicin is the defining characteristic of airway C-fibers. C-fibers in dogs, rats, and mice that are not activated by lung capsaicin challenge have been described (Coleridge and Coleridge 1984; Ho et al. 2001; Kollarik et al. 2003). Moreover, perhaps secondary to the end organ effects associated with C-fiber activation (mucus secretion, vascular engorgement, airway smooth muscle contraction, altered respiratory pattern, and cough), other afferent nerve subtypes, especially intrapulmonary RARs, can be activated by capsaicin challenge (Bergren 1997; Mohammed et al. 1993; Morikawa et al. 1997). A lack of responsiveness to mechanical stimulation and basal activity may also fail to differentiate C-fibers from other subtypes of bronchopulmonary afferent nerves. While C-fibers are generally less responsive to mechanical stimulation, they can be activated by punctate mechanical stimulation or lung inflation, and can have basal activity comparable to that of some RARs (Coleridge and Coleridge 1984; Fox et al. 1993; Lee and Pisarri 2001; Ricco et al. 1996).

C-fibers are found throughout the airways and lungs of all species. The extensively branched terminals of C-fibers in guinea pig and rat tracheae can be immunohistochemically labeled for the neuropeptides calcitonin gene-related peptide (CGRP), substance P, and neurokinin A (Baluk et al. 1992; Hunter and Undem 1999; Kummer et al. 1992; McDonald et al. 1988; Yamamoto et al. 2007). Comparable structures can be found in the airways of other species and in the peripheral airways of guinea pigs (Dey et al. 1990; Lamb and Sparrow 2002; Watanabe et al. 2006; Yamamoto et al. 1998). C-fiber terminals can also be found in the airway microvasculature and airway smooth muscle layer, and comprise at least a portion of Paintal's J-receptors, suggesting peripheral/interstitial lung terminations (Baluk et al. 1992; McDonald et al. 1988; Paintal 1973). To date, however, very little specific information about the intrapulmonary airway and lung terminations of C-fibers is available.

The chemical stimuli most effective at activating bronchopulmonary C-fibers, including capsaicin, bradykinin, and acid, are similarly very effective at initiating cough in conscious human subjects and in conscious animals (Dicpinigaitis 2007; Forsberg et al. 1988; Jia et al. 2002; Karlsson and Fuller 1999; Laude et al. 1993; Trevisani et al. 2004). These stimuli work entirely or partly through TRPV1, and immunohistochemical and single-cell PCR confirms expression of TRPV1 in airway C-fibers (Groneberg et al. 2004; Kwong et al. 2008; Myers et al. 2002; Watanabe et al. 2006). Prior capsaicin desensitization prevents citric acid induced coughing in awake guinea pigs, as does pretreatment with TRPV1 receptor antagonists (Bolser et al. 1991; Forsberg et al. 1988; Gatti et al. 2006; Lalloo et al. 1995; Leung et al. 2007; Trevisani et al. 2004). Taken together, these and other observations argue strongly for a role of bronchopulmonary C-fibers in cough (Canning et al. 2006b).

Some controversy about the role of C-fibers in cough has arisen from the utter inability of C-fiber-selective stimulants to evoke coughing in anesthetized animals (Canning et al. 2004, 2006a; Karlsson et al. 1993; Tatar et al. 1988, 1994) (Fig. 2). Anesthesia has no effect on coughing evoked by mechanical or acid stimulation of the airway mucosa and does not prevent C-fiber activation or other C-fiber-dependent reflexes, and yet capsaicin and bradykinin have been consistently ineffective at evoking cough in anesthetized animals (Canning et al. 2006a; Coleridge and Coleridge 1984; Tatar et al. 1988). But perhaps it should be expected that C-fiber-selective stimulants would fail to evoke coughing in anesthetized animals. Airway and lung C-fibers share many characteristics with somatosensory nociceptors, and it is the objective of general anesthesia to prevent the sensations and reflexes associated with nociceptor activation.

While the effects of anesthesia on nociceptor signaling may explain the inability of C-fiber-selective stimulants to evoke coughing in anesthetized animals, anesthesia cannot account for the known acute inhibitory effects C-fiber activation may have on cough in anesthetized animals, or the inability of some C-fiber stimuli to evoke coughing in conscious animals and in conscious human subjects (Tatar et al. 1988, 1994). We speculated that C-fiber subtypes might account for these opposing effects on cough. Subtypes have been described in several species (Coleridge and Coleridge 1984; Kollarik et al. 2003; Undem et al. 2004). In guinea pigs, airway vagal C-fiber subtypes can be differentiated by their ganglionic origin, distribution in the airways, and responsiveness to ATP, adenosine, and serotonin 5-HT3 receptor agonists (Chuaychoo et al. 2005, 2006; Undem et al. 2004). The ability of C-fiber activation to evoke coughing in awake guinea pigs is reasonably well established, and we also reported a facilitating effect of C-fiber activation on cough (Canning et al. 2006b; Mazzone et al. 2005). In these latter studies, capsaicin or bradykinin applied topically to the tracheal mucosa greatly enhanced sensitivity to subsequent tussive stimuli. On the basis of the location of these bradykinin and capsaicin challenges, C-fibers arising from the jugular ganglia likely promote coughing. By inference, then, we further speculated that nodose C-fiber activation might acutely inhibit coughing. Consistent with this hypothesis, we found that selective activation of nodose C-fibers with adenosine or 2-methyl-5-hydroxytryptamine did not evoke coughing but greatly reduced the ability of citric acid to evoke coughing in anesthetized animals (Fig. 4). Prior adenosine inhalation also inhibited capsaicin-induced coughing in conscious guinea pigs.

The results of studies carried out in other species are at least consistent with the notion that C-fiber subtypes may have opposing effects on cough. In anesthetized dogs and cats, C-fiber activation by either capsaicin or phenyldiguanide (a 5-HT3 receptor agonist) can inhibit cough (Tatar et al. 1988, 1994). There is no published evidence that capsaicin, bradykinin, or citric acid can evoke coughing in awake dogs and cats, but if they cannot, these would be the only species studied that are known to cough but do not cough in response to these stimuli when awake. Similarly, in rats, a species in which we (and others) have been unable to evoke cough, it is unclear from the published literature whether the jugular type (which are insensitive to ATP, adenosine, and serotonin receptor activation), cough-promoting C-fibers that can be found in guinea pigs innervate the airways of rats, a species that may have little or no capacity to cough (Belvisi and Bolser 2002; Lee and Pisarri 2001; Ohi et al. 2004; Tatar et al. 1996, 1997). In rabbits, a species in which cough can be evoked by citric acid aerosol inhalation (at least suggestive of a TRPV1 and C-fiber-dependent mechanism; Adcock et al. 2003; Tatar et al. 1997), it has also been reported that acute sulfur dioxide inhalation is acutely inhibitory on cough (Hanacek et al. 1984). These data have been reasonably interpreted as evidence for a permissive effect of SARs in cough, given the peculiar sensitivity of rabbit SARs to sulfur dioxide inhalation. But sulfur dioxide is also known to activate lung C-fibers (Ho et al. 2001). Adcock and colleagues even speculated that the inhibitory effects of the compound RSD931 in cough induced in rabbits might be due to its ability to activate pulmonary C-fibers (Adcock et al. 2003). In humans, who cough readily in response to capsaicin and bradykinin challenge, serotonin and adenosine challenge may not cause coughing (Burki et al. 2005; Stone et al. 1993). There is also at least one report of serotonin-mediated inhibition of cough in human subjects (Stone et al. 1993). A comparable inability of intravenous capsaicin to evoke coughing has been reported in studies using conscious nonhuman primates (Deep et al. 2001).

6 Central Terminations and Pharmacology of Airway and Lung Afferent Nerves

Studies of airway reflexes in response to stimuli known to be selective for the various airway afferent nerve subtypes largely substantiate the accepted classification schemes for afferent nerves. Implicit in the observation that afferent nerve subtypes subserve distinct reflex functions is that central termination sites of the various afferent nerve subpopulations must diverge to some extent, allowing for reflex specificity. From the little published evidence available, this notion would seem to be substantiated. Most of the work on central terminations of airway sensory nerves has been carried out in cats and rats. Bronchopulmonary C-fibers and RARs terminate extensively and often bilaterally in nucleus tractus solitarius (nTS), particularly in the commissural and medial subnuclei (Bonham and Joad 1991; Davies and Kubin 1986; Ezure et al. 1991; Kalia and Richter 1988; Kubin et al. 1991, 2006; Lipski et al. 1991; Mazzone and Canning 2002; Otake et al. 1992). SARs terminate primarily ipsilateral to their vagal origin, rostral to obex in the lateral and interstitial subnuclei (Bonham and McCrimmon 1990; Davies et al. 1987; Ezure et al. 2002; Kalia and Richter 1985; Kubin et al. 2006). No attempt at differentiating termination sites of RAR, SAR, or C-fiber subtypes has been described.

With respect to the central terminations of the afferent nerve subtypes primarily implicated in cough (cough receptors, tracheal/bronchial C-fibers, as opposed to pulmonary C-fibers), very little specific information derived from physiological studies has been published. The complex multichannel recordings of Shannon and colleagues, and the anatomical results from c-fos and tracing studies provide some insight (Bolser et al. 2006; Gestreau et al. 1997; Jakus et al. 2008; Ohi et al. 2005; Shannon et al. 2004), but to date there have been no electrophysiological or functional studies definitively implicating specific nTS subnuclei as primary sites of cough receptor or tracheal/bronchial C-fiber termination. There are several potential explanations for this gap in our knowledge. For starters, while any brainstem recording in a freely breathing animal is problematic, it would be especially problematic in an animal that was coughing. Similarly, while it is somewhat complicated to deliver stimuli selectively to the airways while an animal is held in a stereotaxic frame, it would be even more difficult to deliver stimuli selectively to the larynx, trachea, and/or mainstem bronchi in such restrained animals. Even if an experimenter could devise a way in which stimuli could be isolated to these large airways and maintain viable recordings while the animals were coughing, the number of afferents targeted would represent only a fraction of those targeted in whole lung challenges (lung inflations or deflations, intravenous capsaicin) to search for nTS termination sites of intrapulmonary RARs, SARs, or C-fibers. In short, an experimenter would have to be both very lucky and very skilled to position an electrode near enough to the relay neurons of the very few afferent nerves targeted by a stimulus confined to the extrapulmonary airways.

As an alternative approach, we have employed a functional assay combined with stereotaxic microinjections to identify the location of the primary synapses of the tracheal afferent nerves regulating cough in guinea pigs (Canning et al. 2006b). What made this approach feasible was the ease with which we could transition from microinjection to subsequent cough challenges in guinea pigs, and the relatively small portion of the airway from which cough is evoked in our studies. We varied the concentration of glutamate receptor antagonists and the location of bilateral microinjection in an attempt to locate the nTS location likely corresponding to the cough receptor termination sites. Bilateral microinjections of a combination of the glutamate receptor antagonists CNQX (6-Cyano-7-nitroquinoxaline-2,3-dione) and AP-5 (D-(-)-2-Amino-5-phosphonopentanoic acid) about 1 mm rostral from obex and 1 mm lateral from midline, near the medial and/or intermediate subnuclei of nTS, readily blocked cough at doses that were without effect when microinjected into adjacent nTS/brainstem locations (0.5–2 mm distal). Importantly, these same microinjections were without effect on basal respiratory rate (largely SAR dependent) or tachypnea evoked by either intravenous histamine (an intrapulmonary RAR-dependent reflex) or intravenous bradykinin (a bronchopulmonary C-fiber-dependent reflex). Subsequent anterograde and retrograde tracing experiments have confirmed a preponderance of cough receptor terminations in this discrete region of nTS.

Establishing this cough model and the stereotaxic locations relevant to cough will guide subsequent electrophysiological analyses. These methods also permit more thorough pharmacological analyses of the central regulation of cough. Initial studies have documented a prominent role of glutamate *N*-methyl-D-aspartate receptors in the regulation of cough. As has been reported by other investigators, γ-aminobutyric acid (inhibiting) and substance P (facilitating) were found to have opposing effects on cough (Bolser et al. 1994; Joad et al. 2004; Mazzone et al. 2005; Mutolo et al. 2007). Previous studies have implicated a role for neurokinins acting centrally in the regulation of cough and other C-fiber-dependent reflexes, even though neurokinins seem to have little or no direct role in RAR-, SAR-, or cough-receptor-dependent signaling in nTS (Bolser et al. 1997; Canning et al. 2001; Mazzone and Canning 2002; Mazzone et al. 2005).

7 Conclusions

Much of what we know about the afferent nerves regulating cough was established in the initial studies of Widdicombe. Unfortunately, these relatively firmly established notions relating to cough have been obscured by an imprecise nomenclature. The role of C-fibers in cough may be the most widely accepted notion, given their unique pharmacological properties and the ability of C-fiber-selective stimulants to initiate coughing in awake animals and human subjects. A role for the laryngeal/tracheal/bronchial mechanoreceptors activated by mechanical stimulation and acid has also been well established. What has not been so well established, however, is an appropriate name for these mechanoreceptors. While they rapidly adapt to a punctate mechanical stimulus, they share few other physiological characteristics with other afferent nerve subtypes called RARs. Cough receptors might be an appropriate name, but only if used to describe these nerves alone. Future studies of the afferent nerves regulating cough should fill gaps in our knowledge about the central pathways regulating cough, and potential targets for cough therapy at the nerve terminals and perhaps also at the central terminations of these vagal afferent nerves.

References

Adcock JJ, Douglas GJ, Garabette M, Gascoigne M, Beatch G, Walker M, Page CP (2003) RSD931, a novel anti-tussive agent acting on airway sensory nerves. Br J Pharmacol 138: 407–416

Adrian ED (1933) Afferent impulses in the vagus and their effect on respiration. J Physiol 79: 332–358

Armstrong DJ, Luck JC (1974) A comparative study of irritant and type J receptors in the cat. Respir Physiol 21:47–60

Baluk P, Nadel JA, McDonald DM (1992) Substance P-immunoreactive sensory axons in the rat respiratory tract: a quantitative study of their distribution and role in neurogenic inflammation. J Comp Neurol 319:586–598

Barry PW, Mason NP, Riordan M, O'Callaghan C (1997) Cough frequency and cough-receptor sensitivity are increased in man at altitude. Clin Sci (Lond) 93:181–186

Belvisi MG, Bolser DC (2002) Summary: animal models for cough. Pulm Pharmacol Ther 15: 249–250

Bergren DR (1997) Sensory receptor activation by mediators of defense reflexes in guinea-pig lungs. Respir Physiol 108:195–204

Bergren DR (2001) Enhanced lung C-fiber responsiveness in sensitized adult guinea pigs exposed to chronic tobacco smoke. J Appl Physiol 91:1645–1654

Bergren DR, Kincaid RJ (1984) Rapidly-adapting receptor activity and intratracheal pressure in guinea pigs. II. Action of aspirin and salicylic acid in antagonizing mediators of allergic asthma. Prostaglandins Leukot Med 16:163–171

Bergren DR, Myers DL (1984) Rapidly-adapting receptor activity and intratracheal pressure in guinea pigs. I. Action of leukotriene C4. Prostaglandins Leukot Med 16:147–161

Bergren DR, Peterson DF (1993) Identification of vagal sensory receptors in the rat lung: are there subtypes of slowly adapting recseptors? J Physiol 464:681–698

Bergren DR, Sampson SR (1982) Characterization of intrapulmonary, rapidly adapting receptors of guinea pigs. Respir Physiol 47:83–95

Bergren DR, Gustafson JM, Myers DL (1984) Effect of prostaglandin F2 alpha on pulmonary rapidly-adapting-receptors in the guinea pig. Prostaglandins 27:391–405

Bolser DC, Aziz SM, Chapman RW (1991) Ruthenium red decreases capsaicin and citric acid-induced cough in guinea pigs. Neurosci Lett 126:131–133

Bolser DC, DeGennaro FC, O'Reilly S, Chapman RW, Kreutner W, Egan RW, Hey JA (1994) Peripheral and central sites of action of GABA-B agonists to inhibit the cough reflex in the cat and guinea pig. Br J Pharmacol 113:1344–1348

Bolser DC, DeGennaro FC, O'Reilly S, McLeod RL, Hey JA (1997) Central antitussive activity of the NK1 and NK2 tachykinin receptor antagonists, CP-99,994 and SR 48968, in the guinea-pig and cat. Br J Pharmacol 121:165–170

Bolser DC, Poliacek I, Jakus J, Fuller DD, Davenport PW (2006) Neurogenesis of cough, other airway defensive behaviors and breathing: a holarchical system? Respir Physiol Neurobiol 152:255–265

Bonham AC, Joad JP (1991) Neurones in commissural nucleus tractus solitarii required for full expression of the pulmonary C fibre reflex in rat. J Physiol 441:95–112

Bonham AC, McCrimmon DR (1990) Neurones in a discrete region of the nucleus tractus solitarius are required for the Breuer-Hering reflex in rat. J Physiol 427:261–280

Bonham AC, Kappagoda CT, Kott KS, Joad JP (1995) Exposing young guinea pigs to side-stream tobacco smoke decreases rapidly adapting receptor responsiveness. J Appl Physiol 78: 1412–1420

Bonham AC, Kott KS, Joad JP (1996) Sidestream smoke exposure enhances rapidly adapting receptor responses to substance P in young guinea pigs. J Appl Physiol 81:1715–1722

Burki NK, Dale WJ, Lee LY (2005) Intravenous adenosine and dyspnea in humans. J Appl Physiol 98:180–185

Canning BJ (2007) Encoding of the cough reflex. Pulm Pharmacol Ther 20:396–401
Canning BJ (2008) The cough reflex in animals: relevance to human cough research. Lung 186:S23–S28
Canning BJ, Reynolds SM, Mazzone SB (2001) Multiple mechanisms of reflex bronchospasm in guinea pigs. J Appl Physiol 91:2642–2653
Canning BJ, Mazzone SB, Meeker SN, Mori N, Reynolds SM, Undem BJ (2004) Identification of the tracheal and laryngeal afferent neurones mediating cough in anaesthetized guinea-pigs. J Physiol 557:543–558
Canning BJ, Farmer DG, Mori N (2006a) Mechanistic studies of acid-evoked coughing in anesthetized guinea pigs. Am J Physiol Regul Integr Comp Physiol 291:R454–R463
Canning BJ, Mori N, Mazzone SB (2006b) Vagal afferent nerves regulating the cough reflex. Respir Physiol Neurobiol 152:223–242
Caterina MJ, Schumacher MA, Tominaga M, Rosen TA, Levine JD, Julius D (1997) The capsaicin receptor: a heat-activated ion channel in the pain pathway. Nature 389:816–824
Chang AB, Phelan PD, Roberts RG, Robertson CF (1996) Capsaicin cough receptor sensitivity test in children. Eur Respir J 9:2220–2223
Chuaychoo B, Lee MG, Kollarik M, Undem BJ (2005) Effect of 5-hydroxytryptamine on vagal C-fiber subtypes in guinea pig lungs. Pulm Pharmacol Ther 18:269–276
Chuaychoo B, Lee MG, Kollarik M, Pullmann R, Jr., Undem BJ (2006) Evidence for both adenosine A1 and A2A receptors activating single vagal sensory C-fibres in guinea pig lungs. J Physiol 575:481–490
Cohn MA, Baier H, Wanner A (1978) Failure of hypoxic pulmonary vasoconstriction in the canine asthma model. Effect of prostaglandin inhibitors. J Clin Invest 61:1463–1470
Coleridge HM, Coleridge JC (1977) Impulse activity in afferent vagal C-fibres with endings in the intrapulmonary airways of dogs. Respir Physiol 29:125–142
Coleridge HM, Coleridge JC, Luck JC (1965) Pulmonary afferent fibres of small diameter stimulated by capsaicin and by hyperinflation of the lungs. J Physiol 179:248–262
Coleridge HM, Coleridge JC, Roberts AM (1983) Rapid shallow breathing evoked by selective stimulation of airway C fibres in dogs. J Physiol 340:415–433
Coleridge JC, Coleridge HM (1984) Afferent vagal C fibre innervation of the lungs and airways and its functional significance. Rev Physiol Biochem Pharmacol 99:1–110
Coleridge JC, Coleridge HM, Roberts AM, Kaufman MP, Baker DG (1982) Tracheal contraction and relaxation initiated by lung and somatic afferents in dogs. J Appl Physiol 52:984–990
Daoui S, Cognon C, Naline E, Emonds-Alt X, Advenier C (1998) Involvement of tachykinin NK3 receptors in citric acid-induced cough and bronchial responses in guinea pigs. Am J Respir Crit Care Med 158:42–48
Davies A, Nadal JA, Weinmann G (1984) Effect of changes in airway pressure on breathing pattern in conscious dogs. Q J Exp Physiol 69:145–153
Davies A, Pirie L, Eyre-Todd RA (1996) Adaptation of pulmonary receptors in the spontaneously breathing anaesthetized rat. Eur Respir J 9:1637–1642
Davies RO, Kubin L (1986) Projection of pulmonary rapidly adapting receptors to the medulla of the cat: an antidromic mapping study. J Physiol 373:63–86
Davies RO, Kubin L, Pack AI (1987) Pulmonary stretch receptor relay neurones of the cat: location and contralateral medullary projections. J Physiol 383:571–585
Deep V, Singh M, Ravi K (2001) Role of vagal afferents in the reflex effects of capsaicin and lobeline in monkeys. Respir Physiol 125:155–168
Delpierre S, Grimaud C, Jammes Y, Mei N (1981) Changes in activity of vagal bronchopulmonary C fibres by chemical and physical stimuli in the cat. J Physiol 316:61–74
De Proost I, Pintelon I, Brouns I, Timmermans JP, Adriaensen D (2007) Selective visualisation of sensory receptors in the smooth muscle layer of ex-vivo airway whole-mounts by styryl pyridinium dyes. Cell Tissue Res 329:421–431
D'Erchia AM, Gissi C, Pesole G, Saccone C, Arnason U (1996) The guinea-pig is not a rodent. Nature 381:597–600

Dey RD, Altemus JB, Zervos I, Hoffpauir J (1990) Origin and colocalization of CGRP- and SP-reactive nerves in cat airway epithelium. J Appl Physiol 68:770–778
Dicpinigaitis PV (2007) Experimentally induced cough. Pulm Pharmacol Ther 20:319–324
El-Hashim AZ, Amine SA (2005) The role of substance P and bradykinin in the cough reflex and bronchoconstriction in guinea-pigs. Eur J Pharmacol 513:125–133
Ezure K, Otake K, Lipski J, She RB (1991) Efferent projections of pulmonary rapidly adapting receptor relay neurons in the cat. Brain Res 564:268–278
Ezure K, Tanaka I, Saito Y, Otake K (2002) Axonal projections of pulmonary slowly adapting receptor relay neurons in the rat. J Comp Neurol 446:81–94
Forsberg K, Karlsson JA, Theodorsson E, Lundberg JM, Persson CG (1988) Cough and bronchoconstriction mediated by capsaicin-sensitive sensory neurons in the guinea-pig. Pulm Pharmacol 1:33–39
Fox AJ, Barnes PJ, Urban L, Dray A (1993) An in vitro study of the properties of single vagal afferents innervating guinea-pig airways. J Physiol 469:21–35
Fox AJ, Barnes PJ, Dray A (1995) Stimulation of guinea-pig tracheal afferent fibres by non-isosmotic and low-chloride stimuli and the effect of frusemide. J Physiol 482 (Pt 1):179–187
Fox AJ, Lalloo UG, Belvisi MG, Bernareggi M, Chung KF, Barnes PJ (1996) Bradykinin-evoked sensitization of airway sensory nerves: a mechanism for ACE-inhibitor cough. Nat Med 2: 814–817
Fujimura M, Sakamoto S, Kamio Y, Matsuda T (1992) Effects of methacholine induced bronchoconstriction and procaterol induced bronchodilation on cough receptor sensitivity to inhaled capsaicin and tartaric acid. Thorax 47:441–445
Fujimura M, Sakamoto S, Kamio Y, Bando T, Kurashima K, Matsuda T (1993) Effect of inhaled procaterol on cough receptor sensitivity to capsaicin in patients with asthma or chronic bronchitis and in normal subjects. Thorax 48:615–618
Gatti R, Andre E, Amadesi S, Dinh TQ, Fischer A, Bunnett NW, Harrison S, Geppetti P, Trevisani M (2006) Protease-activated receptor-2 activation exaggerates TRPV1-mediated cough in guinea pigs. J Appl Physiol 101:506–511
Gaylor JB (1934) The intrinsic nervous mechanism of the human lung. Brain 57:143–160
Gestreau C, Bianchi AL, Grelot L (1997) Differential brainstem Fos-like immunoreactivity after laryngeal-induced coughing and its reduction by codeine. J Neurosci 17:9340–9352
Girard V, Naline E, Vilain P, Emonds-Alt X, Advenier C (1995) Effect of the two tachykinin antagonists, SR 48968 and SR 140333, on cough induced by citric acid in the unanaesthetized guinea pig. Eur Respir J 8:1110–1114
Green JF, Kaufman MP (1990) Pulmonary afferent control of breathing as end-expiratory lung volume decreases. J Appl Physiol 68:2186–2194
Green JF, Schertel ER, Coleridge HM, Coleridge JC (1986) Effect of pulmonary arterial PCO2 on slowly adapting pulmonary stretch receptors. J Appl Physiol 60:2048–2055
Groneberg DA, Niimi A, Dinh QT, Cosio B, Hew M, Fischer A, Chung KF (2004) Increased expression of transient receptor potential vanilloid-1 in airway nerves of chronic cough. Am J Respir Crit Care Med 170:1276–1280
Gu Q, Lee LY (2002) Alveolar hypercapnia augments pulmonary C-fiber responses to chemical stimulants: role of hydrogen ion. J Appl Physiol 93:181–188
Gu Q, Kwong K, Lee LY (2003) Ca2+ transient evoked by chemical stimulation is enhanced by PGE2 in vagal sensory neurons: role of cAMP/PKA signaling pathway. J Neurophysiol 89:1985–1993
Hanacek J, Davies A, Widdicombe JG (1984) Influence of lung stretch receptors on the cough reflex in rabbits. Respiration 45:161–168
Hara J, Fujimura M, Ueda A, Myou S, Oribe Y, Ohkura N, Kita T, Yasui M, Kasahara K (2008) Effect of pressure stress applied to the airway on cough-reflex sensitivity in Guinea pigs. Am J Respir Crit Care Med 177:585–592
Hargreaves M, Ravi K, Kappagoda CT (1993) Effect of bradykinin on respiratory rate in anaesthetized rabbits; role of rapidly adapting receptors. J Physiol 468:501–513

Haxhiu MA, Erokwu B, Dreshaj IA (1997) The role of excitatory amino acids in airway reflex responses in anesthetized dogs. J Auton Nerv Syst 67:192–199
Haxhiu MA, Chavez JC, Pichiule P, Erokwu B, Dreshaj IA (2000) The excitatory amino acid glutamate mediates reflexly increased tracheal blood flow and airway submucosal gland secretion. Brain Res 883:77–86
Ho CY, Gu Q, Lin YS, Lee LY (2001) Sensitivity of vagal afferent endings to chemical irritants in the rat lung. Respir Physiol 127:113–124
House A, Celly C, Skeans S, Lamca J, Egan RW, Hey JA, Chapman RW (2004) Cough reflex in allergic dogs. Eur J Pharmacol 492:251–258
Hunter DD, Undem BJ (1999) Identification and substance P content of vagal afferent neurons innervating the epithelium of the guinea pig trachea. Am J Respir Crit Care Med 159: 1943–1948
Jakus J, Poliacek I, Halasova E, Murin P, Knocikova J, Tomori Z, Bolser DC (2008) Brainstem circuitry of tracheal-bronchial cough: C-FOS study in anesthetized cats. Respir Physiol Neurobiol 160:289–300
Jia Y, McLeod RL, Wang X, Parra LE, Egan RW, Hey JA (2002) Anandamide induces cough in conscious guinea-pigs through VR1 receptors. Br J Pharmacol 137:831–836
Joad JP, Kott KS, Bonham AC (1997) Nitric oxide contributes to substance P-induced increases in lung rapidly adapting receptor activity in guinea-pigs. J Physiol 503(Pt 3):635–643
Joad JP, Munch PA, Bric JM, Evans SJ, Pinkerton KE, Chen CY, Bonham AC (2004) Passive smoke effects on cough and airways in young guinea pigs: role of brainstem substance P. Am J Respir Crit Care Med 169:499–504
Jonzon A, Pisarri TE, Coleridge JC, Coleridge HM (1986) Rapidly adapting receptor activity in dogs is inversely related to lung compliance. J Appl Physiol 61:1980–1987
Kalia M, Richter D (1985) Morphology of physiologically identified slowly adapting lung stretch receptor afferents stained with intra-axonal horseradish peroxidase in the nucleus of the tractus solitarius of the cat. I. A light microscopic analysis. J Comp Neurol 241:503–520
Kalia M, Richter D (1988) Rapidly adapting pulmonary receptor afferents. II. Fine structure and synaptic organization of central terminal processes in the nucleus of the tractus solitarius. J Comp Neurol 274:574–594
Kamei J, Takahashi Y (2006) Involvement of ionotropic purinergic receptors in the histamine-induced enhancement of the cough reflex sensitivity in guinea pigs. Eur J Pharmacol 547: 160–164
Kamei J, Iwamoto Y, Suzuki T, Misawa M, Nagase H, Kasuya Y (1993) Antitussive effects of naltrindole, a selective delta-opioid receptor antagonist, in mice and rats. Eur J Pharmacol 249:161–165
Kamei J, Takahashi Y, Yoshikawa Y, Saitoh A (2005) Involvement of P2X receptor subtypes in ATP-induced enhancement of the cough reflex sensitivity. Eur J Pharmacol 528:158–161
Karlsson JA, Fuller RW (1999) Pharmacological regulation of the cough reflex – from experimental models to antitussive effects in Man. Pulm Pharmacol Ther 12:215–228
Karlsson JA, Zackrisson C, Lundberg JM (1991a) Hyperresponsiveness to tussive stimuli in cigarette smoke-exposed guinea-pigs: a role for capsaicin-sensitive, calcitonin gene-related peptide-containing nerves. Acta Physiol Scand 141:445–454
Karlsson JA, Hansson L, Wollmer P, Dahlback M (1991b) Regional sensitivity of the respiratory tract to stimuli causing cough and reflex bronchoconstriction. Respir Med 85(Suppl A):47–50
Karlsson JA, Sant'Ambrogio FB, Forsberg K, Palecek F, Mathew OP, Sant'Ambrogio G (1993) Respiratory and cardiovascular effects of inhaled and intravenous bradykinin, PGE2, and PGF2 alpha in dogs. J Appl Physiol 74:2380–2386
Kaufman MP, Coleridge HM, Coleridge JC, Baker DG (1980) Bradykinin stimulates afferent vagal C-fibers in intrapulmonary airways of dogs. J Appl Physiol 48:511–517
Knowlton GC, Larrabee MG (1946) A unitary analysis of pulmonary volume receptors. Am J Physiol Lung Cell Mol Physiol 147:100–114
Kollarik M, Undem BJ (2002) Mechanisms of acid-induced activation of airway afferent nerve fibres in guinea-pig. J Physiol 543:591–600

Kollarik M, Dinh QT, Fischer A, Undem BJ (2003) Capsaicin-sensitive and -insensitive vagal bronchopulmonary C-fibres in the mouse. J Physiol 551:869–879
Korpas J, Widdicombe JG, Vrabec M, Kudlicka J (1993) Effect of experimental lung oedema on cough sound creation. Respir Med 87:55–59
Kubin L, Kimura H, Davies RO (1991) The medullary projections of afferent bronchopulmonary C fibres in the cat as shown by antidromic mapping. J Physiol 435:207–228
Kubin L, Alheid GF, Zuperku EJ, McCrimmon DR (2006) Central pathways of pulmonary and lower airway vagal afferents. J Appl Physiol 101:618–627
Kummer W, Fischer A, Kurkowski R, Heym C (1992) The sensory and sympathetic innervation of guinea-pig lung and trachea as studied by retrograde neuronal tracing and double-labelling immunohistochemistry. Neuroscience 49:715–737
Kunz AL, Kawashiro T, Scheid P (1976) Study of CO2 sensitive vagal afferents in the cat lung. Respir Physiol 27:347–355
Kwong K, Carr MJ, Gibbard A, Savage TJ, Singh K, Jing J, Meeker S, Undem BJ (2008) Voltage-gated sodium channels in nociceptive versus non-nociceptive nodose vagal sensory neurons innervating guinea pig lungs. J Physiol 586:1321–1336
Lalloo UG, Fox AJ, Belvisi MG, Chung KF, Barnes PJ (1995) Capsazepine inhibits cough induced by capsaicin and citric acid but not by hypertonic saline in guinea pigs. J Appl Physiol 79: 1082–1087
Lamb JP, Sparrow MP (2002) Three-dimensional mapping of sensory innervation with substance p in porcine bronchial mucosa: comparison with human airways. Am J Respir Crit Care Med 166:1269–1281
Larsell O (1921) Nerve termiunation in the lung of the rabbit. J Comp Neurol 33:105–131
Larsell O (1922) The ganglia, plexuses and nerve-termination of the mammalian lung and pleura pulmonis. J Comp Neurol 35:97–132
Laude EA, Higgins KS, Morice AH (1993) A comparative study of the effects of citric acid, capsaicin and resiniferatoxin on the cough challenge in guinea-pig and man. Pulm Pharmacol 6:171–175
Lee LY, Morton RF (1993) Histamine enhances vagal pulmonary C-fiber responses to capsaicin and lung inflation. Respir Physiol 93:83–96
Lee LY, Pisarri TE (2001) Afferent properties and reflex functions of bronchopulmonary C-fibers. Respir Physiol 125:47–65
Lee MG, Weinreich D, Undem BJ (2005) Effect of olvanil and anandamide on vagal C-fiber subtypes in guinea pig lung. Br J Pharmacol 146:596–603
Leung SY, Niimi A, Williams AS, Nath P, Blanc FX, Dinh QT, Chung KF (2007) Inhibition of citric acid- and capsaicin-induced cough by novel TRPV-1 antagonist, V112220, in guinea-pig. Cough 3:10
Lewis CA, Ambrose C, Banner K, Battram C, Butler K, Giddings J, Mok J, Nasra J, Winny C, Poll C (2007) Animal models of cough: literature review and presentation of a novel cigarette smoke-enhanced cough model in the guinea-pig. Pulm Pharmacol Ther 20:325–333
Lin RL, Gu Q, Lin YS, Lee LY (2005) Stimulatory effect of CO2 on vagal bronchopulmonary C-fiber afferents during airway inflammation. J Appl Physiol 99:1704–1711
Lin YS, Lee LY (2002) Stimulation of pulmonary vagal C-fibres by anandamide in anaesthetized rats: role of vanilloid type 1 receptors. J Physiol 539:947–955
Lipski J, Ezure K, Wong She RB (1991) Identification of neurons receiving input from pulmonary rapidly adapting receptors in the cat. J Physiol 443:55–77
Liu Q, Fujimura M, Tachibana H, Myou S, Kasahara K, Yasui M (2001) Characterization of increased cough sensitivity after antigen challenge in guinea pigs. Clin Exp Allergy 31:474–484
Malik SK, Kher V, Kumar B, Kaur S (1978) Impaired cough-receptor function in leprosy. Lancet 1:1094–1095
Matsumoto S (1997) The changes in pulmonary C-fiber activity and lung mechanics induced by vagal stimulation in rabbits. Life Sci 61:2189–2195

Matsumoto S, Shimizu T, Kanno T, Yamasaki M, Nagayama T (1990) Effects of histamine on slowly adapting pulmonary stretch receptor activities in vagotomized rabbits. Jpn J Physiol 40:737–752

Mazzone SB, Canning BJ (2002) Synergistic interactions between airway afferent nerve subtypes mediating reflex bronchospasm in guinea pigs. Am J Physiol Regul Integr Comp Physiol 283:R86–98

Mazzone SB, McGovern AE (2006) Na+-K+-2Cl- cotransporters and Cl- channels regulate citric acid cough in guinea pigs. J Appl Physiol 101:635–643

Mazzone SB, Mori N, Canning BJ (2005) Synergistic interactions between airway afferent nerve subtypes regulating the cough reflex in guinea-pigs. J Physiol 569:559–573

McAlexander MA, Undem BJ (2000) Potassium channel blockade induces action potential generation in guinea-pig airway vagal afferent neurones. J Auton Nerv Syst 78:158–164

McAlexander MA, Myers AC, Undem BJ (1999) Adaptation of guinea-pig vagal airway afferent neurones to mechanical stimulation. J Physiol 521(Pt 1):239–247

McDonald DM, Mitchell RA, Gabella G, Haskell A (1988) Neurogenic inflammation in the rat trachea. II. Identity and distribution of nerves mediating the increase in vascular permeability. J Neurocytol 17:605–628

McLeod RL, Parra LE, Mutter JC, Erickson CH, Carey GJ, Tulshian DB, Fawzi AB, Smith-Torhan A, Egan RW, Cuss FM, Hey JA (2001) Nociceptin inhibits cough in the guinea-pig by activation of ORL(1) receptors. Br J Pharmacol 132:1175–1178

Mills JE, Sellick H, Widdicombe JG (1969) Activity of lung irritant receptors in pulmonary microembolism, anaphylaxis and drug-induced bronchoconstrictions. J Physiol 203:337–357

Mills JE, Sellick H, Widdicombe JG (1970) Epithelial irritant receptors in the lungs. In: Ruth P (ed) Breathing: Hering-Breuer centenary symposium. Churchill, London, pp 77–99

Mohammed SP, Higenbottam TW, Adcock JJ (1993) Effects of aerosol-applied capsaicin, histamine and prostaglandin E2 on airway sensory receptors of anaesthetized cats. J Physiol 469: 51–66

Morikawa T, Gallico L, Widdicombe JG (1997) Actions of moguisteine on cough and pulmonary rapidly adapting receptor activity in the guinea pig. Pharmacol Res 35:113–118

Mutoh T, Bonham AC, Kott KS, Joad JP (1999) Chronic exposure to sidestream tobacco smoke augments lung C-fiber responsiveness in young guinea pigs. J Appl Physiol 87:757–768

Mutolo D, Bongianni F, Fontana GA, Pantaleo T (2007) The role of excitatory amino acids and substance P in the mediation of the cough reflex within the nucleus tractus solitarii of the rabbit. Brain Res Bull 74:284–293

Myers AC, Kajekar R, Undem BJ (2002) Allergic inflammation-induced neuropeptide production in rapidly adapting afferent nerves in guinea pig airways. Am J Physiol Lung Cell Mol Physiol 282:L775–L781

O'Connell F, Thomas VE, Fuller RW, Pride NB, Karlsson JA (1994) Effect of clonidine on induced cough and bronchoconstriction in guinea pigs and healthy humans. J Appl Physiol 76: 1082–1087

Ohi Y, Yamazaki H, Takeda R, Haji A (2004) Phrenic and iliohypogastric nerve discharges during tussigenic stimulation in paralyzed and decerebrate guinea pigs and rats. Brain Res 1021: 119–127

Ohi Y, Yamazaki H, Takeda R, Haji A (2005) Functional and morphological organization of the nucleus tractus solitarius in the fictive cough reflex of guinea pigs. Neurosci Res 53:201–209

Olgiati R, Birch S, Rao A, Wanner A (1981) Differential effects of methacholine and antigen challenge on gas exchange in allergic subjects. J Allergy Clin Immunol 67:325–329

Otake K, Ezure K, Lipski J, Wong She RB (1992) Projections from the commissural subnucleus of the nucleus of the solitary tract: an anterograde tracing study in the cat. J Comp Neurol 324:365–378

Paintal AS (1969) Mechanism of stimulation of type J pulmonary receptors. J Physiol 203:511–532

Paintal AS (1973) Vagal sensory receptors and their reflex effects. Physiol Rev 53:159–227

Pinto A, Yanai M, Sekizawa K, Aikawa T, Sasaki H (1995) Conditioned enhancement of cough response in awake guinea pigs. Int Arch Allergy Immunol 108:95–98

Plevkova J, Kollarik M, Brozmanova M, Revallo M, Varechova S, Tatar M (2004) Modulation of experimentally-induced cough by stimulation of nasal mucosa in cats and guinea pigs. Respir Physiol Neurobiol 142:225–235

Polacek H, Korpas J, Tatar M, Plank L, Pullmann R (1986) Study of cough in anaesthetized cats with experimental pulmonary oedema. Physiol Bohemoslov 35:481–489

Raj H, Singh VK, Anand A, Paintal AS (1995) Sensory origin of lobeline-induced sensations: a correlative study in man and cat. J Physiol 482 (Pt 1):235–246

Rao SV, Sant'Ambrogio FB, Sant'Ambrogio G (1981) Respiratory reflexes evoked by tracheal distension. J Appl Physiol 50:421–427

Riccio MM, Myers AC, Undem BJ (1996) Immunomodulation of afferent neurons in guinea-pig isolated airway. J Physiol 491(Pt 2):499–509

Ricco MM, Kummer W, Biglari B, Myers AC, Undem BJ (1996) Interganglionic segregation of distinct vagal afferent fibre phenotypes in guinea-pig airways. J Physiol 496(Pt 2):521–530

Sano M, Tsubone H, Sugano S (1992) Vagal afferent activities and respiratory reflexes during drug-induced bronchoconstriction in the guinea pig. J Vet Med Sci 54:989–998

Sellick H, Widdicombe JG (1969) The activity of lung irritant receptors during pneumothorax, hyperpnoea and pulmonary vascular congestion. J Physiol 203:359–381

Sellick H, Widdicombe JG (1971) Stimulation of lung irritant receptors by cigarette smoke, carbon dust, and histamine aerosol. J Appl Physiol 31:15–19

Shannon R, Baekey DM, Morris KF, Nuding SC, Segers LS, Lindsey BG (2004) Production of reflex cough by brainstem respiratory networks. Pulm Pharmacol Ther 17:369–376

Stone RA, Worsdell YM, Fuller RW, Barnes PJ (1993) Effects of 5-hydroxytryptamine and 5-hydroxytryptophan infusion on the human cough reflex. J Appl Physiol 74:396–401

Tatar M, Webber SE, Widdicombe JG (1988) Lung C-fibre receptor activation and defensive reflexes in anaesthetized cats. J Physiol 402:411–420

Tatar M, Sant'Ambrogio G, Sant'Ambrogio FB (1994) Laryngeal and tracheobronchial cough in anesthetized dogs. J Appl Physiol 76:2672–2679

Tatar M, Karcolova D, Pecova R, Brozmanova M (1996) The role of partial laryngeal denervation on the cough reflex in awake guinea-pigs, rats and rabbits. Pulm Pharmacol 9:371–372

Tatar M, Pecova R, Karcolova D (1997) [Sensitivity of the cough reflex in awake guinea pigs, rats and rabbits]. Bratisl Lek Listy 98:539–543

Trevisani M, Milan A, Gatti R, Zanasi A, Harrison S, Fontana G, Morice AH, Geppetti P (2004) Antitussive activity of iodo-resiniferatoxin in guinea pigs. Thorax 59:769–772

Tsubone H, Sant'Ambrogio G, Anderson JW, Orani GP (1991) Laryngeal afferent activity and reflexes in the guinea pig. Respir Physiol 86:215–231

Undem BJ, Chuaychoo B, Lee MG, Weinreich D, Myers AC, Kollarik M (2004) Subtypes of vagal afferent C-fibres in guinea-pig lungs. J Physiol 556:905–917

Vidruk EH, Hahn HL, Nadel JA, Sampson SR (1977) Mechanisms by which histamine stimulates rapidly adapting receptors in dog lungs. J Appl Physiol 43:397–402

Watanabe N, Horie S, Michael GJ, Keir S, Spina D, Page CP, Priestley JV (2006) Immunohistochemical co-localization of transient receptor potential vanilloid (TRPV)1 and sensory neuropeptides in the guinea-pig respiratory system. Neuroscience 141:1533–1543

Widdicombe JG (2001) The race to explore the pathway to cough: who won the silver medal? Am J Respir Crit Care Med 164:729–730

Widdicombe JG (1954a) Receptors in the trachea and bronchi of the cat. J Physiol 123:71–104

Widdicombe JG (1954b) Respiratory reflexes excited by inflation of the lungs. J Physiol 123: 105–115

Widdicombe JG (1954c) Respiratory reflexes from the trachea and bronchi of the cat. J Physiol 123:55–70

Widdicombe JG, Kent DC, Nadel JA (1962) Mechanism of bronchoconstriction during inhalation of dust. J Appl Physiol 17:613–616

Xiang A, Uchida Y, Nomura A, Iijima H, Sakamoto T, Ishii Y, Morishima Y, Masuyama K, Zhang M, Hirano K, Sekizawa K (2002) Involvement of thromboxane A(2) in airway mucous cells in asthma-related cough. J Appl Physiol 92:763–770

Yamamoto Y, Atoji Y, Suzuki Y (1995) Nerve endings in bronchi of the dog that react with antibodies against neurofilament protein. J Anat 187 (Pt 1):59–65
Yamamoto Y, Ootsuka T, Atoji Y, Suzuki Y (1998) Morphological and quantitative study of the intrinsic nerve plexuses of the canine trachea as revealed by immunohistochemical staining of protein gene product 9.5. Anat Rec 250:438–447
Yamamoto Y, Sato Y, Taniguchi K (2007) Distribution of TRPV1- and TRPV2-immunoreactive afferent nerve endings in rat trachea. J Anat 211:775–783
Yu J (2000) Spectrum of myelinated pulmonary afferents. Am J Physiol Regul Integr Comp Physiol 279:R2142–2148
Yu J (2005) Airway mechanosensors. Respir Physiol Neurobiol 148:217–243
Yu J, Coleridge JC, Coleridge HM (1987) Influence of lung stiffness on rapidly adapting receptors in rabbits and cats. Respir Physiol 68:161–176
Yu J, Schultz HD, Goodman J, Coleridge JC, Coleridge HM, Davis B (1989) Pulmonary rapidly adapting receptors reflexly increase airway secretion in dogs. J Appl Physiol 67:682–687
Yu J, Pisarri TE, Coleridge JC, Coleridge HM (1991) Response of slowly adapting pulmonary stretch receptors to reduced lung compliance. J Appl Physiol 71:425–431

Cough Sensors. II. Transient Receptor Potential Membrane Receptors on Cough Sensors

S. Materazzi, R. Nassini, R. Gatti, M. Trevisani, and P. Geppetti(✉)

Contents

Abstract The transient receptor potential (TRP) family of channels is represented by at least six members in primary sensory neurons. These include the TRP vanilloid subtypes 1 (TRPV1), 2, 3, and 4, the cold and menthol receptor TRPM8, and TRPA1. Much interest has been directed to the study of the TRPV1, because capsaicin has been instrumental in discovering the unique role of a subset of primary sensory neurons in causing nociceptive responses, in activating reflex pathways including cough, and in producing neurogenic inflammation. TRPV1 is now regarded as an integrator of diverse sensory modalities because it undergoes marked plasticity and sensitization through a variety of mechanisms, including activation of G-protein-coupled or tyrosine kinase receptors. Evidence in experimental animals and in patients with airway diseases indicates a marked hypersensitivity to cough induced by TRPV1 agonists. Recent studies with newly developed high-affinity and selective TRPV1 antagonists have revealed that TRPV1 inhibition reduces cough induced by citric acid or antigen challenge.

P. Geppetti
Department of Critical Care Medicine and Surgery, University of Florence, Viale Pieraccini, 6, 50139 Florence, Italy
pierangelo.geppetti@unifi.it

K.F. Chung, J.G. Widdicombe (eds.), *Pharmacology and Therapeutics of Cough*,
Handbook of Experimental Pharmacology 187,

1 The Transient Receptor Potential Family of Channels

Few areas of research have known such an intense process of investigation in health and disease like that directed at the identification of the pathophysiological roles of the transient receptor potential (TRP) family of proteins, most likely because these ion channels have been recognized to sense a vast range of stimuli and because of their ubiquitous distribution in different tissues and organs. TRPs putatively consist of six transmembrane-domain proteins that assemble as tetramers to form cation-permeable pores. Three main subclasses of TRP channels have been defined: TRPC, TRPM, and TRPV (V stands for vanilloid) (Clapham 2003; Montell et al. 2002). TRPP, TRPML and TRPN are additional and newly proposed subtypes of TRPs (Clapham 2003; Montell et al. 2002). More recently, a novel TRP-like channel that responds to cold temperature (below 19°C) has been cloned and termed ANKTM1 or TRPA1 (Jaquemar et al. 1999; McKemy et al. 2002; Nilius et al. 2007; Peier et al. 2002).

Uncertainty exists as regard to the precise and multiple roles of TRPs. Their localization at the plasma membranes of neurons (or additional intracellular sites) or other cells and a large body of evidence collected using a plethora of stimuli indicate that these channels are sensors of chemical and physical stimuli. TRPs are the molecular structures used by living organisms from worm to man to sense osmolarity, cell stretching, appreciate sweet and bitter tastes, and discriminate warmth, heat, and cold. However, intracellular localization (e.g., in the endoplasmic reticulum) (Karai et al. 2004) and evidence obtained about the cellular regulation of ion flux has suggested a role as modulators of Ca^{2+} homeostasis (Clapham 2003; Montell 1997), downstream to G-protein-coupled receptors, most probably via the phospholipase C (PLC) pathway. However, this otherwise fascinating hypothesis is not supported yet by the identification of a messenger molecule, which directly binds and activates the channel (Clapham, 2003). Phosphatidylinositol 4,5-bisphosphate (PIP_2) binding and PIP_2 hydrolysis inhibits and activates, respectively, TRP-like (TPRL) in the *Drosophila* (Hardie 2003) and the mammalian TRPV1 (Chuang et al. 2001; Prescott and Julius 2003). However, a major role of PIP_2 as a TRP regulator has been challenged by the observation that constitutive activity of TRPM7 is increased by PIP_2 binding and reduced by PIP_2 hydrolysis (Runnels et al. 2002). For some TRPs, including TRPC3, TPPC6, and TRPC7, diacylglycerol has been proposed to act as the intracellular activator (Clapham 2003); however, this does not seem to be the case for TRPV1 (Hofmann et al. 1999; Hwang et al. 2000). Finally, TRPs have been proposed to regulate the so-called capacitance Ca^{2+} entry or store-operated Ca^{2+} entry. Store-operated Ca^{2+} entries are considered channels that link Ca^{2+} store depletion with Ca^{2+} entry. However, final proof for this hypothesis is still lacking (Clapham 2003). Although there is no evidence for one or more specific and high-affinity endogenous ligands for TRPs, a series of lipid derivatives, including arachidonic acid metabolites, have been claimed to gate TRPs.

2 TRPV1

A 426-unit amino acid protein in the rat (Caterina et al. 1997) uniquely sensitive to vanilloid molecules, including capsaicin, the hot principle contained in the plants of the genus *Capsicum* (Szallasi and Blumberg, 1999), and activated by low extracellular pH (pH 6–5) (Bevan and Geppetti, 1994; Geppetti et al. 1991; Tominaga et al. 1998) and elevated concentrations (in the micromolar range) of the endocannabinoid anandamide (Zygmunt et al. 1999), the lipoxygenase metabolites of arachidonic acid, leukotriene B_4 or 12-hydroperoxyeicosatetraenoic acids (Hwang et al. 2000), and *N*-arachidonoyl dopamine (Huang et al. 2002) has been termed TRPV1. Moderate noxious temperature between 42 and 53°C also stimulates TRPV1 (Caterina et al. 1997), which with the contribution of other TRP channels expressed on sensory neurons (TRPA1, TRPM8, TRPV3, TRPV4, and TRPV2) has been proposed to enable mammals to discriminate different temperatures from noxious cold to noxious heat (Clapham 2003; Montell 1997; Montell et al. 2002). A subset of primary sensory neurons of the trigeminal, vagal, and dorsal root ganglia with C- and A-δ fibers are highly enriched with TRPV1. Because of their ability to detect noxious chemical, thermal, and high-threshold mechanical stimuli these neurons have been defined as polymodal nociceptors. Limbic system, striatum, hypothalamus, thalamic nuclei, substantia nigra, reticular formation, locus coeruleus, and cerebellum are areas of the central nervous system where high levels of messenger RNA (mRNA) for TRPV1 have been reported (Mezey et al. 2000). Nonneuronal cells, including epithelial cells of the urothelium (Birder et al. 2001), keratinocytes (Inoue et al. 2002) and epithelial cells of the palatal rugae (Kido et al. 2003), seem to express significant amounts of TRPV1 mRNA and protein. However, the physiological and pathophysiological role, if any, of TRPV1 either in nonneuronal cells or in nonsensory neuronal cells has not been established yet.

The nocifensor role of sensory neurons, that now we know express TRPV1, was proposed by Sir Thomas Lewis in 1936 following the proposal that stimulation of the nerve terminal results in a depolarization that initiates action potentials, propagated not only orthodromically, but also antidromically. Whereas orthodromic depolarization contributes to reflex responses, including cough, urinary bladder voiding, peristalsis in the gut, and other responses, antidromic diffusion of action potentials to collateral nerve fibers is associated with sensory-neuron-dependent vasomotor responses. There is compelling evidence that neurogenic vasodilatation and plasma protein extravasation result from the local release of neuropeptides, including calcitonin gene-related peptide (CGRP) and the tachykinins substance P (SP) and neurokinin A (NKA) (Geppetti and Holzer, 1996).

TRPV1 does not seem to be required for appropriate temperature sensing, as TRPV1-deficient mice show a normal phenotype for temperature detection (Davis et al. 2000). However, TRPV1 is required to develop thermal hyperalgesia (Davis et al. 2000). Deletion of TRPV1 resulted also in altered urinary bladder function (Birder et al. 2002). The possibility that TRPV1 contributes also to mechanical hyperalgesia derives not from genetic studies (Davis et al. 2000), but from more recent pharmacological evidence obtained with first-generation (capsazepine)

(Walker et al. 2003) and second-generation (Lee et al. 2003; Pomonis et al. 2003) TRPV1 antagonists. The possibility that TRPV1 senses acidic pH and mediates thermal and mechanical hyperalgesia justifies its localization to peripheral terminals of primary sensory neurons. However, TRPV1 expression on central terminals in the dorsal horn of the spinal cord and medulla oblongata raises the question as to whether endogenous ligands activate terminals of primary sensory neurons at this anatomical site and on the pathophysiological role of this activation (Tognetto et al. 2000)

3 TRPA1 and Other TRP Channels

TRPA1 is a recently identified excitatory ion channel (Jaquemar et al. 1999; Nilius et al. 2007), coexpressed with the "capsaicin receptor," TRPV1, by a subpopulation of primary afferent somatosensory neurons, which contain the neuropeptides SP, NKA, and CGRP and mediate pain and neurogenic inflammation (Bautista et al. 2006; Caterina et al. 1997; Geppetti and Holzer, 1996; Story et al. 2003). TRPA1 is activated by some pungent ingredients present in various spices, including mustard, garlic, and cinnamon (Jordt et al. 2004; Nilius et al. 2007), and by environmental irritants such as formaldehyde (McNamara et al. 2007). More recently, TRPA1 has been recognized as the target of a series of endogenous α,β-unsaturated aldehydes, which are produced by lipid peroxidation in response to oxydative stress at sites of inflammation and tissue injury (Bautista et al. 2006; Macpherson et al. 2007; Trevisani et al. 2007). These aldehydes include 4-hydroxy-2-nonenal (HNE), which is produced by peroxidation of omega-6 polyunsaturated fatty acids, such as linoleic acid and arachidonic acid (Benedetti et al. 1980; Esterbauer et al. 1991). HNE is produced in large amounts (1 μM to 5 mM) in various organs in response to oxidative insults (Uchida et al. 1993), as in the lungs of patients with chronic obstructive pulmonary disease (COPD) (Rahman et al. 2002). HNE is a highly diffusible molecule in cells and tissues and has been proposed to mediate many of the toxic effects of reactive oxygen species far from the site of free-radical formation (Esterbauer et al. 1991; Uchida et al. 1993). Acrolein is another α,β-unsaturated aldehyde able to stimulate TRPA1, and is produced endogenously or can be generated from the metabolism of chemotherapeutic agents, and is contained in large amounts in cigarette smoke (Bautista et al. 2006). A variety of known TRPA1 agonists, including acrolein and other α,β-unsaturated aldehydes, possess an electrophilic carbon or sulfur atom that is subjected to nucleophilic attack (Dalle-Donne et al. 2006) by cysteine, lysine, or histidine. Indeed, mutagenesis studies have clarified that such reactivity promotes channel gating through covalent modification of residues within the cytoplasmic N-terminal domain of the channel (Hinman et al. 2006; Macpherson et al. 2007; Trevisani et al. 2007).

The number of agents with a proposed pathophysiological role in airways diseases and that gate TRPA1 is increasing fast. Among these, two constituents of cigarette smoke should be highlighted. Recent reports showed that the saturated aldehyde, acetaldehyde, at millimolar concentrations, was able to gate TRPA1 expressed in mouse neurons and in recombinant systems (Bang et al. 2007) and that H_2O_2 produces a delayed activation of the neuronal wild-type or recombinant TRPA1 channel (Andersson et al. 2008). Much less is known on the other TRP channels expressed by sensory neurons regarding their distribution and function in the airways. Nonneuronal localization in smooth-muscle cells or epithelial cells of TRPV2 and TRPV4 (Dietrich et al. 2006; Jia et al. 2004), although of great interest, is beyond the scope of the present chapter. TRPM8 is not coexpressed in capsaicin-sensitive TRPV1-positive dorsal root ganglion neurons (Story et al. 2003). In addition, TRPM8 does not seem to be expressed in mouse vagal neurons (Nassenstein et al. 2008).

4 The Contribution of Neurogenic Inflammation to Cough

Biological responses produced by neuropeptides released from peripheral endings of capsaicin-sensitive and TRPV1-expressing sensory neurons are collectively referred to as "neurogenic inflammation." In the upper and lower airways, neurogenic inflammatory responses encompass vascular and nonvascular effects. Arterial vasodilatation (CGRP) and plasma protein extravasation and leukocyte adhesion to the vascular endothelium of postcapillary venules (both SP/NKA) are the main vascular components of neurogenic inflammation (Geppetti and Holzer, 1996), whereas nonvascular responses are bronchoconstriction, secretion of mucus from seromucus glands (Geppetti et al. 1993), and excitation of postganglionic cholinergic nerve terminals (Myers et al. 2005). However, it should be underlined that in man although tachykinins, though stimulation of NK2 and NK1 receptors, mediate a robust bronchoconstriction (Amadesi et al. 2001), sensory nerve stimulation by capsaicin or other irritant stimuli does not seem to produce significant bronchoconstriction.

Seromucous secretion or bronchocostriction, effects which are produced by tachykinins through NK1 and NK2 receptor activation, may contribute to cough exacerbation. No evidence supports a direct role of CGRP in cough. However, CGRP levels were found to be significantly higher in coughers with esophagitis than in noncoughers with esophagitis, and elevated CGRP, but not SP or NKA, is associated with chronic cough in children with esophagitis. The level of CGRP in bronchoalveolar lavage appeared to be related to cough sensitivity and at least could be important as a marker of a chronic cough condition (Chang et al. 2007). On the other hand, tachykinins have been proposed to stimulate cough by acting at the peripheral level (Kohrogi et al. 1988). Central localization of tachykinin receptors and a series of pharmacological studies have suggested the possibility that tachykinin released from central endings of airway sensory neurons modulates tussive responses. However, these proposals have not been confirmed by clinical studies, as cough produced by different stimuli, including hypertonic saline (Fahy et al. 1995), was not inhibited by tachykinin NK1 receptor antagonists. Thus, the contribution to cough of the inflammatory neuropeptides that are released in the airways upon stimulation of TRP channels on sensory neurons remains controversial. Nevertheless, the ability of inflammation, and possibly of neurogenic inflammation, to exaggerate responses to TRP channels is supported by robust evidence and, accordingly, the ability to limit inflammation may be a contributing factor to reduce the cough mediated by TRP channels.

5 Localization and Function and Plasticity of TRPV1 in the Airways

In the guinea pig TRPV1-positive nerve fibers are localized within the epithelium of the trachea and around smooth muscle and blood vessels and within the lower airways, in the vicinity of bronchi and bronchioles, and around alveolar tissue, whereas no TRPV1 was found within airway epithelial cells (Watanabe et al. 2005). The physiological relevance of TRPV1 mRNA expression, together with that of other ion channels in bronchial epithelial cells (Agopyan et al. 2003), remains questionable. Similarly, expression of TRPV1 in dendritic cells (Basu and Srivastava, 2005) and the associated hypothesis that TRPV1 orchestrates an early immune response has been challenged by failure to detect any functional TRPV1 in these inflammatory cells (O'Connell et al. 2005). We cannot exclude that in the airways, as in other tissues, TRPV1 occurs in extraneuronal cells, from where it may contribute to homeostasis, inflammation, and ultimately the regulation of the tussive response. However, we underline that this hypothesis lacks conclusive experimental proof.

TRPV1 expression of mRNA/protein and function, as those of other TRP channels, including TRPA1, undergo plasticity by a series of regulatory and inflammatory mediators. Nerve growth factor, which is required for survival of newborn rat dorsal root ganglia neurons and for expression of the TRPV1 phenotype in adult rat dorsal root ganglia neurons in culture (Bevan and Winter 1995), increases TRPV1 protein transportation to the peripheral endings of sensory neurons, a phenomenon associated with an increased neuronal sensitivity (Ji et al. 2002). A major proinflammatory peptide, bradykinin, via the B_2 receptor, sensitizes TRPV1 through different mechanisms, including protein kinase C-ε (Premkumar and Ahern 2000; Sugiura et al. 2002), displacement of PIP_2 from TRPV1 binding (Chuang et al. 2001), and production of 12-lipoxygenase and 5-lipoxygenase metabolites (Carr et al. 2003; Shin et al. 2002). Trypsin and tryptase are signaling molecules that activate protease-activated receptor-2 (PAR-2) via receptor proteolysis. PAR-2 is expressed in TRPV1-positive sensory neurons and its stimulation promotes neurogenic inflammation and sensory neuronal hyperactivity (Steinhoff et al. 2000; Vergnolle et al. 2001) also in airway tissue (Schmidlin et al. 2002; Su et al. 2005). PAR-2 stimulation upregulated the function of TRPV1 through a protein kinase C (PKC) dependent mechanism (Amadesi et al. 2004) (Fig. 1).

Sensitization of TRPV1 by PKC and cyclic AMP dependent protein kinase (PKA) pathways seems to be promiscuously used by different stimuli, including capsaicin, anandamide, heat, and protons (Bhave et al. 2002; De Petrocellis

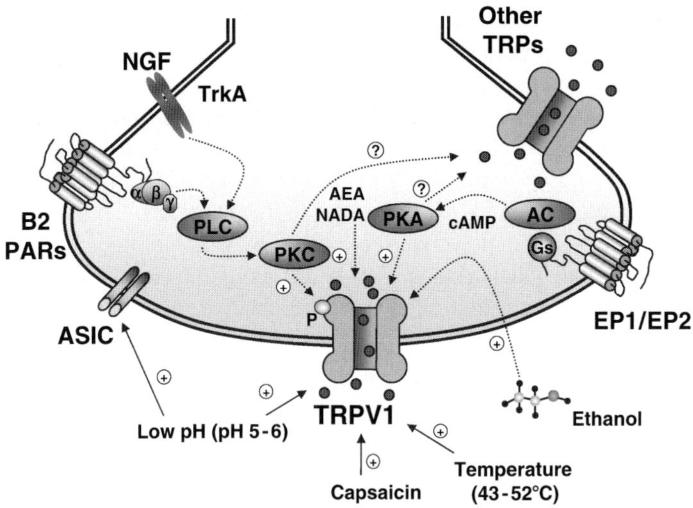

Fig. 1 Transient receptor potential vanilloid 1 (*TRPV1*) and other transient receptor potential channels expressed in primary sensory neurons and their regulation by inflammatory mediators and exogenous or putative endogenous ligands. *NFG* nerve growth factor, *PAR* protease activated receptor 2, *B2* bradykinin B2 receptor, *ASIC* acid-sensing ion channel, *EP1/2* prostanoid E receptor

et al. 2001; Premkumar and Ahern, 2000; Vellani et al. 2001), but these mechanisms are not unique to TRPV1 as also TRPA1 appears to be regulated by similar pathways. Thus, TRPA1 sensitivity is upregulated by PAR-2 via a PLC- and PIP_2-dependent pathway (Dai et al. 2007), and by bradykinin B_2 receptor via PLC- and PKA-dependent pathways (Wang et al. 2008). Neuronal sensitivity to TRPV1 agonists in the cough response is regulated by activation of G-protein-coupled receptors, as indicated by the observation that PAR-2 stimulation exaggerates TRPV1-dependent cough by activation of different mechanism(s), including PKC, PKA, and prostanoid release (Gatti et al. 2006). Clinical findings that treatment of asthmatic patients with prostaglandin I_2 increases cough reflex sensitivity to capsaicin indicate that the mechanism described in guinea pigs may be present in man under specific pathological conditions (Ishiura et al. 2007).

6 Cough and TRPV1

The ability of TRPV1 agonists to cause cough is well established in animal models, in healthy volunteers, and in asthmatic or COPD patients (Chung 2007; Dicpinigaitis and Alva, 2005). Previous review articles have covered these issues (Geppetti et al. 2006; Jia and Lee, 2007; Kollarik and Undem, 2006; Mazzone and McGovern, 2007). Here, we will review more recent data that further underline the importance of TRPV1 in cough pathophysiology and the potential of TRPV1 antagonists as new antitussive drugs.

The demonstration that endogenously produced agonists of TRPV1 can cause cough supports the view that TRPV1 is relevant in the cough mechanism. The endocannabinoid anandamide has been recognized as a TRPV1 agonist (Zygmunt et al. 1999), and an anandamide membrane transporter inhibitor reduced anandamide-induced cough, an effect that is most likely produced through TRPV1 activation (Kamei et al. 2006). Low extracellular pH is often found in the inflammatory milieu and citric acid is a commonly used stimulus to induce cough. Conflicting findings have been reported on the molecular mechanism by which protons cause cough. A recent paper has shown that citric acid evoked coughing in anesthetized guinea pigs is mediated by direct activation of capsaicin-insensitive vagal afferent nerves most likely through sequential activation of acid-sensing ion channels and chloride channels without the contribution of TRPV1 (Canning et al. 2006). However, previous studies using first-generation antagonists (Lalloo et al. 1995; Trevisani et al. 2004) and more recent evidence with second-generation and chemically unrelated TRPV1 antagonists (Bhattacharya et al. 2007; Leung et al. 2007) showed that all these drugs selectively inhibited citric acid induced cough in guinea pigs, thus providing robust circumstantial evidence that TRPV1 contributes to acidic-media-induced cough.

Cough is caused by exposure to environmental pollutants or by endogenous proinflammatory mechanisms. Exposure to the irritant pollutant, sulfur dioxide (SO_2) enhanced TRPV1 receptor function at the level of the nodose ganglia, and this effect occurred in parallel with an increase sensitivity of the cough response to capsaicin (McLeod et al. 2007). The TRPV1 antagonist *N*-(4-Tertiarybutylphenyl)-4(3-cholorphyridin-2-yl)-tetrahydro-pyrazine1(2H)-carboxamide (BCTC) inhibited capsaicin-induced cough and also inhibited antigen-evoked cough responses in ovalbumin-sensitized guinea pigs. Thus, TRPV1 channel activation seems to play a major role in cough mediated by the allergic reaction in guinea pigs (McLeod et al. 2006).

Gender difference has been detected previously in the cough threshold to capsaicin, and more recent findings have confirmed that women appear more sensitive to capsaicin-induced cough (Yamasaki et al. 2008). In addition to women, patients with chronic inflammatory diseases are often characterized by hypersensitivity to, or exhibit a lower threshold for capsaicin or citric acid induced cough. However, little is known about the underlying mechanisms of this hypersensitivity. Considering that the cough stress to airway wall may induce a self-perpetuating cough-reflex cycle it has been speculated that the act of coughing itself may contribute to the release of proinflammatory mediators that, in turn, may exaggerate the cough response (Jatakanon et al. 1999). An intriguing observation reports that in guinea pigs the number of neutrophils in the bronchoalveolar lavage correlated significantly with the number of coughs and that hydroxyurea, which depletes neutrophils, inhibited both the cough-stress-dependent increase in the number of coughs and the associated increase in airway neutrophil accumulation (Hara et al. 2008). It is possible that cough itself by producing a traumatic mechanical stress to the airway wall induces neutrophilic airway inflammation and cough-reflex hypersensitivity to capsaicin.

7 Conclusions

One of the commonest causes of chronic cough is gastroesophageal reflux disease. Selectivity of TRPV1 regulation by acidic-induced inflammation is suggested by the observation that in asthmatics inhibition of gastric acid secretion does not influence bronchial hyperresponsiveness to methacholine, but decreases tussive sensitivity to capsaicin and this effect is related to proximal reflux (Ferrari et al. 2007). The interaction between sensory nerve supply with the inflamed or injured tissue may follow nonobvious anatomical circuitry as acid instillation in the esophagus acutely increases the cough reflex sensitivity to capsaicin in patients with gastroesophageal reflux disease and chronic cough (Javorkova et al. 2008). This phenomenon suggests that cough due to acid reflux does not require the actual presence of acidic media in the airways. The complexity of the pathophysiology of chronic cough has hampered the identification of common underlying mechanisms that may represent the ideal target for efficacious and safe antitussive medicines. The molecular and clinical findings reported in this chapter start from the notion that chronic inflammation determines the tussive phenotype by diverse and sometimes unexpected mechanisms which ultimately regulate the sensitivity sensory neurons to a large variety of protussive endogenous mediators and exogenous irritants. In this scenario, the TRPV1 channel seems to play a role. The introduction of TRPV1 antagonists in clinical investigations should soon answer the question as to whether and to what extent this role is pivotal in chronic cough.

Acknowledgments This work was supported in part by a grant from MURST, Rome, ARCA, Padua, and Fondazione-DEI, Florence, Italy.

References

Agopyan N, Bhatti T, Yu S, et al. (2003) Vanilloid receptor activation by 2- and 10-microm particles induces responses leading to apoptosis in human airway epithelial cells. Toxicol Appl Pharmacol 192:21–35

Amadesi S, Moreau J, Tognetto M, et al. (2001) NK1 receptor stimulation causes contraction and inositol phosphate increase in medium-size human isolated bronchi. Am J Respir Crit Care Med 163:1206–1211

Amadesi S, Nie J, Vergnolle N, et al. (2004) Protease-activated receptor 2 sensitizes the capsaicin receptor transient receptor potential vanilloid receptor 1 to induce hyperalgesia. J Neurosci 24:4300–4312

Andersson DA, Gentry C, Moss S, et al. (2008) Transient receptor potential A1 is a sensory receptor for multiple products of oxidative stress. J Neurosci 28:2485–2494

Bang S, Kim KY, Yoo S, et al. (2007) Transient receptor potential A1 mediates acetaldehyde-evoked pain sensation. Eur J Neurosci 26:2516–2523

Basu S, Srivastava P (2005) Immunological role of neuronal receptor vanilloid receptor 1 expressed on dendritic cells. Proc Natl Acad Sci USA 102:5120–125

Bautista DM, Jordt SE, Nikai T, et al. (2006) TRPA1 mediates the inflammatory actions of environmental irritants and proalgesic agents. Cell 124:1269–1282

Benedetti A, Comporti M, Esterbauer H (1980) Identification of 4-hydroxynonenal as a cytotoxic product originating from the peroxidation of liver microsomal lipids. Biochim Biophys Acta 620:281–296

Bevan S, Geppetti P (1994) Protons: Small stimulants of capsaicin-sensitive sensory nerves. Trends Neurosci 17:509–512

Bevan S, Winter J (1995) Nerve growth factor (NGF) differentially regulates the chemosensitivity of adult rat cultured sensory neurons. J Neurosci 15:4918–4926

Bhattacharya A, Scott BP, Nasser N, et al. (2007) Pharmacology and antitussive efficacy of 4-(3-trifluoromethyl-pyridin-2-yl)-piperazine-1-carboxylic acid (5-trifluoromethyl-pyridin-2-yl)-amide (JNJ17203212), a transient receptor potential vanilloid 1 antagonist in guinea pigs. J Pharmacol Exp Ther 323:665–674

Bhave G, Zhu W, Wang H, et al. (2002) cAMP-dependent protein kinase regulates desensitization of the capsaicin receptor (VR1) by direct phosphorylation. Neuron 35:721–731

Birder LA, Kanai AJ, de Groat WC, et al. (2001) Vanilloid receptor expression suggests a sensory role for urinary bladder epithelial cells. Proc Natl Acad Sci USA 98:13396–13401

Birder LA, Nakamura Y, Kiss S, et al. (2002) Altered urinary bladder function in mice lacking the vanilloid receptor TRPV1. Nat Neurosci 5:856–860

Canning BJ, Farmer DG, Mori N (2006) Mechanistic studies of acid-evoked coughing in anesthetized guinea pigs. Am J Physiol Regul Integr Comp Physiol 291:R454–R463

Carr MJ, Kollarik M, Meeker SN, et al. (2003) A role for TRPV1 in bradykinin-induced excitation of vagal airway afferent nerve terminals. J Pharmacol Exp Ther 304:1275–1279

Caterina MJ, Schumacher MA, Tominaga M, et al. (1997) The capsaicin receptor: a heat-activated ion channel in the pain pathway. Nature 389:816–824

Chang AB, Gibson PG, Ardill J, et al. (2007) Calcitonin gene-related peptide relates to cough sensitivity in children with chronic cough. Eur Respir J 30:66–72

Chuang HH, Prescott ED, Kong H, et al. (2001) Bradykinin and nerve growth factor release the capsaicin receptor from PtdIns(4,5)P2-mediated inhibition. Nature 411:957–962

Chung KF (2007) Chronic cough: future directions in chronic cough: Mechanisms and antitussives. Chron Respir Dis 4:159–165

Clapham DE (2003) TRP channels as cellular sensors. Nature 426:517–524

Dai Y, Wang S, Tominaga M, et al. (2007) Sensitization of TRPA1 by PAR2 contributes to the sensation of inflammatory pain. J Clin Invest 117:1979–1987

Dalle-Donne I, Aldini G, Carini M, et al. (2006) Protein carbonylation, cellular dysfunction, and disease progression. J Cell Mol Med 10:389–406

Davis JB, Gray J, Gunthorpe MJ, et al. (2000) Vanilloid receptor-1 is essential for inflammatory thermal hyperalgesia. Nature 405:183–187

De Petrocellis L, Harrison S, Bisogno T, et al. (2001) The vanilloid receptor (VR1)-mediated effects of anandamide are potently enhanced by the cAMP-dependent protein kinase. J Neurochem 77:1660–1663

Dicpinigaitis PV, Alva RV (2005) Safety of capsaicin cough challenge testing. Chest 128:196–202

Dietrich A, Chubanov V, Kalwa H, et al. (2006) Cation channels of the transient receptor potential superfamily: their role in physiological and pathophysiological processes of smooth muscle cells. Pharmacol Ther 112:744–760

Esterbauer H, Schaur RJ, Zollner H (1991) Chemistry and biochemistry of 4-hydroxynonenal, malonaldehyde and related aldehydes. Free Radic Biol Med 11:81–128

Fahy JV, Wong HH, Geppetti P, et al. (1995) Effect of an NK1 receptor antagonist (CP-99,994) on hypertonic saline-induced bronchoconstriction and cough in male asthmatic subjects. Am J Respir Crit Care Med 152:879–884

Ferrari M, Benini L, Brotto E, et al. (2007) Omeprazole reduces the response to capsaicin but not to methacholine in asthmatic patients with proximal reflux. Scand J Gastroenterol 42:299–307

Gatti R, Andre E, Amadesi S, et al. (2006) Protease-activated receptor-2 activation exaggerates TRPV1-mediated cough in guinea pigs. J Appl Physiol 101:506–511

Geppetti P, Holzer P (1996) Neurogenic inflammation. CRC Press, Boca Raton

Geppetti P, Del Bianco E, Patacchini R, et al. (1991) Low pH-induced release of calcitonin gene-related peptide from capsaicin-sensitive sensory nerves: mechanism of action and biological response. Neuroscience 41:295–301
Geppetti P, Bertrand C, Bacci E, et al. (1993) Characterization of tachykinin receptors in ferret trachea by peptide agonists and nonpeptide antagonists. Am J Physiol 265:L164–L169
Geppetti P, Materazzi S, Nicoletti P (2006) The transient receptor potential vanilloid 1: role in airway inflammation and disease. Eur J Pharmacol 533:207–214
Hara J, Fujimura M, Ueda A, et al. (2008) Effect of pressure stress applied to the airway on cough-reflex sensitivity in Guinea pigs. Am J Respir Crit Care Med 177:585–592
Hardie RC (2003) Regulation of TRP channels via lipid second messengers. Annu Rev Physiol 65:735–759
Hinman A, Chuang HH, Bautista DM, et al. (2006) TRP channel activation by reversible covalent modification. Proc Natl Acad Sci U S A 103:19564–19568
Hofmann T, Obukhov AG, Schaefer M, et al. (1999) Direct activation of human TRPC6 and TRPC3 channels by diacylglycerol. Nature 397:259–263
Huang SM, Bisogno T, Trevisani M, et al. (2002) An endogenous capsaicin-like substance with high potency at recombinant and native vanilloid VR1 receptors. Proc Natl Acad Sci U S A 99:8400–8405
Hwang SW, Cho H, Kwak J, et al. (2000) Direct activation of capsaicin receptors by products of lipoxygenases: endogenous capsaicin-like substances. Proc Natl Acad Sci USA 97:6155–6160
Inoue K, Koizumi S, Fuziwara S, et al. (2002) Functional vanilloid receptors in cultured normal human epidermal keratinocytes. Biochem Biophys Res Commun 291:124–129
Ishiura Y, Fujimura M, Nobata K, et al. (2007) Prostaglandin I2 enhances cough reflex sensitivity to capsaicin in the asthmatic airway. Cough 3:2
Jaquemar D, Schenker T, Trueb B (1999) An ankyrin-like protein with transmembrane domains is specifically lost after oncogenic transformation of human fibroblasts. J Biol Chem 274: 7325–7333
Jatakanon A, Lalloo UG, Lim S, et al. (1999) Increased neutrophils and cytokines, TNF-alpha and IL-8, in induced sputum of non-asthmatic patients with chronic dry cough. Thorax 54:234–237
Javorkova N, Varechova S, Pecova R, et al. (2008) Acidification of the oesophagus acutely increases the cough sensitivity in patients with gastro-oesophageal reflux and chronic cough. Neurogastroenterol Motil 20:119–124
Ji RR, Samad TA, Jin SX, et al. (2002) p38 MAPK activation by NGF in primary sensory neurons after inflammation increases TRPV1 levels and maintains heat hyperalgesia. Neuron 36:57–68
Jia Y, Lee LY (2007) Role of TRPV receptors in respiratory diseases. Biochim Biophys Acta 1772:915–927
Jia Y, Wang X, Varty L, et al. (2004) Functional TRPV4 channels are expressed in human airway smooth muscle cells. Am J Physiol Lung Cell Mol Physiol 287:L272–L278
Jordt SE, Bautista DM, Chuang HH, et al. (2004) Mustard oils and cannabinoids excite sensory nerve fibres through the TRP channel ANKTM1. Nature 427:260–265
Kamei J, Yoshikawa Y, Saitoh A (2006) Effect of *N*-arachidonoyl-(2-methyl-4-hydroxyphenyl) amine (VDM11), an anandamide transporter inhibitor, on capsaicin-induced cough in mice. Cough 2:2
Karai LJ, Russell JT, Iadarola MJ, et al. (2004) Vanilloid receptor 1 regulates multiple calcium compartments and contributes to Ca2+-induced Ca2+ release in sensory neurons. J Biol Chem 279:16377–16387
Kido MA, Muroya H, Yamaza T, et al. (2003) Vanilloid receptor expression in the rat tongue and palate. J Dent Res 82:393–397
Kohrogi H, Graf PD, Sekizawa K, et al. (1988) Neutral endopeptidase inhibitors potentiate substance P- and capsaicin-induced cough in awake guinea pigs. J Clin Invest 82:2063–2068
Kollarik M, Undem BJ (2006) Sensory transduction in cough-associated nerves. Respir Physiol Neurobiol 152:243–254
Lalloo UG, Fox AJ, Belvisi MG, et al. (1995) Capsazepine inhibits cough induced by capsaicin and citric acid but not by hypertonic saline in guinea pigs. J Appl Physiol 79:1082–1087

Lee J, Kang M, Shin M, et al. (2003) *N*-(3-acyloxy-2-benzylpropyl)-*N′*-[4-(methylsulfonylamino)benzyl]thiourea analogues: novel potent and high affinity antagonists and partial antagonists of the vanilloid receptor. J Med Chem 46:3116–3126
Leung SY, Niimi A, Williams AS, et al. (2007) Inhibition of citric acid- and capsaicin-induced cough by novel TRPV-1 antagonist, V112220, in guinea-pig. Cough 3:10
Macpherson LJ, Dubin AE, Evans MJ, et al. (2007) Noxious compounds activate TRPA1 ion channels through covalent modification of cysteines. Nature 445:541–545
Mazzone SB, McGovern AE (2007) Sensory neural targets for the treatment of cough. Clin Exp Pharmacol Physiol 34:955–962
McKemy DD, Neuhausser WM, Julius D (2002) Identification of a cold receptor reveals a general role for TRP channels in thermosensation. Nature 416:52–58
McLeod RL, Fernandez X, Correll CC, et al. (2006) TRPV1 antagonists attenuate antigen-provoked cough in ovalbumin sensitized guinea pigs. Cough 2:10
McLeod RL, Jia Y, McHugh NA, et al. (2007) Sulfur-dioxide exposure increases TRPV1-mediated responses in nodose ganglia cells and augments cough in guinea pigs. Pulm Pharmacol Ther 20:750–757
McNamara CR, Mandel-Brehm J, Bautista DM, et al. (2007) TRPA1 mediates formalin-induced pain. Proc Natl Acad Sci USA. 104:13525–13530
Mezey E, Toth ZE, Cortright DN, et al. (2000) Distribution of mRNA for vanilloid receptor subtype 1 (VR1), and VR1-like immunoreactivity, in the central nervous system of the rat and human. Proc Natl Acad Sci USA 97:3655–3660
Montell C (1997) New light on TRP and TRPL. Mol Pharmacol 52:755–763
Montell C, Birnbaumer L, Flockerzi V, et al. (2002) A unified nomenclature for the superfamily of TRP cation channels. Mol Cell 9:229–231
Myers AC, Goldie RG, Hay DW (2005) A novel role for tachykinin neurokinin-3 receptors in regulation of human bronchial Ganglia neurons. Am J Respir Crit Care Med 171:212–216
Nassenstein C, Kwong K, Taylor-Clark T, et al. (2008) Expression and function of the ion channel TRPA1 in vagal afferent nerves innervating mouse lungs. J Physiol 586:1595–1604
Nilius B, Owsianik G, Voets T, et al. (2007) Transient receptor potential cation channels in disease. Physiol Rev. 87:165–217
O'Connell P J, Pingle SC, Ahern GP (2005) Dendritic cells do not transduce inflammatory stimuli via the capsaicin receptor TRPV1. FEBS Lett 2:2
Peier AM, Moqrich A, Hergarden AC, et al. (2002) A TRP channel that senses cold stimuli and menthol. Cell 108:705–715
Pomonis JD, Harrison JE, Mark L, et al. (2003) *N*-(4-Tertiarybutylphenyl)-4-(3-cholorphyridin-2-yl)tetrahydropyrazine-1(2H)-carbox-amide (BCTC), a novel, orally effective vanilloid receptor 1 antagonist with analgesic properties. II. In vivo characterization in rat models of inflammatory and neuropathic pain. J Pharmacol Exp Ther 306:387–393
Premkumar LS, Ahern GP (2000) Induction of vanilloid receptor channel activity by protein kinase C. Nature 408:985–990
Prescott ED, Julius D (2003) A modular PIP2 binding site as a determinant of capsaicin receptor sensitivity. Science 300:1284–1288
Rahman I, van Schadewijk AA, Crowther AJ, et al. (2002) 4-Hydroxy-2-nonenal, a specific lipid peroxidation product, is elevated in lungs of patients with chronic obstructive pulmonary disease. Am J Respir Crit Care Med 166:490–495
Runnels LW, Yue L, Clapham DE (2002) The TRPM7 channel is inactivated by PIP(2) hydrolysis. Nat Cell Biol 4:329–336
Schmidlin F, Amadesi S, Dabbagh K, et al. (2002) Protease-activated receptor 2 mediates eosinophil infiltration and hyperreactivity in allergic inflammation of the airway. J Immunol 169:5315–5321
Shin J, Cho H, Hwang SW, et al. (2002) Bradykinin-12-lipoxygenase-VR1 signaling pathway for inflammatory hyperalgesia. Proc Natl Acad Sci USA 99:10150–10155
Steinhoff M, Vergnolle N, Young SH, et al. (2000) Agonists of proteinase-activated receptor 2 induce inflammation by a neurogenic mechanism. Nat Med 6:151–158

Story GM, Peier AM, Reeve AJ, et al. (2003) ANKTM1, a TRP-like channel expressed in nociceptive neurons, is activated by cold temperatures. Cell 112:819–829

Su X, Camerer E, Hamilton JR, et al. (2005) Protease-activated receptor-2 activation induces acute lung inflammation by neuropeptide-dependent mechanisms. J Immunol 175:2598–2605

Sugiura T, Tominaga M, Katsuya H, et al. (2002) Bradykinin lowers the threshold temperature for heat activation of vanilloid receptor 1. J Neurophysiol 88:544–548

Szallasi A, Blumberg PM (1999) Vanilloid (Capsaicin) receptors and mechanisms. Pharmacol Rev 51:159–212

Tognetto M, Amadesi S, Harrison S, et al. (2000) Anandamide excites central terminals of dorsal root ganglion neurons via vanilloid receptor-1 (VR-1) activation. J Neurosci 21:1104–1109

Tominaga M, Caterina MJ, Malmberg AB, et al. (1998) The cloned capsaicin receptor integrates multiple pain-producing stimuli. Neuron 21:531–543

Trevisani M, Gazzieri D, Benvenuti F, et al. (2004) Ethanol causes inflammation in the airways by a neurogenic and TRPV1-dependent mechanism. J Pharmacol Exp Ther 309:1167–1173

Trevisani M, Siemens J, Materazzi S, et al. (2007) 4-Hydroxynonenal, an endogenous aldehyde, causes pain and neurogenic inflammation through activation of the irritant receptor TRPA1. Proc Natl Acad Sci USA. 104:13519–13524

Uchida K, Szweda LI, Chae HZ, et al. (1993) Immunochemical detection of 4-hydroxynonenal protein adducts in oxidized hepatocytes. Proc Natl Acad Sci USA 90:8742–8746

Vellani V, Mapplebeck S, Moriondo A, et al. (2001) Protein kinase C activation potentiates gating of the vanilloid receptor VR1 by capsaicin, protons, heat and anandamide. J Physiol 534: 813–825

Vergnolle N, Bunnett NW, Sharkey KA, et al. (2001) Proteinase-activated receptor-2 and hyperalgesia: A novel pain pathway. Nat Med 7:821–826

Walker KM, Urban L, Medhurst SJ, et al. (2003) The VR1 antagonist capsazepine reverses mechanical hyperalgesia in models of inflammatory and neuropathic pain. J Pharmacol Exp Ther 304:56–62

Wang S, Dai Y, Fukuoka T, et al. (2008) Phospholipase C and protein kinase A mediate bradykinin sensitization of TRPA1: a molecular mechanism of inflammatory pain. Brain 20:20

Watanabe N, Horie S, Michael GJ, et al. (2005) Immunohistochemical localization of vanilloid receptor subtype 1 (TRPV1) in the guinea pig respiratory system. Pulm Pharmacol Ther 18: 187–197

Yamasaki M, Ebihara S, Freeman S, et al. (2008) Sex differences in the preference for place of death in community-dwelling elderly people in Japan. J Am Geriatr Soc 56:376

Zygmunt PM, Petersson J, Andersson DA, et al. (1999) Vanilloid receptors on sensory nerves mediate the vasodilator action of anandamide. Nature 400:452–457

Cough Sensors. III. Opioid and Cannabinoid Receptors on Vagal Sensory Nerves

M.G. Belvisi(✉) and D.J. Hele

Contents

Abstract Cough is a persistent symptom of many inflammatory airways' diseases. Cough is mediated by receptors sited on sensory nerves and then through vagal afferent pathways, which terminate in the brainstem respiratory centre. Cough is often described as an unmet clinical need. Opioids are the only prescription-based antitussives currently available in the UK. They possess limited efficacy and exhibit serious unwanted side effects, such as physical dependence, sedation, respiratory depression and gastrointestinal symptoms. There are three classical opioid receptors: the mu, kappa and delta receptors. Peripheral opioid receptors are sited on sensory nerves innervating the airways. A greater understanding of the role of the peripheral and centrally sited opioid receptors is necessary to allow the development of targeted treatments for cough. Because of the limited efficacy and the side-effect profile of the opioids, potential new treatments are sought to alleviate cough. One class of compounds that is currently under examination is the cannabinoids. Like the opioids, cannabinoids have peripheral and centrally sited receptors and also suffer from the blight of unwanted centrally mediated side effects such as sedation, cognitive dysfunction, tachycardia and psychotropic effects. Two cannabinoid receptors have been identified, the CB_1 and CB_2 receptors, and their distribution varies throughout the peripheral and central nervous system. Encouragingly, early studies with these compounds suggest that it may be possible to separate their antitussive

M.G. Belvisi
Respiratory Pharmacology, Airway Diseases, National Heart & Lung Institute, Faculty of Medicine, Imperial College, Guy Scadding Building, Dovehouse Street, London SW3 6LY, UK
m.belvisi@imperial.ac.uk

K.F. Chung, J.G. Widdicombe (eds.), *Pharmacology and Therapeutics of Cough*, Handbook of Experimental Pharmacology 187,

activity from their centrally mediated side effects, with CB_2 agonists showing potential as putative new treatments for cough. In this chapter, we describe the opioid and cannabinoid receptors, their distribution and the effects they mediate. Moreover, we highlight their potential advantages and disadvantages in the treatment of cough.

1 Introduction

Cough is a dominant and persistent symptom of many inflammatory lung diseases, including asthma, chronic obstructive pulmonary disease (COPD), viral infections and bronchiectasis, and is also thought to be induced by gastro-oesophageal disorders (Kastelik et al. 2005). Chronic cough can also be idiopathic in nature, where no obvious causal mechanism is evident. Cough is mediated by the activation of vagal afferent pathways, which converge in brainstem respiratory centres (Canning et al. 2006; Canning 2007). All afferent nerve subtypes innervating the airways can modulate the cough reflex. Rapidly adapting and slowly adapting stretch receptors (RARs and SARs, respectively) innervating the intrapulmonary airways and lung may facilitate coughing. Extrapulmonary C-fibres (sensitive to anaesthesia) and RARs can initiate coughing upon activation. Tracheal and bronchial C-fibres may also interact with other afferents to enhance coughing. Recent studies in anesthetised guinea pigs have identified a myelinated afferent nerve subtype that can be differentiated from intrapulmonary RARs and SARs and plays an essential role in initiating cough. These have been termed 'cough receptors' and are stimulated by low-pH solutions, e.g. citric acid (Canning et al. 2006).

In chronic cough, changes in intrinsic and synaptic excitability at the ganglionic, spinal or brainstem level may be the mechanism by which the cough reflex is enhanced (Chung 2007a). Sensory nerve activity may be enhanced during inflammation such that the protective cough reflex becomes exacerbated and troublesome (Barnes 2001; Belvisi 2002). The cough reflex is often enhanced in airway diseases such as asthma and COPD (Higenbottam 2002). In COPD this can occur owing to increased vagal afferent nerve activity and increased sensitivity to inflammatory mediators (Undem and Kollarik 2005; Carr and Lee 2006).

Cough is the most common complaint for which medical attention is sought, and there are few, if any, effective treatments for cough. Those treatments that exist use opioids, which target receptors on both peripheral sensory nerves and receptors sited in the central nervous system (CNS) and, as such, often display unwanted side effects (Schroeder and Fahey 2002; Eccles 2002; Chung 2007b). Opioids have long provided the mainstay for cough treatment (Sevelius et al. 1971; Chung 2007b), and currently there are three prescription-based medicines available for the treatment of cough in the UK. These are codeine, pholcodeine and the opiate-like dextromethorphan (Widdicombe and Chung 2007). The clinical evidence for the use of opioids to treat cough is sketchy at best, but recently low-dose morphine has been shown to be an effective antitussive in the treatment of intractable chronic cough (Morice et al. 2007). As always with studies conducted with the opioids, it is difficult to achieve beneficial effects while avoiding undesirable side effects, and in this study, an increase in dose produced sedation and drowsiness. It is for these reasons that new and effective treatments for cough are constantly being sought. One of the areas currently being explored to provide a potential new treatment for cough involves the study of the cannabinoids and their receptors (Patel et al. 2003; Chung 2005). One of the potential advantages with the cannabinoids is that by developing receptor-selective treatments it may be possible to separate their antitussive efficacy from their unwanted centrally mediated effects, a problem that has so bedevilled the use of opioids. In this chapter, we will discuss the opioid and cannabinoid receptors involved in the cough response and attempt to highlight their advantages and their failings.

2 Opioid Receptors

Opioids are widely used drugs in the treatment of severe pain, and they have long been advocated for the treatment of cough (Chung 2007b). Agents such as morphine, diamorphine and codeine have all laid claim to efficacy in the treatment of cough but there are few clinical trial data to support these claims. Small, single-dose trials of codeine in chronic bronchitis have shown some benefit (Sevelius and Colmore 1966; Sevelius et al. 1971; Aylward et al. 1984) but a further more recent trial in COPD patients suggested little or no effect of codeine when compared with that of a placebo (Smith et al. 2006). Studies conducted with opiates in acute cough and cough challenge in normal volunteers also suggest a lack of efficacy (Freestone and Eccles 1997). Added to their modest success in controlling cough, opioids suffer from the major drawback that at the dose levels used to treat cough they may cause physical dependence, sedation, respiratory depression and gastrointestinal symptoms such as constipation and colic. However, despite these drawbacks, morphine and diamorphine have proved useful in treating severe distressing cough in patients with a terminal illness such as bronchial carcinoma where dependence is not an issue (Belvisi and Geppetti 2004; Morice et al. 2007).

Originally four opioid receptors were identified, μ, δ, κ and σ (mu, delta, kappa and sigma, respectively). They were initially named after the drugs used to identify them: mu for morphine, kappa for ketocyclazocine and sigma for SKF 10047. The delta receptor was named after deferens for mouse vas deferens (Martin et al. 1976; Lord et al. 1977) (Table 1). However, it was later discovered that the sigma receptor (activated by dextramethorphan) was a non-opioid receptor; therefore, three classical opioid receptors remain (Gundlach et al. 1986). Of these the delta receptor is thought to have two subtypes, delta-1 and delta-2 (Jiang et al. 1991), and recent work has been carried out to determine the role of these subtypes in mediating cough or antitussive effects (Kamei 2002). The three opioid receptors have now been cloned, and there is a high structural homology between the clones. It has also been confirmed that they belong to the superfamily of seven transmembrane G-protein-coupled receptors. More recently, another endogenous opioid

Table 1 Opioid receptor classification: proposed terminology

Proposed nomenclature[a]	Previous nomenclature	Endogenous ligands
μ, mu, MOP, MOR	OP_3	ß-endorphin enkephalins endomorphin-1; endomorphin-2
δ, delta or DOP, DOR	OP_1	Enkephalins ß-endorphin
κ, kappa or KOP, KOR	OP_2	Dynorphin A; dynorphin B; α-neoendorphin
NOP, NOR	OP_4	N/OFQ

[a]Some nomenclature guidelines suggest that when first described in a publication, the receptor should be defined as MOP, DOP, KOP or NOP. Some authors simply use the well-established Greek terminology for μ, δ and κ receptors. Others have suggested that MOR, DOR and KOR should be used to refer to the genes for these receptors

peptide nociceptin/orphanin FQ has also been discovered, which has some homology to the dynorphin family but lacks the N-terminal tyrosine residue, which is essential for binding classical opioid receptors (see Table 1). Nociceptin binds to the G-protein-coupled opioid receptor-like 1 receptor (NOP, formerly ORL1).

Shortly after their first characterisation, it was recognised that opioid receptors are not only expressed in the CNS but also in peripheral tissues, including the lung (Hughes et al. 1975). Autoradiographic binding studies established the presence of dense localisations of opioid receptors in many areas of the brainstem, including the parabrachial nuclei, the superior colliculus, the ventral median raphe nucleus and the nucleus tractus solitarius (Atweh and Kuhar 1977). Some of these areas, such as the raphe nucleus, have been shown to have a role to play in cough (Baekey et al. 2001), and stimulation of laryngeal afferent nerves demonstrated enhanced immunoreactivity in the dorsal raphe nucleus in cat (Ambalavanar et al. 1999). Results from in vivo experiments have suggested that centrally acting opioid antitussives primarily act on the brainstem cough centre. It has also been reported that fictive cough selectively increased immunoreactivity in the nucleus tractus solitarius, the reticular formation, the ambigual complex and the medial parabrachial nucleus in the cat. In all of these areas, codeine significantly reduced the increase in immunoreactivity (Gestreau et al. 1997). These nuclei and the dorsal vagal nerve nucleus show intense expression of mu and kappa opioid receptors. Delta opioid receptors are less abundant in these areas, with the exception of the nucleus parabrachialis, which has a high density of delta receptors. It has been suggested that the nucleus tractus solitarius contains the cough centre, as this region primarily receives sensory input from the airways and its stimulation causes cough-like responses (Takahama and Shirasaki 2007). The nucleus tractus solitarius is also more heavily labelled by a mu ligand than by a kappa ligand in guinea pigs and cats (Dashwood et al. 1988), and microinjection of codeine into the nucleus tractus solitarius inhibited a fictive cough reflex in the guinea pig (Ohi et al. 2005). The raphe nuclei may also be a candidate for the site of action of opioid antitussives, since stimulation of this region depresses the reflex activity caused by stimulation of the superior laryngeal or vagal nerve (Sessle et al. 1981).

Opioid receptors are also located in peripheral sensory nerves, and opioid antitussives have been shown to be effective via these receptors (Adcock 1991). Inhalations of aerosolised codeine, morphine and a peripherally acting specific mu-opioid receptor agonist have demonstrated antitussive effects in guinea pigs (Karlsson et al. 1990; Callaway et al. 1991). It is thought that opioid antitussives delivered directly into the airways exert their effect on sensory nerve endings via activation of mu-opioid receptors in the airways. Further studies by Atweh et al. (1978) established the presence of opioid receptors in sensory nerve fibres in the vagus nerve. Autoradiographic binding studies in rat lung using [^{3}H]morphine demonstrated the presence of high- and low-affinity binding sites with high densities in the alveolar wall and lower densities in the smooth muscle of the trachea and main bronchi (Cabot et al. 1994). The high density of opioid binding sites within the alveolar wall is consistent with the results of early electron micrograph studies, which showed that axons of sensory nerve fibres innervated the alveolar wall of both rat and human lung (Merick and Reid 1971; Fox et al. 1980). However, it is not likely that these receptors play a role in the inhibitory effects of opioids on the cough reflex given cough is not initiated from this site. A further study employing selective ligands for each of the opioid receptors demonstrated receptor binding in the rat trachea, main bronchi, parenchyma and pulmonary artery (Bhargava et al. 1997). Cabot et al. (1996) speculated that the opioid binding sites located within the trachea and main bronchi of the rat airways may be the prejunctional opioid receptors on C-afferent nerve fibres, which modulate the release of potent inflammatory neuropeptides. These are the same nerve fibres known to be involved in bronchoconstriction and the cough reflex. It has been shown that opioids can inhibit non-adrenergic non-cholinergic nerve-mediated bronchoconstrictor responses both in vitro in guinea pig bronchi (Frossard and Barnes 1987) and in vivo in guinea pig airways (Belvisi et al. 1988).

With regard to cough, various studies have shown that all three opioid receptors may have a role to play, and agonists for each of these receptors have been shown to produce antitussive effects (see Table 2 for a list of receptor-selective ligands). Codeine, which is considered to be a mu-opioid receptor agonist, has been shown to inhibit citric acid induced cough in guinea pigs by a peripheral action (Karlsson et al. 1990). Kappa-opioid receptor selective agonists, such as U-50,488H and U-62,066E, have been shown to inhibit cough in animal models by a central action (Kamei et al. 1990), and delta agonists such as SB227122 have been shown to inhibit citric acid induced cough in guinea pigs (Kotzer et al. 2000). As mentioned earlier, work by Kamei (2002) has been carried out to determine the role of the delta receptor subtypes in the cough response and it has been suggested that delta-1 opioid receptors may play an inhibitory role, whereas delta-2 receptors may play a synergistic role, in antitussive processes mediated by delta receptors. Kamei (2002) showed that the delta receptor antagonists naltrindole and 7-benzylidenenaltrexone produce potent antitussive effects and further showed that these effects may be mediated by the antagonism of delta-1, but not delta-2 opioid receptors. They further

Table 2 A selection of the commercially available tools for the opioid receptors

μ	δ	κ	NOP
		Endogenous ligands	
Endomorphin-1	[Met]-enkephalin	Dynorphin A	Nociceptin
Endomorphin-2	[Leu]-enkephalin	Dynorphin B	
		Agonists	
DAMGO-[D-Ala2, N-Me-Phe4, Gly5-ol]-enkephalin	DPDPE-[D-Pen2, d-Pen5]-enkephalin	U-69593	NNC 63–0532
	SB 205607 dihydrobromide (δ_1)	U-50488	Nociceptin (1-13)NH_2
	BW 373U86	BRL 52537 hydrochloride	[Arg^{14},Lys^{15}] nociceptin
	SNC-80	ICI-199,441	Ro-64-6198
	SNC 162	ICI-204,448	
	SB227122		
		Antagonists	
CTAP-Cys2, Tyr3, Arg5, Pen7-amide	Naltrindole	Norbinaltor-phimine	JTC-801
β–Funaltrexamine	ICI 174864		[$Nphe^1$] nociceptin(1-13) NH_2 Trap 101
	ICI-154,129		UFP-101
	N-Benzylnaltrindole hydrochloride (δ_2)		
	Naltriben mesylate (δ_2)		
	BNTX-7-benzylidenenaltrexone maleate (δ_1)		
	SB 244525		

suggest that inhibition of cough seen with delta agonists such as SB227122 (Kotzer et al. 2000) may be via the delta-2 receptor and not the delta-1 receptor. The studies carried out by this group frequently employ mouse or rat models of cough, and as it is thought that mice cannot cough and that rats at best exhibit a weak cough response (Belvisi and Bolser 2002), it would be helpful to confirm these data in an animal such as the guinea pig, which is known to produce a more 'typical' tussive response. Furthermore, in these studies conscious animals were exposed to a nebulised solution of capsaicin using a body plethysmograph. However, it was not clear how cough was assessed and therefore any changes in pressure in the plethysmograph may be counted as a cough. To ensure that what is being counted is actually a cough, the best practice is to detect cough both by pressure change and by sound as recorded by a cough analyser. The cough/exposure chamber can be fitted with a microphone, which is connected to an external speaker, allowing the cough sound to be magnified and the number of coughs to be confirmed by manual counting. A cough can therefore be identified by the characteristic sound coupled with a quick, large abdominal movement. The cough produces a transient increase in airflow over and above the normal flow, which can be detected by computer software and appears on a screen.

More recently, another endogenous opioid peptide, nociceptin/orphanin FQ, has also been discovered which has some homology to the dynorphin family but lacks the N-terminal tyrosine residue, which is essential for binding classical opioid receptors (see Table 1). Nociceptin binds to the G-protein-coupled opioid receptor-like 1 receptor (NOP, formerly ORL1). NOP1 receptors are expressed in capsaicin-sensitive vagal sensory nerves innervating airways, and nociceptin effectively inhibits tachykinin release from vagal C-fibres in guinea pig bronchi (Fischer et al. 1998; Corboz et al. 2001). Intravenous administration of nociceptin also effectively inhibits in conscious animals cough reflexes evoked by mechanical perturbation of the airways or by inhalation of capsaicin (McLeod et al. 2001; Bolser et al. 2001). Although nociceptin may act centrally to inhibit cough, observations that nociceptin can inhibit the function of vagal afferent nerves within the airway indicate that it may also inhibit cough through a peripheral mechanism of action (Fischer et al. 1998; Corboz et al. 2001). In a more recent study, nociceptin effectively inhibited acid-evoked cough in guinea pigs by nearly 70%. Acid (pH 5) increased intracellular free calcium in acutely dissociated vagal jugular ganglionic neurons. The acid-induced increase in intracellular calcium was inhibited by a selective transient receptor potential vanilloid-1 antagonist, 5-iodoresiniferatoxin (1 μM, approximately 80% reduction). The inhibitory effect of 5-iodoresiniferatoxin on acid-induced increases in calcium was mimicked by nociceptin. In gramicidin-perforated patch-clamp recordings on airway-specific capsaicin-sensitive jugular ganglion neurons, acid (pH 5) induced two distinct inward currents. A transient current was evoked that was inhibited by amiloride, and a sustained current was evoked that was inhibited by 5-iodoresiniferatoxin. Nociceptin selectively inhibited only the sustained component of acid-induced inward current, suggesting that the inhibitory effect of nociceptin on acid-induced cough may result from a direct inhibitory effect on peripheral C-fibre activity caused by the selective inhibition of acid-induced transient receptor potential vanilloid-1 activation (Lee et al. 2006). A group at Hoffmann-La Roche reported on the selective, non-peptide NOP agonist Ro 64-6198, which became the most extensively reported non-peptide NOP agonist and a valuable pharmacological tool in determining the potential of the NOP receptor as a therapeutic target. Ro 64-6198 is systemically active and achieves high brain penetration. It has subnanomolar affinity for the NOP receptor and is at least 100 times more selective for the NOP receptor over the classic opioid receptors. Ro 64-6198 ranges from partial to full agonist, depending on the assay. Preclinical data indicate that Ro 64-6198 may have broad clinical uses, such as in treating stress and anxiety, addiction, neuropathic pain, cough and anorexia (Shoblock 2007).

It is now recognised that dextromethorphan, a commonly used non-narcotic antitussive, has a binding K_i of 15 nM at the sigma receptor. Dextromethorphan exhibits a dose-related antitussive activity in a guinea pig model of cough (Callaway et al. 1991; Braga et al. 1994), and it is reduced by treatment with rimcazole, a sigma receptor antagonist (Kamei et al. 1993).

In summary, opioid receptors have a role to play in the cough reflex, and by interfering with them with opioid agonists, we may be able to alleviate cough in animal models and in man. This does not inform us as to where the opioids are exerting their effects. We know that their receptors are sited both centrally at sites in the brain or brainstem and peripherally in nerves that innervate the airways. However, opioids could be exerting their effects at either site or to varying degrees at both sites. In animal models and man the extent to which opioids cross the blood brain barrier is not known, although the fact that they produce centrally mediated effects such as sedation, drowsiness and euphoria confirms that they do cross the blood–brain barrier. The cough centre in the brain has been tentatively identified, and it is probable that drugs like the opioids exert their antitussive actions both centrally on brainstem opioid receptors and peripherally on receptors located on sensory nerve endings in the airways. We have little knowledge of where their specific sites of action reside and, if their site of action is central, we cannot say where in the physiological complex of the cough centre it resides. The signal transduction mechanism of action by which opioid receptors inhibit sensory nerve activation and thereby the cough reflex is not yet known; however, at the neuronal level, mu-opioid and delta-opioid receptor activation causes a decrease in adenylyl cyclase activity, which in turn opens potassium channels (hyperpolarisation) and closes calcium channels. Both these actions may result in decreased sensory nerve activity, thereby suppressing the cough reflex (Pleuvry 2005). Despite these gaps in our knowledge, opioids will remain the mainstay of cough treatment until novel treatments are developed that possess improved efficacy and an improved side-effect profile.

3 Cannabinoid Receptors

Currently there is renewed interest in the therapeutic potential of cannabinoids as antitussives (Chung 2005). Non-selective cannabinoids have been shown to have wide therapeutic applications for a number of important medical conditions, including pain, anxiety, glaucoma, nausea, emesis, muscle spasms and wasting diseases; however, associated side effects such as sedation, cognitive dysfunction, tachycardia and psychotropic effects have hampered the use of these compounds in the clinic (Porter and Felder 2001).

Cannabinoids mediate their effects via at least two specific G-protein-coupled receptors, termed the CB_1 and CB_2 receptors (Matsuda et al. 1990; Munro et al. 1993) (see Table 3 for a list of selective ligands). The cannabinoid receptors were named because of their affinity for the agonist δ^9-tetrahydrocannabinol, a ligand found in organic extracts from *Cannabis sativa* (Howlett 2002). CB_1 receptors are predominantly distributed throughout the brain and spinal cord and are found mainly at the terminals of central neurones and are also expressed at low levels in several peripheral tissues. There is evidence that these receptors can mediate inhibition of ongoing release of a number of different excitatory and inhibitory transmitters, including acetylcholine, noradrenaline, dopamine and 5-hydroxytryptamine (Howlett et al. 2002). It is therefore thought that an important role of the CB_1 receptor is to modulate neurotransmitter release in order to maintain homeostasis in health and disease by preventing the development of excessive neuronal activity in the CNS (Pertwee 2007). In contrast, CB_2 receptors are not commonly expressed in the CNS (Munro et al. 1993; Griffin et al. 1997; Buckley et al. 2000), but primarily on

Table 3 A selection of the commercially available tools for the cannabinoid receptors

	Non-selective	CB_1	CB_2
Endogenous ligands	Anandamide	Docosatetraenylethanolamide (DEA)	
	2-Arachidonylglycerol (2-AG)		
Agonists	WIN 55,212-2 mesylate	Arachidonyl−2′-chloroethylamide (ACEA)	JWH 133
	CP 55940	Arachidonylcyclopropylamide (ACPA)	JWH 015
	HU 210		CB 65 L-759,633 L-759,656
Antagonists	O-2050[a]	AM 251[b] LY320135[b] SR 141716A	AM 630[c] JTE 907[c] SR 144,528

[a]Silent receptor antagonist
[b]CB_1 receptor antagonist/inverse agonist
[c]Selective CB_2 receptor antagonist/inverse agonist

immune tissues such as the spleen and tonsils and also on lymphocytes (Galiegue et al. 1995). It is thought that CB_2 receptor activation alters the release of cytokines from immune cells and may, in addition, affect immune function by modulating immune cell migration both within and outside the CNS (Walter and Stella 2004; Cabral and Staab 2005; Pertwee 2005).

Studies suggest that cannabinoids have diverse effects on sensory nerve function. Activation of spinal CB_1 receptors inhibits nociceptive transmission (Harris et al. 2000), hyperalgesia and neuropeptide release from central primary afferent fibres (Richardson et al. 1998a). The endocannabinoid anandamide has been shown to reduce carrageenan-induced hyperalgesia, oedema, plasma protein extravasation and capsaicin-induced neuropeptide release via peripheral CB_1 receptor activation (Richardson et al. 1998b). Furthermore, a recent study has demonstrated an inhibitory action of the anandamide on the capsaicin-induced cough reflex in the conscious guinea pig, and the authors (Caligano et al. 2000) suggested that anandamide activates CB_1 receptors in this regard. However, CB_1 ligands may have a sedative action in conscious animal models and this activity would inhibit the cough reflex per se. More recently, CB_2 receptor activation has been demonstrated to be sufficient to inhibit acute nociception, inflammatory hyperalgesia and the allodynia and hyperalgesia produced in models of neuropathic pain (Clayton et al. 2002; Malan et al. 2002; Hanus et al. 1999; Elmes et al. 2005). There is currently very little information on CB_2 receptors on peripheral sensory nerves and their involvement in neurogenic inflammatory responses in the airways. However, our group has characterised CB_2 receptor responses on sensory nerves in vitro using a commercially available cannabinoid CB_2 receptor agonist, JWH 133. In these studies, it was demonstrated that activation of the CB_2 receptor subtype inhibits capsaicin-induced depolarisation of the guinea pig and human vagus (Patel et al. 2003). Moreover, it has also been demonstrated that inhibition of sensory nerve function in vivo, also with the CB_2 agonist JWH 133, results in antitussive action in a conscious guinea pig model of cough (Patel et al. 2003). In summary, activation of the CB_2 receptor subtype inhibits sensory nerve activation of guinea pig and human vagus nerve, and importantly it has been shown to inhibit the cough reflex in conscious guinea pigs. CB_2 receptors have been shown to be primarily sited in the periphery, with little evidence for their presence in the CNS. This evidence offers the hope that the development of CB_2 agonists, devoid of CB_1-mediated central effects, may provide new and safe antitussive treatments for cough.

It is not clear yet how cannabinoid receptors may inhibit sensory nerve activation; however, both CB_1 and CB_2 receptors are coupled through Gi/o proteins, negatively to adenylyl cyclase and positively to mitogen-activated protein kinase (Howlett et al. 2002). Furthermore, CB_1 receptors are also thought to be coupled to ion channels through Gi/o proteins, positively to A-type and inwardly rectifying potassium channels and negatively to N-type and P/Q-type calcium channels

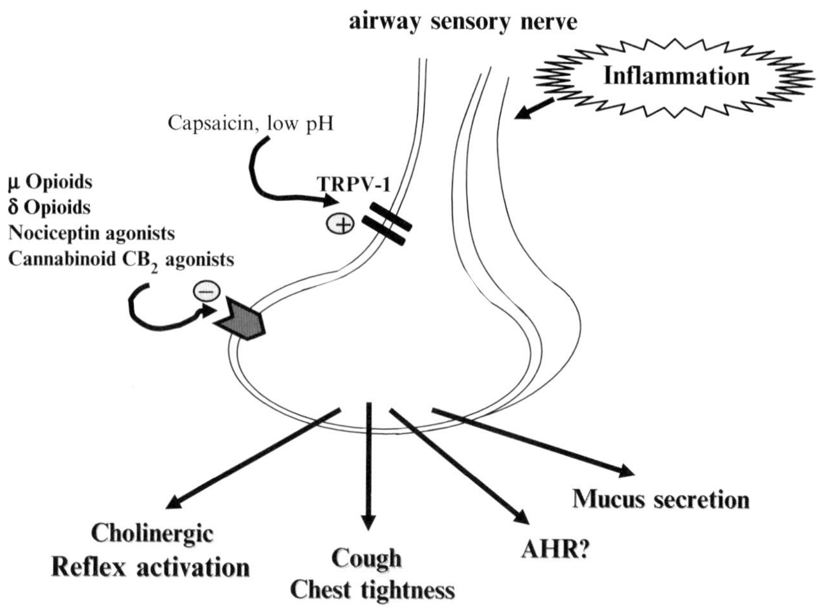

Fig. 1 An airway sensory nerve ending, which can be activated by capsaicin and low-pH stimuli to evoke cough, airway hyperresponsiveness, mucus secretion and cholinergic reflex activation. In inflammatory conditions, these nerves may be sensitised so that they respond at lower thresholds. Opioids, nociceptin and cannabinoid ligands have all been postulated to exert an inhibitory effect on sensory nerve activation

(Howlett 2002; Pertwee 1997). Details of additional signalling mechanisms have also been documented for cannabinoid CB_1 and CB_2 receptors (Howlett et al. 2002).

4 Summary

Opioid receptors and cannabinoid receptors have been characterised and intensively studied. They occur both in the brain or brainstem and in the peripheral tissues, including the airways (Fig. 1). The exact site of these receptors is largely unknown. Cough has long been identified as a debilitating symptom of various airway diseases and can also exist without any obvious cause. Cough is mediated by receptors on sensory nerves sited in the airways via vagal afferent nerves with terminals in the centrally located cough centres. Current treatment for cough is largely unsatisfactory, consisting of over-the-counter cough medicines with limited efficacy and prescription-based opioids, which also have limited efficacy and cause undesirable side effects (Chung 2007a, b). The cannabinoids are a class of compounds that are being evaluated in an attempt to provide a viable antitussive treatment, which will target peripheral cannabinoid receptors and be free of the centrally mediated side effects that are also associated with this class of compounds. Some degree of success has been achieved towards this end but further work needs to be carried out to clarify the receptors involved and to determine whether the cannabinoids will provide a viable alternative to opioids in the clinic.

References

Adcock JJ (1991) Peripheral opioid receptors and the cough reflex. Respir Med 85 (Suppl A):43–46

Ambalavanar R, Tanaka Y, Damirjian M et al. (1999) Laryngeal afferent stimulation enhances Fos immunoreactivity in periaqueductal grey in the cat. J Comp Neurol 409:411–423

Atweh SF, Kuhar MJ (1977) Autoradiographic localization of opiate receptors in rat brain. II. The brain stem. Brain Res 129:1–12

Atweh SF, Murrin LC, Kuhar MJ (1978) Presynaptic localization of opiate receptors in the vagal and accessory optic systems: an autoradiographic study. Neurophamacology 17:65–71

Aylward M, Maddock J, Davies DE et al. (1984) Dextromethorphan and codeine: comparison of plasma kinetics and antitussive effects. Eur J Respir Dis 65:283–291

Baekey DM, Morris K, Gestreau C et al. (2001) Medullary respiratory neurones and control of laryngeal motoneurones during fictive eupnoea and cough in the cat. J Physiol 534:565–581

Barnes PJ (2001) Neurogenic inflammation in the airways. Respir Physiol 125:145–154

Belvisi MG (2002) Overview of the innervation of the lung. Curr Opin Pharmacol 2:211–215

Belvisi MG, Bolser D (2002) Summary: animal models for cough. Pulm Pharmacol Ther 15(3):249–250

Belvisi MG, Geppetti P (2004) Cough 7: current and future drugs for the treatment of chronic cough. Thorax 59(5):438

Belvisi MG, Chung KF, Jackson DM et al. (1988) Opioid modulation of non-cholinergic neural bronchoconstriction in guinea-pig in vivo. Br J Pharmacol 95(2):413–418

Bhargava HN, Villar VM, Cortijo J et al. (1997) Binding of [3H][D-Ala2, MePhe4, Gly-ol5] enkephalin, [3H][D-Pen2, D-Pen5]enkephalin, and [3H]U-69,593 to airway and pulmonary tissues of normal and sensitized rats. Peptides 18:1603–1608

Bolser DC, McLeod RL, Tulshian DB, Hey JA (2001) Antitussive action of nociceptin in the cat. Eur J Pharmacol 430:107–111

Braga PC, Fossati A, Vimercati MG, Caputo R, Guffanti EE (1994) Dextrorphan and dextromethorphan: comparative antitussive effects on guinea pigs. Drugs Exp Clin Res 20:199–203

Buckley NE, McCoy KL, Mezey E et al. (2000) Immunomodulation by cannabinoids is absent in mice deficient for the cannabinoid CB_2 receptor. Eur J Pharmacol 396:141–149

Cabot PJ, Crammond T, Smith MT (1996) Quantitative autoradiography of peripheral opioid binding sites in rat lung. Eur J Pharmacol 310:47–53

Cabot PJ, Dodd PR, Crammond T et al. (1994) Characterization of non-conventional opioid binding sites in rat and human lung. Eur J Pharmacol 268(2):247–255

Cabral GA, Staab A (2005) Effects on the immune system. In: Pertwee RG (ed) Cannabinoids. Handbook of experimental pharmacology, vol. 168. Springer, Berlin, pp 385–423

Caligano A, Katano I, Desarnaud F et al. (2000) Bidirectional control of airway responsiveness by endogenous cannabinoids. Nature 408(6808):96–101

Callaway JK, King RG, Boura AL (1991) Evidence for peripheral mechanisms mediating the antitussive actions of opioids in the guinea pig. Gen Pharmacol 22(6):1103–1108

Canning BJ (2007) Encoding of the cough reflex. Pulm Pharmacol Ther 20:396–401

Canning BJ, Nori N, Mazzone SB (2006) Vagal afferent nerves regulating the cough reflex. Respir Physiol Neurobiol 152(3):223–242

Carr MJ, Lee LY (2006) Plasticity of peripheral mechanisms of cough. Respir Physiol Neurobiol 152(3):298–311

Chung KF (2005) Drugs to suppress cough. Expert Opin Investig Drugs 14(1):19–27

Chung KF (2007a) Review series: chronic cough: future directions in chronic cough: mechanisms and antitussives. Chron Respir Dis 4(3):159–165

Chung KF (2007b) Effective antitussives for the cough patient an unmet need. Pulm Pharmacol Ther 20(4):438–445

Clayton N, Marshall FH, Bountra C et al. (2002) CB_1 and CB_2 cannabinoid receptors are implicated in inflammatory pain. Pain 96:253–260

Corboz MR, Fernandez X, Egan RW, Hey JA (2001) Inhibitory activity of nociceptin/orphanin FQ on capsaicin-induced bronchoconstriction in the guinea-pig. Life Sci 69:1203–1211

Dashwood MR, Muddle JR, Spyer KM (1988) Opiate receptor subtypes in the nucleus tractus solitarii of the cat: the effect of vagal section. Eur J Pharmacol 155(1–2):85–92

Eccles R (2002) The powerful placebo in cough studies? Pulm Pharmacol Ther 15(3):303–308

Elmes SJ, Winyard LA, Medhurst SJ et al. (2005) Activation of CB_1 and CB_2 receptors attenuates the induction and maintenance of inflammatory pain in the rat. Pain 118:327–335

Fischer A, Forssmann WG, Undem BJ. (1998). Nociceptin-induced inhibition of tachykinergic neurotransmission in guinea pig bronchus. J Pharmacol Exp Ther 285:902–907

Fox B, Bull TB, Guz A (1980) Innervation of alveolar walls in the human lung: an electron microscopic study. J Anat 131:683–692

Freestone C, Eccles R (1997) Assessment of the antitussive efficacy of codeine in cough associated with common cold. J Pharm Pharmacol 49:1045–1049

Frossard N, Barnes PJ (1987) Mu-opioid receptors modulate non-cholinergic constrictor nerves in guinea-pig airways. Eur J Pharmacol 141(3):519–522

Galiegue S, Mary S, Marchland J et al. (1995) Expression of central and peripheral cannabinoid receptors in human immune tissues and leukocyte subpopulations. Eur J Biochem 232:54–61

Gestreau C, Bianchi AL, Grelot L (1997) Differential brainstem Fos-like immunoreactivity after laryngeal-induced coughing and its reduction by codeine. J Neurosci 17:9340–9352

Griffin G, Fernando SR, Ross RA et al. (1997) Evidence for the presence of CB_2 like cannabinoid receptors on peripheral nerve terminals. Eur J Pharmacol 339:53–61

Gundlach A L, Largent BL, Snyder SH (1986) Phencyclidine (PCP) receptors: autoradiographic localization in brain with the selective ligand, [3H]TCP. Brain Res 386:266–279

Hanus L, Breuer A, Tchilibon S et al. (1999) HU-308: a specific agonist for CB(2), a peripheral cannabinoid receptor. Proc Natl Acad Sci USA 96:14228–14333
Harris J, Drew LJ, Chapman V (2000) Spinal anandamide inhibits nociceptive transmission via cannabinoid receptor activation in vivo. Neuroreport 11:2817–2819
Higenbottam T (2002) Chronic cough and the cough reflex in common lung diseases. Pulm Pharmacol Ther 15(3):241–247
Howlett AC (2002) The cannabinoid receptors. Prostaglandins Other Lipid Mediat 68–69:619–631
Howlett AC, Barth F, Bonner TI, Cabral G, Casellas P, Devane WA, Felder CC, Herkenham M, Mackie K, Martin BR, Mechoulam R, Pertwee RG (2002) International Union of Pharmacology. XXVII. Classification of cannabinoid receptors. Pharmacol Rev 54:161–202
Hughes J, Smith TW, Kosterlitz et al. (1975) Identification of two related pentapeptides from the brain with potent opiate agonist activity. Nature 258:577–580
Jiang Q, Takemori AE, Sultana M et al. (1991) Differential antagonism of opioid delta antinociception by [D-Ala2,Leu5,Cys6]enkephalin and naltrindole 5/-isothiocyanate: evidence for delta receptor subtypes. J Pharmacol Exp Ther 257(3):1069–1075
Kamei J (2002) Delta-opioid receptor antagonists as a new concept for central acting antitussive drugs. Pulm Pharmacol Ther 15(3):235–240
Kamei J, Tanihara H, Kasuya Y (1990) Antitussive effects of two specific kappa-opioid agonists, U-50,488H and U-62,066E, in rats. Eur J Pharmacol 187:281–286
Kamei J, Iwamoto Y, Misawa M, Kasuya Y (1993) Effects of rimcazole, a specific antagonist of sigma sites, on the antitussive effects of non-narcotic antitussive drugs. Eur J Pharmacol 242:209–211
Karlsson JA, Lanner AS, Persson CG (1990) Airway opioid receptors mediate inhibition of cough and reflex bronchoconstriction in guinea pigs. J Pharmacol Exp Ther 252:863–868
Kastelik JA, Aziz I, Ojoo JC et al. (2005) Investigation and management of chronic cough using a probability-based algorithm. Eur Respir J 25(2):235–243
Kotzer CJ, Haw DW, Dondio G et al. (2000) The antitussive activity of delta-opioid stimulation in guinea pigs. J Pharmacol Exp Ther 292(2):803–809
Lee MG, Undem BJ, Brown C, Carr MJ. (2006) Effect of nociceptin in acid-evoked cough and airway sensory nerve activation in guinea pigs. Am J Respir Crit Care Med 173(3):271–275
Lord JA, Waterfield AA, Hughes J et al. (1977) Endogenous opioid peptides: multiple agonists and receptors. Nature 267:495–499
Malan TP, Ibrahim MM, Vanderah TW et al. (2002) Inhibition of pain responses by activation of CB_2 cannabinoid receptors. Chem Phys Lipids 121:191–200
Martin WR, Eades CG, Thompson JA et al. (1976) The effects of morphine- and nalorphine-like drugs in nondependent and morphine-dependent chronic spinal dog. J Pharmacol Exp Ther 197:517–532
Matsuda LA, Lolait SJ, Brownstein MJ et al. (1990) Structure of a cannabinoid receptor and functional expression of the cloned cDNA. Nature 346:561–564
McLeod RL, Parra LE, Mutter JC, Erickson CH, Carey GJ, Tulshian DB, Fawzi AB, Smith-Torhan A, Egan RW, Cuss FM et al. (2001) Nociceptin inhibits cough in the guinea-pig by activation of ORL_1 receptors. Br J Pharmacol 132:1175–1178
Merick B, Reid L (1971) Nerves in rat intra-acinar alveoli: an electron microscope study. Respir Physiol 11:367–377
Morice AH, Menon MS, Mulrennan SA et al. (2007) Opiate therapy in chronic cough. Am J Respir Crit Care Med 175(4):312–315
Munro S, Thomas KL, Abu-Shaar M (1993) Molecular characterisation of a peripheral receptor for cannabinoids. Nature 365:61–65
Ohi Y, Yamazaki H, Takeda R et al. (2005) Functional and morphological organization of the nucleus tractus solitarius in the fictive cough reflex of guinea pigs. Neurosci Res 53(2): 201–209
Patel HJ, Birrell MA, Crispino N et al. (2003) Inhibition of guinea pig and human sensory nerve activity and the cough reflex in guinea pigs by cannabinoid (CB_2) receptor activation. Br J Pharmacol 140:261–268

Pertwee RG (1997) Pharmacology of cannabinoid CB1 and CB2 receptors. Pharmacol Ther 74(2):129–180
Pertwee RG (2005) Pharmacological actions of cannabinoids. In: Pertwee RG (ed). Cannabinoids. Handbook of experimental pharmacology, vol. 168. Springer, Berlin, pp 1–51
Pertwee RG (2007) The diverse CB1 and CB2 receptor pharmacology of three plant cannabinoids: D9-tetrahydrocannabinol, cannabidiol and D9-tetrahydrocannabivarin. Br J Pharmacol (Sep 10 Epub ahead of print)
Pleuvry BJ (2005) Opioid mechanisms and opioid drugs. Anaesth Intensive Care Med 6(1):30–34
Porter AC, Felder CC (2001) The endocannabinoid nervous system: unique opportunities for therapeutic intervention. Pharmacol Ther 90:45–60
Richardson JD, Aanonsen L, Hargreaves KM (1998a) Antihyperalgesic effects of spinal cannabinoids. Eur J Pharmacol 345:145–153
Richardson JD, Kilo S, Hargreaves KM (1998b) Cannabinoids reduce hyperalgesia and inflammation via interaction with peripheral CB_1 receptors. Pain 75:111–119
Schroeder K, Fahey T (2002) Systematic review of randomised controlled trials of over the counter cough medicines for acute cough in adults. Br Med J 324:1–6
Sessle BJ, Ball GJ, Lucier GE (1981) Suppressive influences from periaqueductal gray and nucleus raphe magnus on respiration and related reflex activities and on solitary tract neurons, and effect of naloxone. Brain Res 216(1):145–161
Sevelius H, Colmore JP (1966) Objective assessment of antitussive agents in patients with chronic cough. J New Drugs 6:216–223
Sevelius H, McCoy JF, Colmore JP (1971) Dose response to codeine in patients with chronic cough. Clin Pharmacol Ther 12:449–455
Shoblock JR (2007) The pharmacology of Ro 64-6198, a systemically active, nonpeptide NOP receptor (opiate receptor-like 1, ORL-1) agonist with diverse preclinical therapeutic activity. CNS Drug Rev (Spring) 13(1):107–136
Smith J, Owen E, Earis J et al. (2006) Effect of codeine on objective measurement of cough in chronic obstructive pulmonary disease. J Allergy Clin Immunol 117:831–835
Takahama K, Shirasaki T (2007) Central and peripheral mechanisms of narcotic antitussives: codeine-sensitive and -resistant coughs. Cough 3:1–8
Undem BJ, Kollarik M (2005) The role of vagal afferent nerves in chronic obstructive pulmonary disease. Proc Am Thorac Soc 2(4):355–360
Walter L, Stella N (2004) Cannabinoids and neuroinflammation. Br J Pharmacol 141:775–785
Widdicombe JG, Chung KF (2007) Cough. Pulm Pharmacol Ther 20(4):305–306

Cough Sensors. IV. Nicotinic Membrane Receptors on Cough Sensors

L.-Y. Lee(✉) and Q. Gu

Contents

Abstract Cigarette smoke is undoubtedly one of the most common inhaled irritants in the human respiratory tract, and invariably evokes coughing in both smokers and nonsmokers. Results obtained from the studies in human volunteers and from single-fiber recording of vagal bronchopulmonary afferents in animals clearly indicate that nicotine is primarily responsible for the airway irritation and coughing caused by inhalation of cigarette smoke. Furthermore, both nicotine and acetylcholine can evoke inward current, membrane depolarization, and action potentials in isolated pulmonary sensory neurons, and these responses are blocked by hexamethonium. Taken together, these findings suggest that the tussive effect of nicotine is probably mediated through an activation of nicotinic acetylcholine receptors (nAChRs) expressed on the sensory terminals of cough receptors located in the airway mucosa. Indeed, the expressions of $\alpha 4 - \alpha 7$ and $\beta 2 - \beta 4$ subunits of nAChR transcripts in

L.-Y. Lee
Department of Physiology, University of Kentucky, Lexington, KY 40536-0298, USA
lylee@uky.edu

K.F. Chung, J.G. Widdicombe (eds.), *Pharmacology and Therapeutics of Cough*, Handbook of Experimental Pharmacology 187,

pulmonary sensory neurons have lent further support to this conclusion. The specific subtypes of the neuronal nAChRs and their subunit compositions expressed on the cough sensors remain to be determined.

1 Cough Evoked by Inhaled Cigarette Smoke

One of the most important functions of the cough reflex is to protect the lung and the rest of body from the hazardous effects of inhaled irritants and environmental air pollutants. There is a long list of common inhaled irritants known to elicit cough reflex in humans, including tobacco smoke, dust, automobile exhausts, acid aerosols, noxious gases, allergens, and a variety of airborne chemical irritants. These irritant substances can activate cough sensors either by directly acting on certain ion channels or receptor proteins expressed on the membrane of sensory terminals or indirectly via an action on other target cells in the airways (e.g., bronchial smooth muscles, immune cells such as macrophages or mast cells, etc.) which can in turn evoke stimulatory effects on the cough sensors. Among these tussive substances, cigarette smoke is undoubtedly one of the most common inhaled irritants in human airways.

In nonsmokers, inhaling only a small puff of cigarette smoke can evoke vigorous cough responses and airway irritation. Similar irritant effects can also be generated by brief exposure to the side-stream smoke, the smoke released from a smoldering cigarette (Tarlo 2006). In smokers, cigarette smoking is the major cause of chronic cough (Braman 2006). Indeed, cough not only serves as an important defense function, but is also a common symptom of various airway diseases associated with chronic smoking (Braman 2006); in conjunction with the mucociliary system, cough can expel from the respiratory tract the cigarette smoke constituents that are deposited in the airway lumen as well as the excessive airway secretion produced by smoke inhalation.

1.1 Role of Nicotine in Cigarette Smoke-Induced Airway Irritation

Despite the well-documented irritant effects of inhaled cigarette smoke, the underlying mechanism and smoke constituent(s) that are responsible for generating the airway irritation and cough were not fully understood. In addition to cough, inhaled cigarette smoke can also elicit reflex bronchoconstriction. For a long time, it was suggested that particulates in the cigarette smoke were primarily responsible for the irritant effects and for eliciting reflex bronchoconstriction (Nadel and Comroe 1961). This notion appeared to be consistent with the finding that the irritant receptors in the lung could be activated not only by cigarette smoke but also by “inert” carbon dust (Sellick and Widdicombe 1971). On the other hand, nicotine, a major constituent of cigarette smoke, was known to activate sensory receptors in various organs, including skin, mesentery, stomach, lung, and carotid body (Douglas 1952; Douglas and Gray 1953; Paintal 1954, 1955). Whether nicotine delivered in the smoke could activate cough sensors was not known.

In a psychometric study of the role of nicotine in cigarette smoking, Henningfield et al. (1985) found that nicotine administered by intravenous injection caused coughing and airway irritation in human smokers. Whether the irritant effects reported in that study were due to an excitatory effect of nicotine on cough receptors in the airways and lung was not known since nicotine injected via the intravenous route may also activate other sensory receptors that can modulate respiratory responses (e.g., cardiac afferents, carotid body chemoreceptors, etc.); a similar circulatory lag for the injected nicotine to reach airway receptors (via bronchial circulation) and these other receptors located in the systemic circulation made it difficult to identify the origin of its irritant effect.

In a study aimed at determining the role of nicotine in cigarette smoke-evoked airway irritation, the responses to inhalation of a single puff of cigarette smoke with different nicotine content were compared in healthy young nonsmokers (Lee et al. 1993). After the upper airways had been locally anesthetized and the view of the smoke delivery system had been blocked, the subjects breathed quietly through a mouthpiece when a bolus of cigarette smoke of 30-ml volume (approximately the average puff volume of smokers) was delivered without notice and immediately inhaled at the front portion of the next inspired volume, mimicking the smoke inhalation pattern in smokers (Tobin and Sackner 1982). Subjects were instructed to signal the detection and the intensity of airway irritation with a push-button device. In eight of the 12 subjects, inhalation of high-nicotine cigarette smoke immediately triggered bouts of coughs (total number of coughs 4.13 ± 0.64). By contrast, low-nicotine (0.38 ± 0.26; $P < 0.001$) and gas-phase smoke (0 ± 0) did not cause cough in the same subjects (Fig. 1a). The intensity of airway irritation measured by the total number of push-button signals after inhalation of high-nicotine cigarette smoke was 6.61 ± 0.87 ($n = 12$), whereas inhalation of low-nicotine and gas-phase smoke either was not detected or caused only very mild irritation (0.89 ± 0.4 and 0.36 ± 0.22, respectively). Furthermore, the intensities of airway irritation and coughs evoked by high-nicotine smoke were markedly reduced after premedication with aerosolized hexamethonium (10% in isotonic saline), a nicotinic acetylcholine receptor antagonist; number of coughs was 4.75 ± 1.02 after administration of a placebo, and 1.26 ± 0.28 after administration of hexamethonium ($P < 0.001$, $n = 8$; Fig. 1b). The irritant effect was clearly attenuated but not completely abolished, largely owing to a marginal blocking effect of the hexamethonium in two of the eight subjects. However, after a higher concentration (20%) of hexamethonium aerosol had been administered in these two subjects, the smoke-evoked airway irritation and cough were completely blunted (Lee et al. 1993). In addition, inhalation of aerosolized nicotine (2% in isotonic saline) also evoked vigorous coughs (total number of coughs 7.36 ± 1.68, $n = 5$) and intense airway irritation (Fig. 1c). Taken together, these studies have provided strong evidence indicating that nicotine is the primary agent in cigarette smoke-evoked cough and airway irritation in nonsmokers. The short latency time of the detection signal (0.93 ± 0.11 s) further suggested

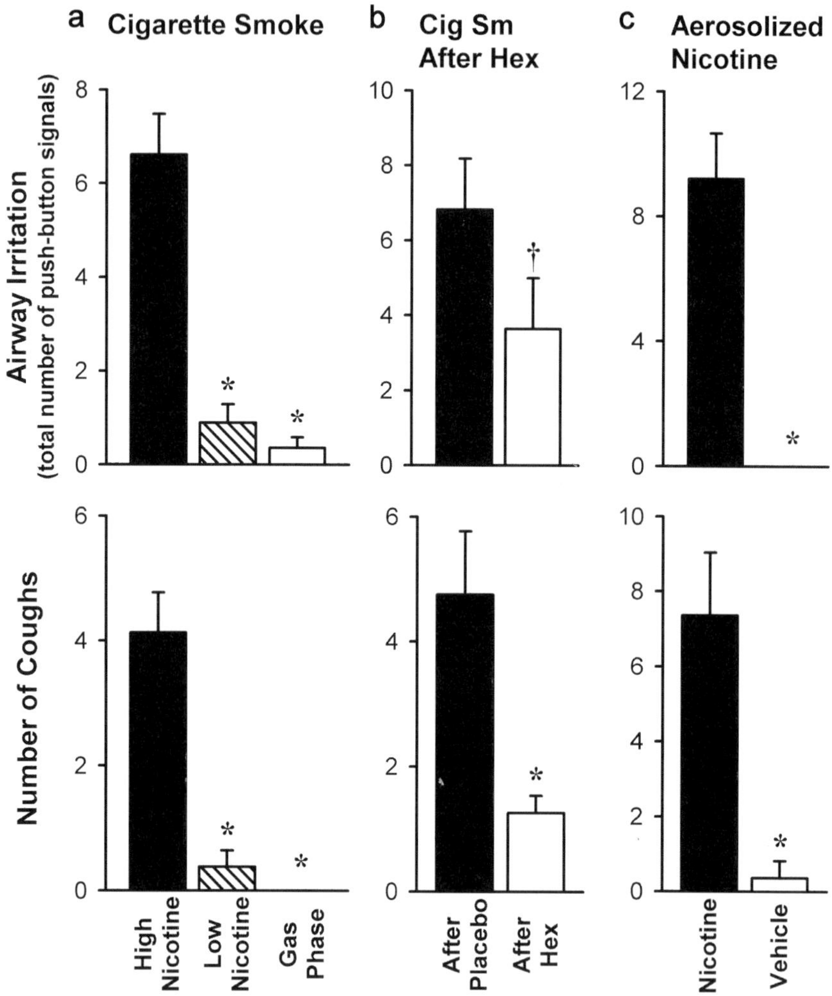

Fig. 1 **a** Airway irritation and number of coughs evoked by inhalation of three different types of cigarette smoke. The intensity of airway irritation was measured by the number of push-button signals generated in the first 5 s following inhalation of a puff of cigarette smoke. Data represent the mean ± the standard error of the mean (SEM) of 12 subjects; data obtained from three consecutive days were averaged for each subject. *, significantly different ($P < 0.001$) from the response to high-nicotine cigarette smoke. **b** A comparison of airway irritation and cough responses to inhalation of high-nicotine cigarette smoke after premedications with aerosolized saline placebo and hexamethonium (10% concentration). † and *, significant difference ($P < 0.01$ and $P < 0.001$, respectively; $n = 8$) between the responses after hexamethonium and placebo administration. **c** Airway irritation and number of coughs evoked by inhalation of aerosolized nicotine or its vehicle. *, significant difference ($P < 0.001$; $n = 5$) between the responses to nicotine and vehicle. (Modified from Lee et al. 1993, with permission)

that the sensory receptors involved in detecting the irritant effect of smoke and eliciting the cough reflex are located in the respiratory tract and the afferent signals are probably conducted by vagus nerves.

1.2 Cough Receptors in the Lung and Airways

The "cough receptor" that initiates the reflex response consists of multiple types of airway afferents with different sensory modalities (chemical vs. mechanical), as characterized by different types of receptor proteins expressed on the neuronal membrane of the sensory endings (Geppetti 2008; Belvisi 2008; Mazzone and Undem 2008). These cough receptors are located in different regions of the respiratory tract, including larynx and tracheobronchial tree. The anatomical locations, neural pathways, and pharmacological properties of the cough receptors are described in more detail elsewhere (Canning 2008). The afferent activities that arise from various cough sensors located in the airways are conducted almost exclusively by branches of vagus nerves (Coleridge and Coleridge 1986; Lee and Pisarri 2001; Sant'Ambrogio 1982; Widdicombe 1998). These vagal afferent fibers innervate the entire respiratory tract diffusively, and project to the "cough centers" located in the medulla (Takahama 2008).

On the basis of the general properties and classically defined criteria, vagal afferents arising from endings located in the lower airways and lungs can be classified into three major categories: slowly adapting receptors (stretch receptors), rapidly adapting receptors (RARs or irritant receptors), and bronchopulmonary C-fibers (Coleridge and Coleridge 1986; Widdicombe 1981). Afferent activities generated from the first two types of receptors are conducted by myelinated A-fibers, whereas those generated by the last one are conducted by nonmyelinated (C-) fibers. For many decades, investigators have been searching for the identities and classifications of a subgroup(s) of sensory receptors in the airways that are specifically responsible for eliciting the cough reflex. RARs, particularly those located in the major airways and the carina, were believed to be primarily responsible for detecting the inhaled irritants and signaling the central nervous system to elicit the cough response in larger mammals (e.g., dogs, cats) (Widdicombe 1998). The evidence in support of an important role of bronchopulmonary C-fiber afferents in eliciting the cough reflex has also been extensively documented. Stimulation of C-fiber afferents by inhalation of capsaicin or acid aerosol consistently produces coughs in a dose-dependent manner in both animals and humans (Bolser et al. 1995; Dicpinigaitis 2007; Groneberg et al. 2004; Hansson et al. 1992; Karlsson and Fuller 1999; Liu et al. 2001; Winning et al. 1986). Recent studies by Canning et al. (2004) and Mazzone (2004) have demonstrated that the cough reflex elicited by the upper-airway irritation in guinea pigs seems to be mediated through the activation of a subset of Aδ afferents innervating the trachea and larynx. These cough sensors are activated by acid solution and punctate mechanical stimulation applied to the upper airways. However, they are not activated by either capsaicin or bradykinin, and are, therefore, unlikely to be involved in eliciting the capsaicin-evoked cough reflex. Furthermore, they appear to be different from the RARs in the airways because they conduct action potentials at a much slower velocity than RARs, and they are not activated by either lung inflation or bronchoconstriction (Canning et al. 2004; Mazzone 2004). Whether they are sensitive to nicotine is not known.

It remains to be determined whether cough receptors belong to a subgroup of these classically defined bronchopulmonary afferents, or represent a distinctly different group of airway afferents whose primary function is to detect tussive stimuli. Furthermore, whether the cough receptor mediating the protective function in healthy lungs against inhaled irritants such as cigarette smoke is also the same sensor that is responsible for triggering the acute or chronic cough in various airway diseases is also not known. Many of these important and yet unanswered questions have been discussed in recent reviews and conference proceedings (Widdicombe and Chung 2004, 2007; Widdicombe and Undem 2006). This chapter will focus solely on the vagal bronchopulmonary afferents that exhibit sensitivity to nicotine and may function as sensors for triggering the cigarette smoke-evoked cough responses.

1.3 Nicotine-Sensitive Afferents in the Lung and Airways

1.3.1 Bronchopulmonary C-Fiber Afferents

Bronchopulmonary C-fibers represent more than 75% of the afferent fibers in the pulmonary branch of the vagus nerve (Agostoni et al. 1957; Jammes et al. 1982), and play an important role in regulating the airway functions in both physiological and pathological conditions (Coleridge and Coleridge 1984; Lee and Pisarri 2001). These afferents can be broadly subdivided into two major groups on the basis of their anatomic locations in the respiratory tract and circulatory accessibility (Coleridge and Coleridge 1984). Pulmonary C-fibers arise from the endings located in small intrapulmonary airways and the lung parenchyma, receiving blood supply from the pulmonary circulation, whereas bronchial C-fiber endings are located in the medium-sized and large airways, receiving blood perfusion primarily from bronchial circulation. It appears that pulmonary C-fibers are more sensitive to mechanical stimuli (e.g., lung inflation), whereas bronchial C-fibers display greater chemosensitivity (Coleridge and Coleridge 1984). A recent study has revealed that bronchial and pulmonary C-fiber afferents arise from different ganglion origins and represent different phenotypes, which may offer a plausible explanation for the noticed difference in the sensory modality between these two groups of C-fibers (Undem et al. 2004). It was further suggested that stimulated bronchial C-fibers are more likely to function as cough sensors than pulmonary C-fibers (Canning et al. 2006), but definite evidence remains to be established (Hansson et al. 1992).

In a study aimed to determine whether vagal bronchopulmonary C-fibers can be activated by cigarette smoke, afferent activities of the vagal C-fibers arising from endings in the lung were recorded using the single-fiber recording technique in anesthetized and mechanically ventilated dogs when cigarette smoke (120 ml) was delivered by the respirator in a single ventilator cycle (Lee et al. 1989). To determine the role of nicotine, the effects of low- and high-nicotine cigarette smoke on the same C-fiber afferents were compared; the volume of smoke generated from the low- and high-nicotine cigarettes was estimated to contain approximately 0.032 and 0.162 mg of nicotine, respectively. The delivery of smoke generated by high-nicotine cigarettes triggered an abrupt and intense burst of activity in the majority (approximately 91%) of the pulmonary C-fibers tested; the activity increased from 0.3 ± 0.1 impulses per second to a peak of 12.6 ± 1.3 impulses per second ($n = 60$) within 1-2 s of the smoke delivery and lasted for 3-5 s. In comparison, smoke generated by low-nicotine cigarettes evoked a milder stimulation in only 33% of pulmonary C-fibers, but did not significantly affect the overall firing frequency. Furthermore, hexamethonium (0.7-1.2 mg/kg, intravenously) prevented C-fiber stimulation by high-nicotine cigarette smoke, but did not alter the response to right atrial injection of capsaicin in the same fibers, indicating a specific stimulatory effect of nicotine on these C-fiber endings. Indeed, Kou et al. (1989) subsequently demonstrated that nicotine solutions injected directly into right atrium stimulated pulmonary C-fibers in a dose-dependent manner in anesthetized dogs (Fig. 2). A direct stimulatory effect of nicotine and cigarette smoke on bronchial

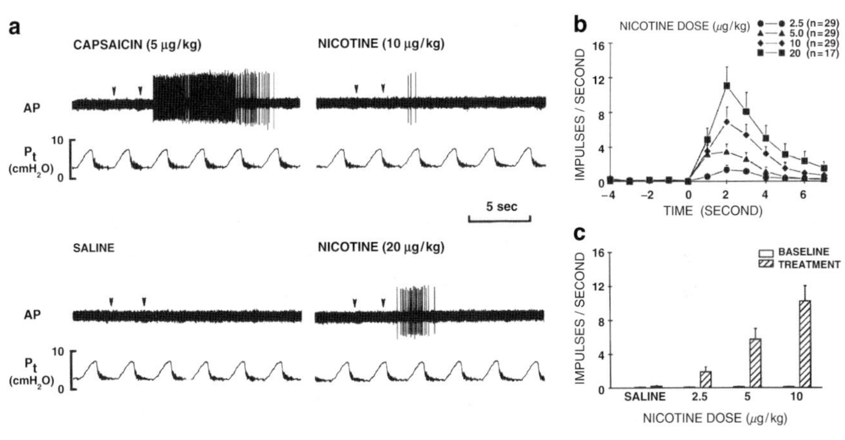

Fig. 2 **a** Effects of capsaicin, saline and nicotine on a vagal pulmonary C-fiber arising from an ending in the right diaphragmatic lobe of an anesthetized dog. *AP* action potentials, P_t tracheal pressure. In each panel, solution was injected into a catheter at the first arrow and then flushed into the right atrium by a bolus of isotonic saline at the second arrow. **b** Effects of injections of nicotine into the right atrium on vagal pulmonary C-fibers. Data are the mean ± SEM; *n* number of fibers studied. Nicotine solution was injected into the right atrium at time zero. **c** Peak responses of vagal pulmonary C-fibers to injections into the right atrium of saline and increasing doses of nicotine ($n = 29$). The effect of treatment was measured as the peak discharge frequency within the first 3 s after each injection in each fiber. Baseline activity was determined as the mean discharge frequency for 10 s immediately preceding each injection. (From Kou et al. 1989, with permission)

and pulmonary C-fibers has also been demonstrated in other species (Ho et al. 2001; Pisarri et al. 1991; Xu et al. 2007).

It should be pointed out that cough sensitivity mediated by C-fibers is susceptible to suppression by anesthesia (Canning 2007; Canning et al. 2006; Mazzone et al. 2005). In fact, an inhibitory effect of pulmonary C-fiber stimulation on cough reflex elicited by mechanical irritation of the tracheobronchial tree and larynx has been reported (Tatar et al. 1988). However, extensive and convincing evidence in the studies of awake animals and humans indicates a distinct role of the bronchopulmonary C-fibers in eliciting the cough reflex. The fact that they are sensitive to nicotine lends additional support to the involvement of these afferents in the cough response to cigarette smoke.

1.3.2 Bronchopulmonary RARs

RARs are located along the entire respiratory tract, with a higher density in the larger airways, and can be identified by the rapid adaptation of their response to lung inflation (Sant'Ambrogio 1982; Widdicombe 1954). The majority of RARs exhibit polymodal sensitivity; they can be activated by certain chemical irritants, lung inflation and deflation, and an increase in lung stiffness (Coleridge and Coleridge 1986; Sant'Ambrogio 1982; Sant'Ambrogio and Widdicombe 2001; Widdicombe 1981).

Delivery of a single breath of high-nicotine cigarette smoke into the lung generated an immediate stimulatory effect on RARs in anesthetized, mechanically ventilated dogs (Kou and Lee 1990); activity increased from a baseline of 0.9 ± 0.2 to a peak of 9.9 ± 1.2 impulses per breath ($n = 58$). After three to six breaths when the receptor discharge returned toward the baseline, a delayed increase of activity emerged (peak activity 3.4 ± 0.6 impulses per breath). By contrast, only a mild stimulatory effect of low-nicotine cigarette smoke was found, either immediately or after a delay, in 28% of the receptors studied. To further understand the mechanisms underlying the responses, studies were repeated after a pretreatment with aerosolized hexamethonium (10%) or isoproterenol (2%), a β-adrenergic agonist and bronchodilator. The initial response of RARs was totally prevented by pretreatment with hexamethonium but was not affected by isoproterenol. The delayed RAR discharge was invariably accompanied by an increase of total lung resistance following smoke delivery; both were abolished by pretreatment with isoproterenol. These results suggest that a direct effect of nicotine on these receptors may be responsible for the immediate stimulation, while the bronchoconstrictive effects of absorbed nicotine may be involved in evoking the delayed stimulation. The fact that cigarette smoke exerted an immediate stimulatory effect on RARs (latency less than 1 s) before any change of bronchomotor tone took place (Kou and Lee 1990, 1991) further supports a direct action of nicotine on RARs. A similar stimulatory effect of smoke on RARs associated with bronchoconstriction has also been reported in anesthetized rabbits (Sellick and Widdicombe 1971), guinea pigs (Bergren and Sampson 1982) and less consistently in dogs (Sampson and Vidruk 1975).

RARs are known to be activated by an increase of fluid filtration into the interstitial space in the lung (Kappagoda and Ravi 2006). An increase in the interstitial fluid volume can also result from the action of nitric oxide, an endogenous free-radical gas, that increases the microvascular permeability and fluid flux into the perivascular space in the airway tissue. Since substance P and other tachykinins are known to trigger the release of nitric oxide from endothelial cells (Nguyen et al. 1995), activation of pulmonary C-fibers by nicotine can produce a secondary stimulatory effect on RARs (Joad et al. 1997). Whether this indirect effect plays a part in the cigarette smoke-induced stimulation of RARs is not known.

1.3.3 Laryngeal Irritant Receptors

The larynx, located in the front position of the respiratory tract, plays a critical role in the protective functions against inhaled irritants. Not surprisingly, the larynx is extremely sensitive to chemical and mechanical stimulations in eliciting the cough reflex (Karlsson et al. 1991; Widdicombe 1954). The larynx is supplied by the largest number of sensory nerves per unit of lumenal surface area in the entire respiratory tract (Sant'Ambrogio and Widdicombe 2001), and is innervated by multiple afferent branches: the recurrent laryngeal nerves and the internal and external branches of the superior laryngeal nerves (Sant'Ambrogio et al. 1995). The cell bodies of these laryngeal afferents are located in the vagal sensory ganglia, with the majority of them in the nodose ganglion (Widdicombe et al. 1988).

A small percentage of the laryngeal afferents exhibit no or irregular spontaneous activity, but possess distinct sensitivity to chemical and mechanical stimulations (Boushey et al. 1974; Lee et al. 1987). They consist of both myelinated and nonmyelinated afferents, and are believed to be responsible for eliciting the protective reflex responses such as cough (Sant'Ambrogio et al. 1995). To examine whether these laryngeal afferents are affected by cigarette smoke, neural activity was recorded from the superior laryngeal nerve in anesthetized dogs. A box-balloon system, connected to the breathing circuit, allowed smoke to be inhaled spontaneously through the isolated upper airway while preserving its normal respiratory flow and pressure. Inhalation of one or two breaths of cigarette smoke (25–50% smoke in air) caused a marked increase in activity of laryngeal irritant receptors (Lee et al. 1987). Interestingly, in sharp contrast to their counterparts in the lower airways and lung, there was no difference between the responses of these laryngeal afferents to high- and low-nicotine cigarette smoke.

2 Neuronal Nicotinic Acetylcholine Receptors on Pulmonary Sensory Neurons

The studies in both humans and animals described in the previous sections suggest that the tussive effect of nicotine is probably mediated through an activation of neuronal nicotinic acetylcholine receptors (NnAChRs) expressed on the sensory terminals of cough receptors. However, direct evidence of their presence on pulmonary sensory nerves was not established in those studies.

2.1 Pharmacological Properties of NnAChRs

The NnAChRs are pentameric ligand-gated ion channels formed by five subunits; each subunit comprises four transmembrane domains, M1–M4, of which M2 is believed to form the ion channel (Gotti and Clementi 2004; Fig. 3a, b). To date, nine α (α2 – α10) and three β (β2 – β4) neuronal subunit genes have been cloned, and these subunits assemble in numerous combinations to form functional NnAChR subtypes (Jensen et al. 2005; Kalamida et al. 2007). The α2 – α6 and β2 – β4 subunits form heteromeric receptors, usually in an $(\alpha_x)_2(\beta_y)_3$ stoichiometry and arranged as αβαββ; the α7 – α9 subunits can form homomeric receptor complexes (Fig. 3c); and the α10 subunit is not able to form a functional homomeric receptor, but it can form a functional NnAChR together with $\alpha9$ (Jensen et al. 2005). It is thought that the homomeric NnAChRs have five identical binding sites for acetylcholine and other orthosteric ligands per receptor molecule (one at each α subunit interface), whereas heteromeric NnAChRs have two orthosteric binding sites per receptor molecule located at the interface between an α and a β subunit (Gotti and Clementi 2004; Gotti et al. 2006; Jensen et al. 2005; see Fig. 3c).

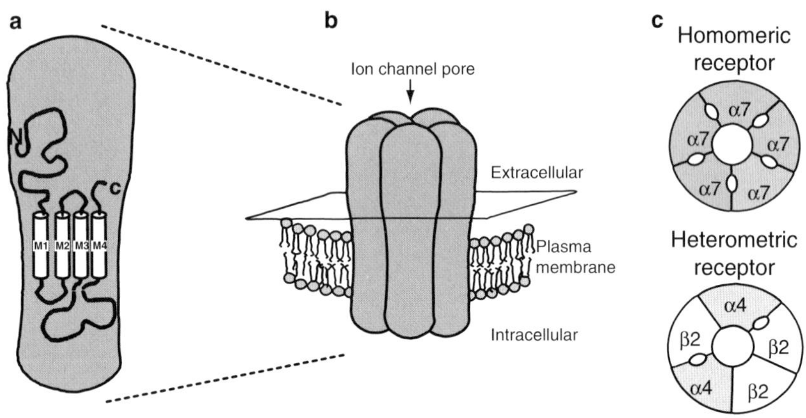

Fig. 3 a–c Organization and structure of neuronal nicotinic acetylcholine receptors (NnAChRs). **a** The putative transmembrane topology of NnAChR subunits. The model shows the extracellular amino terminal portion, followed by three hydrophobic transmembrane domains (M1–M3), a large intracellular loop, and then a fourth hydrophobic transmembrane domain (M4). **b** Pentameric arrangement of NnAChR subunits in an assembled receptor. **c** Subunit arrangement in the homomeric $\alpha7$ and heteromeric $\alpha4\beta2$ subtypes, and localization of the acetylcholine (ACh) binding site. (From Gotti and Clementi 2004, with permission)

Recent heterologous expression studies have demonstrated that both α and β subunits are involved in characterizing the functional and pharmacological properties of the expressed NnAChR subtypes (Gotti and Clementi, 2004). For example, when expressed with the $\beta 2$ subunit, the $\alpha 2$, $\alpha 3$, $\alpha 4$, and $\alpha 6$ subunits all form functional receptors that differ in their channel open time, single channel conductance and agonist/antagonist sensitivity (Gotti et al. 1997). However, not all subunit combinations form functional NnAChRs. The $\beta 3$ and $\alpha 5$ subunits cannot form functional receptors when expressed alone or in paired combinations with other subunits; they only form operational channels when coexpressed with other functional subunit combinations and act as "regulatory" subunits (Gotti et al. 1997; Grinevich et al. 2005). In addition, expression of the $\alpha 7$-$\alpha 9$ subunits alone produces homomeric receptors activated by acetylcholine and blocked by nanomolar concentrations of α-bungarotoxin with dramatically larger Ca^{2+} permeabilities and faster desensitization rates than those of the heteromeric NnAChRs (Gotti and Clementi 2004; Jensen et al. 2005).

The properties of NnAChRs are best described by an allosteric model (Changeux and Edelstein 1998; Jensen et al. 2005). According to this model, NnAChR fluctuates between at least three functional states in the absence of agonists: a resting state, an active state, and a desensitization state (often divided into a fast-onset state and a slow-onset state). Binding of an agonist or a competitive antagonist to the orthosteric sites of NnAChR stabilizes its active and resting states, respectively; whereas a noncompetitive antagonist inhibits the receptor function without interacting with the orthosteric binding sites or directly affecting the binding of an agonist (Hogg and Bertrand 2004; Jensen et al. 2005). The agonists for NnAChRs such as acetylcholine, nicotine, cytisine, epibatidine, and their derived compounds have limited selectivity and cannot discriminate among the different subtypes; however, their different affinity for a certain α–β or α–α interface may be used to characterize various subtypes (Gotti et al. 2006; Jensen et al. 2005). The peptide α-bungarotoxin and methyllycaconitine are competitive and highly selective antagonists to the $\alpha 7, \alpha 9$, and $\alpha 9/\alpha 10$ NnAChRs; other classical competitive NnAChR antagonists such as dihydro-β-erythroidine and D-tubocurarine are far less discriminative between different NnAChR subtypes (Jensen et al. 2005); whereas noncompetitive antagonists of NnAChRs, including mecamylamine and hexamethonium, demonstrate relatively low affinity for NnAChRs (Hirota and McKay 2006). A thorough discussion of NnAChR agonists and antagonists has been given in detail in recent reviews (Arias 2000; Bunnelle et al. 2004; Gotti et al. 2006; Jensen et al. 2005; Romanelli and Gualtieri 2003).

2.2 *Nicotine Sensitivity in Isolated Pulmonary Sensory Neurons*

The studies described in Sect. 1.3 demonstrated a dominant role of nicotine in the stimulatory effect of cigarette smoke on vagal bronchopulmonary afferents. However, NnAChRs are also known to be expressed on other cell types in the airways, including smooth muscles and cholinergic ganglion neurons. Indeed, nicotine has been shown to induce airway smooth-muscle contraction (Hartiala et al. 1985; Hong et al. 1995; Lee et al. 1995). Therefore, the stimulatory effect of nicotine on RARs may have also resulted from an indirect action generated by the nicotine-induced bronchoconstriction because the delayed stimulatory effect of cigarette smoke on RARs was attenuated when bronchoconstriction was prevented by pretreatment with a bronchodilator in canine lungs (Kou and Lee 1991; Sampson and Vidruk 1975).

As mentioned already, NnAChRs are pentameric transmembrane proteins that form cationic channels permeable to both Na^{+} and Ca^{2+} (Gotti and Clementi 2004; Jensen et al. 2005). To investigate the direct effect of nicotine, Xu et al. (2007) measured the change in intracellular Ca^{2+} concentration in response to nicotine in isolated pulmonary sensory neurons that were identified by retrograde labeling from the lungs with a fluorescent tracer (Kwong and Lee 2002). Their results showed that application of nicotine (10, 30, and 100 μM, 20 s) evoked an abrupt and transient increase in intracellular Ca^{2+} concentration in a concentration-dependent manner. Pretreatment with hexamethonium (100 μM, 5 min) almost completely blocked the Ca^{2+} transient evoked by the same concentration of nicotine (100 μM, 20 s), and the blocking effect could be reversed after 15–20-min washout. In 522 pulmonary sensory neurons tested with both capsaicin and nicotine, 100 μM nicotine and 0.1 μM capsaicin evoked a pronounced Ca^{2+} transient in 33.5 and 57.5% of cells, respectively. In the nicotine-sensitive neurons, 59.7% were also activated by capsaicin challenge.

A recent whole-cell patch-clamp recording study (Gu et al. 2008; Fig. 4) has further demonstrated that a brief application of acetylcholine (3, 10, 30, and 100 μM, 2–6 s) concentration-dependently evoked an inward current in pulmonary sensory neurons. The response was reversibly blocked by hexamethonium (100 μM, 5 min) pretreatment, indicating the involvement of activation of NnAChRs. Nicotine and dimethylphenylpiperazinium also evoked inward currents in these neurons in voltage-clamp configuration, and induced membrane depolarization and action potential generation in current-clamp recordings. In addition, the results of the study have shown that the sensitivities to nicotine and the phenotypes of inward current evoked by nicotine varied among different neurons, which may suggest activation of different NnAChR subtypes (Genzen et al. 2001).

2.3 Expression of NnAChRs on Pulmonary Sensory Neurons

The expression of NnAChR subunit messenger RNAs (mRNAs) in pulmonary sensory nerves has been detected recently. Gu et al. (2008) performed Reverse transcriptase-PCR and a nested PCR by using the cytoplasm collected from approximately 20 isolated rat pulmonary sensory neurons. The results revealed the expression of mRNA encoding for the α4, α5, α6, α7, β2, β3, and β4 subunits of the NnAChR, whereas α2 and α3 subunits were not detected (Fig. 5), and α9 and α10 subunits were not tested. Since the results were based on the analysis of mRNA expression from a pool of multiple neurons, they do not provide the definite evidence

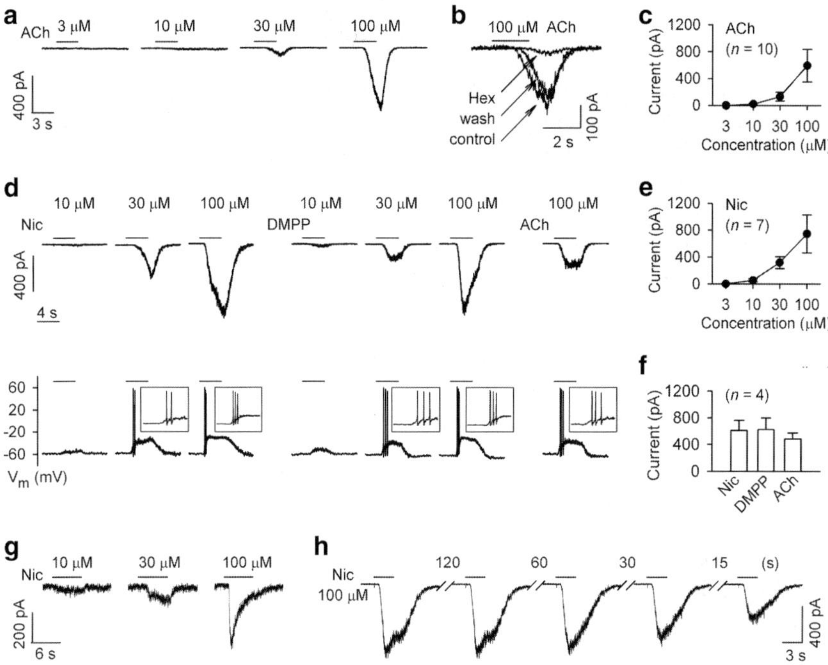

Fig. 4 a–h ACh-, nicotine- and dimethylphenylpiperazinium-evoked whole-cell responses in rat vagal pulmonary sensory neurons. **a** ACh (3, 10, 30, and 100 μM, 3-s duration) concentration-dependently evoked an inward current in a pulmonary sensory neuron (jugular; capacitance 24.9 pF). **b** ACh (100 μM, 2 s)-evoked inward current was reversibly inhibited by hexamethonium (100 μM, 5 min) in a nodose neuron (19.6 pF). **c** Group data showing the inward current evoked by increasing concentration of ACh (3 – 100 μM, 2–6 s). **d** Whole-cell responses (*upper panel* voltage-clamp recording; *lower panel* current-clamp recording) to nicotine (Nic; 10, 30, and 100 μM, 4 s), dimethylphenylpiperazinium (DMPP; 10, 30, and 100 μM, 4 s), and ACh (100 μM, 4 s) in a single nodose neuron (27.2 pF). *Insets*: Action potentials shown on a larger scale. **e** Group data showing the inward current evoked by different concentrations of nicotine (3 – 100 μM, 3–6 s). **f** Comparison of the inward currents evoked by the same concentration (100 μM, 3–4 s) of nicotine, dimethylphenylpiperazinium and ACh recorded from the same neurons. **g** A different phenotype of inward currents evoked by nicotine (10, 30, and 100 μM, 6 s) in a jugular neuron (28.6 pF). **h** Inward currents evoked by 100 μM nicotine (3 s) at different intervals between two consecutive challenges in a nodose neuron (21.2 pF). (From Gu et al. 2008, with permission)

to prove the existence of specific subtypes of NnAChR and their subunit compositions in individual pulmonary sensory neurons. To better understand the role of NnAChRs in regulating the excitability of pulmonary sensory nerve and cough sensitivity in the chronic airway diseases associated with smoking, further study with the employment of additional technical approaches, such as single-cell RT-PCR and real-time quantitative PCR, and use of more specific NnAChR agonists/antagonists will be required.

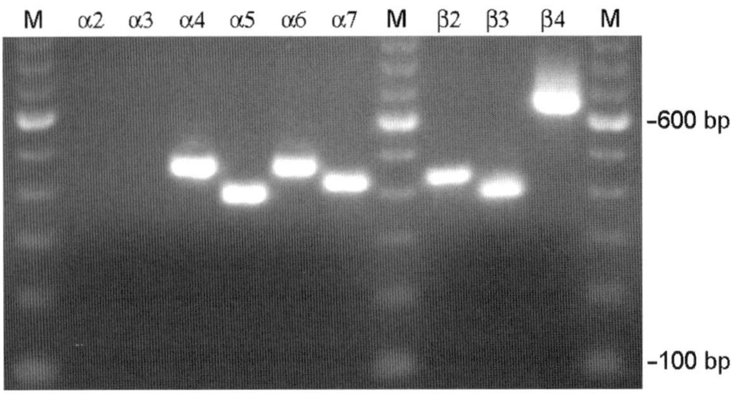

Fig. 5 Reverse transcriptase-PCR analysis of NnAChR messenger RNA (mRNA) in rat vagal pulmonary sensory neurons. Cytoplasm of 20 isolated pulmonary sensory neurons was collected into a single PCR tube. One-step Reverse transcriptase-PCR and a nested PCR were carried out to detect the presence of nicotinic mRNA subunits. The nested PCR products were run on a 1% agarose gel and stained with ethidium bromide. *M* 100-bp DNA marker. (From Gu et al. 2008, with permission)

3 Other Physiological Functions of NnAChRs on Cough Sensors

3.1 Other Physiological Responses Evoked by Activation of NnAChRs on Pulmonary Afferents

In addition to eliciting the cough reflex via the activation by nicotine, what are other potential roles of the NnAChRs expressed on these pulmonary afferents? Reflex responses and neural pathways of the two major types of nicotine-sensitive afferents innervating the lung and airways, C-fibers and RARs, have been described in detail in several extensive reviews (Coleridge and Coleridge 1986; Lee and Pisarri 2001; Sant'Ambrogio and Widdicombe 2001; Widdicombe 1981). Briefly, when these afferents are activated by chemical irritants inhaled into the lung (e.g., nicotine in the inhaled cigarette smoke), action potentials are conducted through the vagus nerves to the nucleus tractus solitarius and immediately elicit a wide range of reflex responses, including reflex bronchoconstriction and mucus secretion, in addition to cough.

Activation of NnAChRs can trigger an abrupt increase in intracellular calcium concentration in the neuron resulting from direct passage of calcium through the receptor channel (Colquhoun and Patrick 1997). Intracellular calcium is an important signal transduction molecule in neurons and plays a critical role in the control of neuronal membrane excitability (Kostyuk and Verkhratsky 1994), synaptic activity, and neurotransmitter release (Llinas et al. 1992; Robitaille et al. 1993). Transient changes in intracellular calcium concentration are known to contribute to short- or long-term alterations in ion channels and gene expression, and can modulate the overall function of these afferents (Ghosh and Greenberg 1995; Simpson et al. 1995).

Several proinflammatory neuropeptides (e.g., tachykinins and calcitonin gene-related peptide) are synthesized primarily in the cell bodies of bronchopulmonary C-fiber neurons (to a distinctly lesser extent in Aδ-fiber neurons), and are stored in large-granular vesicles in the sensory terminals innervating the airways of various species, including human (Barnes and Lundberg 1991; Joos et al. 2000; Lundberg and Saria 1987). Superficial location of these tachykinin-containing nerve endings in the airway mucosa in the human respiratory tract (Komatsu et al. 1991) suggests their susceptibility to inhaled irritants such as nicotine in cigarette smoke. When airway C-fiber endings are stimulated by nicotine, nerve action potentials are transmitted to the central nervous system via the vagus nerves, evoking the irritating sensation, eliciting the defense respiratory reflex responses such as cough and, at the same time, triggering a surge of calcium influx and local release of tachykinins from the sensory endings in the airways (Hong et al. 1995; Lee et al. 1987). Tachykinins (substance P, neurokinins A and B, hemokinins, and endokinins) are a family of structurally related small peptides (Page 2004). Inhalation of aerosolized substance P can elicit the cough reflex in guinea pigs (Kohrogi et al. 1988; Sekizawa et al. 1996), and in patients with respiratory tract infection but not in healthy human subjects (Katsumata et al. 1989). Interestingly, in patients with chronic nonproductive cough, the increased cough sensitivity was associated with an increase in substance P content in the nasal lavage fluid (Cho et al. 2003).

The potent effects of these neuropeptides on airway functions have been well documented (Barnes et al. 1991; Joos et al. 2000; Lundberg and Saria 1987; Solway and Leff 1991). Moreover, these neuropeptides can activate certain target cells in the airways (e.g., mast cells) and lead to the release of various inflammatory mediators (e.g., histamine and bradykinin) (Barnes et al. 1991; Foreman et al. 1982; Joos et al. 2000). Thus, sustained and intense stimulation of these endings can lead to the development of "neurogenic inflammatory reaction" in the airways, characterized by pronounced local responses such as bronchoconstriction, protein extravasation, airway mucosal edema, and inflammatory cell chemotaxis (Barnes et al. 1991; Joos et al. 2000; Lundberg and Saria 1987); these effects seem to be particularly potent in rodent (e.g., guinea pig and rat) airways. Whether and to what extent the repetitive assaults generated by the acute irritant effects of inhaled nicotine contribute to the progressive development of chronic airway diseases in human smokers remains to be determined.

3.2 Interaction with Other Ion Channels Expressed on Cough Sensors

In the study using the single-fiber recording technique in an anesthetized dog preparation, the majority of the pulmonary C-fibers identified by their sensitivity to capsaicin were also activated by right atrial injection of nicotine. Since the action of capsaicin is known to be mediated through an activation of the transient receptor potential vanilloid type 1 receptor (TRPV1), a polymodal transducer for nociception (Caterina et al. 1997), these findings suggest the presence of both TRPV1 and NnAChR in the same sensory terminals (Kou et al. 1989; Xu et al. 2007). Indeed, this finding was further confirmed in the study of isolated neurons; a high percentage (approximately 60%) of the nicotine-sensitive pulmonary neurons were activated by capsaicin (Xu et al. 2007). Interestingly, a recent study in dorsal root ganglion (DRG) neurons showed that activation of NnAChRs by nicotine attenuated the current amplitude evoked by subsequent capsaicin challenges, accompanied by a faster desensitization (Fucile et al. 2005). Although the possible mechanisms underlying the interaction between these two channel proteins in DRG sensory neurons are not fully understood, the Ca^{2+} transient resulting from activation of NnAChRs is believed to be involved (Fig. 6). Furthermore, an inhibitory effect of nicotine on the tetrodotoxin-resistant sodium currents, independent of NnAChR activation, has also been demonstrated in trigeminal neurons by Liu et al. (2004). These recent studies in the somatic nociceptive neurons have provided evidence in support of the previous findings that the analgesic action of nicotine is also in part mediated through the peripheral sensory nerves and that the antinociceptive effects can be induced by activation of the nicotinic cholinergic pathways (Jain 2004). The potential implication of the role of nicotine in regulating the cough receptor sensitivity to capsaicin in the airways remains to be explored. Interestingly, a diminished cough sensitivity to capsaicin has recently been reported in chronic smokers (Dicpinigaitis 2003; Millqvist and Bende 2001).

4 Conclusion

In summary, a substantial amount of evidence has demonstrated that nicotine is a potent tussive agent in human airways, and is primarily responsible for the cigarette smoke-evoked cough response. Nicotine stimulates vagal C-fiber afferents and RARs in the lung and airways of anesthetized animals when it is inhaled as aerosol or cigarette smoke. Nicotine also evokes an inward current with a calcium transient in isolated pulmonary sensory neurons in a concentration-dependent manner. The tussive effect of nicotine is believed to result from an activation of NnAChRs expressed on the sensory terminals of cough receptors located in the airway mucosa because both the cough reflex in humans and responses recorded in sensory nerves and isolated neurons can be prevented by pretreatment with hexamethonium. The expressions of $\alpha4-\alpha7$ and $\beta2-\beta4$ subunits of NnAChR transcripts in pulmonary sensory neurons have lent further support to this conclusion.

There are a number of interesting but unanswered question related to the normal physiological function of NnAChRs on the cough sensors. Acetylcholine applied at relatively high concentrations can stimulate isolated pulmonary sensory neurons via activation of NnAChRs (Gu et al. 2008). The possibility that acetylcholine can act as an endogenous activator of NnAChRs in the airway sensory nerves cannot be

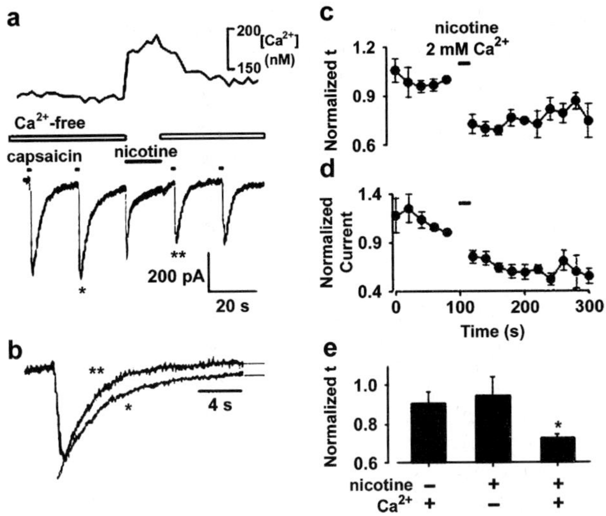

Fig. 6 Effects of acute stimulation of NnAChRs on the capsaicin-induced whole-cell currents in adult dorsal root ganglion (DRG) neurons. **a** Simultaneous recordings of whole-cell currents (*bottom*) and of intracellular Ca^{2+} concentration changes (*top*) in a single DRG neuron. Capsaicin pulses (0.5 μM, 0.5 s) were applied every 20 s in a Ca^{2+}-free medium (as indicated). Note the prolonged nicotine application (500 μM, 15 s) in the presence of normal external Ca^{2+}, inducing inward current (gray line) and Ca^{2+} entry, and modifying the subsequent capsaicin-evoked currents. **b** Capsaicin-induced currents recorded before and after nicotine application. Current amplitudes were normalized to allow a better comparison of desensitization kinetics. Superimposed lines represent single exponential functions best fitting the data, with τ values of 3.70 s (before nicotine) and 2.69 s (after nicotine). Asterisks indicate the corresponding currents in **a** and **b**. **c** Time course of normalized τ of capsaicin-induced currents before and after a 15-s nicotine application (horizontal line, mean τ of the current preceding nicotine application, 3.0 ± 1.0 s). Note absence of τ recovery after nicotine washout. **d** Time course of normalized amplitudes of capsaicin-induced currents before and after nicotine application (mean amplitude of the current preceding nicotine application, 480 ± 70 pA). Note again absence of current recovery. **e** Effects of different treatments on τ of capsaicin-evoked currents. Same protocol as **a**. Values were obtained by averaging the normalized τ from the first five currents following the treatment. $^{*}P < 0.05$. (From Fucile et al. 2005, with permission)

completely ruled out, particularly in light of the finding that substance P-containing afferent nerve profiles, presumably C-fiber endings, have been identified in the close vicinity of cholinergic ganglion neurons in the ferret trachea (Dey et al. 1996).

The specific subtypes of the NnAChRs mediating the response of pulmonary sensory neurons to nicotine and their subunit compositions require further investigation. In view of the recent discoveries and developments of selective ligands targeting specific subunits of NnAChRs as the potential therapeutic interventions for neurodegenerative diseases (e.g., dementia) (Gatto et al. 2004; Ray et al. 2004), their possible side effects on pulmonary sensory nerves and airway irritation certainly merit further investigation.

Acknowledgements This work was supported in part by NIH grants HL40369 and HL67379 to L.-Y.L. Q.G. is a Parker B. Francis Fellow in Pulmonary Research.

References

Agostoni E, Chinnock JE, De Daly MB, Murray JG (1957) Functional and histological studies of the vagus nerve and its branches to the heart, lungs and abdominal viscera in the cat. J Physiol 135:182-205

Arias HR (2000) Localization of agonist and competitive antagonist binding sites on nicotinic acetylcholine receptors. Neurochem Int 36:595-645

Barnes PJ, Lundberg JM (1991) Airway neuropeptides and asthma. In Kaliner MA, Barnes PJ, Persson CGA (eds) Lung Biology in Health and Disease Series, Vol. 49: Asthma: Its Pathology and Treatment. Dekker, New York

Barnes PJ, Baraniuk JN, Belvisi MG (1991) Neuropeptides in the respiratory tract. I. Am Rev Respir Dis 144:1187-1198

Belvisi M (2008) Cough sensors. II. Cannabinoid and opioid membrane receptors on cough sensors. In Chung KF, Widdicombe JG (eds) Pharmacology and Therapeutics of Cough. Springer, Berlin

Bergren DR, Sampson SR (1982) Characterization of intrapulmonary, rapidly adapting receptors of guinea pigs. Respir Physiol 47:83-95

Bolser DC, DeGennaro FC, O'Reilly S, Hey JA, Chapman RW (1995) Pharmacological studies of allergic cough in the guinea pig. Eur J Pharmacol 277:159-164

Boushey HA, Richardson PS, Widdicombe JG, Wise JC (1974) The response of laryngeal afferent fibres to mechanical and chemical stimuli. J Physiol 240:153-175

Braman SS (2006) Chronic cough due to chronic bronchitis: ACCP evidence-based clinical practice guidelines. Chest 129:104S-115S

Bunnelle WH, Dart MJ, Schrimpf MR (2004) Design of ligands for the nicotinic acetylcholine receptors: The quest for selectivity. Curr Top Med Chem 4:299-334

Canning BJ (2007) Encoding of the cough reflex. Pulm Pharmacol Ther 20:396-401

Canning BJ (2008) Mechanisms. II. Sensory and afferent pathways for cough. In Chung KF, Widdicombe JG (eds) Pharmacology and Therapeutics of Cough. Springer, Berlin

Canning BJ, Mazzone SB, Meeker SN, Mori N, Reynolds SM, Undem BJ (2004) Identification of the tracheal and laryngeal afferent neurones mediating cough in anaesthetized guinea-pigs. J Physiol 557:543-558

Canning BJ, Mori N, Mazzone SB (2006) Vagal afferent nerves regulating the cough reflex. Respir Physiol Neurobiol 152:223-242

Caterina MJ, Schumacher MA, Tominaga M, Rosen TA, Levine JD, Julius D (1997) The capsaicin receptor: a heat-activated ion channel in the pain pathway. Nature 389:816-824

Changeux JP, Edelstein SJ (1998) Allosteric receptors after 30 years. Neuron 21:959-980

Cho YS, Park SY, Lee CK, Yoo B, Moon HB (2003) Elevated substance P levels in nasal lavage fluids from patients with chronic nonproductive cough and increased cough sensitivity to inhaled capsaicin. J Allergy Clin Immunol 112:695-701

Coleridge HM, Coleridge JCG (1986) Reflexes evoked from tracheobronchial tree and lungs. In Cherniak NS, Widdicombe JG (eds) Handbook of Physiology, Section 3: The Respiratory System, Vol. II: Control of Breathing, Part I. American Physiological Society, Washington, DC, pp 395-429

Coleridge JCG, Coleridge HM (1984) Afferent vagal C fibre innervation of the lungs and airways and its functional significance. Rev Physiol Biochem Pharmacol 99:1-110

Colquhoun LM, Patrick JW (1997) Pharmacology of neuronal nicotinic acetylcholine receptor subtypes. Adv Pharmacol 39:191-220

Dey RD, Altemus JB, Rodd A, Mayer B, Said SI, Coburn RF (1996) Neurochemical characterization of intrinsic neurons in ferret tracheal plexus. Am J Respir Cell Mol Biol 14:207-216
Dicpinigaitis PV (2003) Cough reflex sensitivity in cigarette smokers. Chest 123:685-688
Dicpinigaitis PV (2007) Experimentally induced cough. Pulm Pharmacol Ther 20:319-324
Douglas WW (1952) The effect of a ganglion-blocking drug, hexamethonium, on the response of the cat's carotid body to various stimuli. J Physiol 118:373-383
Douglas WW, Gray JA (1953) The excitant action of acetylcholine and other substances on cutaneous sensory pathways and its prevention by hexamethonium and D-tubocurarine. J Physiol 119:118-128
Foreman JC, Jordan CC, Piotrowski W (1982) Interaction of neurotensin with the substance P receptor mediating histamine release from rat mast cells and the flare in human skin. Br J Pharmacol 77:531-539
Fucile S, Sucapane A, Eusebi F (2005) Ca^{2+} permeability of nicotinic acetylcholine receptors from rat dorsal root ganglion neurones. J Physiol 565:219-228
Gatto GJ, Bohme GA, Caldwell WS, Letchworth SR, Traina VM, Obinu MC, Laville M, Reibaud M, Pradier L, Dunbar G, Bencherif M (2004) TC-1734: An orally active neuronal nicotinic acetylcholine receptor modulator with antidepressant, neuroprotective and long-lasting cognitive effects. CNS Drug Rev 10:147-166
Genzen JR, Van Cleve W, McGehee DS (2001) Dorsal root ganglion neurons express multiple nicotinic acetylcholine receptor subtypes. J Neurophysiol 86:1773-1782
Geppetti P (2008) Cough sensors. I. TRP membrane receptors on cough sensors. In Chung KF, Widdicombe JG (eds) Pharmacology and Therapeutics of Cough. Springer, Berlin
Ghosh A, Greenberg ME (1995) Calcium signaling in neurons: Molecular mechanisms and cellular consequences. Science 268:239-247
Gotti C, Clementi F (2004) Neuronal nicotinic receptors: From structure to pathology. Prog Neurobiol 74:363-396
Gotti C, Fornasari D, Clementi F (1997) Human neuronal nicotinic receptors. Prog Neurobiol 53:199-237
Gotti C, Zoli M, Clementi F (2006) Brain nicotinic acetylcholine receptors: Native subtypes and their relevance. Trends Pharmacol Sci 27:482-491
Grinevich VP, Letchworth SR, Lindenberger KA, Menager J, Mary V, Sadieva KA, Buhlman LM, Bohme GA, Pradier L, Benavides J, Lukas RJ, Bencherif M (2005) Heterologous expression of human {alpha}6{beta}4{beta}3{alpha}5 nicotinic acetylcholine receptors: Binding properties consistent with their natural expression require quaternary subunit assembly including the {alpha}5 subunit. J Pharmacol Exp Ther 312:619-626
Groneberg DA, Niimi A, Dinh QT, Cosio B, Hew M, Fischer A, Chung KF (2004) Increased expression of transient receptor potential vanilloid-1 in airway nerves of chronic cough. Am J Respir Crit Care Med 170:1276-1280
Gu Q, Ni D, Lee LY (2008) Expression of neuronal nicotinic acetylcholine receptors in rat vagal pulmonary sensory neurons. Respir Physiol Neurobiol 161:87-91
Hansson L, Wollmer P, Dahlback M, Karlsson JA (1992) Regional sensitivity of human airways to capsaicin-induced cough. Am Rev Respir Dis 145:1191-1195
Hartiala JJ, Mapp C, Mitchell RA, Gold WM (1985) Nicotine-induced respiratory effects of cigarette smoke in dogs. J Appl Physiol 59:64-71
Henningfield JE, Miyasato K, Jasinski DR (1985) Abuse liability and pharmacodynamic characteristics of intravenous and inhaled nicotine. J Pharmacol Exp Ther 234:1-12
Hirota CL, McKay DM (2006) Cholinergic regulation of epithelial ion transport in the mammalian intestine. Br J Pharmacol 149:463-479
Ho CY, Gu Q, Lin YS, Lee LY (2001) Sensitivity of vagal afferent endings to chemical irritants in the rat lung. Respir Physiol 127:113-124
Hogg RC, Bertrand D (2004) Nicotinic acetylcholine receptors as drug targets. Curr Drug Targets CNS Neurol Disord 3:123-130
Hong JL, Rodger IW, Lee LY (1995) Cigarette smoke-induced bronchoconstriction: Cholinergic mechanisms, tachykinins, and cyclooxygenase products. J Appl Physiol 78:2260-2266

Jain KK (2004) Modulators of nicotinic acetylcholine receptors as analgesics. Curr Opin Invest Drugs 5:76-81
Jammes Y, Fornaris E, Mei N, Barrat E (1982) Afferent and efferent components of the bronchial vagal branches in cats. J Auton Nerv Syst 5:165-176
Jensen AA, Frolund B, Liljefors T, Krogsgaard-Larsen P (2005) Neuronal nicotinic acetylcholine receptors: Structural revelations, target identifications, and therapeutic inspirations. J Med Chem 48:4705-4745
Joad JP, Kott KS, Bonham AC (1997) Nitric oxide contributes to substance P-induced increases in lung rapidly adapting receptor activity in guinea-pigs. J Physiol 503:635-643
Joos GF, Germonpre PR, Pauwels RA (2000) Role of tachykinins in asthma. Allergy 55:321-337
Kalamida D, Poulas K, Avramopoulou V, Fostieri E, Lagoumintzis G, Lazaridis K, Sideri A, Zouridakis M, Tzartos SJ (2007) Muscle and neuronal nicotinic acetylcholine receptors. Structure, function and pathogenicity. FEBS J 274:3799-3845
Kappagoda CT, Ravi K (2006) The rapidly adapting receptors in mammalian airways and their responses to changes in extravascular fluid volume. Exp Physiol 91:647-654
Karlsson JA, Fuller RW (1999) Pharmacological regulation of the cough reflex – from experimental models to antitussive effects in man. Pulm Pharmacol Ther 12:215-228
Karlsson JA, Hansson L, Wollmer P, Dahlback M (1991) Regional sensitivity of the respiratory tract to stimuli causing cough and reflex bronchoconstriction. Respir Med 85(Suppl A):47-50
Katsumata U, Sekizawa K, Inoue H, Sasaki H, Takishima T (1989) Inhibitory actions of procaterol, a beta-2 stimulant, on substance P-induced cough in normal subjects during upper respiratory tract infection. Tohoku J Exp Med 158:105-106
Kohrogi H, Graf PD, Sekizawa K, Borson DB, Nadel JA (1988) Neutral endopeptidase inhibitors potentiate substance P- and capsaicin-induced cough in awake guinea pigs. J Clin Invest 82:2063-2068
Komatsu T, Yamamoto M, Shimokata K, Nagura H (1991) Distribution of substance P-immunoreactive and calcitonin gene-related peptide-immunoreactive nerves in normal human lungs. Int Arch Allergy Appl Immunol 95:23-28
Kostyuk P, Verkhratsky A (1994) Calcium stores in neurons and glia. Neuroscience 63:381-404
Kou YR, Lee LY (1990) Stimulation of rapidly adapting receptors in canine lungs by a single breath of cigarette smoke. J Appl Physiol 68:1203-1210
Kou YR, Lee LY (1991) Mechanisms of cigarette smoke-induced stimulation of rapidly adapting receptors in canine lungs. Respir Physiol 83:61-75
Kou YR, Frazier DT, Lee LY (1989) The stimulatory effect of nicotine on vagal pulmonary C-fibers in dogs. Respir Physiol 76:347-356
Kwong K, Lee LY (2002) PGE(2) sensitizes cultured pulmonary vagal sensory neurons to chemical and electrical stimuli. J Appl Physiol 93:1419-1428
Lee LY, Pisarri TE (2001) Afferent properties and reflex functions of bronchopulmonary C-fibers. Respir Physiol 125:47-65
Lee LY, Sant'Ambrogio FB, Mathew OP, Sant'Ambrogio G (1987) Acute effect of cigarette smoke on laryngeal receptors. J Appl Physiol 62:1575-1581
Lee LY, Kou YR, Frazier DT, Beck ER, Pisarri TE, Coleridge HM, Coleridge JC (1989) Stimulation of vagal pulmonary C-fibers by a single breath of cigarette smoke in dogs. J Appl Physiol 66:2032-2038
Lee LY, Gerhardstein DC, Wang AL, Burki NK (1993) Nicotine is responsible for airway irritation evoked by cigarette smoke inhalation in men. J Appl Physiol 75:1955-1961
Lee LY, Lou YP, Hong JL, Lundberg JM (1995) Cigarette smoke-induced bronchoconstriction and release of tachykinins in guinea pig lungs. Respir Physiol 99:173-181
Liu L, Zhu W, Zhang ZS, Yang T, Grant A, Oxford G, Simon SA (2004) Nicotine inhibits voltage-dependent sodium channels and sensitizes vanilloid receptors. J Neurophysiol 91:1482-1491
Liu Q, Fujimura M, Tachibana H, Myou S, Kasahara K, Yasui M (2001) Characterization of increased cough sensitivity after antigen challenge in guinea pigs. Clin Exp Allergy 31:474-484
Llinas R, Sugimori M, Silver RB (1992) Microdomains of high calcium concentration in a presynaptic terminal. Science 256:677-679

Lundberg JM, Saria A (1987) Polypeptide-containing neurons in airway smooth muscle. Annu Rev Physiol 49:557-572

Mazzone SB (2004) Sensory regulation of the cough reflex. Pulm Pharmacol Ther 17:361-368

Mazzone SB, Undem BJ (2008) Cough sensors V: Pharmacological modulation of cough sensors. In Chung KF, Widdicombe JG (eds) Pharmacology and Therapeutics of Cough. Springer, Berlin

Mazzone SB, Mori N, Canning BJ (2005) Synergistic interactions between airway afferent nerve subtypes regulating the cough reflex in guinea-pigs. J Physiol 569:559-573

Millqvist E, Bende M (2001) Capsaicin cough sensitivity is decreased in smokers. Respir Med 95:19-21

Nadel JA, Comroe JH Jr. (1961) Acute effects of inhalation of cigarette smoke on airway conductance. J Appl Physiol 16:713-716

Nguyen LS, Villablanca AC, Rutledge JC (1995) Substance P increases microvascular permeability via nitric oxide-mediated convective pathways. Am J Physiol 268:R1060-R1068

Page NM (2004) Hemokinins and endokinins. Cell Mol Life Sci 61:1652-1663

Paintal AS (1954) The response of gastric stretch receptors and certain other abdominal and thoracic vagal receptors to some drugs. J Physiol 126:271-285

Paintal AS (1955) Impulses in vagal afferent fibres from specific pulmonary deflation receptors: The response of these receptors to phenyl diguanide, potato starch, 5-hydroxytryptamine and nicotine, and their role in respiratory and cardiovascular reflexes. Q J Exp Physiol Cogn Med Sci 40:89-111

Pisarri TE, Lee LY, Coleridge HM, Coleridge JCG (1991) Stimulation of vagal bronchial C-fibers by cigarette smoke in dogs (Abstract). FASEB J 5:A1480

Ray M, Bohr I, McIntosh JM, Ballard C, McKeith I, Chalon S, Guilloteau D, Perry R, Perry E, Court JA, Piggott M (2004) Involvement of alpha6/alpha3 neuronal nicotinic acetylcholine receptors in neuropsychiatric features of Dementia with Lewy bodies: [(125)I]-alpha-conotoxin MII binding in the thalamus and striatum. Neurosci Lett 372:220-225

Robitaille R, Garcia ML, Kaczorowski GJ, Charlton MP (1993) Functional colocalization of calcium and calcium-gated potassium channels in control of transmitter release. Neuron 11:645-655

Romanelli MN, Gualtieri F (2003) Cholinergic nicotinic receptors: Competitive ligands, allosteric modulators, and their potential applications. Med Res Rev 23:393-426

Sampson SR, Vidruk EH (1975) Properties of 'irritant' receptors in canine lung. Respir Physiol 25:9-22

Sant'Ambrogio G (1982) Information arising from the tracheobronchial tree of mammals. Physiol Rev 62:531-569

Sant'Ambrogio G, Widdicombe JG (2001) Reflexes from airway rapidly adapting receptors. Respir Physiol 125:33-45

Sant'Ambrogio G, Tsubone H, Sant'Ambrogio FB (1995) Sensory information from the upper airway: Role in the control of breathing. Respir Physiol 102:1-16

Sekizawa K, Jia YX, Ebihara T, Hirose Y, Hirayama Y, Sasaki H (1996) Role of substance P in cough. Pulm Pharmacol 9:323-328

Sellick H, Widdicombe JG (1971) Stimulation of lung irritant receptors by cigarette smoke, carbon dust, and histamine aerosol. J Appl Physiol 31:15-19

Simpson PB, Challiss RA, Nahorski SR (1995) Neuronal Ca^{2+} stores: Activation and function. Trends Neurosci 18:299-306

Solway J, Leff AR (1991) Sensory neuropeptides and airway function. J Appl Physiol 71: 2077-2087

Takahama K (2008) Central mechanisms. III. Mechanisms of action of centrally-acting antitussives. In Chung KF, Widdicombe JG (eds) Pharmacology and Therapeutics of Cough. Springer, Berlin

Tarlo SM (2006) Cough: Occupational and environmental considerations: ACCP evidence-based clinical practice guidelines. Chest 129:186S-196S

Tatar M, Webber SE, Widdicombe JG (1988) Lung C-fibre receptor activation and defensive reflexes in anaesthetized cats. J Physiol 402:411-420

Tobin MJ, Sackner MA (1982) Monitoring smoking patterns of low and high tar cigarettes with inductive plethysmography. Am Rev Respir Dis 126:258-264

Undem BJ, Chuaychoo B, Lee MG, Weinreich D, Myers AC, Kollarik M (2004) Subtypes of vagal afferent C-fibres in guinea-pig lungs. J Physiol 556:905-917

Widdicombe JG (1954) Receptors in the trachea and bronchi of the cat. J Physiol 123:71-104

Widdicombe JG (1981) Nerve receptors in the respiratory tree and lungs. In Hornbein T (ed) Lung Biology in Health and Disease. Regulation of Breathing. Dekker, New York, chap 6

Widdicombe JG (1998) Afferent receptors in the airways and cough. Respir Physiol 114:5-15

Widdicombe JG, Chung KF (eds) (2004) The third international symposium on cough: Acute and chronic. Pulm Pharmacol Ther 17:327-474

Widdicombe JG, Undem BJ (eds) (2006) Cough and its regulation. Respir Physiol Neurobiol 152:221-372

Widdicombe JG, Chung KF (eds) (2007) The fourth international symposium on cough: Acute and chronic. Pulm Pharmacol Ther 20:305-452

Widdicombe JG, Sant'Ambrogio G, Mathew OP (1988) Nerve receptors of the upper airway. In Mathew OP, Sant'Ambrogio G (eds) Respiratory Function of the Upper Airway. Dekker, New York

Winning AJ, Hamilton RD, Shea SA, Guz A (1986) Respiratory and cardiovascular effects of central and peripheral intravenous injections of capsaicin in man: Evidence for pulmonary chemosensitivity. Clin Sci (Lond) 71:519-526

Xu J, Yang W, Zhang G, Gu Q, Lee LY (2007) Calcium transient evoked by nicotine in isolated rat vagal pulmonary sensory neurons. Am J Physiol Lung Cell Mol Physiol 292:L54-L61

Cough Sensors. V. Pharmacological Modulation of Cough Sensors

S.B. Mazzone(✉) and B.J. Undem

Contents

Abstract Several airway afferent nerve subtypes have been implicated in coughing. These include bronchopulmonary C-fibers, rapidly adapting airway mechanoreceptors and touch-sensitive tracheal Aδ-fibers (also called cough receptors). Although the last two afferent nerve subtypes are primarily sensitive to mechanical stimuli, all can be acted upon by one or more different chemical stimuli. In this review we catalogue the chemical agents that stimulate and/or modulate the activity of the airway afferent nerves involved in cough, and describe the specific mechanisms involved in these effects. In addition, we describe the mechanisms of action of a number of chemical inhibitors of these afferent nerve subtypes, and attempt to relate this information to the regulation of coughing in health and disease.

S.B. Mazzone
School of Biomedical Sciences, The University of Queensland, St. Lucia QLD 4072, Australia
s.mazzone@uq.edu.au

K.F. Chung, J.G. Widdicombe (eds.), *Pharmacology and Therapeutics of Cough*,
Handbook of Experimental Pharmacology 187,

1 Introduction

The mammalian airways are innervated by a variety of afferent nerve subtypes that play differing roles in the regulation of cough. This topic has been recently reviewed elsewhere (Mazzone 2004a, Mazzone 2005) and is also covered in detail by Canning and Chou in this volume. Sensory innervation to the airways and lungs is derived from vagal and spinal sources, although there are only data to support the role of vagal afferents in the cough reflex. There are two broad classes of airway vagal sensory nerve types: one class is predominately mechanically sensitive, and includes rapidly and slowly adapting airway mechanosensors, while the other class is predominately chemosensitive, and includes bronchopulmonary C-fibers and chemosensitive Aδ-fibers (Canning et al. 2001; Ho et al. 2001). Accordingly, airway sensory nerve subtypes are differentiated by their physiological, anatomical, and neurochemical characteristics and the reflexes that they initiate. Although there is debate over which nerve subtypes are involved in initiating and/or regulating coughing, a general consensus suggests that rapidly adapting receptors (RARs) and bronchopulmonary C-fibers, are directly responsible for the initiation of cough, whereas slowly adapting receptors (SARs) play a more modulatory role in shaping the reflex response (Mazzone 2005). More recently, tracheal touch-sensitive Aδ-fibers (also known as cough receptors) have been shown to be involved in coughing in guinea pigs (Canning et al. 2004a). These fibers display characteristics that are distinct from those of other classically described pulmonary mechanoreceptors.

In this chapter we review the current knowledge relating to the chemical activation, modulation and inhibition of airway afferent nerves, focusing particularly on touch-sensitive Aδ-fibers, RARs, and bronchopulmonary C-fibers as these are the likely afferents nerves that are primarily involved in coughing. Although the most up-to-date information is presented, it is readily apparent from reading the proceeding chapters that a great many more substances are known to modulate the activity of bronchopulmonary C-fibers compared with the other types of airway afferent nerves. This regrettably necessitates that a larger portion of the chapter be devoted to C-fiber pharmacology. It is also worth noting that distinct subtypes of bronchopulmonary C-fibers have been described (e.g., pulmonary and bronchial C-fibers in dogs, placodal and neural crest C-fibers in guinea pigs) and there is some debate about the role of each of these afferent subclasses in the cough reflex. This matter has not been addressed in the current chapter, but is dealt with in detail in the chapter by Canning and Chou. Rather, the discussion has been limited to the pharmacological modulation of each afferent nerve type.

2 Increasing Activity

2.1 Touch-Sensitive Aδ-Fibers (Cough Receptors)

Studies in guinea pigs have identified a distinct airway afferent nerve subtype, termed the “cough receptor,” which appears to play an essential role in regulating the cough reflex in this species (Canning et al. 2004a; reviewed in Mazzone 2005). Cough receptors are characterized as extrapulmonary, low-threshold Aδ-fiber mechanosensors that are highly sensitive to touch. The functional and structural properties of cough receptors are considered quite distinct from those of the classic intrapulmonary RARs and SARs. Their conduction velocity is about 3–5 times slower than that of the intrapulmonary RARs and SARs, but about 3–5 times faster than that of C-fibers in this species (Canning et al. 2004a). Although this sensory nerve subtype adapts rapidly to mechanical stimuli (McAlexander et al. 1999), touch-sensitive Aδ-fibers are not sensitive to airway stretch or bronchospasm (stimuli that activate classic RARs). Rather touch-sensitive Aδ-fibers respond to punctate mechanical stimuli (touch) and rapid changes in pH (Canning et al. 2004a; .Kollarik and Undem 2002). These are two stimuli that readily evoke cough in conscious and anesthetized animals and humans. Their location within the airways and functional and structural profiles makes cough receptors the likely sensory nerve subtype involved in defensive coughing in guinea pigs (Canning et al. 2004a; Mazzone 2004a, Mazzone 2005). Circumstantial evidence also suggests that similar nerve fibers may regulate defensive coughing in other species, including humans.

2.1.1 Acid

Touch-sensitive Aδ-fibers in the guinea pig extrapulmonary airways (cough receptors) are largely insensitive to many chemical stimuli. However, they are activated in response to sudden reductions in pH via a mechanism that does not involve stimulation of the capsaicin-sensitive channel, TRPV1 (involved in the acid sensitivity of other airway afferent nerve subtypes; see later) (Kollarik and Undem 2002). Indeed, cough evoked by citric acid activation of touch-sensitive Aδ-fibers in anesthetized guinea pigs is actually potentiated in the presence of the TRPV1 antagonist capsazepine (Canning et al. 2006), albeit via an unknown mechanism since this afferent nerve subtype does not express TRPV1 channels (Myers et al. 2002) and capsazepine has no effect on acid-evoked firing of individual cough receptors in vitro (Kollarik and Undem 2002).

The mechanism by which acid activates tracheal touch-sensitive Aδ-fibers is presently unclear, although possible insights may be gained by assessing the unique kinetics of the acid-evoked responses. For example, acid only activates touch-sensitive Aδ-fibers in the guinea pig extrapulmonary airways when the reduction in pH occurs very rapidly (from pH 7.4 to 5.0 in 1–2 s) (Kollarik and Undem 2002). A comparable reduction in pH over a much slower time scale (e.g., 60 s) does not activate touch-sensitive Aδ-fibers (Kollarik and Undem 2002) and does not cause cough in anesthetized guinea pigs (Mazzone, unpublished observations). Furthermore, following an initial activation, there is a rapid inactivation of the response to acid application. This time course resembles that of certain acid-sensing ion channels (ASICs). The ASICs are a family of channels that are related to the epithelial/sodium (degenerin) channels (ENaCs). There are several subtypes and splice variants, including ASIC1a, ASIC1b, ASIC2a, ASIC2b, ASIC3, and ASIC4. Of these subtypes,

ASIC3 has the characteristics that best reflect the acid-evoked responses in touch-sensitive Aδ-fibers. Consistent with this hypothesis, using a single-cell PCR strategy, we have identified ASIC3 messenger RNA (mRNA) within individual nodose ganglion neurons (Undem, unpublished observations). It is also of note that expression of the ASICs in sensory neurons is increased by inflammatory mediators (Mamet et al. 2002). In theory, this may translate into enhanced acid responsiveness of the sensory nerves. Nevertheless, amiloride and other inhibitors of ASICs are only modestly effective, at best, at reducing acid-evoked coughing mediated via touch-sensitive Aδ-fibers in guinea pigs (Canning et al. 2006). Given that there are many other types of ion channels sensitive to decreases in pH, it remains unknown whether an ASIC or some other ionic mechanism is involved in acid-induced activation of touch-sensitive Aδ-fibers.

2.2 Rapidly Adapting Receptors

RARs are sporadically active throughout the respiratory cycle, excited by the dynamic mechanical forces accompanying lung inflation and deflation and becoming more active as the rate and volume of lung inflation increase (Ho et al. 2001; Pack and DeLaney 1983). Indeed a wide range of mechanically related stimuli can modify the activity of RARs. Conversely, reports of chemical stimuli that can directly activate RAR nerve fibers are limited. RARs are insensitive to the direct effects of serotonin (5-hydroxytryptamine, 5-HT), capsaicin, bradykinin, and adenosine, stimuli that readily excite bronchopulmonary C-fibers and other chemosensitive nerve subtypes (see later) (Canning et al. 2001; Ho et al. 2001). Nevertheless, some chemical stimuli may activate RARs indirectly, secondary to producing mechanical-like perturbations in the local lung tissue environment surrounding individual RAR nerve terminals.

2.2.1 Adenosine Triphosphate

In guinea pigs, RARs may be activated directly by some purinergic receptor agonists (Canning et al. 2004a). For example, intrapulmonary RARs in vitro are robustly excited by adenosine triphosphate (ATP) and the selective P2X receptor agonist α,β-methylene ATP. Responses to both ATP and α,β-methylene ATP are abolished by the P2X receptor antagonist pyridoxal phosphate 6-azophenyl-2′,4′-disulfonic acid (Canning et al. 2004a). It is not known if purinergic mechanisms are involved in the endogenous excitation of RARs in vivo, nor is it known which specific subtype of P2X receptor is expressed by RARs. With respect to the latter, given the pharmacological profile of α,β-methylene ATP and on the basis of studies assessing the expression of purinergic receptors in the vagal (nodose) sensory ganglia, it is reasonable to speculate that receptors incorporating $P2X_2$ and $P2X_3$ subunits may be involved.

Inhaled ATP causes dyspnea and cough in asthmatic subjects. Whether this is secondary to activation of RAR or C-fibers (see later) is not known (Basoglu et al. 2005). However, inhaled α,β-methylene ATP does not evoke cough in conscious guinea pigs (at concentrations up to 1 mM; Canning et al. 2004a) or anaesthetized guinea pigs (at concentrations up to 10 mM; Mazzone, unpublished observations) despite these same animals responding readily to citric acid. Purinergic agonists may, however, potentiate cough evoked by other stimuli (Kamei et al. 2005), and endogenous ATP may be involved in the sensitizing effects of histamine on the cough reflex (Kamei and Takahashi 2006).

2.2.2 Agents That Evoke Smooth-Muscle Contraction and Vascular Congestion

Although RARs are insensitive to the direct actions of many chemicals, the sensitivity of RARs to stretch and other mechanical perturbations means that their activity can be increased by any chemical stimulus that evokes bronchospasm or obstruction resulting from mucus secretion or edema in the airways (Bergren and Sampson 1982; Bonham et al. 1996; Canning et al. 2001; Jonzon et al. 1986; Pack and DeLaney 1983; Mohammed et al. 1993). Bronchospastic agents such as histamine, methacholine, and leukotirene C4, and agents that evoke plasma extravasation, including substance P (or capsaicin, via neurogenic release of substance P), produce robust activation of RARs, not necessarily because RARs express receptors for these agents, but rather owing to the secondary pulmonary end organ effects (Canning et al. 2001). Not surprisingly then, RAR activation by chemicals such as histamine, capsaicin and substance P is markedly inhibited or abolished by preventing the local end organ effects that these stimuli evoke. This is perhaps best highlighted by the fact that pretreatment with the bronchodilator isoproteronol prevents bronchospasm-evoked activation of RARs, whereas blocking endogenous β-adrenoceptors with propranolol potentiates bronchospasm-evoked activation of RARs (Bergren 1997; Canning et al. 2001).

Indirect activation of RARs by chemical stimuli may confound attempts to ascribe a sensory nerve subtype to a particular reflex response. For example, the well-known cough-evoking effects of capsaicin could be presumed to be mediated via airway capsaicin-sensitive C-fibers. However, peripherally acting neurokinin receptor antagonists effectively reduce capsaicin-evoked cough in several models, suggesting that axon reflex mediated end organ effects, and the subsequent activation of RARs, may also be involved (reviewed in Mazzone 2004b). Nevertheless, although lung inflation may be associated with cough under certain circumstances, many stimuli that produce robust activation of RARs (e.g., thromboxane, leukotriene C_4, histamine, neurokinins, methacholine) are ineffective or only modestly effective at evoking cough (Barnes et al. 1984; Canning et al. 2004a; Joos et al. 1987). Moreover, in some coughing species (e.g., guinea pigs) many RARs are spontaneously active throughout the respiratory cycle and yet cough is only induced in response to very specific stimuli (Bergren and Sampson 1982; Canning et al. 2001; Mazzone and Canning 2002).

2.3 Bronchopulmonary C-Fibers

Afferent C-fibers are thin unmyelinated nerves that conduct action potentials at a velocity typically less than 1 m/s. It should be kept in mind, however, that there is more than one subtype of afferent C-fibers. The respiratory tract is innervated mainly by afferent C-fibers arising from the vagus nerve, but also by a substantial number of spinal afferent C-fibers with cell bodies situated in the thoracic dorsal root ganglia (DRG). In addition to vagal and spinal C-fiber subtypes, the Coleridges and their colleagues noted in a series of elegant studies that there are subtypes of C-fibers arising from the vagus nerve (Coleridge and Coleridge 1984). The vagal C-fibers innervating the large airways (referred to as bronchial C-fibers) may be pharmacologically distinguishable from those situated in more peripheral lung tissue (pulmonary C-fibers). Part of the phenotypic distinction between different vagal C-fiber subtypes may be based on whether their cell bodies are located in the nodose or jugular vagal ganglia. Neurons in the nodose ganglia are derived embryonically from the epibranchial placodes, whereas the neurons in the jugular ganglia are derived from the neural crest (Baker 2005). In this sense, jugular ganglia neurons are similar to spinal DRG neurons. Several studies have demonstrated pharmacological and neurochemical differences between the neural crest and placodal C-fibers innervating the respiratory tract (Kollarik and Undem 2006).

C-fibers respond to a variety of chemical irritants and endogenous inflammatory mediators, and inhalation of chemicals known to stimulate bronchopulmonary C-fibers is effective at evoking cough in conscious animals, including guinea pigs, rabbits, cats, dogs, and humans. Unlike mechanical stimulation of the extrapulmonary airways, selective C-fiber stimulants are less effective or ineffective at evoking cough in anesthetized animals (Canning et al. 2004a). Whether the placodal C-fibers, neural crest C-fibers or both subtypes evoke cough when activated is not yet understood.

Chemicals activate (evoke action potential discharge) in the C-fiber terminals via one of two general mechanisms: they either act directly on ligand-gated ion channels that result in an inward current upon stimulation, or they act on metabotropic receptors (e.g., G-protein-coupled receptors) that evoke inward currents in the nerve terminal secondary to various signal transduction schemes that ultimately alter ion channel kinetics.

2.3.1 Activation via Ligand-Gated Ion Channels

Nicotinic Receptor Agonists

The nicotinic cholinergic receptor is perhaps the most thoroughly studied ligand-gated ion channel. The functional receptor takes the form of a pentomere comprising various combinations of α and β subunits (Lukas et al. 1999). There are ten identified α subunits (α1–α10) and four β subunits (β1–4). Nine (α2–α10) of the α subunits have been identified in the vertebrate nervous system, although α8 is limited to avians.

Inhalation of cigarette smoke, in naïve smokers, causes intense airway irritation and cough. A series of studies revealed that this is largely accounted for by nicotinic receptor activation (Lee et al. 1993). The most parsimonious explanation of how inhaled nicotine could lead to sensations of irritation and urge to cough is that nicotine directly activates C-fibers near the airway epithelium. This hypothesis has been addressed using extracellular recording of C-fibers in vivo (Lee et al. 2007). Nicotine and cigarette smoke both lead to action potential discharge in a population of airway C-fibers, and this can be blocked by the nicotininc receptor antagonist hexamethonium. A population of RAR fibers is also activated upon inhalation of nicotine (Lee et al. 2007). From these in vivo studies one cannot be certain whether the nerve activation is due to a direct effect of nicotine on the nerve terminals, or secondary to other effects in the lungs (vascular effects, bronchoconstriction, etc.; see earlier).

That nicotine directly interacts with receptors on the afferent nerves is supported by studies showing that nodose neurons express nicotinic receptors (Mao et al. 2006). Furthermore, in functional studies that assess increases in cytosolic calcium as a measure of neuronal activation, nicotine and other nicotinic receptor agonists cause activation of a subset of capsaicin-sensitive neurons retrogradely labeled from the lungs. Nicotine also activated a subset of capsaicin-insensitive lung afferent nodose neurons (perhaps RAR neurons) (Xu et al. 2007).

The nicotinic receptor subtype responsible for stimulation of pulmonary afferent nerves is not known. Immunoprecipitation and immunhistochemical studies provide evidence of $\alpha2$, $\alpha3$, $\alpha4$, and $\alpha5$ subunits along with $\beta2$ and $\beta4$ in the nodose ganglion neurons (Mao et al. 2006). The $\alpha3$ and $\alpha5$ subunits may be particularly prominent in the capsaicin-sensitive (i.e., C-fiber) population (Spies et al. 2006). It is also not known whether the nicotinic receptors on afferent nerve terminals in the lungs are ever stimulated by endogenous agonists. There is little evidence that cholinergic parasympathetic nerves synapse with afferent C-fibers. There, are however, potential nonneuronal sources of acetylcholine in the airways that may be juxtaposed to afferent nerve terminals, including the airway epithelial cells (Klapproth et al. 1997). Moreover, choline, a rather ubiquitous chemical, is an effective agonist at certain subtypes of nicotinic receptors (Lukas et al. 1999).

5-Hydroxytryptamine

Among the dozen or so 5-HT receptor subtypes only 5-HT3 is an ionotropic receptor. The functional 5-HT3 receptor, like the nicotinic receptor, takes the form of a pentomer (Hoyer et al. 1994). So far three 5-HT3 subunits have been described (5-HT3A, 5-HT3B, and 5HT-3C). In the 1970s and 1980s several laboratories were using phenyldiguanide to stimulate pulmonary C-fibers (or J-receptors) in classical studies of the pulmonary chemoreflex (Coleridge and Coleridge 1984). These studies predated the discovery that phenyldiguanide is a very selective 5-HT3 receptor agonist; nevertheless, in retrospect, they provided the first evidence that 5-HT3 receptor activation can lead to stimulation of pulmonary C-fibers. Incidentally, phenyldiguanide is still a useful 5-HT3-selective agonist, but owing to slight 5HT3 receptor sequence differences among species, it fails to stimulate 5-HT3 receptors in some mammals, including guinea pigs (Lankiewicz et al. 1998). 2-Methyl-5-hydroxytryptamine is a selective 5-HT3 receptor agonist that is more universally effective among species.

5-HT effectively activates C-fibers (but not RARs or SARs) in rabbit lungs and this is blocked by a selective 5-HT3 antagonist. Interestingly, the same 5-HT3 receptor antagonist prevents C-fiber activation caused by experimentally induced pulmonary embolism (Kay and Armstrong 1991). In guinea pigs, 5-HT was not effective at stimulating all C-fiber subtypes. The nodose (placodal) C-fibers in the lungs were effectively stimulated by 5-HT (and 5-HT3-selective agonists), whereas the jugular (neural-crest)-derived C-fibers were not stimulated by 5-HT (Chuaychoo et al. 2005). Consistent with these finding, 5-HT evoked an inward current in patch-clamped pulmonary nodose neurons, but failed to evoke such a current in pulmonary jugular neurons. Like the jugular neurons, the dorsal root ganglion C-fiber neurons labeled from the lungs are also unresponsive to 5-HT. In dogs, 5-HT was found to evoke a larger response in bronchial than in pulmonary C-fibers.

Adenosine Triphosphate

ATP binds to P2Y and P2X purinergic receptors, but only P2X receptors are ionotropic. There are seven subtypes of P2X receptors (P2X1–P2X7) (Chizh and Illes 2001). The functional receptors are trimeric and can be either homopolymers or heteropolymers. As described earlier, the P2X receptors in nodose neurons are generally thought to be P2X2, P2X3, and heteromeric P2X2,3 receptors (Cockayne et al. 2005).

Pelleg and Hurt (1996) were first to show ATP and P2X-selective agonists cause action potential discharge in dog lungs studied in vivo. The P2X receptors seem to define the placodal population of afferent neurons innervating the respiratory tract (Undem et al. 2004; Canning et al. 2004a). In patch-clamp studies, P2X currents can be seen on virtually all nodose neurons (Cockayne et al. 2005). By contrast, jugular ganglion neurons and lung-specific dorsal root ganglion neurons (neural crest in origin) do not respond to selective P2X agonists. In extracellular recording studies, ATP, or the P2X3-selective agonist $\alpha s \beta$-methyl ATP, evokes action potential discharge in nodose C-fiber terminals, but not jugular C-fibers (Undem et al. 2004). Given that inhaled α,β-methyl ATP (up to 1 mM) does not evoke cough in conscious guinea pigs, a question could be raised as to the specific role in cough played by nodose-derived C-fiber afferent nerves (Canning et al. 2004a).

Stimuli of TRPV1 Channels

Capsaicin is commonly used as a pharmacological tool to activate visceral and somatosensory C-fibers. It had long been suspected that capsaicin may act via a specific receptor, but it was not until 1997 that the capsaicin receptor was identified (Caterina et al. 1997). It was found to be a ligand (capsaicin)-gated ion channel originally termed vanilloid receptor-1, but later renamed TRPV1. TRPV1 is a member of a large family of transient receptor potential (TRP) channels that is thoroughly reviewed in Nilius et al. (2007).

The relevance of TRPV1 in C-fiber biology was amplified by the findings that this ion channel can be gated by mechanisms independently of chemicals found in hot peppers (Tominaga et al. 2008). Temperature itself can open TRPV1 and lead to action potential discharge, but this typically requires temperature in excess of what would be found in the lower respiratory tract. A decrease in pH can also activate C-fibers in the airways via TRPV1 gating (see later). This is almost certainly relevant to acid-provoked cough, and may be relevant in inflammatory diseases in general inasmuch as the pH of inflamed tissues is typically acidic. TRPV1 can also be gated by signaling mechanisms downstream of stimulation of metabotropic receptors. Chuang et al. (2001) reported that TRPV1 is tonically inhibited by the membrane phospholipid phosphatidylinositol 4,5-bisphosphate (PIP_2). Activation of phospholipase C results in the conversion of PIP_2 to diacylglycerol and inositol 1,4,5-trisphosphate, thereby removing the tonic inhibition. Another way in which phospholipase C activation can lead to activation is through the downstream production of lipoxygenase products of arachidonic acid (Shin et al. 2002). Lipoxygenase metabolites such as 12-hydroxyeicosatetraenoic acid or 15-hydroxyeicosatetraenoic acid are structurally similar to capsaicin and are effective at intracellularly gating the TRPV1 channel (Hwang et al. 2000) and evoking TRPV1-dependent reflex bronchospasm in guinea pigs (Mazzone and Canning 2002). These findings explain why bradykinin-induced action potential discharge in airway C-fibers is inhibited in a nonredundant manner by TRPV1 antagonists (or in TRPV1 knockout mice) and 12-lipoxygenase inhibitors (Carr et al. 2003; Kollarik and Undem 2003).

Stimuli of TRPA1 Channels

TRPA1 is another member of the TRP family of ion channels and is named for its prominent ankyrin repeats in the N-terminal domain (Nilius et al. 2007). In preliminary studies we found that all lung-specific C-fibers in the mouse express both TRPV1 and TRPA1 mRNA, and ligands known to activate either TRPV1 or TRPA1 lead to action potential discharge in mouse and guinea pig bronchopulmonary C-fibers. TRPA1 agonists also evoke neurogenic (tachykinin-mediated) bronchospasm in anesthetized guinea pigs (Mazzone, unpublished observations). TRPA1 is not activated by capsaicin, but is stimulated by cinnamaldahyde and allyl isothiocyante (active ingredient in mustard oil, wasabi, horseradish, etc.) (Jordt et al. 2004; Bandell et al. 2004). Like TRPV1, TRPA1 may be gated by autacoids that act through certain G-protein-coupled receptors, e.g., bradykinin acting via B2 receptors. The mechanism(s) underlying G-protein-coupled-receptor-dependent activation of TRPA1, however, appears to be different from those involved in TRPV1 activation. TRPA1 has an N-terminal EF-hand calcium-binding domain making it sensitive to any stimulus that increases intracellular calcium. These mechanisms provide a pathway through which inflammatory mediators acting via metabotropic receptors may increase the activation rate of TRPA1 containing nerves (Doerner et al. 2007). Stimuli that lead to calcium influx through other ligand-gated ion channels, including TRPV1, may also indirectly activate TRPA1 via this mechanism. Decreases in temperature can activate TRPA1 and this may have relevance in cold-air-induced reflex activity in the airways. In addition, certain relevant environmental irritants such as isothiocyantes and acrolein may directly activate TRPA1, providing a mechanism by which air-pollutants can lead to nasal and bronchial C-fiber activation (Bautista et al. 2006).

Acid

Acid is an effective stimulator of bronchopulmonary C-fibers (Fox et al. 1995a; Hong et al. 1997; Kollarik and Undem 2002). The mechanism by which C-fibers are activated by acid depends on both the acid concentration and the rate of change in pH. Sustained acidification of the airway mucosal surface in the guinea pig evokes sustained action potential discharge in vagal C-fiber nerves. The pH threshold for activation of these nerves is about 6.0 (Fox et al. 1995a; Kollarik and Undem 2002). Pharmacological and TRPV1 knockout studies revealed that this activation is in part mediated by TRPV1; however, up to 50% of this activation is TRPV1-independent (Kollarik and Undem 2004).

An analysis of the characteristics of the TRPV1-independent acid-induced response in the nerve terminals of the touch-sensitive Aδ-fibers and C-fibers predicts an acid transducer with low pH threshold (pH $>$ 6.7) that rapidly (within seconds) inactivates in the presence of acid (Kollarik and Undem 2002). It also predicts that inactivation may occur in the absence of activation if the rate of pH change is slow. As described earlier for tracheal touch-sensitive Aδ-fibers, ASIC3 most closely matches the properties of the TRPV1-independent acid-induced response in vagal afferent C-fibers.

2.3.2 Activation via G-Protein-Coupled "Metabotropic" Receptors

There are a large number of G-protein-coupled-receptor agonists that through various second messenger systems can affect the afferent function of airway sensory nerves. With a few notable exceptions, the ligands acting via G-protein-coupled receptors do not overtly activate the C-fiber terminals. Rather, they alter the excitability of the nerve such that the net response of any given stimulus is enhanced or inhibited. This change in excitability, or "neuromodulation," typically involves phosphorylation of various ion channels or pumps in the nerve terminal. In some cases, ligands acting via G-protein-coupled receptors have been found to directly activate the C-fibers, causing action potential discharge (Table 1).

Table 1 Overview of the known chemicals that activate, modulate, or inhibit the airway sensory nerves involved in cough

Agent	Pharmacological Mechanism of action	Touch sensitive Aδ-fibers	Nodose C-fibers	Jugular C-fibers and Aδ-fibers	RARs[a]
		Activators			
Adenosine	A1/A2A receptors	No	Yes	No	No
Acid	TRPV1 channels	No	Yes	Yes	No
	ASICs[b]	Yes	Yes	Yes	Unknown
Acetylcholine/ nicotine	nAChR[c]receptors	No	Yes[d]	Yes[d]	Some[d]
ATP	P2X[e]receptors	No	Yes	No	Yes
Bradykinin	B2 receptors	No	Yes	Yes	No
Capsaicin, heat, lipoxygenase products	TRPV1 channels	No	Yes	Yes	No
Thrombin	PAR1 receptors	No	Yes	Yes	No
5-HT	5-HT3 receptors	No	Yes	No	No
Cinnamaldehyde, mustard oil, extreme cold	TRPA1 channels	No	Yes[f]	Yes[f]	No
		Modulators			
Histamine	H1 receptors	No	Yes	Unknown	No
Cysteinyl leukotrienes	cys-LT1 receptors	No	Yes	Yes	No
Neurokinins	NK1/NK2 receptors	No	Yes	Unknown	No
Thrombin	PAR2[g]receptors	No	Yes	Yes	No
PGE2	EP3 receptors	No	Yes	Yes	No
		Inhibitors			
Cl^- channel blockers	Cl^- channels	*Possibly*[h]	Yes	Yes	Unknown
Loop diuretics	NKCC1	*Possibly*[h]	Yes	Yes	*Possibly*[h]
Local anesthetics	Na^+ channels	Yes	Yes	Yes	Yes
Menthol[i]	TRPM8/TRPA1 channels	Unknown	Unknown	Unknown	Unknown
Nociceptin	NOP1 receptors	No	Yes	Yes	Unknown
Ouabain	α3-ATPase pumps	Yes	Unknown	No	Unknown
Opioids	Mu receptors	No	Yes	Yes	Unknown
	Kappa receptors	Unknown	Unknown	Unknown	Unknown
	Delta receptors	Unknown	Unknown	Unknown	Unknown

Activators are agents that evoke action potential formation; modulators are agents that alter the membrane potential thereby altering neuronal excitability, and inhibitors are only those compounds that reduce nerve excitability (i.e., although specific antagonists of the known activators will effectively inhibit that activator from evoking afferent responses, such compounds do not reduce the overall excitability of the nerve and are therefore not included in the list). The mechanism of action is the pharmacological receptor, ion channel or ion transporter that the listed compound directly interacts with. All afferent nerves can be activated to punctuate mechanical stimuli and large changes in extracellular osmolarity, albeit with varying sensitivities. The mechanisms of these activations are presently unknown

Table 1 Continued

RARs rapidly adapting receptors, *ASICs*, acid-sensing ion channels, *nAChR* nicotinic acetylcholine receptor, *5-HT* 5- hydroxytryptamine, *PGE2* prostaglandin E2, *PAR2* protease-activated receptor-2, *NKCC1* furosemide-sensitive $Na^{+}/K^{+}/2Cl^{-}$ cotransporter

[a]Only agents that directly activate RARs are listed

[b]Acid activates airway afferent nerves in part via a TRPV1-independent mechanism. Although it is not clear yet, this likely involves activation of one or more ASICs.

[c]The subunits that comprise the nAChR are unknown

[d]The evidence for C-fiber and RAR fiber activation comes from studies in dogs. The precise fibers activated are unclear

[e]The subunits that comprise the P2X receptor are not known, although they are likely to be P2X2 and P2X3

[f]Endogenous activators of TRPA1 are unknown at present. Also, the exact fibers activated by TRPA1 stimulants (cinnamaldehyde, mustard oil, etc.) have not been fully defined, although TRPV1-expressing nociceptors are the likely candidates

[g]Sensitization of airway nociceptors via PAR2 mechanisms is probably secondary to PGE2 release from the epithelium

[h]Functional studies provide some evidence for a role of NKCC1 and/or Cl^{-}channels in regulating cough receptor excitability

[i]Menthol inhibits cough; however, the mechanism via which it does so has not been investigated

Bradykinin

Bradykinin B2 receptors have been localized on neurons in human nodose ganglia, and bradykinin causes sneezing and coughing when applied to appropriate sites in human airways (Choudry et al. 1989; Riccio and Proud 1996). Bradykinin stimulates action potential discharge in airway C-fiber afferent nerves in several species studied in vivo (Coleridge et al. 1989; Hargreaves et al. 1993). In the guinea pig isolated airway preparation, bradykinin is one of the few G-protein-coupled-receptor agonists that consistently leads to action potential discharge when applied directly to the receptive fields of afferent nerves in the airway wall (Fox et al. 1993; Kajekar et al. 1999). This effect of bradykinin is blocked by B2 receptor antagonists, and is selective for C-fibers and Aδ nociceptive fibers (Kajekar et al. 1999).

The question remains as to the ionic mechanism that leads to the depolarizing generator potential following bradykinin B2 receptor activation. At least two mechanisms appear to be involved. As mentioned above, it is known that bradykinin B2 receptor stimulation can lead to the opening of TRPV1. Supporting this idea are the findings that the bradykinin-induced action potential discharge in airway C-fibers is inhibited by TRPV1 antagonist, and reduced in fibres from TRPV1 knockout mice (Carr et al. 2003; Kollarik and Undem 2004). Blocking TRPV1 does not, however, abolish the bradykinin response. The balance of the response may be due to the opening of certain calcium-activated chloride channels. The efflux of chloride down its concentration gradient leads to the membrane depolarization and action potential discharge. This hypothesis is supported by the observations that the bradykinin-induced inward current in nodose and jugular neurons is dependent on chloride channel activity (Oh and Weinreich 2004; Lee et al. 2005), and chloride channel blockers inhibit the action potential discharge in guinea pig tracheal C-fiber terminals (Lee et al. 2005). Another potential ion channel that may be involved in the response to bradykinin is TRPA1. In this case, as mentioned earlier, TRPA1 may be gated secondary to an elevation in intracellular calcium.

Adenosine

Adenosine can evoke sensations of dyspnea and cough in humans subjects (Burki et al. 2005; Basoglu et al. 2005). Adenosine increases action potential discharge in rat pulmonary C-fibers (Hong et al. 1998) studied in vivo. This effect is unaffected by adenosine A2 antagonists, but is inhibited by A1 antagonists. In the isolated innervated guinea pig lung preparation, adenosine effectively evokes action potential discharge in C-fiber terminals. The effect of adenosine can be mimicked by adenosine A1 or adenosine A2A receptor selective agonists, and is blocked entirely only with a combination of A2A and A1 receptor antagonists. Single-cell Reverse transcriptase-PCR analysis of guinea pig lung-specific C-fiber neurons revealed that both A1 and A2A receptors are expressed. These data suggest that signaling mechanisms downstream of either A1 or A2A receptor are capable of causing inward currents that activate guinea pig lung C-fiber terminals (Chuaychoo et al. 2006). The precise signaling mechanisms have not yet been studied. The effect of adenosine on guinea pig lung C-fibers is relatively selective for nodose (placodal) C-fibers with the response of jugular C-fibers being either trivial or absent.

Protease-Activated Receptor-1

Among the four known protease-activated receptors (PARs) (PAR1–PAR4), PAR1 and PAR2 appear to the types most often linked to C-fiber activity. In the vagally innervated, isolated mouse lung preparation, thrombin and selective PAR1 activators robustly evoke action potential discharge in bronchopulmonary C-fibers. The PAR1-selective activating peptide also causes a rapid and large inward current in lung-specific C-fiber neurons. In addition, single neuron Reverse transcriptase-PCR analysis of mouse lung-specific C-fiber neurons revealed that the neurons routinely express PAR1 and PAR3 with little evidence of either PAR2 or PAR4 expression (Undem, unpublished observations).

2.3.3 Modulation of Bronchopulmonary C-Fiber Excitability

As mentioned already, most inflammatory mediators fail to directly evoke action potential discharge when applied to C-fibers ex vivo. Within our narrow definition of the term, we may say that these mediators fail to *activate* the nerve. Many of these mediators, however, increase, or in some cases decrease, the excitability of the nerve.

It is often difficult to determine if a given chemical should be classified as a sensory nerve activator or modulator (or both). Confusing the issue are experiments where a nerve modulator is associated with action potential discharge. Bronchopulmonary C-fibers often reveal little background activity when studied in vivo, but this should not be interpreted as the complete absence of a potential stimulus. In the complex environment of the tissue, one can envisage that bronchopulmonary C-fibers are constantly exposed to potentially activating stimuli (e.g., changes in osmolarity, mechanical forces, acid concentration, ATP, adenosine, choline, bradykinin), but the stimuli are subthreshold for action potential discharge. In the presence of a neuromodulating autacoid that increases excitability, one may observe action potential discharge when a complex system is studied where these subliminal stimuli are present (e.g., in vivo). A modulator would not be expected to evoke action potentials in nerves studied in a more controlled or reduced environment. At the level of the cell soma, studied using patch-clamp techniques, a neuromodulator may influence active and passive electrophysiological properties of nerves, but generally speaking will not cause a large inward current (i.e., it is not sufficient to evoke an action potential).

Histamine

Histamine H1 receptor activation depolarizes a subpopulation of vagal sensory neurons in several species. Unlike the ligand-gated ion channels, and bradykinin, the membrane depolarization evoked by histamine is small and typically associated with a decrease in ion conductance (Undem and Weinreich 1993; Jafri et al. 1997). Histamine inhibits the resting or so-called leak potassium current in nodose ganglion neurons, and in some species inhibits voltage-gated calcium currents and the calcium-activated potassium current that subserves afterspike hyperpolarizations (Jafri et al. 1997). These effects on sensory neurons predict that histamine alone will not directly evoke action potential discharge in airway afferent endings, but will increase the excitability of the nerve. Consistent with this prediction, histamine does not evoke action potential discharge in the guinea pig isolated airway preparation (Fox et al. 1993; Riccio et al. 1996), but it has been shown to increase the mechanical and chemical (capsaicin) sensitivity of afferent C-fibers in dog airways (Lee and Morton 1993).

Eicosanoids

Prostaglandins have long been known to increase the excitability of sensory nerves. Electrophysiological studies on vagal sensory ganglion neurons have demonstrated excitatory affects of several prostaglandins, including prostaglandin E2 (PGE2), prostaglandin D2 (PGD2), and prostaglandin I2 (PGI2; prostacyclin) (Fowler et al. 1985; Undem and Weinreich 1993). PGE2 increases the sodium current through voltage-gated tetrodotoxin (TTX) resistant sodium channels (likely NaV1.8) in lung-specific nodose neurons (Kwong and Lee 2005). This effect appears to be dependent on EP3 receptors. In addition to enhancing sodium currents, certain prostaglandins (e.g., PGE2, PGI2, and PGD2) have been found to inhibit calcium-activated potassium currents involved in the afterspike hyperpolarization (Weinreich and Wonderlin 1987; Undem and Weinreich 1993), and can lead to an increase in the hyperpolarization-activated cation current (Ingram and Williams 1996). If these effects occur at the nerve terminals they would likely lead to an increase in the peak frequency of action potential discharge. Consistent with this prediction, in studies on vagal ganglion neurons Ho et al. (2000) found that low concentrations of PGE2 do not cause action potential discharge in airway afferent nerves, but effectively increased action potential discharge in pulmonary C-fiber afferents induced mechanically by lung inflation, or chemically with capsaicin. Others have noted that thromboxane, PGE2, PGI2, and prostaglandin F2a (PGF2a) increased the rate of baseline discharge in airway C-fibers (Coleridge et al. 1976; Roberts et al. 1985; Bergren 2006; Karla et al. 1992). In support of a role for prostaglandins in increasing the excitability of human airway afferent C-fibers are the findings that PGE2 and PGF2a inhalation increased the sensitivity of the cough response in human volunteers (Stone et al. 1992).

There has been little investigation into the potential role of cysteinyl leukotrienes (cys-LTs) on airway vagal afferent nerve activity. Cyst-LTs inhibit the afterspike hyperpolarization of vagal sensory ganglion neurons, and cause membrane depolarization of identified airway C-fiber neurons in sensory ganglia (McAlexander et al. 1998; Undem and Weinreich 1993). This latter effect is due to an inhibition of a resting potassium current, and is blocked by cys-LT1 receptor antagonists. Other lipoxygenase products, as discussed earlier, may interact directly with the TRPV1 channel on nociceptive nerve terminals in the airways. Inhibitors of leukotreine cys-LT1 receptors inhibit the heightened cough response in animal models and human subjects (Nishitsuji et al. 2007; Brozmanova et al. 2005; Kopriva et al. 2004; Dicpinigaitis et al. 2002).

Bradykinin

Bradykinin is a mediator that may be considered as both an activator (as discussed earlier) and a modulator. One mechanism by which bradykinin modulates C-fiber excitability is blockade of calcium-dependent potassium currents responsible for an afterspike hyperpolarization (Weinreich et al. 1995). This effect is mediated by bradykinin B2 receptors, although the latter effect on the afterspike hyperpolarization appears to be secondary to prostacyclin production by the neuron.

Neurokinins

Neurokinin 1 receptor agonists have been found to depolarize (Oh et al. 2000) or hyperpolarize (Jafri and Weinreich 1998) the membrane potential of nodose ganglion neurons depending on the species. Membrane hyperpolarization would be expected to inhibit the excitability of the C-fibers. The hyperpolarization of ferret nodose ganglion neurons is secondary to activation of a calcium-gated potassium current. Neurokinin 2 receptor agonists depolarize guinea pig nodose ganglion neurons secondary to an increase in nonselective cation current (Moore et al. 2000). Interestingly, this effect is "unmasked" by inflammatory mediators. Thus, whereas the membrane potential of none out of 156 guinea pig isolated nodose ganglion neurons depolarized in response to neurokinin 2 receptor activation normally, within a day of allergen challenge of the airways more than 80% of the nodose ganglion neurons responded to neurokinin 2 agonists with membrane depolarization (Moore et al. 2000). Substance P has been associated with an increase in C-fiber activity in guinea pig lungs. This effect was inhibited by indomethacin, suggesting a role for prostaglandins in the response.

Bombesin-Like Peptides

Bombesin and gastrin-releasing peptides have been shown to sensitize C-fiber neurons to subsequent activation by capsaicin and ATP. In addition, these peptides enhance the pulmonary chemoreflex induced by C-fiber stimulants. The relevance of these observations may be heightened in small-cell lung cancers, as the cancer cells are known to secrete various peptides, including bombesin-like peptides (Gu and Lee 2005).

Protease-Activated Receptor-2

PAR2 activating peptides failed to stimulate rat C-fiber neurons, but were effective in enhancing the C-fiber responses to subsequent exposure to capsaicin (Gu and Lee 2006). Inhalation of PAR2-selective activating peptides increases C-fiber mediated reflexes in rats (Gu and Lee 2006) and cough sensitivity in guinea pigs (Gatti et al. 2006). The effect of PAR2 activators on cough is mainly secondary to prostaglandin production (Gatti et al. 2006), likely owing to PAR2-mediated prostaglandin E release from the airway epithelium (Cocks et al. 1999).

3 Decreasing Activity

Relatively few pharmacological inhibitors of airway sensory nerve activity have been described, and even fewer have been investigated carefully to assess their specific modes of action. Whereas some inhibitors exert actions primarily on a single afferent nerve subtype (e.g., bronchopulmonary C-fibers), others, such as the local anesthetics, may act less specifically. With the relatively recent discovery of tracheal touch-sensitive Aδ-fibers, there is a paucity of knowledge about selective pharmacological inhibition of this afferent nerve subtype. Accordingly, the following section highlights only the inhibitors that have been studied in some detail or where the mechanism of action can be predicted from work on other systems. Specific references to the afferent nerve subtypes targeted by these agents are made where appropriate.

3.1 Opioids

Often considered the "gold standard" antitussive therapy, opioids acting at mu-, kappa- or delta-opioid receptors have all been shown to be antitussive, although the mechanisms via which opioids inhibit cough are debatable (Takahama and Shirasaki 2007). Given that effective doses of opioids for cough suppression also produce sedation and can lead to addiction, as well as the reported limited antitussive actions of inhaled or peripherally acting compounds in humans, it seems that the site of action is likely primarily somewhere within the central nervous system. Nevertheless, opioid receptors are expressed on the peripheral terminals of sensory nerves and there is evidence that activation of peripheral opioid receptors can suppress sensory nerve activity. Mu-opioid agonists inhibit calcium currents in cultured rat nodose neurons, albeit this inhibitory effect is more frequently observed in neurons from neonatal animals than in those from adults (Hamra et al. 1999). Morphine and mu-opioid receptor agonists inhibit neurogenic bronchospasm and airway plasma extravasation evoked by antidromic electrical stimulation of the vagus nerves in guinea pigs, an effect that is likely due to prejunctional inhibition of tachykinin release from vagal C-fibers in the airways (Belvisi et al. 1988, 1989; Shankley et al. 1992). Consistent with this, mu-opioid receptor transport in rat vagus nerve is greatly reduced by capsaicin pretreatment (Laduron 1984) and immunohistochemical studies have shown that mu-opioid receptors are expressed in substance P-containing vagal ganglia neurons (Ding et al. 1998).

Despite the effect of morphine and related compounds on neurogenic effects in the airways, there have been relatively few studies that have looked specifically at whether opioids inhibit orthodromic activity in pulmonary vagal sensory nerves. In guinea pigs, cough evoked by inhaled citric acid has been shown to be inhibited by inhaled morphine and selective mu-opioid receptor agonists, perhaps indicating a potential inhibitory effect of opioids on centrally directed activity in airway afferent nerves (Adcock et al. 1988; Callaway et al. 1991; Karlsson et al. 1990). Furthermore, in bullfrogs, morphine and dihydrocodeine have been shown to reduce the number of spontaneous action potential discharges in vagal pulmonary afferents (Kontani and Koshiura 1985). There are also no data confirming the presence of either delta-opioid or kappa-opioid receptors on the peripheral terminals of airway vagal afferent nerves, although vagal afferents innervating the stomach are inhibited by selective kappa-opioid receptor agonists, suggesting vagal afferent terminals can express other opioid receptors (Ozaki et al. 2000).

3.2 Nociceptin

Nociceptin (also known as orphanin FQ) is an endogenous peptide that was originally shown to activate the "orphan" opioid receptor-like receptor (ORL1, now known as NOP1). Both nociceptin and NOP1 receptors are expressed by airway vagal sensory neurons (Fischer et al. 1998) and exogenously administered nociceptin, and related agonists, inhibit tachykinergic-mediated airway neurogenic bronchospasm and plasma extravasation in guinea pigs and rabbits (Corboz et al. 2000; D'Agostino et al. 2005; Fischer et al. 1998). Furthermore, intravenously administered nociceptin inhibits cough in guinea pigs and cats, perhaps in part via an action at peripheral NOP1 receptors in the airways (Bolser et al. 2001; Lee et al. 2006; McLeod et al. 2001, 2004).

Recent studies have revealed insights into how activation of NOP1 receptor activation may inhibit vagal sensory nerve activity. Inhibition of capsaicin- and acid-evoked coughing in conscious guinea pigs is associated with an inhibition of TRPV1-dependent increases in intracellular calcium in dissociated guinea pig vagal ganglia sensory neurons (Jia et al. 2002; Lee et al. 2006). Jia et al. (2002) showed that prior treatment with an inhibitor of inward-rectifier potassium channels, but not with an inhibitor of voltage-gated calcium channels, prevented capsaicin-evoked calcium response in dissociated neurons and prevented nociceptin inhibition of capsaicin-evoked contractions of the guinea pig bronchus. These data would suggest that neuronal hyperpolarization mediated via inward-rectifier potassium channels may underlie how nociceptin regulates cystolic calcium responses (and hence peripheral neurotransmitter release) in vagal afferent nerves. In contrast, however, Lee et al. (2006) suggest that this mechanism may not be involved in how nociceptin regulates all TRPV1-dependent responses in vagal sensory neurons. In patch-clamped airway-projecting jugular ganglia neurons, under voltage-clamp conditions that limit the contribution of inward-rectifier potassium channel activity, Lee et al. (2006) showed that inward-rectifier potassium channels contribute little, if any, to nociceptin inhibition of acid-evoked TRPV1-mediated inward currents. Rather they speculate that a secondary signaling event (perhaps a reduction in cytosolic cyclic AMP formation) likely mediates the effect. They also showed that TRPV1-independent acid responses (presumably ASIC-mediated currents) are not affected by nociceptin. Identifying the exact mechanisms via which nociceptin selectively inhibits TRPV1 channel activity may provide novel approaches for sensory nerve activity suppression.

3.3 Sodium Channel Blockers

Local anesthetics such as lignocaine and mexiletine, which block sensory neuron voltage-gated Na^+ channels, are effective antitussive agents in humans and animals. Blockade of neuronal voltage-gated Na^+ channels reduces action potential formation in sensory neurons evoked by a variety of stimuli (Adcock et al. 2003;

Carr 2006). However the effects of local anesthetic drugs are typically transient and are limited by acute tachyphylaxis with repeated administration. Insights into the specific sites of action of local anesthetics have been made in the hope of discovering novel anesthetic-like molecules which more effectively inhibit sensory nerve activity. For example, Carr (2006) showed that mexilitine can inhibit action potential formation in guinea pig tracheal touch-sensitive Aδ-fibers (cough receptors) at concentrations that do not block action potential conduction along the sensory nerve axon. This observation would suggest that cough receptor nerve terminals express Na^+ channel characteristics that are different from those along the axons. The observation that the novel local anesthetic RSD931 inhibits RAR-evoked cough in guinea pigs and rabbits via a mechanism that is distinct from that of lidocaine (Adcock et al. 2003) also suggests heterogeneity in Na^+ channels on sensory nerves. Understanding the nature of this heterogeneity could lead to the development of more selective Na^+ channel blockers (perhaps modified local anesthetic-based drugs) for inhibiting sensory-neuron-mediated events, including cough.

In addition to playing a role in the basic generation of action potentials, Na^+ channels are also likely involved in altered sensory neuron excitability in disease. For example, the TTX-resistant Na^+ channel is expressed by specific subsets of sensory nerve subtypes, specifically those displaying the characteristics of capsaicin-sensitive nociceptors (Kwong and Lee 2005). In guinea pigs, neurons that originate in the jugular ganglia and project to the trachea (but not those from the nodose ganglia) specifically express a TTX-resistant Na^+ current (Carr, personal communication). As discussed, this population of nerves consists exclusively of capsaicin-sensitive C-fibers and Aδ-fibers, and is distinct from the nodose derived touch-sensitive Aδ-fibers in this species (Canning et al. 2004a). Capsaicin-sensitive neurons are robustly sensitized by inflammatory mediators such as PGE2, a response that is likely mediated in part by an increase in TTX-resistant Na^+ currents (Kwong and Lee 2005). Physiological studies in guinea pigs have suggested that capsaicin-sensitive nociceptors are likely quiescent in the normal airways, but are recruited during airway inflammation (Mazzone and Canning 2002; Mazzone et al. 2005). Given that nociceptor recruitment is likely to contribute to cough hyperreflexia, selective inhibitors of TTX-resistant Na^+ channels may prove useful cough suppressants in hypertussive states.

3.4 NKCC1 and Chloride Channel Blockers

Furosemide and other loop diuretics reduce cough evoked by citric acid, low-chloride solutions and angiotensin-converting enzyme inhibitors in guinea pigs and humans (Franova 2001; Karlsson et al. 1992; Mazzone and McGovern 2006; Stone et al. 1993; Ventresca et al. 1990), and prevent the activation of laryngeal and tracheobronchial sensory nerves by solutions that are deficient in chloride ions (Fox et al. 1995b; Sant'Ambrogio et al. 1993). This reduction in sensory nerve activity is likely due to inhibition of a neuronally expressed chloride transporter in the airways (Mazzone and McGovern 2006). The furosemide-sensitive $Na^+/K^+/2Cl^-$ cotransporter (NKCC1) is expressed by the majority of neurons in the vagal sensory ganglia and by the peripheral terminals of touch-sensitive Aδ-fibers in the guinea pig trachea (Mazzone and McGovern 2006). NKCC1 functions to accumulate intracellular chloride ions above the electrochemical equilibrium. Opening of membrane chloride channels results in a depolarizing chloride current that contributes to the activation of sensory fibers by some stimuli. The identity of the chloride channels involved in carrying this depolarizing current has received some attention. Lee et al. (2005) recently showed that bradykinin-evoked depolarization of bronchopulmonary C-fibers is inhibited with niflumic acid, a selective inhibitor of calcium-activated chloride channels. Consistent with this, niflumic acid reduces citric acid evoked cough in guinea pigs, although other chloride channel subtypes may also be involved (Mazzone and McGovern 2006). These findings encourage additional studies to further assess whether compounds selective for either NKCC1 or neuronally expressed chloride channels are likely to be useful inhibitors of specific sensory neuron subtypes in the airways.

3.5 α3 Na^+/K^+ ATPase Inhibitors

Immunohistochemical analyses have revealed that a population of somatic and visceral myelinated mechanosensitive afferent nerve fibers are characterized by the expression of an isozyme of the sodium pump containing the alpha3 subunit (Brouns et al. 2006; Canning et al. 2004b; Dobretsov et al. 2003; Mazzone and McGovern 2006; Yu et al. 2003). In the airways, α3 Na^+/K^+ ATPase is expressed by SARs and RARs in rabbits and rats (Yu et al. 2003), myelinated afferent fibers associated with neuroepithelial bodies in rats (Brouns et al. 2006) and touch-sensitive Aδ-fibers (cough receptors) in guinea pigs (Canning et al. 2004b; Mazzone and McGovern 2006), and is distinctly absent from capsaicin-sensitive C-fibers (Canning et al. 2004b; Mazzone and McGovern 2006). Although the exact role of α3 Na^+/K^+ ATPase is unclear, pharmacological studies suggest an essential role for this isozyme in regulating the excitability of mechanosensory nerves. The Na^+/K^+ ATPase inhibitor ouabain, at doses that are reportedly selective for inhibiting the α3 subunit, selectively inhibits cough receptor activation and coughing evoked by citric acid, mechanical stimulation or electrical stimulation of the tracheal mucosa in anesthetized guinea pigs, but has little effect on C-fiber-dependent reflexes evoked from the trachea (Canning et al. 2004b). This inhibitory effect of ouabain on the cough receptors contrasts with the findings of other studies showing that ouabain enhances the excitability of baroreceptors, SARs in the lung and mechanoreceptors innervating the renal artery (Chapleau et al. 1993; Kopp et al. 1994; Winner et al. 2005). Preliminary electrophysiological studies suggest that unique kinetic properties of the α3 Na^+/K^+ ATPase isozyme may facilitate high-frequency action potential conduction along the cough receptor nerve terminal (Mazzone and Canning, unpublished results). Indeed, functional studies have confirmed that cough can only be evoked when cough receptors are activated at high firing frequencies (Canning et al. 2004b; Mazzone et al. 2005). Although these data are promising, it is too early at present to speculate whether selective inhibition of $\alpha 3$ Na^{+}/K^{+} ATPase is a viable option for cough suppression.

3.6 Menthol

Menthol, a natural compound extracted from peppermint, eucalyptus and citronella oils, has been used for many years to alleviate itch and pain and cough. However, despite its long therapeutic history, insights into the specific pharmacological actions of menthol have only been described in recent years. Menthol, when applied topically to the skin or inhaled, produces a characteristic cooling sensation owing to an agonist activity at the TRPM8 ion channel (reviewed in Patel et al. 2007; Reid 2005). TRPM8 is a distant relative of the TRPV1 channel (capsaicin receptor), and like TRPV1, TRPM8 is expressed by subsets of sensory nerves that are involved in thermosensation. However, unlike TRPV1, which has a temperature threshold for activation of 43°C and therefore senses moderate heat, TRPM8 is activated by cool temperatures in the 8–28°C range (McKemy et al. 2002). Although TRPM8 is expressed by subsets of sensory neurons in the mouse nodose ganglia (Zhang et al. 2004), very little is known about the specific characteristics of these visceral sensory neurons. Therefore, only inferences can be made at present about how menthol inhibits cough on the basis of data from somatosensory neurons in the DRG.

TRPM8 is generally considered to be expressed in a population of small-diameter DRG neurons that do not express classic nociceptive markers, like TRPV1 (Kobayashi et al. 2005; McKemy et al. 2002, Peier et al. 2002); however, there is controversy in relation to this point. Several studies have provided evidence for TRPM8 expression in, and/or menthol activation of, TRPV1-expressing neurons (Babes et al. 2004; Reid et al. 2002; Xing et al. 2006). This discrepancy warrants some discussion as understanding the sensory neuron populations that express TRPM8 may provide important insights into the mechanisms that underlie how menthol exerts its antipruritic, analgesic, and antitussive effects (see later). Many experiments that have investigated the expression of TRPM8 and TRPV1 in DRG neurons have used neurons that were cultured in the presence of nerve growth factor, which has been shown to induce TRPV1 expression in neurons that do not normally express the capsaicin receptor (Story et al. 2003). In addition, functional studies that show menthol sensitivity in TRPV1-expressing neurons (that have been cultured without nerve growth factor) must take into account that menthol is not selective for TRPM8 and can activate other TRP channels, including TRPV3 (Macpherson et al. 2006). This is of course not conclusive evidence against the coexpression of TRPM8 and TRPV1 in DRG neurons, but rather highlights some of the issues that confound these studies.

Menthol inhibits cough in both animals and humans (Morice et al. 1994; Laude et al. 1994), although how it does so is not readily apparent. Prolonged or repeated exposure to menthol desensitizes nociceptive responses, including those evoked by subsequent exposure to capsaicin (Cliff and Green 1994; Green and McAuliffe 2000), which may provide a potential peripheral antitussive mechanism if TRPM8 is expressed in neurons responsible for cough (i.e., TRPV1-expressing bronchopulmonary C-fibers). Recent studies have shown that, in addition to activating TRPM8, menthol also inhibits TRPA1 channels (Macpherson et al. 2006). Given that TRPA1 is likely coexpressed with TRPV1 (Story et al. 2003), blocking TRPA1 channels expressed by airway sensory neurons could conceivably reduce the activity of capsaicin-sensitive neurons, in particular the increased activity that occurs during airway inflammation. One study in guinea pigs has showed that menthol can directly relax airway smooth muscle (Wright et al. 1997). This bronchodilator effect could have antitussive effects by reducing RAR activity in response to bronchospastic agents. Indeed, inhaled β-agonists have antitussive effects in humans and animals (Bolser et al. 1995; Horiuchi et al. 1995). Alternatively, menthol may inhibit cough via central mechanisms that do not involve the suppression of primary afferent nerve activity. Consistent with this, menthol-induced suppression of capsaicin evoked pain (when both agents are applied to the skin) has been shown to be mediated via specific glutamatergic mechanisms in the spinal cord, rather than via a peripheral mechanism (Proudfoot et al. 2006). In this scenario, input from TRPM8-expressing neurons reduces spinal synaptic transmission of noxious sensory input from TRPV1-expressing neurons. This is of course consistent with the data showing that different primary afferent nerve subtypes express TRPM8 and TRPV1. Finally, the pleasant cooling sensation of inhaled menthol may simply soothe the airways, reducing the urge to cough via psychotropic mechanisms.

4 Concluding Remarks

There is an obvious utility in identifying chemical mediators (and their mode of action) that activate, sensitize or inhibit the airway sensory nerves involved in coughing as understanding these processes will undoubtedly pave the way forward for the design of new therapeutic strategies for the treatment of cough disorders. Although our current knowledge in this area is rapidly progressing, there is still much work to be done. In reality we presently know relatively little about the transduction processes involved in converting cough stimuli into sensory nerve action potentials and therefore, at this stage, suggesting viable targets for cough suppression is difficult. Nevertheless the progress made thus far is promising and ongoing research in this exciting field should lead to new therapeutic discoveries in the future.

References

Adcock JJ, Douglas GJ, Garabette M, Gascoigne M, Beatch G, Walker M, Page CP (2003) RSD931, a novel anti-tussive agent acting on airway sensory nerves. Br J Pharmacol 138(3):407–16

Adcock JJ, Schneider C, Smith TW (1988) Effects of codeine, morphine and a novel opioid pentapeptide BW443C, on cough, nociception and ventilation in the unanaesthetized guinea-pig. Br J Pharmacol 93(1):93–100

Babes A, Zorzon D, Reid G (2004) Two populations of cold-sensitive neurons in rat dorsal root ganglia and their modulation by nerve growth factor. Eur J Neurosci 20(9):2276–82

Baker CVH (2005) The embryology of vagal sensory neurons. In: Undem BJ, Weinreich D (eds) Advances in vagal afferent neurobiology. Taylor and Francis, Florida

Bandell M, Story GM, Hwang SW, Viswanath V, Eid SR, Petrus MJ, Earley TJ, Patapoutian A (2004) Noxious cold ion channel TRPA1 is activated by pungent compounds and bradykinin. Neuron 41(6):849–57

Barnes NC, Piper PJ, Costello JF (1984) Comparative effects of inhaled leukotriene C4, leukotriene D4, and histamine in normal human subjects. Thorax 39(7):500–4

Basoglu OK, Pelleg A, Essilfie-Quaye S, Brindicci C, Barnes PJ, Kharitonov SA (2005) Effects of aerosolized adenosine 5′-triphosphate vs. adenosine 5′-monophosphate on dyspnea and airway caliber in healthy nonsmokers and patients with asthma. Chest 128(4):1905–9

Bautista DM, Jordt SE, Nikai T, Tsuruda PR, Read AJ, Poblete J, Yamoah EN, Basbaum AI, Julius D (2006) TRPA1 mediates the inflammatory actions of environmental irritants and proalgesic agents. Cell 124(6):1269–82

Belvisi MG, Rogers DF, Barnes PJ (1989) Neurogenic plasma extravasation: inhibition by morphine in guinea pig airways in vivo. J Appl Physiol 66(1):268–72

Belvisi MG, Chung KF, Jackson DM, Barnes PJ (1988) Opioid modulation of non-cholinergic neural bronchoconstriction in guinea-pig in vivo. Br J Pharmacol 95(2):413–18

Bergren DR (2006) Prostaglandin involvement in lung C-fiber activation by substance P in guinea pigs. J Appl Physiol 100(6):1918–27

Bergren DR (1997) Sensory receptor activation by mediators of defense reflexes in guinea-pig lungs. Respir Physiol 108(3):195–204

Bergren DR, Sampson SR (1982) Characterization of intrapulmonary, rapidly adapting receptors of guinea pigs. Respir Physiol 47(1):83–95

Bolser DC, McLeod RL, Tulshian DB, Hey JA (2001) Antitussive action of nociceptin in the cat. Eur J Pharmacol 430(1):107–11

Bolser DC, DeGennaro FC, O'Reilly S, Hey JA, Chapman RW (1995) Pharmacological studies of allergic cough in the guinea pig. Eur J Pharmacol 277(2–3):159–64

Bonham AC, Kott KS, Ravi K, Kappagoda CT, Joad JP (1996) Substance P contributes to rapidly adapting receptor responses to pulmonary venous congestion in rabbits. J Physiol 493 (Part 1): 229–38

Brouns I, Pintelon I, De Proost I, Alewaters R, Timmermans JP, Adriaensen D (2006) Neurochemical characterisation of sensory receptors in airway smooth muscle: comparison with pulmonary neuroepithelial bodies. Histochem Cell Biol 125(4):351–67

Brozmanova M, Plevkova J, Bartos V, Plank L, Tatar M (2005) Antileukotriene treatment and allergic rhinitis-related cough in guinea pigs. J Physiol Pharmacol 56 (Suppl 4):21–30

Burki NK, Dale WJ, Lee LY (2005) Intravenous adenosine and dyspnea in humans. J Appl Physiol 98(1):180–5

Callaway JK, King RG, Boura AL (1991) Evidence for peripheral mechanisms mediating the antitussive actions of opioids in the guinea pig. Gen Pharmacol 22(6):1103–8

Canning BJ, Farmer DG, Mori N (2006) Mechanistic studies of acid-evoked coughing in anesthetized guinea pigs. Am J Physiol Regul Integr Comp Physiol 291(2):R454–R463

Canning BJ, Mazzone SB, Meeker SN, Mori N, Reynolds SM, Undem BJ (2004a) Identification of the tracheal and laryngeal afferent neurones mediating cough in anaesthetized guinea-pigs. J Physiol 557(Part 2):543–58

Canning BJ, Kollarik M, Undem BJ, Mazzone SB (2004b) Demonstration of the essential role of the alpha3-expressing isozyme of the Na^+-K^+-ATPase in regulating cough receptor activation in guinea pigs. Am J Respir Crit Care Med 169(7):A799

Canning BJ, Reynolds SM, Mazzone SB (2001) Multiple mechanisms of reflex bronchospasm in guinea pigs. J Appl Physiol 91(6):2642–53

Carr MJ (2006) Influence of mexiletine on action potential discharge and conduction in nodose Adelta afferent neurons innervating guinea pig isolated trachea. Pulm Pharmacol Ther 19(4):258–63

Carr MJ, Kollarik M, Meeker SN, Undem BJ (2003) A role for TRPV1 in bradykinin-induced excitation of vagal airway afferent nerve terminals. J Pharmacol Exp Ther 304(3):1275–9

Caterina MJ, Schumacher MA, Tominaga M, Rosen TA, Levine JD, Julius D (1997) The capsaicin receptor: a heat-activated ion channel in the pain pathway. Nature 389(6653):816–24

Chapleau MW, Lu J, Hajduczok G, Abboud FM (1993) Mechanism of baroreceptor adaptation in dogs: attenuation of adaptation by the K^+ channel blocker 4-aminopyridine. J Physiol 462: 291–306

Chizh BA, Illes P (2001) P2X receptors and nociception. Pharmacol Rev 53(4):553–68

Choudry NB, Fuller RW, Pride NB (1989) Sensitivity of the human cough reflex: effect of inflammatory mediators prostaglandin E2, bradykinin, and histamine. Am Rev Respir Dis 140(1):137–41

Chuang HH, Prescott ED, Kong H, Shields S, Jordt SE, Basbaum AI, Chao MV, Julius D (2001) Bradykinin and nerve growth factor release the capsaicin receptor from PtdIns(4,5)P2-mediated inhibition. Nature 411(6840):957–62

Chuaychoo B, Lee MG, Kollarik M, Pullmann R Jr, Undem BJ (2006) Evidence for both adenosine A1 and A2A receptors activating single vagal sensory C-fibres in guinea pig lungs. J Physiol 575(Part 2):481–90

Chuaychoo B, Lee MG, Kollarik M, Undem BJ (2005) Effect of 5-hydroxytryptamine on vagal C-fiber subtypes in guinea pig lungs. Pulm Pharmacol Ther 18(4):269–76

Cliff MA, Green BG (1994) Sensory irritation and coolness produced by menthol: evidence for selective desensitization of irritation. Physiol Behav 56(5):1021–9

Cockayne DA, Dunn PM, Zhong Y, Rong W, Hamilton SG, Knight GE, Ruan HZ, Ma B, Yip P, Nunn P, McMahon SB, Burnstock G, Ford AP (2005) P2X2 knockout mice and P2X2/P2X3 double knockout mice reveal a role for the P2X2 receptor subunit in mediating multiple sensory effects of ATP. J Physiol 567Part 2):621–39

Cocks TM, Fong B, Chow JM, Anderson GP, Frauman AG, Goldie RG, Henry PJ, Carr MJ, Hamilton JR, Moffatt JD (1999)A protective role for protease-activated receptors in the airways. Nature 398(6723):156–60

Coleridge HM, Coleridge JC, Schultz HD (1989) Afferent pathways involved in reflex regulation of airway smooth muscle. Pharmacol Ther 42(1):1–63

Coleridge HM, Coleridge JC, Ginzel KH, Baker DG, Banzett RB, Morrison MA (1976) Stimulation of 'irritant' receptors and afferent C-fibres in the lungs by prostaglandins. Nature 264(5585):451–3

Coleridge JC, Coleridge HM (1984) Afferent vagal C fibre innervation of the lungs and airways and its functional significance. Rev Physiol Biochem Pharmacol 99:1–110

Corboz MR, Rivelli MA, Egan RW, Tulshian D, Matasi J, Fawzi AB, Benbow L, Smith-Torhan A, Zhang H, Hey JA (2000) Nociceptin inhibits capsaicin-induced bronchoconstriction in isolated guinea pig lung. Eur J Pharmacol 402(1–2):171–9

D'Agostino B, Marrocco G, De Nardo M, Calo G, Guerrini R, Gallelli L, Advenier C, Rossi F (2005) Activation of the nociceptin/orphanin FQ receptor reduces bronchoconstriction and microvascular leakage in a rabbit model of gastroesophageal reflux. Br J Pharmacol 144(6): 813–20

Dicpinigaitis PV, Dobkin JB, Reichel J (2002) Antitussive effect of the leukotriene receptor antagonist zafirlukast in subjects with cough-variant asthma. J Asthma 39(4):291–7

Ding YQ, Li JL, Lu BZ, Wang D, Zhang ML, Li JS (1998) Co-localization of mu-opioid receptor-like immunoreactivity with substance P-LI, calcitonin gene-related peptide-LI and nitric oxide synthase-LI in vagal and glossopharyngeal afferent neurons of the rat. Brain Res 792(1):149–53

Dobretsov M, Hastings SL, Sims TJ, Stimers JR, Romanovsky D (2003) Stretch receptor-associated expression of alpha 3 isoform of the Na^+, K^+-ATPase in rat peripheral nervous system. Neuroscience 116(4):1069–80

Doerner JF, Gisselmann G, Hatt H, Wetzel CH (2007) Transient receptor potential channel A1 is directly gated by calcium ions. J Biol Chem 282(18):13180–9

Fischer A, Forssmann WG, Undem BJ (1998) Nociceptin-induced inhibition of tachykinergic neurotransmission in guinea pig bronchus. J Pharmacol Exp Ther 285(2):902–7

Fowler JC, Wonderlin WF, Weinreich D (1985) Prostaglandins block a $Ca2^{+}$-dependent slow spike after hyperpolarization independent of effects on $Ca2^{+}$ influx in visceral afferent neurons. Brain Res 345(2):345–9

Fox AJ, Urban L, Barnes PJ, Dray A (1995a) Effects of capsazepine against capsaicin- and proton-evoked excitation of single airway C-fibres and vagus nerve from the guinea-pig. Neuroscience 67(3):741–52

Fox AJ, Barnes PJ, Dray A (1995b) Stimulation of guinea-pig tracheal afferent fibres by non-isosmotic and low-chloride stimuli and the effect of frusemide. J Physiol 482(Part 1):179–87

Fox AJ, Barnes PJ, Urban L, Dray A (1993) An in vitro study of the properties of single vagal afferents innervating guinea-pig airways. J Physiol 469:21–35

Franova S (2001) The influence of inhaled furosemide on adverse effects of ACE-inhibitors in airways. Bratisl Lek Listy 102(7):309–13

Gatti R, Andre E, Amadesi S, Dinh TQ, Fischer A, Bunnett NW, Harrison S, Geppetti P, Trevisani M (2006) Protease-activated receptor-2 activation exaggerates TRPV1-mediated cough in guinea pigs. J Appl Physiol 101(2):506–11

Green BG, McAuliffe BL (2000) Menthol desensitization of capsaicin irritation. Evidence of a short-term anti-nociceptive effect. Physiol Behav 68(5):631–9

Gu Q, Lee LY (2006) Hypersensitivity of pulmonary chemosensitive neurons induced by activation of protease-activated receptor-2 in rats. J Physiol 574(Part 3):867–76

Gu Q, Lee LY (2005) Sensitization of pulmonary chemosensitive neurons by bombesin-like peptides in rats. Am J Physiol Lung Cell Mol Physiol 289(6):L1104–L1112

Hamra M, McNeil RS, Runciman M, Kunze DL (1999) Opioid modulation of calcium current in cultured sensory neurons: mu-modulation of baroreceptor input. Am J Physiol 277(2, Part 2):H705–H713

Hargreaves KM, Roszkowski MT, Swift JQ (1993) Bradykinin and inflammatory pain. Agents Actions Suppl 41:65–73

Ho CY, Gu Q, Lin YS, Lee LY (2001) Sensitivity of vagal afferent endings to chemical irritants in the rat lung. Respir Physiol 127(2–3):113–24

Ho CY, Gu Q, Hong JL, Lee LY (2000) Prostaglandin E(2) enhances chemical and mechanical sensitivities of pulmonary C fibers in the rat. Am J Respir Crit Care Med 162(2, Part 1):528–33

Hong JL, Ho CY, Kwong K, Lee LY (1998) Activation of pulmonary C fibres by adenosine in anaesthetized rats: role of adenosine A1 receptors. J Physiol 508(Part 1):109–18

Hong JL, Kwong K, Lee LY (1997) Stimulation of pulmonary C fibres by lactic acid in rats: contributions of H^{+} and lactate ions. J Physiol 500(Part 2):319–29

Horiuchi H, Nonaka T, Ito M, Kaniwa T, Kishimoto T, Fuchikami J, Kohjimoto Y, Kiyoki M (1995) Suppressive effect of clenbuterol on citric acid-induced cough reflex in guinea pigs. Res Commun Mol Pathol Pharmacol 88(3):293–301

Hoyer D, Clarke DE, Fozard JR, Hartig PR, Martin GR, Mylecharane EJ, Saxena PR, Humphrey PP (1994) International Union of Pharmacology classification of receptors for 5-hydroxytryptamine (serotonin). Pharmacol Rev 46(2):157–203

Hwang SW, Cho H, Kwak J, Lee SY, Kang CJ, Jung J, Cho S, Min KH, Suh YG, Kim D, Oh U (2000) Direct activation of capsaicin receptors by products of lipoxygenases: endogenous capsaicin-like substances. Proc Natl Acad Sci USA 97(11):6155–60

Ingram SL, Williams JT (1996) Modulation of the hyperpolarization-activated current (Ih) by cyclic nucleotides in guinea-pig primary afferent neurons. J Physiol 492(Part 1):97–106

Jafri MS, Weinreich D (1998) Substance P regulates Ih via a NK-1 receptor in vagal sensory neurons of the ferret. J Neurophysiol 79(2):769–77

Jafri MS, Moore KA, Taylor GE, Weinreich D (1997) Histamine H1 receptor activation blocks two classes of potassium current, IK(rest) and IAHP, to excite ferret vagal afferents. J Physiol 503(Part 3):533–46

Jia Y, Wang X, Aponte SI, Rivelli MA, Yang R, Rizzo CA, Corboz MR, Priestley T, Hey JA (2002) Nociceptin/orphanin FQ inhibits capsaicin-induced guinea-pig airway contraction through an inward-rectifier potassium channel. Br J Pharmacol 135(3):764–70

Jonzon A, Pisarri TE, Coleridge JC, Coleridge HM (1986) Rapidly adapting receptor activity in dogs is inversely related to lung compliance. J Appl Physiol 61(5):1980–7

Joos G, Pauwels R, van der Straeten M (1987) Effect of inhaled substance P and neurokinin A on the airways of normal and asthmatic subjects. Thorax 42(10):779–83

Jordt SE, Bautista DM, Chuang HH, McKemy DD, Zygmunt PM, Hogestatt ED, Meng ID, Julius D (2004) Mustard oils and cannabinoids excite sensory nerve fibres through the TRP channel ANKTM1. Nature 427(6971):260–5

Kajekar R, Proud D, Myers AC, Meeker SN, Undem BJ (1999) Characterization of vagal afferent subtypes stimulated by bradykinin in guinea pig trachea. J Pharmacol Exp Ther 289(2):682–7

Kamei J, Takahashi Y (2006) Involvement of ionotropic purinergic receptors in the histamine-induced enhancement of the cough reflex sensitivity in guinea pigs. Eur J Pharmacol 547(1–3): 160–4

Kamei J, Takahashi Y, Yoshikawa Y, Saitoh A (2005) Involvement of P2X receptor subtypes in ATP-induced enhancement of the cough reflex sensitivity. Eur J Pharmacol 528(1–3):158–61

Karla W, Shams H, Orr JA, Scheid P (1992) Effects of the thromboxane A2 mimetic, U46,619, on pulmonary vagal afferents in the cat. Respir Physiol 87(3):383–96

Karlsson JA, Choudry NB, Zackrisson C, Fuller RW (1992) A comparison of the effect of inhaled diuretics on airway reflexes in humans and guinea pigs. J Appl Physiol 72(2):434–8

Karlsson JA, Lanner AS, Persson CG (1990) Airway opioid receptors mediate inhibition of cough and reflex bronchoconstriction in guinea pigs. J Pharmacol Exp Ther 252(2):863–8

Kay IS, Armstrong DJ (1991) MDL 72222 (a selective 5-HT3 receptor antagonist) prevents stimulation of intrapulmonary C fibres by pulmonary embolization in anaesthetized rabbits. Exp Physiol 76(2):213–18

Klapproth H, Reinheimer T, Metzen J, Munch M, Bittinger F, Kirkpatrick CJ, Hohle KD, Schemann M, Racke K, Wessler I (1997) Non-neuronal acetylcholine, a signalling molecule synthezised by surface cells of rat and man. Naunyn Schmiedebergs Arch Pharmacol 355(4): 515–23

Kobayashi K, Fukuoka T, Obata K, Yamanaka H, Dai Y, Tokunaga A, Noguchi K (2005) Distinct expression of TRPM8, TRPA1, and TRPV1 mRNAs in rat primary afferent neurons with adelta/c-fibers and colocalization with trk receptors. J Comp Neurol 493(4):596–606

Kollarik M, Undem BJ (2006) Sensory transduction in cough-associated nerves. Respir Physiol Neurobiol 152(3):243–54

Kollarik M, Undem BJ (2003) Activation of bronchopulmonary vagal afferent nerves with bradykinin, acid and vanilloid receptor agonists in wild-type and TRPV1t/t mice. J Physiol 555(Part 1):115–23

Kollarik M, Undem BJ (2002) Mechanisms of acid-induced activation of airway afferent nerve fibres in guinea-pig. J Physiol 543(Part 2):591–600

Kontani H, Koshiura R (1985) The effects of morphine and morphine-related compounds on the rate of afferent discharges from the pulmonary receptors in bullfrog. Jpn J Pharmacol 38(3):295–303

Kopp UC, Smith LA, Pence AL (1994) Na(+)-K(+)-ATPase inhibition sensitizes renal mechanoreceptors activated by increases in renal pelvic pressure. Am J Physiol 267(4, Part 2):R1109–R1117

Kopriva F, Sobolova L, Szotkowska J, Zapalka M (2004) Treatment of chronic cough in children with montelukast, a leukotriene receptor antagonist. J Asthma 41(7):715–20

Kwong K, Lee LY (2005) Prostaglandin E2 potentiates a TTX-resistant sodium current in rat capsaicin-sensitive vagal pulmonary sensory neurones. J Physiol 564(Part 2):437–50

Laduron PM (1984) Axonal transport of opiate receptors in capsaicin-sensitive neurones. Brain Res 294(1):157–60

Lankiewicz S, Lobitz N, Wetzel CH, Rupprecht R, Gisselmann G, Hatt H (1998) Molecular cloning, functional expression, and pharmacological characterization of 5-hydroxytryptamine3 receptor cDNA and its splice variants from guinea pig. Mol Pharmacol 53(2):202–12
Laude EA, Morice AH, Grattan TJ (1994) The antitussive effects of menthol, camphor and cineole in conscious guinea-pigs. Pulm Pharmacol 7(3):179–84
Lee LY, Burki NK, Gerhardstein DC, Gu Q, Kou YR, Xu J (2007) Airway irritation and cough evoked by inhaled cigarette smoke: role of neuronal nicotinic acetylcholine receptors. Pulm Pharmacol Ther 20(4):355–64
Lee LY, Gerhardstein DC, Wang AL, Burki NK (1993) Nicotine is responsible for airway irritation evoked by cigarette smoke inhalation in men. J Appl Physiol 75(5):1955–61
Lee LY, Morton RF (1993) Histamine enhances vagal pulmonary C-fiber responses to capsaicin and lung inflation. Respir Physiol 93(1):83–96
Lee MG, Undem BJ, Brown C, Carr MJ (2006) Effect of nociceptin in acid-evoked cough and airway sensory nerve activation in guinea pigs. Am J Respir Crit Care Med 173(3):271–5
Lee MG, Macglashan DW Jr, Undem BJ (2005) Role of chloride channels in bradykinin-induced guinea pig airway vagal C-fibre activation. J Physiol 566(Part 1):205–12
Lukas RJ, Changeux JP, Le Novere N, Albuquerque EX, Balfour DJ, Berg DK, Bertrand D, Chiappinelli VA, Clarke PB, Collins AC, Dani JA, Grady SR, Kellar KJ, Lindstrom JM, Marks MJ, Quik M, Taylor PW, Wonnacott S (1999) International Union of Pharmacology. XX. Current status of the nomenclature for nicotinic acetylcholine receptors and their subunits. Pharmacol Rev 51(2):397–401
Macpherson LJ, Hwang SW, Miyamoto T, Dubin AE, Patapoutian A, Story GM (2006) More than cool: promiscuous relationships of menthol and other sensory compounds. Mol Cell Neurosci 32(4):335–43
Mamet J, Baron A, Lazdunski M, Voilley N (2002) Proinflammatory mediators, stimulators of sensory neuron excitability via the expression of acid-sensing ion channels. J Neurosci 22(24):10662–70
Mao D, Yasuda RP, Fan H, Wolfe BB, Kellar KJ (2006) Heterogeneity of nicotinic cholinergic receptors in rat superior cervical and nodose ganglia. Mol Pharmacol 70(5):1693–9
Mazzone SB, McGovern AE (2006) Na^{+}-K^{+}-2Cl- cotransporters and Cl- channels regulate citric acid cough in guinea pigs. J Appl Physiol 101(2):635–43
Mazzone SB, Mori N, Canning BJ (2005) Synergistic interactions between airway afferent nerve subtypes regulating the cough reflex in guinea-pigs. J Physiol 569(Part 2):559–73
Mazzone SB (2005) An overview of the sensory receptors regulating cough. Cough 1:2
Mazzone SB (2004a) Sensory regulation of the cough reflex. Pulm Pharmacol Ther 17(6):361–8
Mazzone SB (2004b) Targeting tachykinins for the treatment of obstructive airways disease. Treat Respir Med 3(4):201–16
Mazzone SB, Canning BJ (2002) Synergistic interactions between airway afferent nerve subtypes mediating reflex bronchospasm in guinea pigs. Am J Physiol Regul Integr Comp Physiol 283(1):R86–R98
McAlexander MA, Myers AC, Undem BJ (1999) Adaptation of guinea-pig vagal airway afferent neurones to mechanical stimulation. J Physiol 521(Part 1):239–47
McAlexander MA, Myers AC, Undem BJ (1998) Inhibition of 5-lipoxygenase diminishes neurally evoked tachykinergic contraction of guinea pig isolated airway. J Pharmacol Exp Ther 285(2):602–7
McKemy DD, Neuhausser WM, Julius D (2002) Identification of a cold receptor reveals a general role for TRP channels in thermosensation. Nature 416(6876):52–8
McLeod RL, Jia Y, Fernandez X, Parra LE, Wang X, Tulshian DB, Kiselgof EJ, Tan Z, Fawzi AB, Smith-Torhan A, Zhang H, Hey JA (2004) Antitussive profile of the NOP agonist Ro-64-6198 in the guinea pig. Pharmacology 71(3):143–9
McLeod RL, Parra LE, Mutter JC, Erickson CH, Carey GJ, Tulshian DB, Fawzi AB, Smith-Torhan A, Egan RW, Cuss FM, Hey JA (2001) Nociceptin inhibits cough in the guinea-pig by activation of ORL(1) receptors. Br J Pharmacol 132(6):1175–8

Mohammed SP, Higenbottam TW, Adcock JJ (1993) Effects of aerosol-applied capsaicin, histamine and prostaglandin E2 on airway sensory receptors of anaesthetized cats. J Physiol 469: 51–66

Moore KA, Undem BJ, Weinreich D (2000) Antigen inhalation unmasks NK-2 tachykinin receptor-mediated responses in vagal afferents. Am J Respir Crit Care Med 161(1):232–6

Morice AH, Marshall AE, Higgins KS, Grattan TJ (1994) Effect of inhaled menthol on citric acid induced cough in normal subjects. Thorax 49(10):1024–6

Myers AC, Kajekar R, Undem BJ (2002) Allergic inflammation-induced neuropeptide production in rapidly adapting afferent nerves in guinea pig airways. Am J Physiol Lung Cell Mol Physiol 282(4):L775–L781

Nilius B, Owsianik G, Voets T, Peters JA (2007) Transient receptor potential cation channels in disease. Physiol Rev 87(1):165–217

Nishitsuji M, Fujimura M, Oribe Y, Nakao S (2007) Effect of montelukast in a guinea pig model of cough variant asthma. Pulm Pharmacol Ther 2008;21(1):142-5.

Oh EJ, Weinreich D (2004) Bradykinin decreases K(+) and increases Cl(−) conductances in vagal afferent neurones of the guinea pig. J Physiol 558(Part 2):513–26

Oh EJ, Thompson LP, Weinreich D (2000) Sexually dimorphic regulation of NK-1 receptor-mediated electrophysiological responses in vagal primary afferent neurons. J Neurophysiol 84(1):51–6

Ozaki N, Sengupta JN, Gebhart GF (2000) Differential effects of mu-, delta-, and kappa-opioid receptor agonists on mechanosensitive gastric vagal afferent fibers in the rat. J Neurophysiol 83(4):2209–16

Pack AI, DeLaney RG (1983) Response of pulmonary rapidly adapting receptors during lung inflation. J Appl Physiol 55(3):955–63

Patel T, Ishiuji Y, Yosipovitch G (2007) Menthol: a refreshing look at this ancient compound. J Am Acad Dermatol 57(5):873–8

Peier AM, Moqrich A, Hergarden AC, Reeve AJ, Andersson DA, Story GM, Earley TJ, Dragoni I, McIntyre P, Bevan S, Patapoutian A (2002) A TRP channel that senses cold stimuli and menthol. Cell 108(5):705–15

Pelleg A, Hurt CM (1996) Mechanism of action of ATP on canine pulmonary vagal C fibre nerve terminals. J Physiol 490(Part 1):265–75

Proudfoot CJ, Garry EM, Cottrell DF, Rosie R, Anderson H, Robertson DC, Fleetwood-Walker SM, Mitchell R (2006) Analgesia mediated by the TRPM8 cold receptor in chronic neuropathic pain. Curr Biol 16(16):1591–605

Reid G (2005) ThermoTRP channels and cold sensing: what are they really up to? Pflugers Arch 451(1):250–63

Reid G, Babes A, Pluteanu F (2002) A cold- and menthol-activated current in rat dorsal root ganglion neurones: properties and role in cold transduction. J Physiol 545(Part 2):595–614

Riccio MM, Proud D (1996) Evidence that enhanced nasal reactivity to bradykinin in patients with symptomatic allergy is mediated by neural reflexes. J Allergy Clin Immunol 97(6):1252–63

Riccio MM, Kummer W, Biglari B, Myers AC, Undem BJ (1996) Interganglionic segregation of distinct vagal afferent fibre phenotypes in guinea-pig airways. J Physiol 496(Part 2):521–30

Roberts AM, Schultz HD, Green JF, Armstrong DJ, Kaufman MP, Coleridge HM, Coleridge JC (1985) Reflex tracheal contraction evoked in dogs by bronchodilator prostaglandins E2 and I2. J Appl Physiol 58(6):1823–31

Sant'Ambrogio FB, Sant'Ambrogio G, Anderson JW (1993) Effect of furosemide on the response of laryngeal receptors to low-chloride solutions. Eur Respir J 6(8):1151–5

Shankley CE, Anderson CW, Adcock JJ (1992) Effect of BW443C81, a novel opioid, on non-cholinergic bronchoconstrictor responses and neurogenic plasma extravasation in the guinea pig. Agents Actions 36(1–2):22–8

Shin J, Cho H, Hwang SW, Jung J, Shin CY, Lee SY, Kim SH, Lee MG, Choi YH, Kim J, Haber NA, Reichling DB, Khasar S, Levine JD, Oh U (2002) Bradykinin-12-lipoxygenase-VR1 signaling pathway for inflammatory hyperalgesia. Proc Natl Acad Sci USA 99(15):10150–5

Spies M, Lips KS, Kurzen H, Kummer W, Haberberger RV (2006) Nicotinic acetylcholine receptors containing subunits alpha3 and alpha5 in rat nociceptive dorsal root ganglion neurons. J Mol Neurosci 30(1–2):55–6

Stone R, Barnes PJ, Fuller RW (1992) Contrasting effects of prostaglandins E2 and F2 alpha on sensitivity of the human cough reflex. J Appl Physiol 73(2):649–53

Stone RA, Barnes PJ, Chung KF (1993) Effect of frusemide on cough responses to chloride-deficient solution in normal and mild asthmatic subjects. Eur Respir J 6(6):862–7

Story GM, Peier AM, Reeve AJ, Eid SR, Mosbacher J, Hricik TR, Earley TJ, Hergarden AC, Andersson DA, Hwang SW, McIntyre P, Jegla T, Bevan S, Patapoutian A (2003) ANKTM1, a TRP-like channel expressed in nociceptive neurons, is activated by cold temperatures. Cell 112(6):819–29

Takahama K, Shirasaki T (2007) Central and peripheral mechanisms of narcotic antitussives: codeine-sensitive and -resistant coughs. Cough 3:8

Tominaga M, Caterina MJ, Malmberg AB, Rosen TA, Gilbert H, Skinner K, Raumann BE, Basbaum AI, Julius D (1998) The cloned capsaicin receptor integrates multiple pain-producing stimuli. Neuron 21(3):531–43

Undem BJ, Chuaychoo B, Lee MG, Weinreich D, Myers AC, Kollarik M (2004) Subtypes of vagal afferent C-fibres in guinea-pig lungs. J Physiol 556(Part 3):905–17

Undem BJ, Weinreich D (1993) Electrophysiological properties and chemosensitivity of guinea pig nodose ganglion neurons in vitro. J Auton Nerv Syst 44(1):17–33

Ventresca PG, Nichol GM, Barnes PJ, Chung KF (1990) Inhaled furosemide inhibits cough induced by low chloride content solutions but not by capsaicin. Am Rev Respir Dis 142(1):143–6

Weinreich D, Koschorke GM, Undem BJ, Taylor GE (1995) Prevention of the excitatory actions of bradykinin by inhibition of PGI2 formation in nodose neurones of the guinea-pig. J Physiol 483(Part 3):735–46

Weinreich D, Wonderlin WF (1987) Inhibition of calcium-dependent spike after-hyperpolarization increases excitability of rabbit visceral sensory neurones. J Physiol 394:415–27

Winner E, Zhang JW, Proctor M, Yu J (2005) Ouabain stimulates slowly adapting pulmonary stretch receptors. Sheng Li Xue Bao 57(6):689–95

Wright CE, Laude EA, Grattan TJ, Morice AH (1997) Capsaicin and neurokinin A-induced bronchoconstriction in the anaesthetized guinea-pig: evidence for a direct action of menthol on isolated bronchial smooth muscle. Br J Pharmacol 121(8):1645–50

Xing H, Ling J, Chen M, Gu JG (2006) Chemical and cold sensitivity of two distinct populations of TRPM8-expressing somatosensory neurons. J Neurophysiol 95(2):1221–30

Xu J, Yang W, Zhang G, Gu Q, Lee LY (2007) Calcium transient evoked by nicotine in isolated rat vagal pulmonary sensory neurons. Am J Physiol Lung Cell Mol Physiol 292(1):L54–L61

Yu J, Wang YF, Zhang JW (2003) Structure of slowly adapting pulmonary stretch receptors in the lung periphery. J Appl Physiol 95(1):385–93

Zhang L, Jones S, Brody K, Costa M, Brookes SJ (2004) Thermosensitive transient receptor potential channels in vagal afferent neurons of the mouse. Am J Physiol Gastrointest Liver Physiol 286(6):G983–G991

Peripheral Mechanisms I: Plasticity of Peripheral Pathways

M.A. McAlexander and M.J. Carr(✉)

Contents

Abstract Cough plays a vital role in protecting the lower airways from inhaled irritants, pollutants, and infectious agents. The cough reflex exhibits remarkable plasticity, such that in the context of infectious or inflammatory respiratory diseases such as asthma, chronic bronchitis, and idiopathic pulmonary fibrosis the cough reflex can become dysregulated, leading to a chronic cough. A chronic, nonproductive (dry) cough can rob sufferers of quality of life. Plasticity of the cough reflex likely involves multiple intersecting pathways within the airways, the peripheral nerves that supply them, and the central nervous system. While further studies are needed to determine the presence and relevance of many of these specific pathways in cough associated with chronic respiratory disease, the last decade has yielded unprecedented insight into the molecular identity of the ion channels and associated

M.J. Carr
GlaxoSmithKline, 709 Swedeland Rd, King of Prussia, PA 19406, USA
michael.j.carr@gsk.com

K.F. Chung, J.G. Widdicombe (eds.), *Pharmacology and Therapeutics of Cough*,
Handbook of Experimental Pharmacology 187,

proteins that initiate and conduct action potentials in the primary sensory nerves involved in reflexes such as cough. We now know, for instance, that members of the transient receptor potential superfamily of nonselective cation channels function as transducers that convert specific external stimuli into neuronal activation. We also know that certain Na^+ and K^+ channels play specialized roles in regulating action potential discharge in irritant-sensing afferent nerves. In this chapter, we summarize the available information regarding factors that may modulate afferent neuron function acutely, via posttranslational modifications and over the longer term through neurotrophin-dependent alterations of the transcriptional programs of adult sensory neurons.

1 Introduction

In healthy individuals, the cough reflex is vital for preserving airway homeostasis, as the forced expulsions of air that accompany it can clear the upper airways of infectious organisms, particulate matter, and other irritants before they deposit into the distal lung. The cough reflex exhibits remarkable plasticity, such that in the context of respiratory diseases such as asthma, chronic bronchitis, and idiopathic pulmonary fibrosis the cough reflex can become dysregulated, leading to a chronic cough. A chronic, nonproductive (dry) cough not only serves no appreciable protective function, but can rob sufferers of quality of life. Hypothetically, a dysregulated cough reflex may involve inflammation and tissue remodeling in the airways, aberrant central nervous system (CNS) activity and plasticity of the sensitivity/activity of primary afferent neurons that encounter tussive stimuli within the airways. This chapter focuses on recent advances in our understanding of plasticity of sensory neurons and the potential relevance of this to disease-associated cough.

A schematic overview highlighting sensory nerve activation and its role in both acute and chronic sensitization of reflexes such as cough is presented in Fig. 1a. Activation of primary sensory neurons associated with cough occurs following an encounter of the peripheral nerve terminal with tussigenic stimuli. In many cases this is thought to involve direct or indirect gating of transducer ion channels in response to specific chemical or physical stimuli. This leads to sodium and/or calcium influx into the nerve terminal, which depolarizes the membrane potential past a threshold that causes voltage-gated sodium (Na_V) channel activation and the consequent formation of action potentials. Action potentials are conducted along axons within the vagal nerves to synapses within the CNS where they evoke the release of a variety of neurotransmitters, leading to the activation of neuronal pathways responsible for generating an urge to cough.

Figure 1b illustrates several known mechanisms by which the terminals of sensory neurons can become acutely sensitized. Such acute changes in excitability can be driven by G-protein-coupled-receptor-dependent elevation of second messengers such as cyclic AMP (cAMP) and intracellular Ca^{2+}, which increase the activity of kinases such as protein kinase A and protein kinase C. Such kinases, in turn, phosphorylate specific serine and/or threonine residues of ion channels involved in transduction and action potential formation to alter their trafficking and activity. The net result of these changes is a lower threshold for activation of the nerve terminal and/or a greater rate of discharge. These forms of plasticity would be expected to decrease the threshold for the urge to cough. Recent advances in our understanding of the mechanisms underpinning plasticity of sensory neuron transducer mechanisms and the role of voltage-gated ion channels in setting sensory neuron excitability are described in Sects. 2 and 3.

A longer-term process could also drive enhanced cough sensitivity in disease. Over time, tissue remodeling and inflammation cause the release of factors such as neurotrophins that alter adult neurons both acutely, downstream of effectors such as

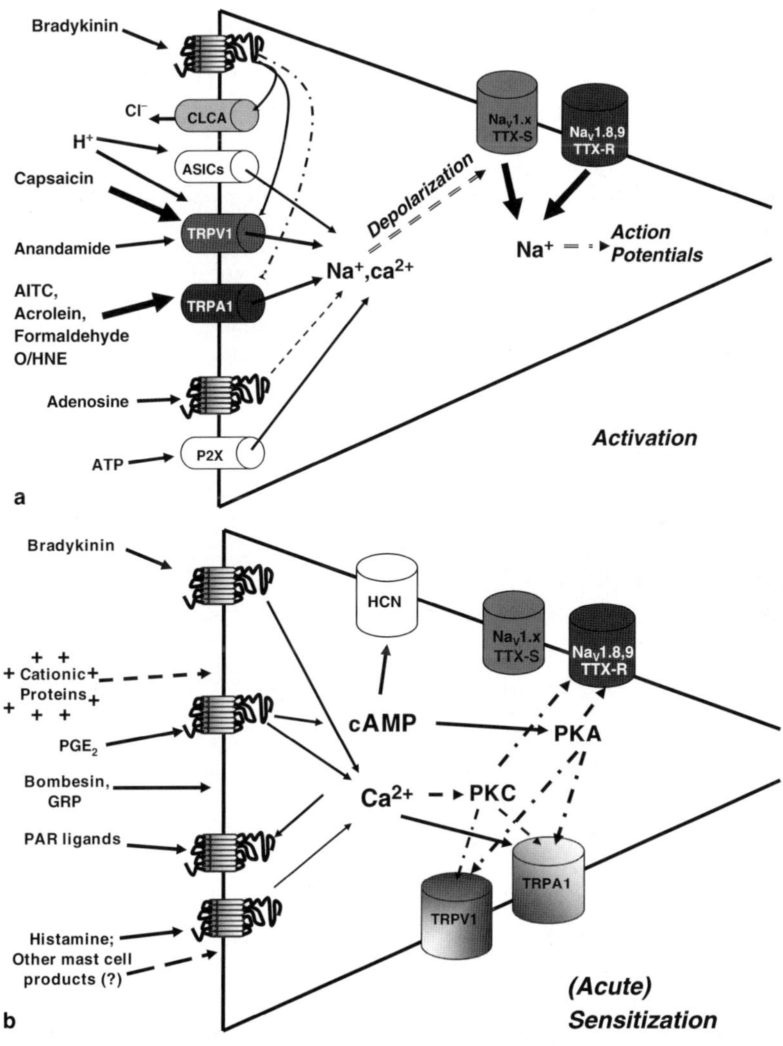

Fig. 1 a, b (Caption see next page)

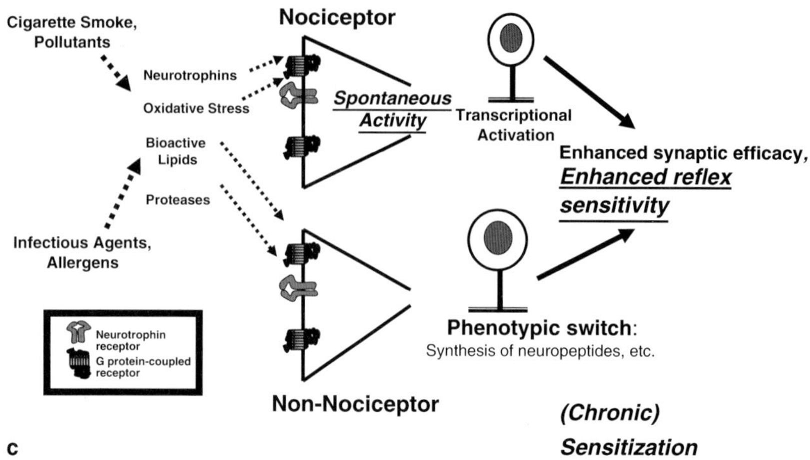

Fig. 1 (**a**) Mediators such as bradykinin and adenosine elicit activation of respiratory nociceptors downstream of their actions on their cognate G-protein-coupled receptors (GPCRs), while protons and ATP can activate respiratory nociceptors via direct activation of ligand-gated ion channels. In addition, numerous irritants and mediators of tissue damage converge upon TRPV1 and/or TRPA1. Bradykinin appears unique, as it first acts on its GPCR to elicit signals that activate TRPA1, TRPV1, and calcium-activated chloride channels. When any of these mechanisms are engaged with an intensity sufficient to depolarize the nerve terminal beyond its threshold, voltage-gated sodium channels are opened, triggering action potentials. (**b**) Known pathways of acute sensory nerve activation occur primarily through GPCR-dependent elevations in the intracellular second messengers Ca^{2+} and cyclic AMP (cAMP), which lead to increases in the activity of protein kinase C (PKC) and protein kinase A, respectively. These kinases phosphorylate their substrates, causing changes in trafficking and/or gating of key ion channels, such as tetrodotoxin-resistant Na^{+} channels and transient receptor potentials, which render the terminal more excitable. (**c**) The continued presence of inflammation and tissue damage can have longer-term effects on airway nerve terminals. Continued exposure to this milieu causes peripheral sensitization, which can lead to spontaneous activity. This activity, combined with neurotrophin-dependent phenotypic alterations in non-nociceptors, increases synaptic efficacy, contributing to central sensitization and a feed-forward loop of enhancement of reflex sensitivity

phospholipase C (PLC), and over the longer term, by altering the transcription of genes for ion channels, peptide neurotransmitters, and other modulatory proteins. An overview of these forms of plasticity is represented in Fig. 1c and described in greater detail in Sect. 4.

2 Transduction

2.1 Mechanotransducers

Mechanical stimulation of the airways is effective at evoking the cough reflex. The exact molecular identity of the mechanosensor(s) present in primary afferent nerve terminals is unknown. However, some insight is emerging. Recently, a mechanically activated current with pharmacological properties distinct from those of multiple other candidate mechanically activated ion currents was identified in dorsal root ganglion (DRG) neurons (Drew 2007) and a peptide blocker (dubbed NMB-1 for "noxious mechanosensation blocker 1") of this current also inhibited behavioral responses to noxious mechanical stimuli. As mechanically evoked responses in distinct airway sensory neuron subtypes differ in threshold and discharge patterns (McAlexander et al. 1999), it seems likely that several mechanotransduction mechanisms may be involved in detecting tussive mechanical stimuli. Indeed, mechanotransduction likely involves a variety of structural proteins, the extracellular matrix, and various cytoskeletal proteins (Christensen and Corey 2007).

Many airway sensory fibers respond to direct mechanical perturbation, and this response is subject to sensitization by inflammation. For example, exposing the isolated trachea of previously sensitized guinea pigs to allergen resulted in resident mast cell activation and mediator release, followed by an approximately fourfold decrease in the amount of mechanical force required to stimulate action potential generation in mechanosensitive fibers (Riccio et al. 1996b). This decrease could be the result of one or multiple mast cell mediators altering the activity of ion channels associated with mechanotransduction and/or voltage-gated channels involved in regulation of action potential characteristics.

Recent studies have confirmed that the sensitivity of neuronal pathways activated by tussive mechanical stimuli of human airways is subject to plasticity. Subjects with upper respiratory tract infections were found to have reduced thresholds for mechanically evoked cough and in addition coughed more in response to mechanical stimuli than did healthy subjects (Lee and Eccles 2004). It has yet to be determined if mechanically evoked chough is similarly sensitized during chronic lung diseases.

2.2 *Transient Receptor Potential Channels: Multimodal Transducers of Tussive Stimuli*

The mammalian transient receptor potential (TRP) superfamily of six transmembrane domain cation channels encompasses 28 members divided into six subfamilies (TRPA, TRPC, TRPM, TRPV, TRPML and TRPP) on the basis of amino acid sequence homology (Clapham et al. 2005; Nilius and Voets 2005; Ramsey et al. 2006).

Currently, the weight of evidence suggests six members of the TRP family of channels are expressed in sensory neurons, where they serve to detect a variety of physical and chemical stimuli. These include the multimodal transducer TRPV1 (Tominaga et al. 1998), the noxious heat sensor TRPV2 (Caterina et al. 1999), the warmth sensor TRPV3 (Smith et al. 2002; Xu et al. 2002), and the warmth, chemo-, and hypo-osmolarity sensor TRPV4 (Suzuki et al. 2003). The cold and menthol receptor TRPM8 (McKemy et al. 2002; Peier et al. 2002) and the nociceptor-specific TRPA1 (Story et al. 2003) are also present in subsets of sensory neurons. While TRPV2–TRPV4 may play important roles in airway nerves (Ni et al. 2006), and the hypothesis that TRPM8 mediates hyperresponsiveness to cold air is intriguing, most work in the field of airway sensory nerve biology has focused on TRPV1 and, more recently, TRPA1.

2.2.1 TRPV1

Capsaicin, a pungent ingredient present in hot peppers, is a potent activator of a subset of sensory neurons that normally only respond to potentially tissue damaging noxious stimuli (nociceptors), including those associated with the cough reflex (Davenport et al. 2007; Fuller et al. 1985; Fuller 1991; Mazzone et al. 2007b). The capsaicin receptor, TRPV1, was the first transducer of nociceptive information to be cloned and characterized (Caterina et al. 1997).

TRPV1 is unusual in that it is gated not only by ligands but also by low pH and heat (Tominaga et al. 1998). At physiological pH, the TRPV1 pore is opened only at temperatures above approximately 42°C. At low pH (6.3), however, the threshold for channel opening is reduced to approximately 30°C; thus, low pH sensitizes TRPV1 such that it may be tonically active at physiological temperatures. Inflammatory pain may be at least partially due to this sensitization, since the pH of inflamed tissues is lower than that of noninflamed tissue, and TRPV1-deficient mice display reduced inflammatory hyperalgesia (Davis et al. 2000). In experimental systems, TRPV1 is also sensitized and/or activated by multiple mediators and intracellular second messengers, including anandamide, lipoxygenase products, serine/threonine protein kinases, and tyrosine phosphorylation (Chuang et al. 2001; Hwang et al. 2000; Zhang et al. 2005; Zygmunt et al. 1999). These findings have lead to the hypothesis that TRPV1 can function not only as a transducer of noxious heat, acidity, and irritants, but also in neuronal metabotropic-receptor-mediated sensitization and activation of afferent nerve endings. This hypothesis is supported by recent studies in which TRPV1-deficient mice display reduced bradykinin-evoked hyperalgesia (Bautista et al. 2006) and histamine-evoked pruritis (Shim et al. 2007).

Within the respiratory tract, capsaicin activates C fibers and a subpopulation of high mechanical threshold Aδ fibers (Coleridge et al. 1965; Fox et al. 1993; Kollarik et al. 2003; Mohammed et al. 1993; Riccio et al. 1996a; Tatar et al. 1988). Depending on the context and extent of their activation, activation of these fibers can evoke several respiratory reflexes, including bronchoconstriction, vascular and glandular secretions, apneas (accompanied by bradycardia and hypotension in the stereotypical "pulmonary chemoreflex"), and rapid, shallow breathing (Coleridge et al. 1965, 1992; Davis et al. 1982; Green et al. 1984; Roberts et al. 1981). Other endogenous and/or inhaled TRPV1 agonists may contribute to vagal reflexes in vivo, a notion supported by findings such as the inhibition of bradykinin-evoked action potential discharge in guinea pig airway nerve terminals by ruthenium red and capsazepine (Carr et al. 2003), the inhibition of hyperthermia-evoked activation of rat vagal neurons by the cinnamide TRPV1 blocker AMG9810 (Gavva et al. 2005) and capsazepine (Ni et al. 2006), and the abolition of anandamide-evoked pulmonary C-fiber discharge in the TRPV1 knockout mouse (Kollarik and Undem 2004).

Inhaled capsaicin evokes cough (and at lower doses, urge to cough) in humans (Davenport et al. 2007; Fuller et al. 1985; Fuller 1991; Mazzone et al. 2007a). Similarly, capsaicin evokes cough in a variety of animals (Forsberg et al. 1988; Tatar et al. 1994) that is inhibited by TRPV1 blockers such as capsazepine (Lalloo et al. 1995) and iodoresiniferatoxin (Trevisani et al. 2004). This indicates that capsaicin is sufficient to cause cough; curiously, capsaicin does not normally evoke cough in anesthetized animals, even at doses that cause robust reflexes (Canning et al. 2004; Tatar et al. 1988). An attractive explanation for this is that the activation of capsaicin-sensitive C fibers does not cause cough per se, but lowers the threshold for coughing secondary to peripheral and/or central sensitization (Canning et al. 2004; Mazzone et al. 2005).

2.2.2 TRPA1

TRPA1 has generated considerable excitement because of its restricted expression and unique biology. In particular, TRPA1 has been identified as a neuronal sensor of multiple respiratory irritants, including acrolein, cinnamaldehyde, and allyl isothiocyanate (AITC), the active ingredient in mustard oil (Bandell et al. 2004; Jordt et al. 2004).

Studies in TRPA1 knockout mice have indicated that the respiratory irritant acrolein activates sensory neurons through a TRPA1-dependent pathway (Bautista et al. 2006). The finding that a chemical as broadly reactive and toxic as acrolein failed to acutely activate sensory neurons in the absence of one single ion channel is intriguing, and not the only example of a highly reactive toxin that activates TRPA1. Our own laboratory has recently reported that the lipid peroxidation product 4-hydroxynonenal (4-HNE), which is produced following oxidative tissue damage and elevated in respiratory disease (Rahman et al. 2002), is an endogenously occurring agonist of TRPA1. Indeed, 4-HNE activated both HEK cells stably transfected with human TRPA1 (Fig. 2) and a subset of nasal-specific, capsaicin-sensitive trigeminal neurons (Taylor-Clark et al. 2007a). These findings are in agreement with recently published studies showing TRPA1 is activated by and necessary for pain behaviors evoked by 4-HNE as well as formalin (Macpherson et al. 2007b; McNamara et al. 2007; Trevisani et al. 2007).

The puzzle of how one ion channel could be activated by such a broad range of irritants (e.g., AITC, formalin, allicin, diallyl disulfide, acrolein, cinnamaldehyde) with minimal structural similarity was reconciled by the discovery that TRPA1 is activated when electrophiles form covalent adducts with three cysteines (and, in human TRPA1, one lysine) within the channel's cytosolic N terminus (Hinman et al. 2006; Macpherson et al. 2007a). This suggests that any cysteine-reactive electrophile that can access these crucial residues has the potential to activate TRPA1 and evoke nociceptive reflexes. In support of this, iodoacetamide, the cysteine-reactive tool molecule often used in biochemical experiments, activates TRPA1 (albeit with very low potency) and evokes nocifensive behaviors in wild-type but not TRPA1 knockout mice (Macpherson et al. 2007b).

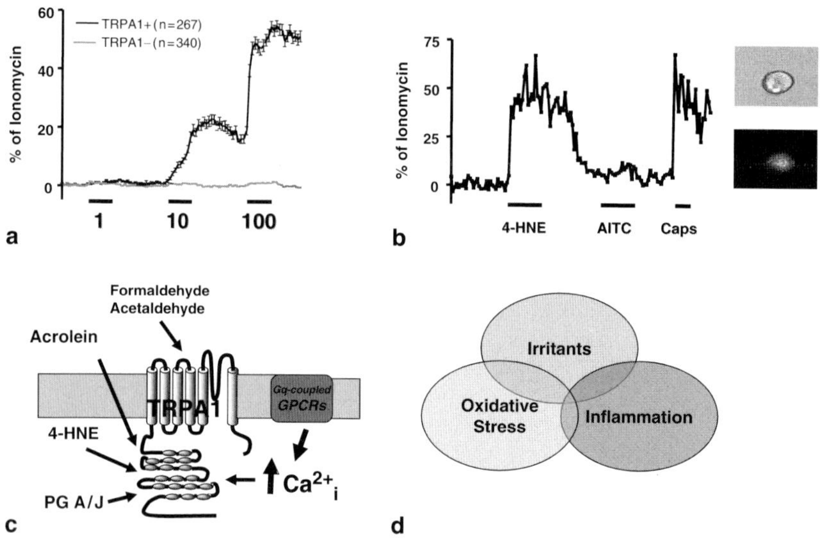

Fig. 2 (**a**) The lipid peroxidation product 4-hydroxynonenal (4-HNE) activates TRPA1, as evidenced by calcium imaging experiments on HEK293 cells stably expressing human TRPA1. Cells were exposed to increasing concentrations (1 – 100 μM) of 4-HNE, 60 s for each concentration (as denoted by bars under traces). (**b**) 4-HNE (100 μM) also activated three out of five mouse trigeminal nasal-specific neurons. *Inset*: Brightfield (*top*) and fluorescence (*bottom*) images of a representative labeled neuron. 4-HNE-responsive neurons responded to capsaicin (1 μM), confirming their nociceptive character, although responses to the known TRPA1 agonist allyl isothiocyanate (AITC; 100 μM) were minimal. This is likely because 4-HNE forms an irreversible adduct with intracellular cysteines within TRPA1 (see reference in the text), preventing further activation by electrophiles such as AITC. (**c**) Proposed role of TRPA1 in airway nociceptors as a polymodal sensor of reactive irritants such as acrolein and formaldehyde; inflammation, through Gq-driven intracellular Ca^{2+} elevations or A- and J-series prostaglandins; and oxidative stress, via responses to lipid peroxidation products such as 4-HNE. (**d**) TRPA1 appears to play a unique role by integrating responses to irritants, oxidative stress, and inflammation, three major factors likely involved in the ongoing activation and sensitization of respiratory sensory nerves (modified from Taylor-Clark et al., 2007a)

Functional evidence in animals and humans suggests that TRPA1 plays a vital role in irritant-induced responses within the respiratory tract. TRPA1 immunoreactivity has been localized to neurons within the mouse nodose ganglion (Nagata et al. 2005). The TRPA1 agonist acrolein is extremely irritating when applied to human mucosal surfaces (Sim and Pattle 1957) and causes profound acute effects in the respiratory tract of rodents, including vasodilation and inhibition of respiration (Lee et al. 1992; Morris et al. 1999). Of note, acrolein-induced reflex apneic responses in rats are blocked by capsaicin desensitization (Lee et al. 1992), but similar responses in mice are unaltered by genetic deletion of TRPV1 (Symanowicz et al. 2004). The most likely interpretation of these results is that acrolein evokes respiratory reflexes by acting on TRPA1 in TRPV1-containing nociceptors.

In addition to its sensitivity to products of lipid peroxidation and exogenous reactive irritants, TRPA1 is also regulated by inflammatory mediators. Bradykinin and muscarinic agonists activate the channel, albeit downstream of Gq/PLC signaling, rather than directly (Bandell et al. 2004; Jordt et al. 2004). Protease-activated receptor-2 agonists also sensitize TRPA1-mediated cellular and behavioral responses in a PLC-dependent manner (Dai et al. 2007), which raises the possibility that activation of any Gq-coupled receptor on TRPA1-containing respiratory sensory neurons may "tune" their excitability during inflammation. Recently our laboratory has also identified a novel mechanism by which inflammatory mediators can directly activate TRPA1. The A- and J-series prostaglandins, metabolites of the primary mast cell prostaglandin D_2 and the nociceptor-sensitizing prostaglandin E_2, activated both heterologously expressed TRPA1 and a subset of mouse capsaicin-sensitive trigeminal neurons (Taylor-Clark et al. 2007b). A representative member of the series, 15-deoxy-$\Delta^{12,14}$-prostaglandin J_2, was shown to act directly since it evoked currents in TRPA1-expressing (but not TRPV1-expressing) HEK cells in the nominal absence of calcium.

Taken together, these findings provide prima facie evidence that TRPA1 is a key transducer of inflammatory stimuli, oxidative tissue damage, and multiple inhaled irritants and that it serves a unique role in activating sensory neurons independent of TRPV1.

3 Ion Channels Modulating Action Potential Threshold and Rate of Discharge

Following the detection of a tussive stimulus, the airway's sensory nerve terminals are depolarized to a critical point where action potentials are produced. Thresholds and pattern of action potential discharge are not fixed but rather change in the presence of inflammatory mediators, thereby altering the sensitivity of the cough reflex. A discussion of several such ion channels and information regarding their function in airway sensory fibers is presented below.

3.1 Na_V Channels

Na_V channels are essential for action potential formation, and thus, neuronal function.

Of the nine Na_Vs, six are highly expressed in neurons. Intriguingly, of these six, three (Na_V1.7, Na_V1.8, and Na_V1.9) are primarily concentrated within sensory neurons (Raymond et al. 2004), where they may each play critical, nonredundant roles in regulating sensory neuron activity. This specialized expression of function of Na_V-channel subtypes highlights the potential of Na_V-subtype-selective channel blockers as a possible therapeutic strategy for diseases involving neuronal hyperexcitability, including pruritis, chronic pain states, and respiratory diseases associated with cough.

3.1.1 Na$_V$1.7

Na$_V$1.7 carries a tetrodotoxin (TTX)-sensitive sodium current, its expression appears concentrated in nociceptors (Djouhri et al. 2003) and sympathetic neurons (Sangameswaran et al. 1997), and nociceptor-specific deletion of Na$_V$1.7 provides protection against hyperalgesia in a variety of inflammatory pain models (Nassar et al. 2004). Consistent with the hypothesis that Na$_V$1.7 plays an important role in pain perception, several gain-of-function mutations in the gene encoding Na$_V$1.7 (*SCN9A*) have been identified in patients with the painful conditions erythermalgia (Drenth et al. 2005; Waxman et al. 2007a, b; Yang et al. 2004) and paroxysmal extreme pain disorder (Fertleman et al. 2006, 2007).

More recently, a separate set of rare *SCN9A* mutations, that result in Na$_V$1.7 channels that are either not expressed or unable to function, have been described (Ahmad et al. 2007; Cox et al. 2006; Goldberg et al. 2007). Remarkably, individuals possessing these mutations appear grossly normal, with the exception that they cannot perceive pain. Recent findings suggest that Na$_V$1.7 is expressed in airway sensory nerves (Kwong et al. 2007), however, its role in regulation or plasticity of the cough reflex is yet to be determined.

3.1.2 Na$_V$1.8

Na$_V$1.8 is selectively expressed in primary afferent nociceptors, where it appears to be the primary contributor to the action potential upstroke (Akopian et al. 1999). Na$_V$1.8 carries a TTX-resistant (TTX-R) sodium current that is considerably different from those carried by TTX-sensitive channels. Na$_V$1.8 currents have a more depolarized voltage dependence of activation and of fast inactivation (Akopian et al. 1999). Thus, Na$_V$1.8 is one of the few Na$_V$s that could contribute to continuous action potential discharge during periods of sustained depolarization (Renganathan et al. 2001).

Preclinical evidence suggests selective Na$_V$1.8 inhibitors may provide relief from inflammatory pain conditions without mechanism-based cardiac and CNS side effects. To date, this hypothesis has been validated in rodent models of pain using targeted genetic disruption (Akopian et al. 1999; Laird et al. 2002), antisense oligonucleotide-mediated depletion (Villarreal et al. 2005), a μ-conotoxin less than tenfold selective for Na$_V$1.8 over other sodium channels (Ekberg et al. 2006), and, most recently, a small molecule with nanomolar potency and more than 100-fold selectivity over other sodium channels (Jarvis et al. 2007).

A variety of evidence suggests a role of a Na$_V$1.8-like TTX-R sodium current in the plasticity of sensory nerves associated with cough. Na$_V$1.8 is expressed in airway sensory neurons (Kwong et al. 2007) and a recent study showed that prostaglandin E (PGE_2) increased the macroscopic conductance of an Na$_V$1.8-like TTX-R sodium current that was predominantly expressed in capsaicin-sensitive pulmonary sensory neurons, and thereby enhanced the overall excitability of these neurons (Kwong and Lee 2005).

3.1.3 $Na_V1.9$

Within the somatosensory sensory nervous system $Na_V1.9$ is found largely in putative nociceptors (Cummins et al. 1999; Dib-Hajj et al. 1998, 1999) and is responsible for a TTX-R current distinct from that carried by $Na_V1.8$. Rodent studies suggest that $Na_V1.9$ carries a persistent current that is activated at potentials close to standard neuronal resting membrane potentials in the range from −60 to −70 mV (Cummins et al. 1999; Herzog et al. 2001). This current is subject to regulation by multiple factors, including glial-cell-derived neurotrophic factor and PGE_V (Cummins et al. 2000; Rush and Waxman 2004), suggesting $Na_V1.9$ may play a significant role in regulating nociceptor excitability. Consistent with this hypothesis, two independent groups have demonstrated reduced peripheral sensitization in response to a number of inflammatory mediators in $Na_V1.9$ knockout mice (Amaya et al. 2006; Priest et al. 2005).

More recently, we have used these same knockout mice (Amaya et al. 2006) to investigate the role of $Na_V1.9$ in vagal neurons (Ghatta et al. 2007). While these mice have a normal distribution of conduction velocities in their vagal compound action potentials and, on the basis of patch-clamp studies of jugular-nodose complex neurons, have only subtle differences in membrane properties, their pulmonary C fibers have marked deficits in irritant (α,β-methylene ATP and bradykinin)-evoked action potential generation. These studies provide the first direct evidence that $Na_V1.9$ plays a critical role in nerve terminal responses to noxious stimuli; however, further studies are necessary to determine whether it also modulates synaptic communication in the brainstem. Thus, while many details of $Na_V1.9$'s function remain arcane, it clearly plays a significant role in rodent sensory neurons, including those within the respiratory tract.

3.2 Hyperpolarization-Activated Cyclic-Nucleotide-Gated Channels

Hyperpolarization-activated, cyclic-nucleotide-gated (HCN) channel subunits HCN1–HCN4 underlie the current frequently referred to as Ih (for hyperpolarization-induced current) or If, for the less scientific, though similarly descriptive "funny current" (Accili et al. 2002). These channels demonstrate a mixed cation conductance that, like for Na_V channels, tends to cause membrane depolarization (Pape 1996). Thus, by depolarizing membranes following its activation by hyperpolarization, Ih facilitates repetitive firing in neurons and cardiac pacing in the sinoatrial node (Accili et al. 2002; DiFrancesco 2006).

All four known HCN channel subunits have been identified in numerous neuronal populations, and several lines of evidence suggest that Ih carried by one or more of the subtypes plays an important role in vagal afferent neurobiology. All neonatal rat nodose neurons possess an Ih involved in regulating resting membrane potential (Doan and Kunze 1999), while roughly a third of adult guinea pig nodose ganglion neurons possess an Ih that is blocked by high micromolar concentrations of the

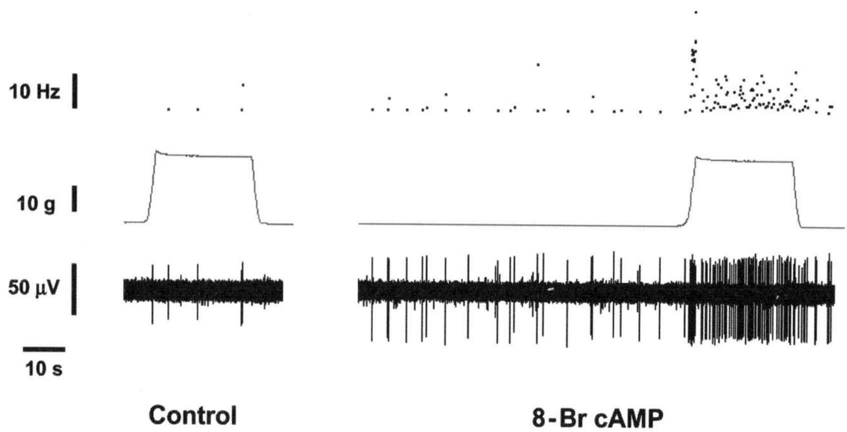

Fig. 3 Under control conditions (*left*), applying 25 g of force with a blunt probe caused only minimal action potential discharge from this guinea pig tracheal jugular C fiber (conduction velocity $0.7\,\mathrm{m\,s^{-1}}$). Addition of the membrane-permeant cAMP analogue 8-bromo-cAMP (1 mM, 15 min) to the buffer perfusing the isolated airways caused a marked sensitization of the fiber to subsequent mechanical stimulation (*right*). Also note that the fiber began to discharge spontaneously in the presence of 8-bromo-cAMP, suggesting that 8-bromo-cAMP-sensitive pathways can modulate both the active and the passive properties of airway sensory fiber terminals. These effects occurred in $3\,\mu\mathrm{M}$ indomethacin–containing buffer, so they are likely independent of cyclo-oxygenase products

prototypical blocker cesium (Undem and Weinreich 1993). More recently, HCN1-, HCN2,- and HCN4-like immunoreactivities were demonstrated in rat nodose neurons and their peripheral terminals (Doan et al. 2004), and the Ih blocker cesium lowered the threshold pressure necessary to evoke aortic baroreceptor activation.

HCN channels may also play a role in regulating the excitability of airway peripheral terminals following inflammation, since PGE_2, which sensitizes jugular ganglion-derived airway C fiber terminals (Carr 2007), also increases Ih in guinea pig nodose neurons (Ingram and Williams 1996), most likely by increasing cAMP levels within the neuron. Since multiple receptor-coupled mechanisms are likely required to produce maximal rises in nerve terminal cAMP (Alessandri-Haber et al. 2006), we have begun using the membrane-permeable analogue 8-bromo-cAMP to study mechanisms of acute peripheral sensitization of airway sensory neurons. As predicted, 8-bromo-cAMP causes profound sensitization of guinea pig tracheobronchial jugular fibers (Fig. 3) and can evoke spontaneous firing in some fibers. How cAMP sensitizes airway fiber terminals, and what role (if any) HCN channels play in the process, is yet to be determined.

3.3 Cl^- Channels

Neuronal chloride channels, exemplified by the ligand-gated $GABA_A$ receptor, generally hyperpolarize membranes and are thus considered inhibitory. A notable exception to this general rule, however, is primary sensory neurons that accumulate relatively large concentrations of intracellular chloride owing to the Na^+–K^+–$2Cl^-$ (NKCC) cotransporter NKCC1 (Sung et al. 2000). In these neurons, the concentration gradient for chloride is shifted such that when chloride channels are open, the negatively charged chloride ions leave the cell, causing a net depolarization. This also occurs in spinal cord neurons following nerve injury, a phenomenon that may contribute to neuropathic pain (Coull et al. 2003). Interestingly, NKCC1 has also been immunolocalized in guinea pig airway fibers, and both low chloride solutions and chloride channel blockers enhance the cough reflex in anesthetized guinea pigs, while the NKCC1 blocker furosemide has antitussive effects (Mazzone and McGovern 2006).

Oh and Weinreich (2004) produced the first evidence of excitatory chloride currents in vagal neurons. They found that bradykinin caused a chloride current in guinea pig nodose neurons that was inhibited by niflumic acid or intracellular calcium chelation with 1,2-bis(2-aminophenoxy)ethane-N, N, N', N'-tetraacetic acid, properties consistent with a calcium-activated chloride current. These channels are likely also involved in guinea pig tracheobronchial nerve terminals, as the TRP blockade-insensitive component of bradykinin-evoked action potential generation in these fibers was largely eliminated by two separate calcium-activated chloride channel inhibitors (Lee et al. 2005).

3.4 Voltage-Gated Ca^{2+} Channels

Voltage-gated calcium (Ca_V) currents are carried by members of the Ca_V families of ion channels. The pore-forming $\alpha 1$ subunits of Ca_V channels are separated into two general categories, based on their current–voltage relationship: high-voltage-activated (HVA) and low-voltage-activated (LVA). The HVA channels comprise $Ca_V1.1$–$Ca_V1.4$ and $Ca_V2.1$–$Ca_V2.3$ and are responsible for L-, N-, P/Q-, and R-type currents, while T-type currents carried by $Ca_V3.1$–$Ca_V3.3$ are LVA (Catterall et al. 2005).

3.4.1 HVA Ca_V

L-type calcium channels play significant roles in central neurons and N-type calcium channels are expressed in multiple populations of neurons, including primary sensory neurons in the DRG, where they regulate the release of neurotransmitters onto secondary dorsal horn neurons. Pharmacological evidence suggests that L- and N-type Ca^{2+} channels are also present in nodose ganglion neurons (Cordoba-Rodriguez et al. 1999), but multiple blockers of Ca_V channels do not alter the response of nodose or jugular ganglion-derived airway fiber terminals to mechanical stimuli (Undem et al. 2003). It is thus surprising in light of this that removal of extracellular Ca^{2+} enhances the mechanical sensitivity of jugular ganglion-derived terminals (Undem et al. 2003). The mechanism(s) involved remain unknown, but may be related to the finding that removal of calcium from the extracellular solution bathing jugular vagal sensory neurons activated a nonselective cation current (Undem et al. 2003) that could cause a depolarization large enough to evoke action potential discharge.

3.4.2 T-Type (Ca_V3)

The pore-forming $\alpha 1$ subunits $Ca_V3.1$–$Ca_V3.3$ are responsible for T-type Ca^{2+} currents and are considered LVA since they are activate at membrane potentials far more hyperpolarized than their HVA counterparts (Catterall et al. 2005). Intriguingly, T-type Ca^{2+} channels are crucial for the normal function of mouse D-hair mechanosensitive fibers (Shin et al. 2003) that adapt rapidly to mechanical stimuli in a manner qualitatively similar to airway nodose Aδ "cough receptor" fibers (McAlexander et al. 1999).

There is evidence that T-type Ca^{2+} channels may play a role in vagal neurobiology, as messenger RNA for all three Ca_V subunits has been detected in rat nodose ganglion neurons and antisense depletion of Ca_V3 channels inhibited a component of Ca^{2+} entry evoked by application of a mock action potential (Lambert et al. 1997). Despite these compelling data, the T-type Ca^{2+} channel blocker Ni^{2+} (300 μM) did not appreciably alter the responses of guinea pig tracheobronchial cough receptor fibers or nociceptors to citric acid or mechanical stimulation (unpublished observations). Still, these negative findings do not rule out the possibility that Ca^{2+} entry through T-type or other Ca_V channels may play a role in the longer-term sensitization of airway sensory fibers.

3.5 K^+ Channels

In addition to their role in setting resting membrane potential, K^+ channels play roles in setting the threshold for action potential discharge and determining action potential discharge frequency and pattern. This versatility results from the vast heterogeneity of channels that can contribute to K^+ currents, the most heterogeneous of any known ion channel superfamily (Gutman et al. 2003). Members of this channel superfamily include the voltage-gated K channels, denoted K_V, and small- and intermediate-conductance Ca^{2+}-activated K^+ channels, denoted K_{Ca}, both of which possess six transmembrane domains. G-protein-coupled channels, inwardly rectifying channels, and ATP-activated channels (K_{ATP}), contain two transmembrane domains, the large-conductance channel contains seven transmembrane domains, and the K_{2P} family contains two transmembrane domains. Further complexity is achieved through differential interactions with modulatory subunits.

Like most neurons, vagal sensory neurons express multiple diverse K^+ conductances (Christian et al. 1994) and, consistent with their role in regulating neuronal excitability, the K^+ channel blockers α-dendrotoxin and 4-aminopyridine evoke action potential discharge in airway nodose ganglion-derived fibers (McAlexander and Undem 2000). These findings suggest that K^+ currents active at rest maintain airway sensory neuron membrane potential below threshold for action potential discharge. A role of modulation of these tonically active K^+ currents in controlling excitability during inflammation is consistent with the finding that antigenic activation of resident mast cells or application of inflammatory mediators such as histamine, bradykinin prostacyclin, or cysteinyl leukotrienes inhibited resting K^+ currents in nodose ganglion neurons and lead to a membrane depolarization and an increase in excitability (Undem and Weinreich 1993).

Several types of K_{Ca} that are present in vagal sensory neurons may play a role in regulating excitability. One of these is responsible for a slowly developing, long-lasting afterspike hyperpolarization (AHP_{slow}). The hyperpolarization caused by K^+ efflux regulates neuronal excitability and the frequency and pattern of neuronal discharge (Cordoba-Rodriguez et al. 1999; Undem and Weinreich 1993). Both the amplitude and the duration of AHP_{slow} in guinea-pig nodose cell bodies are significantly reduced in the presence inflammatory mediators such as bradykinin, prostacylin, prostaglandin E_2, prostaglandin D_2, 5-hydroxytryptamine, or leukotriene C_4 (Cordoba-Rodriguez et al. 1999; Undem and Weinreich 1993). The inhibition of AHP_{slow} in these cell bodies was accompanied by an increase in the frequency at which the neuron could successfully elicit repetitive action potentials (Undem and Weinreich 1993). These findings suggest that autacoids may not need to overtly evoke action potentials to sensitize afferent neurons, but may do so by modulating the rate of neuronal discharge.

Another K_{Ca} that may modulate airway afferent nerve excitability is the large conductance Ca^{2+}-activated channel (BK_{Ca}). An opener of this channel, NS 1619, inhibited citric acid induced cough in conscious guinea pigs and inhibited the generation of action potentials in guinea-pig tracheal Aδ and C fibers in vitro (Fox et al. 1997). The effects of NS 1619 on Aδ and C fibers were prevented by iberiotoxin, a BK_{Ca} channel-selective blocker.

4 Neurotrophins and Plasticity

The expression pattern of transducers, voltage-gated channels, and neuropeptide transmitters present in sensory neurons is not fixed, but is modulated, through mechanisms largely dependent upon the action of various neurotrophins, a family of molecules which are known to act on sensory nerve endings to send growth signals to their remotely located cell bodies (Helke 2005).

Sources of neurotrophins within the airways include the respiratory epithelium, T lymphocytes, alveolar macrophages, and mast cells (Leon et al. 1994; Braun et al. 1999). Neurotrophins initiate their effects in vagal afferent neurons, in part, by binding to high-affinity, tyrosine kinase coupled receptors, followed by uptake and retrograde transport to the cell body in the vagal ganglia (Helke 2005). Among the neurotrophins, most attention has been given to nerve growth factor (NGF) as a potential mediator in afferent plasticity in inflamed airways. NGF is found in human airways and the NGF content of airway lavage fluid is increased following allergen challenge (Sanico et al. 2000; Virchow et al. 1998; Wu et al. 2006). The influence of neurotrophins on the expression of various transducers, ion channels, and neuropeptides may be of particular relevance to plasticity of the airway sensory innervation associated with cough.

4.1 Modulation of Ionotropic and Metabotropic Transducers

A variety of distinct transducers are expressed in cough-evoking sensory neurons (Kollarik and Undem 2004). Although the relative importance or indeed the exact identity of all transducers likely to be involved is yet to be completely identified, there is a growing body of evidence suggesting that transducer plasticity contributes to peripheral sensitization of sensory neurons. While several posttranslational mechanisms can alter the trafficking and/or activity of transducers and other neuronal proteins, their net function may also be enhanced as a result of neurotrophin-induced upregulation of gene expression. Experimental observations indicate that the expression of a variety of metabotropic transducers such as bradykinin B1 (Vellani et al. 2004) and B2 receptors (Lee et al. 2002), as well as the ionotropic transducers TRPA1 (Diogenes et al. 2007; Elitt et al. 2006), TRPV1 (Amaya et al. 2004), P2X3 (Ramer et al. 2001), and various ASICs (Mamet et al. 2003) can all be regulated by neurotrophins.

The potential role of increased expression of transducer ion channels in chronic cough is consistent with increased expression of TRPV1-like immunoreactivity within the airways of patients with chronic cough (Groneberg et al. 2004; Mitchell et al. 2005). Moreover, a significant correlation between capsaicin-evoked tussive response and the number of TRPV1-positive nerves within the airways of patients with cough has been reported (Groneberg et al. 2004). The level of expression of other neuronal transducers in the airways of patients with chronic cough has not been reported to date.

4.2 Modulation of Voltage-Gated Ion Channel Expression

The activation of a particular transducer within the terminal of a cough-evoking sensory nerve is encoded into an action potential discharge via the activation of Na_V and K_V currents. A variety of evidence has recently emerged indicating that the expression levels of the multiple Na_V and K^+ channels in sensory neurons are subject to transcriptional regulation by factors such as NGF and that these changes are in part responsible for altered neuronal activity during inflammation. For example, ulcer formation produced by a single injection of acetic acid in the stomach of rats was associated with a significant increase in excitability as shown by a lower threshold for action potential initiation and an increase in discharge frequencies during prolonged depolarization (Dang et al. 2004). These changes are consistent with neurotrophin-dependent increased expression of Na^+ currents in the cell bodies of sensory neurons within remotely located vagal sensory ganglia (Bielefeldt et al. 2003).

Like Na^+ channels, the level of the expression of distinct K^+ currents within sensory neurons is not fixed but changes in response to tissue injury. The complement of K^+ channels expressed in airway afferent neurons associated with the cough reflex is unknown and, on the basis of studies of isolated vagal sensory neuron cell bodies, it is likely that distinct airway sensory neuron phenotypes express multiple distinct K^+ currents. One type of K^+ current in particular, a rapidly activating transient current known as an A-type current, was decreased in rat DRG and nodose neurons following gastric inflammation (Dang et al. 2004). A-type K^+ currents also appear to be downregulated in bladder sensory neurons following inflammation, resulting in lower thresholds and greater discharge frequencies (Yoshimura and Groat 1999). Notably, intrathecal NGF administration causes similar responses in bladder neurons (Yoshimura et al. 2006).

There is currently no direct evidence addressing the hypothesis that modulation of ion channel expression in airway sensory neurons contributes to the plasticity of the cough reflex. Thus, while by necessity the discussion above was focused on findings in a variety of nonrespiratory tissues, regulation of the expression pattern of various Na_V and K^+ channels appears to be a general feature of sensory nerves; therefore, it is conceivably a relevant form of plasticity contributing to the enhanced urge to cough that can occur in respiratory diseases.

4.3 Regulation of Neuropeptide Expression

The release of the peptide neurotramsmitter substance P and related neurokinins is known to augment synaptic transmission between the central terminals of airway afferent neurons and secondary neurons within the nucleus tractus solitarius (Joad et al. 2004; Mazzone and Geraghty 2000; Mutoh et al. 2000). Normally, these neuropeptides are expressed only in high-threshold and therefore difficult-to-activate nociceptors. As a consequence of this limited expression pattern, under normal conditions the release of sensory neuropeptides in the brainstem from vagal afferent nerves, including those associated with the cough reflex (Mazzone et al. 2005), may be limited to situations where the airways are subjected to intense noxious stimuli. This may change following irritation and inflammation of the airways. In particular, airway inflammation is associated with an increased content of neuropeptides, including within nerves (Fischer et al. 1996; Kwong et al. 2001). A somewhat surprising observation was that following allergen challenge (Chuaychoo et al. 2005; Myers et al. 2002) or viral infection (Carr et al. 2002) tachykinin expression was detectable in neurons whose axons project low-threshold airway mechanosensors.

The full expression of this "phenotypic switch" in nodose ganglion neurons requires intact vagus nerves, but if allergen reached the systemic circulation in sufficient quantities, it could also stimulate substance P synthesis by local activation of vagal ganglionic mast cells (Chuaychoo et al. 2005).

A phenotypic switch as described above may create a scenario in which low-threshold stimuli could result in neuropeptide release in the brainstem. On the basis of these findings, it is conceivable that an increase in neuropeptides released into the brainstem from low-threshold, easily activated afferent fibers may explain at least a portion of the plasticity of the cough reflex that often accompanies lung inflammation.

4.4 Density of the Sensory Innervation of the Airways

Neurotrophins influence the density of the sensory innervation of various tissues. To date, only one study has examined the relationship between sensitivity to inhaled tussive stimuli and nerve fiber density in patients with idiopathic persistent non-productive cough (O'Connell et al. 1995). Concentrations of capsaicin required to elicit at least two and five coughs were significantly lower in patients than in control subjects. In bronchial epithelium taken from the carina of the right upper lobe and a subsegmental carina of the right lower lobe, the total nerve density (protein gene product 9.5 immunoreactivity) was greater in patients than in controls, although this was not statistically significant. However, CGRP-immunoreactive nerve density in the epithelium was significantly higher in patients than in control biopsies taken from the carina of the right upper lobe. Substance P immunoreactive nerves were not significantly different in the two groups.

Although these findings do not support an association of increased nerve density in idiopathic persistent nonproductive cough, they are consistent with the hypothesis of plasticity in peptide neurotransmitter expression with the sensory innervation of the airways.

5 Concluding Remarks

The dramatically heightened sensitivity of the cough reflex and the resulting nonproductive cough are common features of many respiratory diseases. These symptoms are the result of some combination of tissue injury and remodeling, plasticity within airway primary afferent neurons associated with cough, and aberrant CNS activity. Rather than viewing these processes as separate entities, current evidence suggests they are all interconnected, with the primary afferent neuron serving as a critical conduit necessary to drive tissue injury- and tissue remodeling-dependent changes in the CNS. If this view is correct, it is imperative to understand the processes and molecules within primary sensory nerves that perpetuate this deleterious cycle of sensitization and plasticity. Fortunately, many promising targets that appear to have specific functions and restricted tissue distributions have been identified in primary sensory neurons; thus, therapies directed against these ion channels and/or the factors that regulate their expression and activity may provide the most tractable option for relief from disease-associated cough.

References

Accili EA, Proenza C, Baruscotti M, DiFrancesco D (2002) From Funny Current to HCN Channels: 20 Years of Excitation. News Physiol Sci 17:32–37

Ahmad S, Dahllund L, Eriksson AB, Hellgren D, Karlsson U, Lund PE, Meijer IA, Meury L, Mills T, Moody A, Morinville A, Morten J, O'Donnell D, Raynoschek C, Salter H, Rouleau GA, Krupp JJ (2007) A stop codon mutation in SCN9A causes lack of pain sensation. Hum Mol Genet 16:2114–2121

Akopian AN, Souslova V, England S, Okuse K, Ogata N, Ure J, Smith A, Kerr BJ, McMahon SB, Boyce S, Hill R, Stanfa LC, Dickenson AH, Wood JN (1999) The tetrodotoxin-resistant sodium channel SNS has a specialized function in pain pathways. Nat Neurosci 2:541–548

Alessandri-Haber N, Dina OA, Joseph EK, Reichling D, Levine JD (2006) A transient receptor potential vanilloid 4-dependent mechanism of hyperalgesia is engaged by concerted action of inflammatory mediators. J Neurosci 26:3864–3874

Amaya F, Shimosato G, Nagano M, Ueda M, Hashimoto S, Tanaka Y, Suzuki H, Tanaka M (2004) NGF and GDNF differentially regulate TRPV1 expression that contributes to development of inflammatory thermal hyperalgesia. Eur J Neurosci 20(9):2303–2310

Amaya F, Wang H, Costigan M, Allchorne AJ, Hatcher JP, Egerton J, Stean T, Morisset V, Grose D, Gunthorpe MJ, Chessell IP, Tate S, Green PJ, Woolf CJ (2006) The voltage-gated sodium channel Nav1.9 is an effector of peripheral inflammatory pain hypersensitivity. J Neurosci 26:12852–12860

Bandell M, Story GM, Hwang SW, Viswanath V, Eid SR, Petrus MJ, Earley TJ, Patapoutian A (2004) Noxious cold ion channel TRPA1 is activated by pungent compounds and bradykinin. Neuron 41:849–857

Bautista DM, Jordt SE, Nikai T, Tsuruda PR, Read AJ, Poblete J, Yamoah EN, Basbaum AI, Julius D (2006) TRPA1 mediates the inflammatory actions of environmental irritants and proalgesic agents. Cell 124:1269–1282

Bielefeldt K, Ozaki N, Gebhart GF (2003) Role of nerve growth factor in modulation of gastric afferent neurons in the rat. Am J Physiol Gastrointest Liver Physiol 284:G499–G507

Braun A, Lommatzsch M, Mannsfeldt A, Neuhaus-Steinmetz U, Fischer A, Schnoy N, Lewin GW, Renz H (1999) Cellular sources of enhanced brain-derived neurotrophic factor production in a mouse model of allergic inflammation. Am J Respir Cell Mol Biol. 21(4):537–546

Canning BJ, Mazzone SB, Meeker SN, Mori N, Reynolds SM, Undem BJ (2004) Identification of the tracheal and laryngeal afferent neurones mediating cough in anaesthetized guinea-pigs. J Physiol (Lond) 557:543–558

Carr MJ (2007) Plasticity of the afferent innervation of the airways: The role of ion channels. Pulm Pharmacol Ther 20:412–415

Carr MJ, Hunter DD, Jacoby DB, Undem BJ (2002) Expression of tachykinins in nonnociceptive vagal afferent neurons during respiratory viral infection in guinea pigs. Am J Respir Crit Care Med 165:1071–1075

Carr MJ, Kollarik M, Meeker SN, Undem BJ (2003) A role for TRPV1 in bradykinin-induced excitation of vagal airway afferent nerve terminals. J Pharmacol Exp Ther 304:1275–1279

Caterina MJ, Schumacher MA, Tominaga M, Rosen TA, Levine JD, Julius D (1997) The capsaicin receptor: A heat-activated ion channel in the pain pathway. Nature 389:816–824

Caterina MJ, Rosen TA, Tominaga M, Brake AJ, Julius D (1999) A capsaicin-receptor homologue with a high threshold for noxious heat. Nature 398:436–441

Catterall WA, Perez-Reyes E, Snutch TP, Striessnig J (2005) International Union of Pharmacology. XLVIII. Nomenclature and structure-function relationships of voltage-gated calcium channels. Pharmacol Rev 57:411–425

Christensen AP, Corey DP (2007) TRP channels in mechanosensation: Direct or indirect activation? Nat Rev Neurosci 8:510–521

Christian EP, Togo J, Naper KE (1994) Guinea pig visceral C-fiber neurons are diverse with respect to the K+ currents involved in action-potential repolarization. J Neurophysiol 71:561–574

Chuang Hh, Prescott ED, Kong H, Shields S, Jordt SE, Basbaum AI, Chao MV, Julius D (2001) Bradykinin and nerve growth factor release the capsaicin receptor from PtdIns(4,5)P2-mediated inhibition. Nature 411:957–962

Chuaychoo B, Hunter DD, Myers AC, Kollarik M, Undem BJ (2005) Allergen-induced substance P synthesis in large-diameter sensory neurons innervating the lungs. J Allergy Clin Immunol 116:325–331

Clapham DE, Julius D, Montell C, Schultz G (2005) International Union of Pharmacology. XLIX. Nomenclature and structure-function relationships of transient receptor potential channels. Pharmacol Rev 57:427–450

Coleridge HM, Coleridge JC, Luck JC (1965) Pulmonary afferent fibres of small diameter stimulated by capsaicin and by hyperinflation of the lungs. J Physiol (Lond) 179:248–262

Coleridge HM, Coleridge JC, Green JF, Parsons GH (1992) Pulmonary C-fiber stimulation by capsaicin evokes reflex cholinergic bronchial vasodilation in sheep. J Appl Physiol 72:770–778

Cordoba-Rodriguez R, Moore KA, Kao JPY, Weinreich D (1999) Calcium regulation of a slow post-spike hyperpolarization in vagal afferent neurons. PNAS 96:7650–7657

Coull JAM, Boudreau D, Bachand K, Prescott SA, Nault F, Sik A, Koninck PD, Koninck YD (2003) Trans-synaptic shift in anion gradient in spinal lamina I neurons as a mechanism of neuropathic pain. Nature 424:938–942

Cox JJ, Reimann F, Nicholas AK, Thornton G, Roberts E, Springell K, Karbani G, Jafri H, Mannan J, Raashid Y, Al Gazali L, Hamamy H, Valente EM, Gorman S, Williams R, McHale DP, Wood JN, Gribble FM, Woods CG (2006) An SCN9A channelopathy causes congenital inability to experience pain. Nature 444:894–898

Cummins TR, Dib-Hajj SD, Black JA, Akopian AN, Wood JN, Waxman SG (1999) A novel persistent tetrodotoxin-resistant sodium current in SNS-null and wild-type small primary sensory neurons. J Neurosci 19:RC43

Cummins TR, Black JA, Dib-Hajj SD, Waxman SG (2000) Glial-derived neurotrophic factor upregulates expression of functional SNS and NaN sodium channels and their currents in axotomized dorsal root ganglion neurons. J Neurosci 20:8754–8761

Dai Y, Wang S, Tominaga M, Yamamoto S, Fukuoka T, Higashi T, Kobayashi K, Obata K, Yamanaka H, Noguchi K (2007) Sensitization of TRPA1 by PAR2 contributes to the sensation of inflammatory pain. J Clin Invest 117:1979–1987

Dang K, Bielefeldt K, Gebhart GF (2004) Gastric ulcers reduce A-type potassium currents in rat gastric sensory ganglion neurons. Am J Physiol Gastrointest Liver Physiol 286:G573–579

Davenport PW, Bolser DC, Vickroy T, Berry RB, Martin AD, Hey JA, Danzig M (2007) The effect of codeine on the Urge-to-Cough response to inhaled capsaicin. Pulm Pharmacol Ther 20: 338–346

Davis B, Roberts AM, Coleridge HM, Coleridge JC (1982) Reflex tracheal gland secretion evoked by stimulation of bronchial C-fibers in dogs. J Appl Physiol 53:985–991

Davis JB, Gray J, Gunthorpe MJ, Hatcher JP, Davey PT, Overend P, Harries MH, Latcham J, Clapham C, Atkinson K, Hughes SA, Rance K, Grau E, Harper AJ, Pugh PL, Rogers DC, Bingham S, Randall A, Sheardown SA (2000) Vanilloid receptor-1 is essential for inflammatory thermal hyperalgesia. Nature 405:183–187

Dib-Hajj SD, Tyrrell L, Black JA, Waxman SG (1998) NaN, a novel voltage-gated Na channel, is expressed preferentially in peripheral sensory neurons and down-regulated after axotomy. Proc Natl Acad Sci USA 95:8963–8968

Dib-Hajj SD, Tyrrell L, Cummins TR, Black JA, Wood PM, Waxman SG (1999) Two tetrodotoxin-resistant sodium channels in human dorsal root ganglion neurons. FEBS Lett 462:117–120

DiFrancesco D (2006) Serious workings of the funny current. Prog Biophys Mol Biol 90:13–25

Diogenes A, Akopian AN, Hargreaves KM (2007) NGF Up-regulates TRPA1: Implications for orofacial pain. J Dent Res 86:550–555

Djouhri L, Newton R, Levinson SR, Berry CM, Carruthers B, Lawson SN (2003) Sensory and electrophysiological properties of guinea-pig sensory neurones expressing Nav1.7 (PN1) Na^{+} channel alphasubunit protein. J Physiol (Lond) 546:565–576

Doan TN, Kunze DL (1999) Contribution of the hyperpolarization-activated current to the resting membrane potential of rat nodose sensory neurons. J Physiol (Lond) 514:125–138

Doan TN, Stephans K, Ramirez AN, Glazebrook PA, Andresen MC, Kunze DL (2004) Differential distribution and function of hyperpolarization-activated channels in sensory neurons and mechanosensitive fibers. J Neurosci 24:3335–3343

Drenth JPH, te Morsche RHM, Guillet G, Taieb A, Kirby RL, Jansen JBMJ (2005) SCN9A mutations define primary erythermalgia as a neuropathic disorder of voltage gated sodium channels. J Investig Dermatol 124:1333–1338

Drew LJ (2007) High-threshold mechanosensitive ion channels blocked by a novel conopeptide mediate pressure-evoked pain. PLoS One 2:e515

Ekberg J, Jayamanne A, Vaughan CW, Aslan S, Thomas L, Mould J, Drinkwater R, Baker MD, Abrahamsen B, Wood JN, Adams DJ, Christie MJ, Lewis RJ (2006) {micro}O-conotoxin MrVIB selectively blocks Nav1.8 sensory neuron specific sodium channels and chronic pain behavior without motor deficits. PNAS 103:17030–17035

Elitt CM, McIlwrath SL, Lawson JJ, Malin SA, Molliver DC, Cornuet PK, Koerber HR, Davis BM, Albers KM (2006) Artemin overexpression in skin enhances expression of TRPV1 and TRPA1 in cutaneous sensory neurons and leads to behavioral sensitivity to heat and cold. J Neurosci 26:8578–8587

Fertleman CR, Baker MD, Parker KA, Moffatt S, Elmslie FV, Abrahamsen B, Ostman J, Klugbauer N, Wood JN, Gardiner RM, Rees M (2006) SCN9A mutations in paroxysmal extreme pain disorder: Allelic variants underlie distinct channel defects and phenotypes. Neuron 52: 767–774

Fertleman CR, Ferrie CD, Aicardi J, Bednarek NAF, Eeg-Olofsson O, Elmslie FV, Griesemer DA, Goutieres F, Kirkpatrick M, Malmros INO, Pollitzer M, Rossiter M, Roulet-Perez E, Schubert R, Smith VV, Testard H, Wong V, Stephenson JBP (2007) Paroxysmal extreme pain disorder (previously familial rectal pain syndrome). Neurology 69:586–595

Fischer A, McGregor GP, Saria A, Philippin B, Kummer W (1996) Induction of tachykinin gene and peptide expression in guinea pig nodose primary afferent neurons by allergic airway inflammation. J Clin Invest 98:2284–2291

Forsberg K, Karlsson J-A, Theodorsson E, Lundberg JM, Persson CGA (1988) Cough and bronchoconstriction mediated by capsaicin-sensitive sensory neurons in the guinea-pig. Pulm Pharmacol 1:33–39

Fox AJ, Barnes PJ, Urban L, Dray A (1993) An in vitro study of the properties of single vagal afferents innervating guinea-pig airways. J Physiol (Lond) 469:21–35

Fox AJ, Barnes PJ, Venkatesan P, Belvisi MG (1997) Activation of large conductance potassium channels inhibits the afferent and efferent function of airway sensory nerves in the Guinea pig. J Clin Invest 99:513–519

Fuller RW (1991) Pharmacology of inhaled capsaicin in humans. Respir Med 85:31–34

Fuller RW, Dixon CM, Barnes PJ (1985) Bronchoconstrictor response to inhaled capsaicin in humans. J Appl Physiol 58:1080–1084

Gavva NR, Tamir R, Qu Y, Klionsky L, Zhang TJ, Immke D, Wang J, Zhu D, Vanderah TW, Porreca F, Doherty EM, Norman MH, Wild KD, Bannon AW, Louis JC, Treanor JJS (2005) AMG 9810 [(E)-3-(4-t-Butylphenyl)-N-(2,3-dihydrobenzo[b][1,4] dioxin-6-yl)acrylamide], a novel vanilloid receptor 1 (TRPV1) antagonist with antihyperalgesic properties. J Pharmacol Exp Ther 313:474–484

Ghatta S, McAlexander MA, Carr MJ (2007) Characterization of vagal afferent neurons from NaV1.9 -/- and NaV1.9 +/+ mice. Neuroscience Meeting Planner, Society for Neuroscience, 2007 Online

Goldberg YP, MacFarlane J, MacDonald ML, Thompson J, Dube MP, Mattice M, Fraser R, Young C, Hossain S, Pape T, Payne B, Radomski C, Donaldson G, Ives E, Cox J, Younghusband HB, Green R, Duff A, Boltshauser E, Grinspan GA, Dimon JH, Sibley BG, Andria G, Toscano E, Kerdraon J, Bowsher D, Pimstone SN, Samuels ME, Sherrington R, Hayden MR (2007) Loss-of-function mutations in the Nav1.7 gene underlie congenital indifference to pain in multiple human populations. Clin Genet 71:311–319

Green JF, Schmidt ND, Schultz HD, Roberts AM, Coleridge HM, Coleridge JC (1984) Pulmonary C-fibers evoke both apnea and tachypnea of pulmonary chemoreflex. J Appl Physiol 57: 562–567

Groneberg DA, Niimi A, Dinh QT, Cosio B, Hew M, Fischer A, Chung KF (2004) Increased expression of transient receptor potential vanilloid-1 in airway nerves of chronic cough. Am J Respir Crit Care Med 170:1276–1280

Gutman GA, Chandy KG, Adelman JP, Aiyar J, Bayliss DA, Clapham DE, Covarriubias M, Desir GV, Furuichi K, Ganetzky B, Garcia ML, Grissmer S, Jan LY, Karschin A, Kim D, Kuperschmidt S, Kurachi Y, Lazdunski M, Lesage F, Lester HA, McKinnon D, Nichols CG, O'Kelly I, Robbins J, Robertson GA, Rudy B, Sanguinetti M, Seino S, Stuehmer W, Tamkun MM, Vandenberg CA, Wei A, Wulff H, Wymore RS (2003) International Union of Pharmacology. XLI. Compendium of voltage-gated ion channels: Potassium channels. Pharmacol Rev 55:583–586

Helke CJ (2005) Vagal afferent neurons: Neurotrophic factors and epigenetic influences. In: Undem BJ, Weinreich D (eds) Advances in vagal afferent neurobiology. CRC, Boca Raton, pp 27–74

Herzog RI, Cummins TR, Waxman SG (2001) Persistent TTX-resistant Na + current affects resting potential and response to depolarization in simulated spinal sensory neurons. J Neurophysiol 86:1351–1364

Hinman A, Chuang Hh, Bautista DM, Julius D (2006) TRP channel activation by reversible covalent modification. PNAS 103:19564–19568

Hwang SW, Cho H, Kwak J, Lee SY, Kang CJ, Jung J, Cho S, Min KH, Suh YG, Kim D, Oh U (2000) Direct activation of capsaicin receptors by products of lipoxygenases: Endogenous capsaicin-like substances. PNAS 97:6155–6160

Ingram SL, Williams JT (1996) Modulation of the hyperpolarization-activated current (Ih) by cyclic nucleotides in guinea-pig primary afferent neurons. J Physiol (Lond) 492:97–106

Jarvis MF, Honore P, Shieh CC, Chapman M, Joshi S, Zhang XF, Kort M, Carroll W, Marron B, Atkinson R, Thomas J, Liu D, Krambis M, Liu Y, McGaraughty S, Chu K, Roeloffs R, Zhong C, Mikusa JP, Hernandez G, Gauvin D, Wade C, Zhu C, Pai M, Scanio M, Shi L, Drizin I, Gregg R, Matulenko M, Hakeem A, Gross M, Johnson M, Marsh K, Wagoner PK, Sullivan JP, Faltynek CR, Krafte DS (2007) A-803467, a potent and selective Nav1.8 sodium channel blocker, attenuates neuropathic and inflammatory pain in the rat. PNAS 104:8520–8525

Joad JP, Munch PA, Bric JM, Evans SJ, Pinkerton KE, Chen CY, Bonham AC (2004) Passive smoke effects on cough and airways in young guinea pigs: Role of brainstem substance P. Am J Respir Crit Care Med 169:499–504

Jordt SE, Bautista DM, Chuang Hh, McKemy DD, Zygmunt PM, Hogestatt ED, Meng ID, Julius D (2004) Mustard oils and cannabinoids excite sensory nerve fibres through the TRP channel ANKTM1. Nature 427:260–265

Kollarik M, Undem BJ (2004) Activation of bronchopulmonary vagal afferent nerves with bradykinin, acid and vanilloid receptor agonists in wild-type and TRPV1-/- mice. J Physiol (Lond) 555:115–123

Kollarik M, Dinh QT, Fischer A, Undem BJ (2003) Capsaicin-sensitive and -insensitive vagal bronchopulmonary C-fibres in the mouse. J Physiol (Lond) 551:869–879

Kwong K, Lee LY (2005) Prostaglandin E2 potentiates a TTX-resistant sodium current in rat capsaicin-sensitive vagal pulmonary sensory neurones. J Physiol (Lond) 564:437–450

Kwong K, Wu ZX, Kashon ML, Krajnak KM, Wise PM, Lee LY (2001) Chronic smoking enhances tachykinin synthesis and airway responsiveness in Guinea pigs. Am J Respir Cell Mol Biol 25:299–305

Kwong K, Carr MJ, Undem BJ (2007) TTX-R INa expression in sensory neurons innervating the lungs correlates with their embryonic origin in adult Guinea pigs. Am J Respir Crit Care Med 175:A666

Laird JMA, Souslova V, Wood JN, Cervero F (2002) Deficits in visceral pain and referred hyperalgesia in Nav1.8 (SNS/PN3)-null mice. J Neurosci 22:8352–8356

Lalloo UG, Fox AJ, Belvisi MG, Chung KF, Barnes PJ (1995) Capsazepine inhibits cough induced by capsaicin and citric acid but not by hypertonic saline in guinea pigs. J Appl Physiol 79: 1082–1087

Lambert RC, Maulet Y, Mouton J, Beattie R, Volsen S, De Waard M, Feltz A (1997) T-type $Ca2^+$ current properties are not modified by $Ca2^+$ channel beta ásubunit depletion in nodosus ganglion neurons. J Neurosci 17:6621–6628

Lee P, Eccles R (2004) Cough induced by mechanical stimulation of the upper airway in humans. Acta Oto-Laryngol 124:720–725

Lee BP, Morton RF, Lee LY (1992) Acute effects of acrolein on breathing: Role of vagal bronchopulmonary afferents. J Appl Physiol 72:1050–1056

Lee YJ, Zachrisson O, Tonge DA, McNaughton PA (2002) Upregulation of bradykinin B2 receptor expression by neurotrophic factors and nerve injury in mouse sensory neurons. Mol Cell Neurosci 19(2):186–200

Lee MG, MacGlashan DW, Jr., Undem BJ (2005) Role of chloride channels in bradykinin-induced guinea pig airway vagal C-fibre activation. J Physiol (Lond) 566:205–212

Leon A, Buriani A, Dal Toso R, Fabris M, Romanello S, Aloe L, Levi-Montalcini R (1994) Mast cells synthesize, store and release nerve growth factor. Proc Natl Acad Sci U S A 91(9):3739–3743

Macpherson LJ, Dubin AE, Evans MJ, Marr F, Schultz PG, Cravatt BF, Patapoutian A (2007a) Noxious compounds activate TRPA1 ion channels through covalent modification of cysteines. Nature 445:541–545

Macpherson LJ, Xiao B, Kwan KY, Petrus MJ, Dubin AE, Hwang S, Cravatt B, Corey DP, Patapoutian A (2007b) An ion channel essential for sensing chemical damage. J Neurosci 27:11412–11415

Mamet J, Lazdunski M, Voilley N (2003) How nerve growth factor drives physiological and inflammatory expressions of acid-sensing ion channel 3 in sensory neurons. J Biol Chem 278: 48907–48913

Mazzone SB, Geraghty DP (2000) Respiratory actions of tachykinins in the nucleus of the solitary tract: Characterization of receptors using selective agonists and antagonists. Br J Pharmacol 129:1121–1131

Mazzone SB, McGovern AE (2006) Na^+-K^+-2Cl- cotransporters and Cl- channels regulate citric acid cough in guinea pigs. J Appl Physiol 101:635–643

Mazzone SB, Mori N, Canning BJ (2005) Synergistic interactions between airway afferent nerve subtypes regulating the cough reflex in guinea-pigs. J Physiol 569:559–573

Mazzone SB, McLennan L, McGovern AE, Egan GF, Farrell MJ (2007a) Representation of capsaicin-evoked Urge-to-Cough in human brain using functional MRI. Am J Respir Crit Care Med 176:327–332

Mazzone SB, McLennan L, McGovern AE, Egan GF, Farrell MJ (2007b) Representation of Capsaicin-evoked Urge-to-Cough in the human brain using functional magnetic resonance imaging. Am J Respir Crit Care Med 176:327–332

McAlexander MA, Undem BJ (2000) Potassium channel blockade induces action potential generation in guinea-pig airway vagal afferent neurones. J Auton Nerv Syst 78:158–164

McAlexander MA, Myers AC, Undem BJ (1999) Adaptation of guinea-pig vagal airway afferent neurones to mechanical stimulation. J Physiol (Lond) 521:239–247

McKemy DD, Neuhausser WM, Julius D (2002) Identification of a cold receptor reveals a general role for TRP channels in thermosensation. Nature 416:52–58

McNamara CR, Mandel-Brehm J, Bautista DM, Siemens J, Deranian KL, Zhao M, Hayward NJ, Chong JA, Julius D, Moran MM, Fanger CM (2007) TRPA1 mediates formalin-induced pain. PNAS 104:13525–13530

Mitchell JE, Campbell AP, New NE, Sadofsky LR, Kastelik JA, Mulrennan SA, Compton SJ, Morice AH (2005) Expression and Characterization of the intracellular vanilloid receptor (TRPV1) in bronchi from patients with chronic cough. Exp Lung Res 31:295–306

Mohammed SP, Higenbottam TW, Adcock JJ (1993) Effects of aerosol-applied capsaicin, histamine and prostaglandin E2 on airway sensory receptors of anaesthetized cats. J Physiol (Lond) 469:51–66

Morris JB, Stanek J, Gianutsos G (1999) Sensory nerve-mediated immediate nasal responses to inspired acrolein. J Appl Physiol 87:1877–1886

Mutoh T, Bonham AC, Joad JP (2000) Substance P in the nucleus of the solitary tract augments bronchopulmonary C fiber reflex output. Am J Physiol Regul Integr Comp Physiol 279: R1215–R1223

Myers AC, Kajekar R, Undem BJ (2002) Allergic inflammation-induced neuropeptide production in rapidly adapting afferent nerves in guinea pig airways. Am J Physiol Lung Cell Mol Physiol 282:L775–L781

Nagata K, Duggan A, Kumar G, Garcia-Anoveros J (2005) Nociceptor and Hair Cell Transducer Properties of TRPA1, a Channel for Pain and Hearing. J Neurosci 25:4052–4061

Nassar MA, Stirling LC, Forlani G, Baker MD, Matthews EA, Dickenson AH, Wood JN (2004) Nociceptor-specific gene deletion reveals a major role for Nav1.7 (PN1) in acute and inflammatory pain. PNAS 101:12706–12711

Ni D, Gu Q, Hu HZ, Gao N, Zhu MX, Lee LY (2006) Thermal sensitivity of isolated vagal pulmonary sensory neurons: Role of transient receptor potential vanilloid receptors. Am J Physiol Regul Integr Comp Physiol 291:R541–R550

Nilius B, Voets T (2005) TRP channels: A TR(I)P through a world of multifunctional cation channels. Pflugers Archiv Eur J Physiol 451:1–10

O'Connell F, Springall DR, Moradoghli-Haftvani A, Krausz T, Price D, Fuller RW, Polak JM, Pride NB (1995) Abnormal intraepithelial airway nerves in persistent unexplained cough? Am J Respir Crit Care Med 152:2068–2075

Oh EJ, Weinreich D (2004) Bradykinin decreases K+ and increases Cl- conductances in vagal afferent neurones of the guinea pig. J Physiol (Lond) 558:513–526

Pape HC (1996) Queer current and pacemaker: The hyperpolarization-activated cation current in neurons. Annu Rev Physiol 58:299–327

Peier AM, Moqrich A, Hergarden AC, Reeve AJ, Andersson DA, Story GM, Earley TJ, Dragoni I, McIntyre P, Bevan S, Patapoutian A (2002) A TRP channel that senses cold stimuli and menthol. Cell 108:705–715

Priest BT, Murphy BA, Lindia JA, Diaz C, Abbadie C, Ritter AM, Liberator P, Iyer LM, Kash SF, Kohler MG, Kaczorowski GJ, MacIntyre DE, Martin WJ (2005) Contribution of the tetrodotoxin-resistant voltage-gated sodium channel NaV1.9 to sensory transmission and nociceptive behavior. PNAS 102:9382–9387

Rahman I, van Schadewijk AAM, Crowther AJL, Hiemstra PS, Stolk J, MacNee W, De Boer WI (2002) 4-Hydroxy-2-Nonenal, a specific lipid peroxidation product, is elevated in lungs of patients with chronic obstructive pulmonary disease. Am J Respir Crit Care Med 166:490–495

Ramer MS, Bradbury EJ, McMahon SB (2001) Nerve growth factor induces P2X(3) expression in sensory neurons. J Neurochem 77(3):864–875

Ramsey IS, Delling M, Clapham DE (2006) An introduction to TRP channels. Annu Rev Physiol 68:619–647

Raymond CK, Castle J, Garrett-Engele P, Armour CD, Kan Z, Tsinoremas N, Johnson JM (2004) Expression of alternatively spliced sodium channel {alpha}-subunit genes: Unique splicing patterns are observed in dorsal root ganglia. J Biol Chem 279:46234–46241

Renganathan M, Cummins TR, Waxman SG (2001) Contribution of Na(v)1.8 sodium channels to action potential electrogenesis in DRG neurons. J Neurophysiol 86:629–640

Riccio MM, Kummer W, Biglari B, Myers AM, Undem BJ (1996a) Interganglionic segregation of distinct vagal afferent fibre phenotypes in guinea-pig airways. J Physiol (Lond) 496:521–530
Riccio MM, Myers AM, Undem BJ (1996b) Immunomodulation of afferent neurons in guinea-pig isolated airway. J Physiol (Lond) 491(Part 2):499–509
Roberts AM, Kaufman MP, Baker DG, Brown JK, Coleridge HM, Coleridge JC (1981) Reflex tracheal contraction induced by stimulation of bronchial C-fibers in dogs. J Appl Physiol 51: 485–493
Rush AM, Waxman SG (2004) PGE2 increases the tetrodotoxin-resistant Nav1.9 sodium current in mouse DRG neurons via G-proteins. Brain Res 1023:264–271
Sangameswaran L, Fish LM, Koch BD, Rabert DK, Delgado SG, Ilnicka M, Jakeman LB, Novakovic S, Wong K, Sze P, Tzoumaka E, Stewart GR, Herman RC, Chan H, Eglen RM, Hunter JC (1997) A novel tetrodotoxin-sensitive, voltage-gated sodium channel expressed in rat and human dorsal root ganglia. J Biol Chem 272:14805–14809
Sanico AM, Stanisz AM, Gleeson TD, Bora S, Proud D, Bienenstock J, Koliatsos VE, Togias A (2000) Nerve growth factor expression and release in allergic inflammatory disease of the upper airways. Am J Respir Crit Care Med 161:1631–1635
Shim WS, Tak MH, Lee MH, Kim M, Kim M, Koo JY, Lee CH, Kim M, Oh U (2007) TRPV1 mediates histamine-induced itching via the activation of phospholipase A2 and 12-Lipoxygenase. J Neurosci 27:2331–2337
Shin JB, Martinez-Salgado C, Heppenstall PA, Lewin GR (2003) A T-type calcium channel required for normal function of a mammalian mechanoreceptor. Nat Neurosci 6:724–730
Sim VM, Pattle RE (1957) Effect of possible smog irritants on human subjects. J Am Med Assoc 165:1908–1913
Smith GD, Gunthorpe MJ, Kelsell RE, Hayes PD, Reilly P, Facer P, Wright JE, Jerman JC, Walhin JP, Ooi L, Egerton J, Charles KJ, Smart D, Randall AD, Anand P, Davis JB (2002) TRPV3 is a temperature-sensitive vanilloid receptor-like protein. Nature 418:186–190
Story GM, Peier AM, Reeve AJ, Eid SR, Mosbacher J, Hricik TR, Earley TJ, Hergarden AC, Andersson DA, Hwang SW (2003) ANKTM1, a TRP-like channel expressed in nociceptive neurons, is activated by cold temperatures. Cell 112:819–829
Sung KW, Kirby M, McDonald MP, Lovinger DM, Delpire E (2000) Abnormal GABAA receptor-mediated currents in dorsal root ganglion neurons isolated from Na-K-2Cl cotransporter null mice. J Neurosci 20:7531–7538
Suzuki M, Mizuno A, Kodaira K, Imai M (2003) Impaired pressure sensation in mice lacking TRPV4. J Biol Chem 278:22664–22668
Symanowicz PT, Gianutsos G, Morris JB (2004) Lack of role for the vanilloid receptor in response to several inspired irritant air pollutants in the C57Bl/6J mouse. Neurosci Lett 362:150–153
Tatar M, Webber SE, Widdicombe JG (1988) Lung C-fibre receptor activation and defensive reflexes in anaesthetized cats. J Physiol (Lond) 402:411–420
Tatar M, Sant'Ambrogio G, Sant'Ambrogio FB (1994) Laryngeal and tracheobronchial cough in anesthetized dogs. J Appl Physiol 76:2672–2679
Taylor-Clark T, Undem BJ, Ghatta S, McAlexander MA (2007a) 4-Hydroxynonenal: A potential endogenous activator of transient receptor potential A1 (TRPA1) channels. Am J Respir Crit Care Med 175:667
Taylor-Clark TE, Undem BJ, MacGlashan J, Ghatta S, Carr MJ, McAlexander MA (2007b) Prostaglandin-induced activation of nociceptive neurons via direct interaction with TRPA1. Mol Pharmacol Nov 13; [Epub ahead of print]
Tominaga M, Caterina MJ, Malmberg AB, Rosen TA, Gilbert H, Skinner K, Raumann BE, Basbaum AI, Julius D (1998) The cloned capsaicin receptor integrates multiple pain-producing stimuli. Neuron 21:531–543
Trevisani M, Milan A, Gatti R, Zanasi A, Harrison S, Fontana G, Morice AH, Geppetti P (2004) Antitussive activity of iodo-resiniferatoxin in guinea pigs. Thorax 59:769–772
Trevisani M, Siemens J, Materazzi S, Bautista DM, Nassini R, Campi B, Imamachi N, Andre E, Patacchini R, Cottrell GS, Gatti R, Basbaum AI, Bunnett NW, Julius D, Geppetti P (2007)

4-Hydroxynonenal, an endogenous aldehyde, causes pain and neurogenic inflammation through activation of the irritant receptor TRPA1. PNAS 104:13519–13524

Undem BJ, Weinreich D (1993) Electrophysiological properties and chemosensitivity of guinea pig nodose ganglion neurons in vitro. J Auton Nerv Syst 44:17–33

Undem BJ, Oh EJ, Lancaster E, Weinreich D (2003) Effect of extracellular calcium on excitability of guinea pig airway vagal afferent nerves. J Neurophysiol 89:1196–1204

Vellani V, Zachrisson O, McNaughton PA (2004) Functional bradykinin B1 receptors are expressed in nociceptive neurones and are upregulated by the neurotrophin GDNF. J Physiol 560:391–401

Villarreal CF, Sachs D, Queiroz Cunha F, Parada CA, Ferreira SH (2005) The role of NaV1.8 sodium channel in the maintenance of chronic inflammatory hypernociception. Neurosci Lett 386:72–77

Virchow J, Julius P, Lommatzsch M, Luttman W, Renz H, Braun A (1998) Neurotrophins are increased in bronchoalveolar lavage fluid after segmental allergen provocation. Am J Respir Crit Care Med 158:2002–2005

Waxman SG (2007a) Channel, neuronal and clinical function in sodium channelopathies: From genotype to phenotype. Nat Neurosci 10:405–409

Waxman SG (2007b) Nav1.7, its mutations, and the syndromes that they cause. Neurology 69: 505–507

Wu X, Myers AC, Goldstone AC, Togias A, Sanico AM (2006) Localization of nerve growth factor and its receptors in the human nasal mucosa. J Allergy Clin Immunol 118:428–433

Xu H, Ramsey IS, Kotecha SA, Moran MM, Chong JA, Lawson D, Ge P, Lilly J, Silos-Santiago I, Xie Y, DiStefano PS, Curtis R, Clapham DE (2002) TRPV3 is a calcium-permeable temperature-sensitive cation channel. Nature 418:181–186

Yang Y, Wang Y, Li S, Xu Z, Li H, Ma L, Fan J, Bu D, Liu B, Fan Z, Wu G, Jin J, Ding B, Zhu X, Shen Y (2004) Mutations in SCN9A, encoding a sodium channel alpha subunit, in patients with primary erythermalgia. J Med Genet 41:171–174

Yoshimura N, de Groat WC (1999) Increased excitability of afferent neurons innervating rat urinary bladder after chronic bladder inflammation. J Neurosci 19:4644–4653

Yoshimura N, Bennett NE, Hayashi Y, Ogawa T, Nishizawa O, Chancellor MB, de Groat WC, Seki S (2006) Bladder overactivity and hyperexcitability of bladder afferent neurons after intrathecal delivery of nerve growth factor in rats. J Neurosci 26:10847–10855

Zhang X, Huang J, McNaughton PA (2005) NGF rapidly increases membrane expression of TRPV1 heat-gated ion channels. EMBO J 24:4211–4223

Zygmunt PM, Petersson J, Andersson DA, Chuang Hh, Sorgard M, Di Marzo V, Julius D, Hogestatt ED (1999) Vanilloid receptors on sensory nerves mediate the vasodilator action of anandamide. Nature 400:452–457

Peripheral Mechanisms II: The Pharmacology of Peripherally Active Antitussive Drugs

D. Spina, I. McFadzean, F.K.R. Bertram, and C.P. Page(✉)

Contents

Abstract Cough is an indispensable defensive reflex. Although generally beneficial, it is also a common symptom of diseases such as asthma, chronic obstructive pulmonary disease, upper respiratory tract infections, idiopathic pulmonary fibrosis and lung cancer. Cough remains a major unmet medical need and although the centrally acting opioids have remained the antitussive of choice for decades, they have

C.P. Page
Sackler Institute of Pulmonary Pharmacology, Division of Pharmaceutical Sciences, School of Biomedical and Health Sciences, King's College, London SE1 1UL, UK
clive.page@kcl.ac.uk

K.F. Chung, J.G. Widdicombe (eds.), *Pharmacology and Therapeutics of Cough*, Handbook of Experimental Pharmacology 187,

many unwanted side effects. However, new research into the behaviour of airway sensory nerves has provided greater insight into the mechanisms of cough and new avenues for the discovery of novel non-opioid antitussive drugs. In this review, the pathophysiological mechanisms of cough and the development of novel antitussive drugs are reviewed.

1 Introduction

Cough is a reflex mechanism protecting the airways by ejecting obstructive or harmful substances. The afferent component of the reflex involves vagal sensory nerves with bodies in the jugular or nodose ganglia and terminals within the airway wall, activated by a combination of chemical and/or mechanical stimuli. Up to five subtypes of sensory airway afferents have been characterized (Canning 2006), all synapsing in the nucleus tractus solitarius (nTS) of the brainstem, from where second-order neurones project to the medullary respiratory pattern generator to initiate cough. Whilst acute cough (lasting a few weeks) is not a serious clinical problem, it is the most common respiratory ailment for which patients seek medical help and around $1 billion is spent per annum in the USA on cough remedies (McGarvey and Morice 2006). Chronic cough (persisting for more than 8 weeks) is a serious debilitating condition estimated to affect some 14% of the population (Cullinan 1992). It can be a symptom of both pulmonary conditions, such as asthma (Dicpinigaitis 2006b), chronic obstructive pulmonary disease (COPD) (Braman 2006), idiopathic pulmonary fibrosis (Hope-Gill et al. 2003), upper-airway cough syndrome (postnasal drip) (Pratter 2006) and lung cancer (Kvale 2006), and extrapulmonary conditions, such as gastro-oesophageal reflux (Irwin 2006). It has an adverse effect on quality of life, being associated with complications ranging from frequent retching through incontinence and insomnia to social embarrassment (French et al. 1998, 2004, 2005). Treatment of the underlying disease often resolves chronic cough (Morice et al. 2004; Pratter et al. 2006), but this can take several months, during which time effective antitussive treatment would be desirable. Furthermore, there remains a significant cohort ($\sim$40% of patients) (O'Connell et al. 1994) in which either the underlying cause is known but is irreversible or there is no identifiable underlying cause. In these individuals, relief from chronic cough depends upon effective antitussive drug treatment. Available drugs are of either limited effectiveness (Schroeder and Fahey 2002, 2004) or have problematic side effects and consequently there is a pressing need for safer, more effective agents (Dicpinigaitis 2004; Morice and Geppetti 2004; Reynolds et al. 2004).

Current antitussives are broadly divided according to their site of action as either central or peripheral, although many act to some extent at both locations. Centrally acting antitussives act within the central nervous system (CNS) to suppress central cough pathways and comprise the majority of currently used drugs. However, opioids, which are gold standard antitussive agents, suffer from numerous unwanted side effects, including sedation, constipation and drug dependence. There is therefore a clear and unmet need for the development of novel drugs which target airway afferent nerves directly and/or that target the processes that give rise to central sensitization of the cough reflex, possibly at the level of the nTS.

2 Mechanisms of Hypertussive States

In chronic, hypertussive cough states, normally innocuous stimuli that would not normally initiate a response, evoke cough (Bonham et al. 2004). There are clear parallels here with alterations during chronic pain states, characterized by hyperalgesia and allodynia. The mechanisms underlying such alterations have been well studied (Ji and Woolf 2001; Melzack et al. 2001; Woolf and Salter 2000) and can be subdivided into 'peripheral' and 'central' sensitization.

2.1 Peripheral Sensitization

Peripheral sensitization involves hyperexcitability of sensory neurones due to, for example, changes in voltage-gated sodium channel properties (Lai et al. 2004). Under inflammatory or disease conditions, many pathological changes can occur around and within sensory nerve fibres, leading to increased excitability as well as phenotypic changes in receptor and neurotransmitter expression. For instance, airway mechanosensitive Aδ-fibres do not contain neuropeptides under physiological conditions, but following viral and/or allergen challenge begin to synthesize neuropeptides (Carr et al. 2002; Myers et al. 2002). In addition, the excitability of airway Aδ-fibres and nTS neurones can be increased by antigen stimulation of the lungs of sensitized animals (Chen et al. 2001; Undem et al. 2002). Similarly, chronic exposure of guinea pigs to sidestream tobacco smoke augments pulmonary C-fibre responses to capsaicin and to lung hyperinflation (Mutoh et al. 1999, 2000). In addition, allergic airway inflammation has been reported to increase neuropeptide expression (substance P, neurokinin A and calcitonin gene related peptide, CGRP) in sensory neurones originating in the guinea pig airway (Fischer and Hoffmann 1996; Myers et al. 2002). Together, these experimental studies suggested that sensory nerve plasticity can occur during inflammatory insults to the lung.

This view is also supported by clinical studies which have reported an increase in the number of TRPV1 immunopositive neurones (Groneberg et al. 2004) and the density of neuropeptide immunoreactive nerves (Lee et al. 2003; O'Connell et al. 1995) in the airways of individuals with chronic cough syndromes. This plasticity of the sensory neurones mediating cough provides a plausible mechanism leading to the hypertussive state, or 'sensitization'. The mechanisms contributing to the sensitization of airway nerves may be a more rational target for antitussive drug development rather than the 'hard-wired' cough reflex since it is generally accepted that inhibition of cough per se is probably of limited clinical benefit as this protective mechanism is of critical importance for clearing unwanted airway secretions.

2.2 Central Sensitization

In pain pathways, central sensitization involves enhanced efficiency of excitatory synaptic transmission pathways within the CNS, notably in the dorsal horn of the spinal cord at the synapse between the sensory nociceptor fibres and second-order relay neurones (Ji et al. 2003). It has been suggested that central sensitization in the cough reflex loop might also underlie hypertussive states (Bonham et al. 2004, 2006a, b), but to date, there are few data confirming this.

In the spinal cord, central sensitization manifests itself as increased postsynaptic excitability of second-order relay neurones following a period of high-intensity stimulation of nociceptor afferents (Melzack et al. 2001; Woolf and Salter 2000). The relay neurones normally have a high threshold, responding only to the high-intensity stimulation associated with the 'pain' response (Woolf and King 1989). Following central sensitization, the threshold is lowered such that normally innocuous stimuli now activate the relay neurones (Cook et al. 1987; Woolf and King 1990; Woolf and Wall 1986). The primary neurotransmitter at the synapse is glutamate, acting postsynaptically on both ionotropic (*N*-methyl-D-aspartate, NMDA; and α-amino-3-hydroxy-5-methyl-4-isoxazolepropionate, AMPA) receptors and metabotropic receptors (Andresen and Yang 1995; Chen et al. 1999; Drewe et al. 1990; Fortin and Champagnat 1993; Mifflin and Felder 1990; Smith et al. 1998; Wang and Bradley 1995), whilst γ-aminobutyric acid (GABA) acting on both $GABA_A$ and $GABA_B$ receptors mediates inhibitory synaptic responses (Brooks et al. 1992; Smith et al. 1998; Wang and Bradley 1995). There is also evidence that substance P is released (Colin et al. 2002; Duggan et al. 1990) and acts on neurokinin 1 (NK_1) receptors to cause postsynaptic excitation and presynaptic modulation of glutamate and GABA release (Bailey et al. 2004; Sekizawa et al. 2003) thereby modulating the responsiveness of the postsynaptic cells.

Several mechanisms have been reported to contribute to the increase in synaptic efficacy (Ji et al. 2003; Ji and Woolf 2001; Malcangio and Lessmann 2003; Park and Vasko 2005). Relatively low frequency ($\sim$5 Hz) stimulation of nociceptors can cause 'windup' whereby successive stimuli cause increasing degrees of postsynaptic depolarization owing to removal of the voltage-dependent magnesium block of NMDA receptors. Windup is short-lived, dissipating on removal of the synaptic input, but the calcium entry associated with NMDA receptor activation, allied to that through voltage-gated channels, activates intracellular signalling cascades that lead to longer-lasting changes in the expression and function of receptors and ion channels (Caudle et al. 2005; Gao et al. 2005; Guo et al. 2002; Ji et al. 1999; Pezet et al. 2005; Suzuki et al. 2005; Yang et al. 2004; Zou et al. 2000).

There is good evidence to suggest that a similar scenario might operate in the nTS following stimulation of airway afferent nerves during an inflammatory insult to the airways. Insults that can induce a hypertussive state produce changes in the presynaptic inputs to the nTS, but whether they also produce postsynaptic changes indicative of central sensitization is less clear. This view is supported by various experimental findings. The increased C-fibre responsiveness to capsaicin in guinea pigs exposed to tobacco smoke is paralleled by increased firing of nTS relay neurones (Bonham et al. 2001; Mutoh et al. 2000); extended allergen exposure and episodic exposure to ozone can also cause an increase in excitability of primate nTS neurones (Chen et al. 2001, 2003). Furthermore, the intracerebroventricular administration of neurokinin antagonists inhibited the ability of airway C-fibre stimulation to increase the sensitivity of the cough response to mechanical irritation of the airway mucosa (Mazzone et al. 2005). The release of sensory neuropeptides centrally into the brainstem might facilitate synaptic input from mechanosensitive afferent nerves in the lung that are involved in cough.

Clearly many sensory afferent fibre types contribute to or modulate the cough reflex. The integration of their signals occurs at the level of the nTS, found in the dorsal medulla. Here, both pulmonary and extrapulmonary (from other vagally innervated organs) afferent fibres terminate and provide polysynaptic input to second-order neurones (Jordan 2001). Although subject to substantial cortical control, these second-order neurones likely alter the activity of the respiratory neurones typically responsible for normal breathing to produce cough (Pantaleo et al. 2002). Each of the synapses in this 'cough centre' is a potential pharmacological target for centrally acting antitussives.

3 Role of Sensory Nerves in the Cough Reflex

The cough reflex is known to include a 'hard-wiring' circuit as recently proposed (Canning et al. 2004; Canning 2006). The stimuli initiating the cough reflex stimulate sensory nerve fibres that have been broadly divided in to three main groups: Aδ-fibres, C-fibres and slowly adapting stretch receptors (SARs). These fibres have been differentiated on the basis of their neurochemistry, anatomical location, conduction velocity, physiochemical sensitivity and adaptation to lung inflation (Canning et al. 2004; Canning 2006; Reynolds et al. 2004).

3.1 Aδ-Fibres

Rapidly adapting receptors (RARs) are thought to terminate within or slightly beneath the epithelium throughout the intrapulmonary airways and respond to changes in airway mechanics to regulate normal breathing (Widdicombe 2003). These fibres are sensitive to most tussive stimuli and it is likely their stimulation is of primary importance in the elicitation of the cough reflex (Widdicombe 2003). In general, their activity is increased by mechanical stimuli such as mucus secretion or oedema, but they are insensitive to many chemical stimuli that provoke cough, including bradykinin and capsaicin (Lee et al. 2001). However, in the guinea pig at least, the variability in mechanical and chemical sensitivities of Aδ-fibres is sufficient to believe there may be as many as three subdivisions: RAR-like, nociceptive and polymodal Aδ-fibres (Mazzone 2005).

The RAR-like Aδ-fibres are very responsive to mechanical stimulation, unresponsive to direct chemical stimuli like bradykinin and capsaicin, and have their cell bodies in the nodose ganglia (Hunter and Undem 1999; Riccio et al. 1996). By comparison, nociceptive Aδ-fibres are sensitive to capsaicin and bradykinin, are 15 times less responsive to mechanical stimulation, have their cell bodies in the jugular ganglia and are mainly carried in the superior laryngeal nerve. The role of RARs as primary initiators of cough has been disputed following a number of observations (Mazzone and McGovern 2007). For example, cough sensitivity in healthy subjects is not altered by bronchoconstrictor agents which would be expected to activate RARs (Fujimura et al. 1996); exogenous administration of neuropeptides which would be anticipated to activate RARs indirectly owing to a combination of bronchoconstriction, mucus secretion and oedema was ineffective at evoking cough in healthy subjects, although cough to this stimulus is observed in patients with inflamed lungs (Hope-Gill et al. 2003; Sekizawa et al. 1996); and sensory nerve activation and, by extension, release of sensory neuropeptides failed to initiate cough following indirect activation of RAR's in anaesthetized guinea pigs (Canning et al. 2004; Mazzone et al. 2005). It seems likely that activation of RARs resulting in cough may be of greater relevance in disease. Nonetheless, these inconsistencies have led investigators to further probe the role of Aδ-fibres in cough.

Only recently have the polymodal Aδ-fibres been identified. These fibres are similar to RAR-like fibres in that they originate in the nodose ganglia, are activated by mechanical stimulation and acid but are unresponsive to capsaicin, bradykinin, smooth-muscle contraction or stretching of the airways. Interestingly, the conduction velocity of these polymodal Aδ-fibres is much slower than the conduction velocity of RARs or SARs and is much faster than that of C-fibres. Furthermore, severing the recurrent laryngeal nerves which supply innervation to the trachea abolished cough in response to mechanical irritation of the tracheal mucosal surface in guinea pigs (Canning et al. 2004). It has been proposed that the primary function of these fibres is the elicitation of cough and as such they may be regarded as the 'hard-wiring' of the cough reflex (Canning et al. 2004; Mazzone et al. 2005).

3.2 C-Fibres

C-fibres play an important role in airway defensive reflexes. They respond to both mechanical (though with a higher threshold than Aδ-fibres) and chemical stimuli, including sulphur dioxide, capsaicin and bradykinin (Lee et al. 2001). In certain species they evoke the peripheral release of neuropeptides such as substance P, neurokinin A and CGRP via an axon reflex which leads to bronchoconstriction and neurogenic inflammation. Neuropeptide-dependent airway smooth-muscle contraction, oedema and mucus secretion can activate RARs. However, human airways contain very few substance P-containing nerve fibres and at present there is a lack of evidence indicating that these nerves correspond to the terminals of capsaicin-sensitive C-fibres (Lundberg et al. 1984). Consistent with this observation, human airways from non-diseased individuals only contract modestly in response to capsaicin when compared with those of guinea pigs (Spina et al. 1998), although TRPV1-positive nerves have been described in the lung which do not contain sensory neuropeptides (Kagaya et al. 2002; Watanabe et al. 2005, 2006), suggesting that capsaicin-mediated effects may be able to occur independently of the release of neuropeptides. Similarly, TRPV1 receptors have been identified on Aδ nociceptive fibres (Myers et al. 2002), which under normal physiological conditions do not synthesize neuropeptides, but can be activated by capsaicin. If these fibres were involved in the sensitization of the cough reflex, then this might explain the inability of neurokinin antagonists to modify cough clinically (Fahy et al. 1995). However, it is also likely that changes in C-fibre activity may occur in disease and hence these nerves would be anticipated to play a role in hypertussive states. Indeed, cough sensitivity to capsaicin is heightened in respiratory disease (Hope-Gill et al. 2003; Nakajima et al. 2006; O'Connell et al. 1996; Weinfeld et al. 2002) and the density of TRPV1- and neuropeptide-containing nerves is increased in chronic cough states (Groneberg et al. 2004; Lee et al. 2003; O'Connell et al. 1995).

A critical role for C-fibres in cough has been proposed since this response can be induced by citric acid, capsaicin and bradykinin, all agents known to be stimulants of C-fibres in humans and animals (Karlsson 1996). In animal studies, chronic pretreatment with capsaicin to deplete C-fibres of their sensory neuropeptides abolished cough as a response to citric acid in conscious animals, without affecting mechanically induced cough (Forsberg et al. 1988). It is plausible that high doses of capsaicin might also impair the function of a subpopulation of Aδ nociceptive fibres which express TRPV1 (Myers et al. 2002) and therefore might have contributed to the functional impairment to cough in response to citric acid. Furthermore cough induced by capsaicin and citric acid was abolished when animals were pretreated with neurokinin antagonists (Bolser et al. 1997); whilst cough induced by capsaicin, cigarette smoke and bronchospasm in guinea pigs is inhibited with neurokinin antagonists (Yasumitsu et al. 1996) and by promoting degradation of endogenously released neuropeptides following treatment with neutral endopeptidase (Kohrogi et al. 1989), suggesting a role for neuropeptides in this cough response, at least in guinea pigs.

However, there is evidence to show that activation of C-fibres does not evoke cough. Anaesthetized animals failed to cough in response to capsaicin and bradykinin, yet the topical application of citric acid or mechanical irritation of the mucosal surface still evoked a cough response (Canning et al. 2004). This suggested that activation of C-fibres does not incite cough per se. In several studies, the systemic administration of capsaicin actually inhibited cough following mechanical stimulation of the upper airways in anaesthetized animals (Tatar et al. 1988, 1994). In contrast to these findings, the subthreshold stimulation of C-fibres by capsaicin or bradykinin administered topically to the lung increased cough sensitivity to mechanical irritation of the upper airways (Mazzone et al. 2005) and could therefore help to explain why direct stimulation of these nerves is not necessary to induce cough per se, but is likely to 'sensitize' the cough reflex.

3.3 Slowly Adapting Stretch Receptors

The role of SARs in cough has received very little attention, although they are not believed to be directly involved, as their activity is not altered by stimuli evoking cough. However, experimental evidence in cats and rabbits suggested that SARs may facilitate the cough reflex, via interneurones called 'pump cells', by mechanisms that remain to be established, which either permit or augment the cough reflex due to RAR activity (Shannon et al. 2000).

4 Antitussive Drugs

Most cough treatments are designed to target the underlying disease which can cause cough as a symptom. However, there are a range of drugs available which directly target neuronal pathways (Fig. 1).

4.1 Sodium Channel Blockers

At least nine subtypes of sodium channels have been characterized (Alexander et al. 2007) and of particular interest are the tetrodotoxin-insensitive sodium channels ($Na_V 1.8$, Na_V 1.9) expressed on pulmonary C-fibres (Kwong and Lee 2005).

Local anaesthetics such as lidocaine, benzonatate and mexiletine are the most consistently effective peripherally acting antitussive drugs used to treat cough that is resistant to other treatments; however, their effects are transient and their repeated use is associated with tachyphylaxis, making high doses necessary, which can lead to unwanted side effects (Yukioka et al. 1985). Their mechanism of action is believed to be through use-dependent inhibition of voltage-gated sodium channels, thereby reducing action potential generation and transmission in afferent nerves. However, whilst topical administration of lidocaine to the airways has been reported to inhibit cough induced by capsaicin in healthy subjects (Choudry et al. 1990), the orally acting local anaesthetic mexiletine only attenuated cough induced by tartaric acid and not by capsaicin (Fujimura et al. 2000). The explanation for these observations might be explained by pharmacodynamic differences due to the routes of drug administration, although this remains to be determined.

RSD 931 is a quaternary ammonium compound (carcainium chloride) and a structural analogue of lidocaine, that exhibits antitussive activity in both guinea

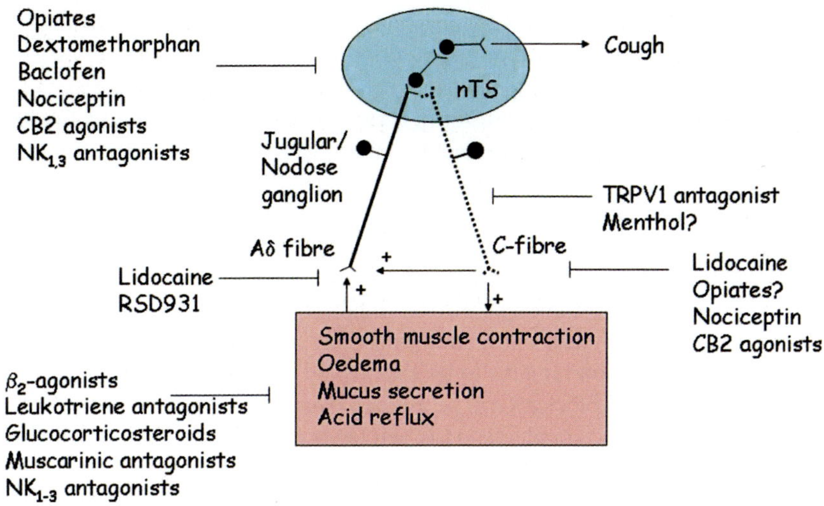

Fig. 1 The cough reflex and sites of action of some antitussive agents. Airway sensory nerves activated in response to a protussive stimulus travel though the vagus nerve to the medulla, where they terminate in the nucleus tractus solitarius (nTS). Second-order neurons relay the message to the respiratory pattern generator, which modifies the activity of the inspiratory and expiratory motor neurons and leads to cough. Antitussives can act peripherally at the level of the airway receptors or on nerve conduction. They can also act centrally, both presynaptically and postsynaptically. C-fibres are depicted as playing a role in facilitating nTS activity. The activation of their terminal processes by inflammatory mediators such as 15-hydroperoxyeicosatetraenoic acid, bradykinin, *N*-arachidonoyl dopamine or low pH stimulates C-fibre input into the nTS, thereby facilitating central processing of afferent information received from Aδ-fibres innervating the proximal airways. Also note that the Aδ-fibres shown consist of nociceptive Aδ-fibres, 'cough' receptor and rapidly adapting receptors. It is the lattermost subset which can be activated by sensory neuropeptides either directly or indirectly via bronchoconstriction, oedema and mucus secretion

pigs and rabbits (Adcock et al. 2003). Whilst RSD 931 is a weak local anaesthetic, in the rabbit it only inhibited spontaneous and histamine-evoked discharges from Aδ-fibres, whilst modestly activating pulmonary C-fibres, which is distinctive from the effect of lidocaine, which suppressed all nerve fibres in the airway (Adcock et al. 2003). Such results suggest that RSD931 is antitussive via a selective effect on Aδ-fibres independent of local anaesthetic effects. Another structurally related analogue of lidocaine, JMF2-1, has also been described to have anti-inflammatory activity and to be able to inhibit bronchospasm, whilst being 2 orders of magnitude weaker than lidocaine as a Na^+ channel blocker (da Costa et al. 2007), although it is unclear whether this agent has antitussive activity.

Citric acid is a tussive stimulus that is thought to cause cough by direct activation of a recently described subset of Aδ fibres referred to as 'cough' receptors (Canning et al. 2004). The precise molecular target for citric acid on these nerves has not been identified, but pharmacological studies suggested the potential involvement of acid-sensing ion channels (ASICs) (Canning et al. 2006). ASICs are members of the Na^+ channel superfamily that includes the epithelium Na^+ channel (ENaC) and the degenerins (DEG) that support Na^+ conductance in response to lowering of extracellular pH. The activation of these ion channels on afferent nerves either in the oesophagus or following microaspiration of acid into the respiratory tract has been implicated in the cough resulting from gastro-oesophageal reflux (Kollarik et al. 2007). Amiloride is a non-selective inhibitor of ASICs and partially inhibited cough induced by citric acid in an anaesthetized-guinea-pig model (Canning et al. 2006), although an effect on other Na^+ channels in this response cannot be ruled out. These experimental data are also consistent with a report showing the ability of amiloride to suppress cough induced by acetic acid in asthmatic children (Mochizuki et al. 1995), although whether inhibition of ASICs on epithelial cells resulting in changes in local concentration of ions (e.g. chloride) was responsible for this antitussive action remains to be established. The development of subtype-selective ASIC channel blockers will be useful for further investigation of the role of these ion channels in the cough response, and to ascertain their utility as antitussive drugs.

4.2 Tachykinin Receptor Antagonists

Tachykinins are a group of peptides, including substance P, neurokinin A and neurokinin B, which are located in the peripheral endings of capsaicin-sensitive primary afferent neurones (C-fibres). Neuropeptides have been implicated in cough as their release from C-fibres may stimulate RARs either directly, or indirectly as a consequence of airway smooth-muscle contraction and increased vascular permeability leading to mucosal oedema and/or mucus secretion (Widdicombe 1995). Hence, tachykinin antagonists are most likely to act in the peripheral lung to inhibit the actions of endogenously released neuropeptides and thereby ameliorate cough to tussive agents which also stimulate the activation and release of sensory neuropeptides from sensory C-fibres (e.g. citric acid).

However, direct stimulation of the ‘cough receptor’ in anaesthetized guinea pigs by citric acid was unaffected by systemic or topical application of neurokinin antagonists, thereby ruling out a role for the release of sensory neuropeptides into the airways and subsequent modulation of the activity of these Aδ-fibres (Canning et al. 2006; Mazzone et al. 2005). In contrast, administration of neurokinin antagonists directly into the nTS, whilst not preventing cough per se, did attenuate the sensitization of the cough reflex following local application of capsaicin to the airways (Mazzone et al. 2005). Therefore, a central site of action cannot be entirely ruled out in view of the purported role of brainstem neuropeptides in central sensitization (Mazzone et al. 2005).

A number of potent tachykinin antagonists have been developed as antitussive drugs. The NK_1 antagonists FK 888, CP-99,994 and NKP608 inhibited cough induced by tobacco smoke and citric acid in guinea pigs, and mechanical stimulation of the trachea in anesthetized cats (Bolser et al. 1997; Chapman et al. 2004; El-Hashim et al. 2004; Yasumitsu et al. 1996). CP-99,994 can cross the blood–brain barrier and interference with neuropeptide transmission within the nTS and/or other regions of the CNS involved in cough may have accounted for the antitussive activity of this drug (Bolser et al. 1997). However, in healthy human subjects CP-99,994 did not inhibit bronchoconstriction or cough induced by hypertonic saline (Fahy et al. 1995), which is consistent with the view that the cough receptor is distinct from nociceptors.

The neurokinin 2 (NK_2) receptor antagonist (SR 48968) suppressed the cough reflex and was more potent than codeine in the guinea pig (Girard et al. 1995). Interestingly both SR48968 and codeine only partially inhibited the cough reflex when administered by the inhaled route (Advenier and EmondsAlt 1996; Yasumitsu et al. 1996), possibly reflecting differences in the concentration of these antagonists at receptor sites following the different routes of administration.

Neurokinin 3(NK_3) receptors have received very little attention in relation to the cough reflex. Nevertheless the NK_3 receptor antagonist SR142,801 inhibited cough induced by citric acid in guinea pigs and in pigs (Daoui et al. 1998; Moreaux et al. 2000). Another high-binding-affinity, non-peptide NK_3 receptor antagonist, SB 235375 (Giardina et al. 1996), having low CNS penetration, was also effective against cough in the guinea pig (Hay et al. 2002) and could be a useful compound as a novel peripherally acting antitussive drug.

4.3 NOP Receptor Agonists

Nociceptin/orphanin FQ (NOP) is an opioid-like peptide and is the endogenous ligand for the NOP1 receptor (Meunier et al. 1995). NOP1 receptors are $G_{i/o}$-coupled receptors that mediate presynpatic inhibition of neurotransmitter release (Meis and Pape 2001) and are widely distributed in the CNS and on peripheral airway nerves (Fischer et al. 1998). Nociceptin has been found to inhibit the release of sensory neuropeptides following depolarization of C-fibres (Shah et al. 1998), but interestingly did not prevent neuropeptide release from C-fibre terminal endings following direct activation of TRPV1 (Fischer et al. 1998; Shah et al. 1998). However, the antidromic stimulation of C-fibres by capsaicin is sensitive to nociceptin (Corboz et al. 2000), presumably reflecting a preferential inhibition of the opening of N-type calcium channels induced by this stimulus (Buchan and Adcock 1992).

Nociceptin inhibited cough induced by mechanical stimuli and capsaicin in guinea pigs and cats (McLeod et al. 2002) and induced by citric acid in guinea-pigs (Lee et al. 2006), and the antitussive action of the non-peptide NOP agonist Ro-64–6198 demonstrates the possibility of developing novel drugs free from side effects normally associated with opioid treatment (McLeod et al. 2004). Hence, NOP1 receptors are involved in the modulation of the cough reflex and selective NOP1 agonists may have potential as novel peripherally acting antitussives, although to date there have been no studies with such drugs in humans.

4.4 Cannabinoids

The psychotropic effects of smoking cannabis are due to the pharmacological action of the active constituent (-)-Δ^9-tetrahydrocannabinol (THC) within the CNS. In the airways, THC has been shown to cause both bronchodilation and bronchoconstriction in non-diseased subjects and asthmatic patients (Tashkin et al. 1977). However, an early report suggested that THC may have antitussive properties in an experimental model of cough induced by mechanical irritation of the tracheal mucosal surface in cats (Gordon et al. 1976). The discovery of cannabinoid (CB) receptors has renewed interest in the possible antitussive efficacy of CBs.

CB receptors are $G_{i/o}$-coupled receptors that mediate presynpatic inhibition of neurotransmitter release (Freiman and Szabo 2005; Melis et al. 2004; Soya et al. 2005). CB_1 receptors are predominantly located within the CNS but are also expressed by non-neuronal cells (e.g. lymphocytes), whilst CB_2 receptors are primarily found on immune cells (Pacher et al. 2006). It was reported that within the airways, CB_1 receptors were found on sensory nerves and cough induced by capsaicin was augmented by the CB_1 receptor antagonist SR14167A, but not by the CB_2 selective antagonist SR144528 (Calignano et al. 2000). These findings could be explained by the well-known inverse agonist activity of SR14167A, although this agent did not cause cough per se when administered alone. It was concluded that the increased cough response following administration of SR14167A was most likely a consequence of the antagonism of the antitussive action of endogenously released endocannabinoids in the airways (Calignano et al. 2000). However, this conclusion is complicated by the fact that cough was elicited in guinea pigs following aerosol exposure to the endogenous endocannabinoid anandamide (Jia et al. 2002). Anandamide was initially described as an endogenous activator of CB receptors, but it can also activate TRPV1 receptors (Smart et al. 2000; Zygmunt et al. 1999). As a consequence, its effect on cough is variable and may be dependent on the balance between CB and TRPV1 activity (Calignano et al. 2000; Jia et al. 2002).

In stark contrast to these findings, an agonist selective for the CB_2 receptors was reported to suppress cough induced by citric acid in guinea pigs (Patel et al. 2003). A peripheral mechanism of action was implied since the CB_2 selective agonist JWH 133 inhibited the depolarization of isolated vagus nerve from both guinea pigs and humans, in response to capsaicin in vitro (Patel et al. 2003). Moreover, the antitussive action of CBs is not likely to be mediated by suppression of neuropeptide release from the peripheral terminals of C-fibres in the airways, since the non-selective CB agonist CP 55940 was without significant effect on the contraction of airway smooth muscle, owing to the release of sensory neuropeptides, following electrical stimulation of C-fibres (Tucker et al. 2001). However, it is not known whether the antitussive action of JWH-133 also occurred within the cough circuitry of the nTS as the nucleus is rich in CB receptors (Mailleux and Vanderhaeghen 1992).

4.5 Drugs Targeting Transient Receptor Potential Channels

Transient receptor potential (TRP) channels play an important role in a diversity of sensory function, including taste, smell, hearing and perception of pain. This family comprises proteins with six transmembrane-spanning domains that are assembled either as homotetramers or heterotetramers to form non-selective cation channels permeable to calcium and sodium ions. There are at least six subfamilies, including TRPC (canonical), TRPM (melastatin), TRPV (vanilloid), TRPA (ankyrin), TRPML (mucolipin) and TRPP (polycystin) (Nilius et al. 2007).

One particular member of this superfamily (TRPV1) is predominantly localized to small-diameter peptidergic and nonpeptidergic neurones in dorsal and vagal sensory ganglia (Szallasi and Blumberg 1999) and activation of this channel by capsaicin gives rise to feelings of warmth and pain. The cloning and expression of TRPV1 has led to a greater appreciation of the role this channel plays in various pathophysiological pain states (Caterina et al. 1997). TRPV1 receptors are polymodal, activated by protons, heat (Caterina et al. 1997) and a range of lipid mediators such as 15-hydroperoxyeicosatetraenoic acid (15-HPETE) (Hwang et al. 2000), *N*-arachidonoyl dopamine (NADA) (Huang et al. 2002) and anandamide (Smart et al. 2000; Zygmunt et al. 1999) and indirectly by bradykinin (Chuang et al. 2001; Shin et al. 2002) and nerve growth factor (NGF) (Chuang et al. 2001). The release of these mediators during an inflammatory insult can result in the activation of TRPV1, thereby potentially leading to the cascade of events culminating in the sensitization of the cough reflex (Fig. 1).

An upper respiratory tract infection can increase cough sensitivity to citric acid in humans (Empey et al. 1976). Similarly, cough induced by capsaicin is increased in subjects with respiratory inflammatory diseases such as asthma, COPD and idiopathic cough (Doherty et al. 2000; Hathaway et al. 1993; Hope-Gill et al. 2003; O'Connell et al. 1996), indicating that mediators released during an inflammatory response within the respiratory tract increase cough sensitivity. It is speculated that in chronic diseases this process may result in plasticity changes in sensory function and pattern of innervation within the respiratory tract. The density of TRPV1-immunoreactive nerves in the airways is elevated in bronchial biopsies from individuals with chronic cough (Groneberg et al. 2004) and it has been reported that the density of substance P (Lee et al. 2003) and CGRP (O'Connell et al. 1995) immunoreactive nerves is increased in cough-variant asthma and chronic cough of unknown cause, respectively.

These clinical observations suggest that neuronal innervation patterns and perhaps function might underlie the physiological manifestation of hypertussive cough responses in these pathological conditions. Interference with the function of these afferent nerves might offer a selective and novel therapeutic approach to the treatment of cough. A number of studies have shown that TRPV1 antagonists, including ruthenium red (Bolser 1991), capsazepine (Lalloo et al. 1995), 5′-iodoresiniferatoxin (Trevisani et al. 2004) and *N*-(4-*tertiary*-butylphenyl)-4-(3-cholrophyridin-2-yl)-tetrahydropyrazine-1(2*H*)-carboxamide (McLeod et al. 2006) reduced cough in response to a variety of tussive stimuli (e.g. capsaicin, citric acid and allergen).

Similarly, the chemical inactivation of sensory C-fibres suppressed the ability of guinea pigs to cough in response to capsaicin and citric acid (Forsberg et al. 1988), confirming the importance of C-fibres in this response. However, the role of C-fibres in causing cough per se has been questioned (see above) and in some experimental models stimulation of C-fibres does not evoke cough (Canning et al. 2004), and may even inhibit cough (Tatar et al. 1988, 1994), whilst subacute stimulation of C-fibres can lead to an increase in the cough response to tussive stimuli such as citric acid. This has led to the proposal that activation of C-fibres can alter the threshold for cough, an interaction that is likely to be modulated at the level of the nTS (Mazzone et al. 2005). Hence, drugs targeting TRPV1 may be of considerable utility as they might not inhibit defensive cough, but might serve to normalize the sensitized cough response.

Targeting the TRPV1 receptor may offer a new approach to the treatment of cough, since defensive cough might be unaffected. For example, chemical ablation of sensory C-fibres did not result in a loss of the tussive response following mechanical irritation of the airways (Forsberg et al. 1988), nor did capsazepine affect cough induced by hypertonic saline (Lalloo et al. 1995). This would imply that TRPV1 antagonists would not affect the 'hard-wiring' pathways mediating defensive cough, but may be of potential use under conditions where there is sensitization of the cough reflex via activation of TRPV1-sensitive afferents. In contrast to these observations, one study has shown that whilst capsazepine did not inhibit cough in response to citric acid per se (Canning et al. 2006), it did inhibit the TRPV1-dependent sensitization of the cough reflex (Mazzone et al. 2005). Therefore, TRPV1 antagonists may actually reduce the sensitization of the cough reflex that occurs as a consequence of the activation of these receptors on afferent nerve terminals following release of endogenous activators (e.g. 15-HPETE, bradykinin, NADA, NGF, pH).

One study has suggested that TRPV1 antagonism may not be a useful approach in suppressing all forms of hypertussive cough. In a model of hypertussive cough following cigarette exposure, neither capsazepine nor 5′-iodoresiniferatoxin inhibited cough in response to citric acid or capsaicin, despite the fact that a dual $NK_{1,2}$ receptor antagonist was effective in this model (Lewis et al. 2007). It was unclear from this study whether the two TRPV1 antagonists inhibited cough to these tussive stimuli in air-exposed animals, but if this is the case, then sensory activation occurred independently of TRPV1 during smoke exposure. The implication of these studies is that other ion channels (e.g. ASICs) play a complementary role in modulating cough under certain pathological conditions and, therefore, development of TRPV1 antagonists may not be applicable to all forms of hypertussive cough; and desensitization of sensory C-fibres with partial TRPV1 agonists may offer a better approach to suppress sensory input into the nTS than TRPV1 antagonists.

The role of other TRP channels in regulating sensory nerve function has been investigated, although to what extent these channels contribute towards cough remains to be established. Of particular interest is the recognition that menthol, an ingredient of many over-the-counter cough remedies, is antitussive in humans (Morice et al. 1994) and guinea pigs (Laude et al. 1994), and, furthermore, is now appreciated to be an activator of TRPM8 (McKemy et al. 2002), a channel that plays a major role in cold thermosensation (Bautista et al. 2007). Interestingly, menthol is also an antagonist of TRPA1 channels, which are activated by mustard oil and a range of lipid mediators derived from oxidative stress (Macpherson et al. 2007). Whether targeting these channels on sensory nerves will yield novel drugs with antitussive activity remains to be established.

4.6 Ion Channel Modulators

NS1619 is a selective opener of calcium-activated potassium channels resulting in hyperpolarization and reduced cell excitability. In the airways, topical application of NS1619 inhibited cough induced by citric acid and bradykinin, the mechanism of action of which has been suggested to be via the suppression of impulses in guinea pig tracheal Aδ- and C-fibres (Fox et al. 1997); however, clinical studies have not yet been reported with such drugs.

Moguisteine is a peripherally acting antitussive that may act as an ATP-sensitive K^+channel opener and in clinical trials was shown to reduce the frequency of cough in patients with lung cancer to a level comparable to that resulting from administration of codeine (Barnabe et al. 1995).

Furosemide is a loop diuretic, and suppressed cough induced by low-chloride solutions but not by capsaicin (Foresi et al. 1996; Karlsson et al. 1992; Mochizuki et al. 1995; Stone et al. 1993a; Ventresca et al. 1990). This drug blocks the $Na^+/K^+/2Cl^-$ cotransporter expressed in the peripheral terminals of the 'cough' receptor in guinea pigs (Mazzone and McGovern 2006) and laryngeal RARs in the dog (Sant'Ambrogio et al. 1993). The cotransporter facilitates the accumulation of chloride ions into the peripheral terminations of the cough receptor. It has been proposed that furosemide inhibits the effective accumulation of chloride ions into these terminations, thereby preventing nerve depolarization in response to opening of chloride ion channels in the membrane. Inhibitors of chloride channels, including 4,4′-diisothiocyanatostilbene-2,2′-disulfonic acid (a non-selective anion-channel inhibitor) and niflumic acid (a selective inhibitor of calcium activated chloride channels), also inhibited cough in response to citric acid (Mazzone and McGovern 2006). These channels offer attractive targets for the development of novel antitussive agents, although to target receptors and ion channels present on the 'cough' receptor which triggers an essential part of the 'hard-wired' cough reflex would probably be unwise.

4.7 Leukotriene Receptor Antagonists

Zafirlukast, a cysteinyl leukotriene antagonist, has shown therapeutic efficacy in cough-variant asthma, even in patients unresponsive to inhaled bronchodilators and corticosteroids (Dicpinigaitis et al. 2002), whist a similar drug, montelukast has also been demonstrated to be of utility in various cough syndromes (Bisgaard 2003; Kopriva et al. 2004), although there are too few clinical trials to determine whether this drug should be routinely used in patients with cough (Chang et al. 2006). The mechanism of action of these drugs as antitussives could stem from their ability to reduce mucus secretion, oedema and bronchoconstriction, which are all stimulated by leukotrienes and thereby indirectly facilitate action potential generation in sensitized afferent nerves. Alternatively, there is evidence supporting the view that leukotrienes may directly increase the excitability of afferent nerves (McAlexander et al. 1998; Undem and Weinreich 1993).

4.8 Bradykinin Antagonists

Captopril and enalapril are angiotensin converting enzyme (ACE) inhibitors and are often used in the treatment of hypertension. Cough is a side effect of these drugs in patients with irritable airways and is a result of inhibition of ACE, leading to a decreased metabolism and subsequent accumulation of bradykinin in the airways (Dicpinigaitis 2006a). Bradykinin caused cough when administered topically to the airways of healthy subjects (Choudry et al. 1989). Similarly, cough sensitivity to capsaicin, but not other tussive agents such as citric acid or distilled water, was increased if endogenous levels of bradykinin were elevated following treatment with an ACE inhibitor (Morice et al. 1987). Furthermore, experimental models have also demonstrated that bradykinin increased cough sensitivity to tussive stimuli (Mazzone et al. 2005) and pharmacological treatment with HOE-140, a B_2 receptor antagonist, inhibited citric acid induced cough in the guinea pig (Featherstone et al. 1996). However, it remains to be seen whether B_2 receptor antagonists are antitussive in humans.

4.9 Opioids

Currently there are few effective cough suppressants and it has been concluded that patients requiring symptomatic relief in chronic cough derive a very low level of benefit from available drugs (Bolser 2006). Most effective are opioids such as codeine which produce a reduction in the frequency of cough of 19–60% (Bolser 2006; Morice et al. 2007). However, the usefulness of opioids is limited by their sedative properties and dependence liability (Bolser 2006).

Opioids such as codeine, morphine and dihydrocodone are all good antitussives primarily via their action at central μ-opioid-receptors (Kamei 1996). It is not known, however, whether this occurs within the cough circuitry of the nTS, even though it is recognized that this location is rich in this receptor (Monteillet-Agius et al. 1998). These receptors are $G_{i/o}$-coupled and mediate presynaptic inhibition of neurotransmitter release (Emmerson and Miller 1999). Recently, both selective δ-opioid receptor antagonists (TRK-851) and agonists (SB-227122) have been found to be antitussive in animal models (Chung 2003). Additionally, κ-opioid receptor agonists have been reported to be anti-tussive (Kamei 1996). Therefore, modulation of previously untargeted central opioid receptors appears a promising avenue for the development of novel antitussive drugs.

Opioid receptors have also been located in the periphery, but their ability to exert considerable antitussive action at this level is debatable. Aerosol administration of codeine is not antitussive in humans challenged with capsaicin (Kamei 2002), and although the peripherally acting peptide BW443C is antitussive in animal studies (Adcock et al. 1988), there is only limited clinical evidence to support this antitussive effect in humans (Pavord et al. 1994).

4.10 Sigma Receptor Ligands

Dextromethorphan, the dextrorotatory optical isomer of the opioid levomethorphan, retains its antitussive action despite being inactive at opioid receptors. It has been reported that this drug is an agonist at the 'sigma-1' receptor (Guitart et al. 2004; Monnet 2005) and a non-competitive antagonist (channel blocker) of NMDA receptors (Franklin and Murray 1992; Netzer et al. 1993), actions that could have consequences for excitatory glutamatergic neurotransmission in the nTS.

It is thought that dextromethorphan and noscapine act on sigma receptors (centrally and peripherally) rather than at classic opioid receptors (Brown et al. 2004). However, the inhibitory effect of noscapine on the bradykinin B_2 receptor (Ebrahimi et al. 2003) and dextromethorphan's activity at NMDA receptors (Brown et al. 2004) may also contribute to the antitussive activity of these drugs. Some research has been undertaken with newer, more efficacious sigma agonists such as SKF-10,047 (Brown et al. 2004), although plenty of opportunities remain for further investigation in this area.

4.11 GABA Receptor Ligands

GABA is an inhibitory neurotransmitter present centrally and in the peripheral nervous system. The $GABA_B$ receptor agonist baclofen has been shown to be antitussive centrally in animal studies and several clinical trials have proven its efficacy as an antitussive drug in humans (Dicpinigaitis and Gayle 2003a). An analogue of baclofen that does not cross the blood–brain barrier has also been shown experimentally to be antitussive (Bolser et al. 1994), suggesting that peripherally acting $GABA_B$ agonists may also be useful in the treatment of cough, particularly since the $GABA_A$ agonist muscimol administered locally to the mucosal surface augmented cough induced by citric acid in guinea pigs (Mazzone and McGovern 2006).

5 Challenges to the Development of Novel Antitussive Drugs

The development of novel antitussives is a challenging area, not least because of the lack of good predictive models, both clinically and preclinically. For instance, it is highly debatable whether capsaicin challenge in healthy volunteers is a good cough model when many drugs which improve pathological cough fail to inhibit capsaicin-induced cough (Table 1). Similarly, the demonstration of antitussive activity of novel substances against stimuli used to induce tussive responses in humans rather than determining their effect on the natural history of cough is likely to lead to false-negatives. For example, opioids have been shown to reduce the frequency of cough in subjects with chronic cough, without altering cough sensitivity to citric acid (Morice et al. 2007). Therefore, non-invasive ambulatory monitoring of cough should be urgently validated as a method to assess the efficacy of novel antitussive agents.

Much of the preclinical work has also relied on using capsaicin-induced or citric acid induced cough in healthy guinea pigs and there are many examples of drugs that are antitussive in these preclinical models that do not show antitussive effects in humans. For example, NK_1 antagonists had promising preclinical antitussive data experimentally that were not confirmed in clinical studies in humans (Joos et al. 2003). Following a similar pattern, NK_1 antagonists attenuated nociceptive responses in animals but failed as analgesics in humans (Hill 2000). Furthermore, a dual dopamine D2 and β_2-adrenoceptor agonist was recently found not to improve the symptom of cough in a large clinical trial of patients with COPD despite exhibiting antitussive activity in animals (Laursen et al. 2003). The value and limitations of animal models of cough has been discussed elsewhere (Karlsson and Fuller 1999) and it would seem sensible to ensure that novel drugs exhibit antitussive activity in at least two species before being tried in humans and perhaps more importantly should exhibit antitussive activity in hypertussive models since the ultimate aim of antitussive therapy in the clinic is to reduce excess cough (sensitization of cough reflexes), rather than inhibiting cough altogether (i.e. the 'hard-wiring' cough reflex). Clearly there is an urgent need to develop more realistic models of cough preclinically and as importantly to consider the best way of evaluating novel antitussives in early clinical development for proof of concept before entering into larger clinical studies.

Furthermore, given the place of the nTS in the afferent arm of the cough reflex, we hypothesize that drugs that inhibit excitatory synaptic transmission in this nucleus will possess antitussive activity. Additionally this region of the brain lacks a complete blood–brain barrier and is therefore accessible to blood-borne drugs (Gross et al. 1990). Agonists acting at a range of receptors have been shown to act presynaptically to inhibit excitatory neurotransmission in the nTS (Brooks et al. 1992; Chen et al. 2002; Glatzer and Smith 2005; Kato and Shigetomi 2001; Sekizawa et al. 2003), though not in the context of the cough response.

Table 1 Results of clinical trials investigating the efficacy of selected antitussive compounds

Class/drug	Capsaicin	Low Cl^- concentration	Citric acid	Angiotensin converting enzyme inhibitor	Pathological cough[a]	Postinfectious cough	Clinical use in cough treatment
Opioids	±(Davenport et al. 2007; Fuller et al. 1988)		±(Empey et al. 1979; Morice et al. 2007)		✓(Bolser 2006; Morice et al. 2007)	×(Eccles 1996; Herbert and Brewster 2000)	Gold standard
Dextromethorphan			✓(Grattan et al. 1995; Karttunen et al. 1987)		✓(Braman 2006)	×(Lee et al. 2000)	Yes
Baclofen	✓(Dicpinigaitis and Rauf 1998)			✓(Dicpinigaitis 1996)	✓(Dicpinigaitis and Rauf 1998)		Yes
Noscapine			×(Empey et al. 1979)	✓(Mooraki et al. 2005)			
H_1-antagonists	×(Dicpinigaitis and Gayle 2003b; Studham and Fuller 1992)		✓(Packman et al. 1991)		±(Danzon et al. 1988; Lilienfield et al. 1976)	±(Dicpinigaitis and Gayle 2003b)	Primarily used as an antiallergic drug

Table 1 Continued

Class/drug	Capsaicin	Low Cl^- concentration	Citric acid	Angiotensin converting enzyme inhibitor	Pathological cough[a]	Postinfectious cough	Clinical use in cough treatment
Furosemide	×(Ventresca et al. 1990)	✓(Foresi et al. 1996; Stone et al. 1993b; Ventresca et al. 1990)					
Glaucine			×(Rees and Clark 1983)		✓(Ruhle et al. 1984)	✓(Gastpar et al. 1984)	
Local anaesthetics	✓(Choudry et al. 1990; Hansson et al. 1994)	✓(Sheppard et al. 1983)	✓(Midgren et al. 1992)		✓(Chong et al. 2005; Fuller and Jackson 1990; Trochtenberg 1994)		
β_2−agonists	×(Smith et al. 1991)	✓(Higenbottam 1987; Lowry et al. 1987)	✓(Pounsford et al. 1985)		± (Chong et al. 2005; Mulrennan et al. 2004; Nichol et al. 1990)	×(Pillay and Swingler 2003)	Primarily used as a bronchodilator
Sodium cromoglycate	×(Collier and Fuller 1984)	×(Fuller and Collier 1984; Sheppard et al. 1983)		±(Hargreaves and Benson 1995)	±(Chang et al. 2004)		Primarily used as an antiallergic drug

Expectorants and mucolytics					✓(Ishiura et al. 2003)	✓(Dicpinigaitis and Gayle 2003a)	Over-the-counter ‘cough remedies’
Muscarinic receptor antagonists	×(Smith et al. 1991)	±(Fuller and Collier 1984; Lowry et al. 1987; Sheppard et al. 1983)	✓(Pounsford et al. 1985)		✓(Braman 2006)	✓(Holmes et al. 1992)	Primarily used in the treatment of chronic obstructive respiratory disease
LT-receptor antagonists	×(Dicpinigaitis and Dobkin 1999)				±(Chang et al. 2006; Dicpinigaitis et al. 2002; Kopriva et al. 2004)	✓(Bisgaard 2003)	Primarily used as an antiasthma drug
Moguisteine					±(Aversa et al. 1993; Barnabe et al. 1995)		

✓, significant antitussive effect; × , no significant effect; ±, conflicting data

[a]Includes cancer, chronic obstructive respiratory disease, chronic non-productive cough, cough associated with other respiratory diseases

6 Conclusion

Our understanding of the reflexes involved in modulating cough and hypertussive responses is increasing, but there is still much to learn. Cough remains a relatively poorly studied area of pulmonary research when compared with bronchospasm and airway inflammation. However, there remains a clear need to develop a non-opioid antitussive drug that ideally modulates abnormal heightened cough reflexes, whilst leaving the normal cough reflex unaltered (Table 1, Fig. 1). Nonetheless, significant progress is being made to develop novel antitussive drugs and the growing recognition of cough as an unmet medical need will hopefully ensure there is more research into this important clinical problem.

References

Adcock JJ, Schneider C, Smith TW (1988) Effects of codeine, morphine and a novel opioid pentapeptide BW443C, on cough, nociception and ventilation in the unanaesthetized guinea-pig. Br J Pharmacol 93:93–100

Adcock JJ, Douglas GJ, Garabette M, Gascoigne M, Beatch G, Walker M, Page CP (2003) RSD931, a novel anti-tussive agent acting on airway sensory nerves. Br J Pharmacol 138: 407–416

Advenier C, EmondsAlt X (1996) Tachykinin receptor antagonists and cough. Pulm Pharmacol 9:329–333

Alexander SP, Mathie A, Peters JA (2007) Ion channels. Br J Pharmacol 150(Suppl 1):S96–S121

Andresen MC, Yang M (1995) Dynamics of sensory afferent synaptic transmission in aortic baroreceptor regions on nucleus tractus solitarius. J Neurophysiol 74:1518–1528

Aversa C, Cazzola M, Clini V, Dal NR, Maiorano V, Tana F, Allegra L (1993) Clinical trial of the efficacy and safety of moguisteine in patients with cough associated with chronic respiratory diseases. Drugs Exp Clin Res 19:273–279

Bailey CP, Maubach KA, Jones RS (2004) Neurokinin-1 receptors in the rat nucleus tractus solitarius: Pre- and postsynaptic modulation of glutamate and GABA release. Neuroscience 127: 467–479

Barnabe R, Berni F, Clini V, Pirrelli M, Pisani CA, Robuschi M, Rossi M, Sestini P, Tana F, Vaghi A (1995) The efficacy and safety of moguisteine in comparison with codeine phosphate in patients with chronic cough. Monaldi Arch Chest Dis 50:93–97

Bautista DM, Siemens J, Glazer JM, Tsuruda PR, Basbaum AI, Stucky CL, Jordt SE, Julius D (2007) The menthol receptor TRPM8 is the principal detector of environmental cold. Nature 448:204–208

Bisgaard H (2003) A randomized trial of montelukast in respiratory syncytial virus postbronchiolitis. Am J Respir Crit Care Med 167:379–383

Bolser DC (1991) Fictive cough in the cat. J Appl Physiol 71:2325–2331

Bolser DC (2006) Cough suppressant and pharmacologic protussive therapy: ACCP evidence-based clinical practice guidelines. Chest 129:238S–249S

Bolser DC, DeGennaro FC, O'Reilly S, Chapman RW, Kreutner W, Egan RW, Hey JA (1994) Peripheral and central sites of action of GABA-B agonists to inhibit the cough reflex in the cat and guinea pig. Br J Pharmacol 113:1344–1348

Bolser DC, DeGennaro FC, O'Reilly S, McLeod RL, Hey JA (1997) Central antitussive activity of the NK1 and NK2 tachykinin receptor antagonists, CP-99,994 and SR 48968, in the guinea-pig and cat. Br J Pharmacol 121:165–170

Bonham AC, Chen CY, Mutoh T, Joad JP (2001) Lung C-fiber CNS reflex: Role in the respiratory consequences of extended environmental tobacco smoke exposure in young guinea pigs. Environ Health Perspect 109(Suppl 4):573–578
Bonham AC, Sekizawa SI, Joad JP (2004) Plasticity of central mechanisms for cough. Pulm Pharmacol Ther 17:453–457
Bonham AC, Chen CY, Sekizawa S, Joad JP (2006a) Plasticity in the nucleus tractus solitarius and its influence on lung and airway reflexes. J Appl Physiol 101:322–327
Bonham AC, Sekizawa S, Chen CY, Joad JP (2006b) Plasticity of brainstem mechanisms of cough. Respir Physiol Neurobiol 152:312–319
Braman SS (2006) Chronic cough due to chronic bronchitis: ACCP evidence-based clinical practice guidelines. Chest 129:104S–115S
Brooks PA, Glaum SR, Miller RJ, Spyer KM (1992) The actions of baclofen on neurones and synaptic transmission in the nucleus tractus solitarii of the rat in vitro. J Physiol 457:115–129
Brown C, Fezoui M, Selig WM, Schwartz CE, Ellis JL (2004) Antitussive activity of sigma-1 receptor agonists in the guinea-pig. Br J Pharmacol 141:233–240
Buchan P, Adcock JJ (1992) Capsaicin-induced bronchoconstriction in the guinea-pig: Contribution of vagal cholinergic reflexes, local axon reflexes and their modulation by BW443C. Br J Pharmacol 105:448–452
Calignano A, Katona I, Desarnaud F, Giuffrida A, La RG, Mackie K, Freund TF, Piomelli D (2000) Bidirectional control of airway responsiveness by endogenous cannabinoids. Nature 408: 96–101
Canning BJ (2006) Anatomy and neurophysiology of the cough reflex: ACCP evidence-based clinical practice guidelines. Chest 129:33S–47S
Canning BJ, Mazzone SB, Meeker SN, Mori N, Reynolds SM, Undem BJ (2004) Identification of the tracheal and laryngeal afferent neurones mediating cough in anaesthetized guinea-pigs. J Physiol 557:543–558
Canning BJ, Farmer DG, Mori N (2006) Mechanistic studies of acid-evoked coughing in anesthetized guinea pigs. Am J Physiol Regul Integr Comp Physiol 291:R454–R463
Carr MJ, Hunter DD, Jacoby DB, Undem BJ (2002) Expression of tachykinins in nonnociceptive vagal afferent neurons during respiratory viral infection in guinea pigs. Am J Respir Crit Care Med 165:1071–1075
Caterina MJ, Schumacher MA, Tominaga M, Rosen TA, Levine JD, Julius D (1997) The capsaicin receptor: A heat-activated ion channel in the pain pathway. Nature 389:816–824
Caudle RM, Perez FM, Del Valle-Pinero AY, Iadarola MJ (2005) Spinal cord NR1 serine phosphorylation and NR2B subunit suppression following peripheral inflammation. Mol Pain 1:25
Chang A, Marchant JM, McKean M, Morris P (2004) Inhaled cromones for prolonged non-specific cough in children. Cochrane Database Syst Rev CD004436
Chang AB, Winter D, Acworth JP (2006) Leukotriene receptor antagonist for prolonged non-specific cough in children. Cochrane Database Syst Rev CD005602
Chapman RW, House A, Liu F, Celly C, Mei H, Hey JA (2004) Antitussive activity of the tachykinin NK1 receptor antagonist, CP-99994, in dogs. Eur J Pharmacol 485:329–332
Chen CY, Horowitz JM, Bonham AC (1999) A presynaptic mechanism contributes to depression of autonomic signal transmission in NTS. Am J Physiol 277:H1350–H1360
Chen CY, Bonham AC, Schelegle ES, Gershwin LJ, Plopper CG, Joad JP (2001) Extended allergen exposure in asthmatic monkeys induces neuroplasticity in nucleus tractus solitarius. J Allergy Clin Immunol 108:557–562
Chen CY, Ling Eh EH, Horowitz JM, Bonham AC (2002) Synaptic transmission in nucleus tractus solitarius is depressed by group II and III but not group I presynaptic metabotropic glutamate receptors in rats. J Physiol 538:773–786
Chen CY, Bonham AC, Plopper CG, Joad JP (2003) Neuroplasticity in nucleus tractus solitarius neurons after episodic ozone exposure in infant primates. J Appl Physiol 94:819–827
Chong CF, Chen CC, Ma HP, Wu YC, Chen YC, Wang TL (2005) Comparison of lidocaine and bronchodilator inhalation treatments for cough suppression in patients with chronic obstructive pulmonary disease. Emerg Med J 22:429–432

Choudry NB, Fuller RW, Pride NB (1989) Sensitivity of the human cough reflex: Effect of inflammatory mediators prostaglandin E2, bradykinin, and histamine. Am Rev Respir Dis 140: 137–141

Choudry NB, Fuller RW, Anderson N, Karlsson JA (1990) Separation of cough and reflex bronchoconstriction by inhaled local anaesthetics. Eur Respir J 3:579–583

Chuang HH, Prescott ED, Kong H, Shields S, Jordt SE, Basbaum AI, Chao MV, Julius D (2001) Bradykinin and nerve growth factor release the capsaicin receptor from PtdIns(4,5)P2-mediated inhibition. Nature 411:957–962

Chung KF (2003) Current and future prospects for drugs to suppress cough. IDrugs 6:781–786

Colin I, Blondeau C, Baude A (2002) Neurokinin release in the rat nucleus of the solitary tract via NMDA and AMPA receptors. Neuroscience 115:1023–1033

Collier JG, Fuller RW (1984) Capsaicin inhalation in man and the effects of sodium cromoglycate. Br J Pharmacol 81:113–117

Cook AJ, Woolf CJ, Wall PD, McMahon SB (1987) Dynamic receptive field plasticity in rat spinal cord dorsal horn following C-primary afferent input. Nature 325:151–153

Corboz MR, Rivelli MA, Egan RW, Tulshian D, Matasi J, Fawzi AB, Benbow L, Smith-Torhan A, Zhang H, Hey JA (2000) Nociceptin inhibits capsaicin-induced bronchoconstriction in isolated guinea pig lung. Eur J Pharmacol 402:171–179

Cullinan P (1992) Persistent cough and sputum: Prevalence and clinical characteristics in south east England. Respir Med 86:143–149

da Costa JC, Olsen PC, de Azeredo SR, de FC, V, Serra MF, Alves LA, Faria RX, Xisto DG, Rocco PR, Cordeiro RS, Rodrigues E Silva PM, Martins MA (2007) JMF2–1, a lidocaine derivative acting on airways spasm and lung allergic inflammation in rats. J Allergy Clin Immunol 119:219–225

Danzon A, Lacroix J, Infante-Rivard C, Chicoine L (1988) A double-blind clinical trial on diphenhydramine in pertussis. Acta Paediatr Scand 77:614–615

Daoui S, Cognon C, Naline E, Emonds-Alt X, Advenier C (1998) Involvement of tachykinin NK3 receptors in citric acid-induced cough and bronchial responses in guinea pigs. Am J Respir Crit Care Med 158:42–48

Davenport PW, Bolser DC, Vickroy T, Berry RB, Martin AD, Hey JA, Danzig M (2007) The effect of codeine on the urge-to-cough response to inhaled capsaicin. Pulm Pharmacol Ther 20: 338–346

Dicpinigaitis PV (1996) Use of baclofen to suppress cough induced by angiotensin- converting enzyme inhibitors. Ann Pharmacother 30:1242–1245

Dicpinigaitis PV (2004) Potential new cough therapies. Pulm Pharmacol Ther 17:459–462

Dicpinigaitis PV (2006a) Angiotensin-converting enzyme inhibitor-induced cough: ACCP evidence-based clinical practice guidelines. Chest 129:169S–173S

Dicpinigaitis PV (2006b) Potential future therapies for the management of cough: ACCP evidence-based clinical practice guidelines. Chest 129:284S–286S

Dicpinigaitis PV, Dobkin JB (1999) Effect of zafirlukast on cough reflex sensitivity in asthmatics. J Asthma 36:265–270

Dicpinigaitis PV, Rauf K (1998) Treatment of chronic, refractory cough with baclofen. Respiration 65:86–88

Dicpinigaitis PV, Dobkin JB, Rauf K, Aldrich TK (1998) Inhibition of capsaicin-induced cough by the gamma-aminobutyric acid agonist baclofen. J Clin Pharmacol 38:364–367

Dicpinigaitis PV, Dobkin JB, Reichel J (2002) Antitussive effect of the leukotriene receptor antagonist zafirlukast in subjects with cough-variant asthma. J Asthma 39:291–297

Dicpinigaitis PV, Gayle YE (2003a) Effect of guaifenesin on cough reflex sensitivity. Chest 124:2178–2181

Dicpinigaitis PV, Gayle YE (2003b) Effect of the second-generation antihistamine, fexofenadine, on cough reflex sensitivity and pulmonary function. Br J Clin Pharmacol 56:501–504

Doherty MJ, Mister R, Pearson MG, Calverley PM (2000) Capsaicin responsiveness and cough in asthma and chronic obstructive pulmonary disease. Thorax 55:643–649

Drewe JA, Miles R, Kunze DL (1990) Excitatory amino acid receptors of guinea pig medial nucleus tractus solitarius neurons. Am J Physiol 259:H1389–H1395

Duggan AW, Hope PJ, Jarrott B, Schaible HG, Fleetwood-Walker SM (1990) Release, spread and persistence of immunoreactive neurokinin A in the dorsal horn of the cat following noxious cutaneous stimulation. Studies with antibody microprobes. Neurosci 35:195–202

Ebrahimi SA, Zareie MR, Rostami P, Mahmoudian M (2003) Interaction of noscapine with the bradykinin mediation of the cough response. Acta Physiol Hung 90:147–155

Eccles R (1996) Codeine, cough and upper respiratory infection. Pulm Pharmacol 9:293–297

El-Hashim AZ, Wyss D, Lewis C (2004) Effect of a novel NK1 receptor selective antagonist (NKP608) on citric acid induced cough and airway obstruction. Pulm Pharmacol Ther 17: 11–18

Emmerson PJ, Miller RJ (1999) Pre- and postsynaptic actions of opioid and orphan opioid agonists in the rat arcuate nucleus and ventromedial hypothalamus in vitro. J Physiol 517(Part 2): 431–445

Empey DW, Laitinen LA, Jacobs L, Gold WM, Nadel JA (1976) Mechanisms of bronchial hyperreactivity in normal subjects after upper respiratory tract infection. Am Rev Respir Dis 113:131–139

Empey DW, Laitinen LA, Young GA, Bye CE, Hughes DT (1979) Comparison of the antitussive effects of codeine phosphate 20 mg, dextromethorphan 30 mg and noscapine 30 mg using citric acid-induced cough in normal subjects. Eur J Clin Pharmacol 16:393–397

Fahy JV, Wong HH, Geppetti P, Reis JM, Harris SC, Maclean DB, Nadel JA, Boushey HA (1995) Effect of an NK1 receptor antagonist (CP-99,994) on hypertonic saline-induced bronchoconstriction and cough in male asthmatic subjects. Am J Respir Crit Care Med 152:879–884

Featherstone RL, Parry JE, Church MK (1996) The effects of a kinin antagonist on changes in lung function and plasma extravasation into the airways following challenge of sensitized guinea-pigs. Clin Exp Allergy 26:235–240

Fischer A, Hoffmann B (1996) Nitric oxide synthase in neurons and nerve fibers of lower airways and in vagal sensory ganglia of man. Correlation with neuropeptides. Am J Respir Crit Care Med 154:209–216

Fischer A, Forssmann WG, Undem BJ (1998) Nociceptin-induced inhibition of tachykinergic neurotransmission in guinea pig bronchus. J Pharmacol Exp Ther 285:902–907

Foresi A, Cavigioli G, Pelucchi A, Mastropasqua B, Marazzini L (1996) Effect of acetazolamide on cough induced by low-chloride-ion solutions in normal subjects: Comparison with furosemide. J Allergy Clin Immunol 97:1093–1099

Forsberg K, Karlsson JA, Theodorsson E, Lundberg JM, Persson CG (1988) Cough and bronchoconstriction mediated by capsaicin-sensitive sensory neurons in the guinea-pig. Pulm Pharmacol 1:33–39

Fortin G, Champagnat J (1993) Spontaneous synaptic activities in rat nucleus tractus solitarius neurons in vitro: Evidence for re-excitatory processing. Brain Res 630:125–135

Fox AJ, Barnes PJ, Venkatesan P, Belvisi MG (1997) Activation of large conductance potassium channels inhibits the afferent and efferent function of airway sensory nerves in the guinea pig. J Clin Invest 99:513–519

Franklin PH, Murray TF (1992) High affinity [3H]dextrorphan binding in rat brain is localized to a noncompetitive antagonist site of the activated N-methyl-D-aspartate receptor-cation channel. Mol Pharmacol 41:134–146

Freiman I, Szabo B (2005) Cannabinoids depress excitatory neurotransmission between the subthalamic nucleus and the globus pallidus. Neurosci 133:305–313

French CL, Irwin RS, Curley FJ, Krikorian CJ (1998) Impact of chronic cough on quality of life. Arch Intern Med 158:1657–1661

French CT, Fletcher KE, Irwin RS (2004) Gender differences in health-related quality of life in patients complaining of chronic cough. Chest 125:482–488

French CT, Fletcher KE, Irwin RS (2005) A comparison of gender differences in health-related quality of life in acute and chronic coughers. Chest 127:1991–1998

Fujimura M, Kamio Y, Hashimoto T, Matsuda T (1996) Effect of capsaicin inhalation on methacholine responsiveness in normal subjects. Clin Auton Res 6:45–48

Fujimura M, Kamio Y, Myou S, Hashimoto T (2000) Effect of oral mexiletine on the cough response to capsaicin and tartaric acid. Thorax 55:126–128

Fuller RW, Collier JG (1984) Sodium cromoglycate and atropine block the fall in FEV1 but not the cough induced by hypotonic mist. Thorax 39:766–770

Fuller RW, Jackson DM (1990) Physiology and treatment of cough. Thorax 45:425–430

Fuller RW, Karlsson JA, Choudry NB, Pride NB (1988) Effect of inhaled and systemic opiates on responses to inhaled capsaicin in humans. J Appl Physiol 65:1125–1130

Gao X, Kim HK, Chung JM, Chung K (2005) Enhancement of NMDA receptor phosphorylation of the spinal dorsal horn and nucleus gracilis neurons in neuropathic rats. Pain 116:62–72

Gastpar H, Criscuolo D, Dieterich HA (1984) Efficacy and tolerability of glaucine as an antitussive agent. Curr Med Res Opin 9:21–27

Giardina GA, Sarau HM, Farina C, Medhurst AD, Grugni M, Foley JJ, Raveglia LF, Schmidt DB, Rigolio R, Vassallo M, Vecchietti V, Hay DW (1996) 2-Phenyl-4-quinolinecarboxamides: A novel class of potent and selective non-peptide competitive antagonists for the human neurokinin-3 receptor. J Med Chem 39:2281–2284

Girard V, Naline E, Vilain P, Emonds-Alt X, Advenier C (1995) Effect of the two tachykinin antagonists, SR 48968 and SR 140333, on cough induced by citric acid in the unanaesthetized guinea pig. Eur Respir J 8:1110–1114

Glatzer NR, Smith BN (2005) Modulation of synaptic transmission in the rat nucleus of the solitary tract by endomorphin-1. J Neurophysiol 93:2530–2540

Gordon R, Gordon RJ, Sofia D (1976) Antitussive activity of some naturally occurring cannabinoids in anesthetized cats. Eur J Pharmacol 35:309–313

Grattan TJ, Marshall AE, Higgins KS, Morice AH (1995) The effect of inhaled and oral dextromethorphan on citric acid induced cough in man. Br J Clin Pharmacol 39:261–263

Groneberg DA, Niimi A, Dinh QT, Cosio B, Hew M, Fischer A, Chung KF (2004) Increased expression of transient receptor potential vanilloid-1 in airway nerves of chronic cough. Am J Respir Crit Care Med 170:1276–1280

Gross PM, Wall KM, Pang JJ, Shaver SW, Wainman DS (1990) Microvascular specializations promoting rapid interstitial solute dispersion in nucleus tractus solitarius. Am J Physiol 259:R1131–R1138

Guitart X, Codony X, Monroy X (2004) Sigma receptors: Biology and therapeutic potential. Psychopharmacology (Berl) 174:301–319

Guo W, Zou S, Guan Y, Ikeda T, Tal M, Dubner R, Ren K (2002) Tyrosine phosphorylation of the NR2B subunit of the NMDA receptor in the spinal cord during the development and maintenance of inflammatory hyperalgesia. J Neurosci 22:6208–6217

Hansson L, Midgren B, Karlsson JA (1994) Effects of inhaled lignocaine and adrenaline on capsaicin- induced cough in humans. Thorax 49:1166–1168

Hargreaves MR, Benson MK (1995) Inhaled sodium cromoglycate in angiotensin-converting enzyme inhibitor cough. Lancet 345:13–16

Hathaway TJ, Higenbottam TW, Morrison JF, Clelland CA, Wallwork J (1993) Effects of inhaled capsaicin in heart-lung transplant patients and asthmatic subjects. Am Rev Respir Dis 148:1233–1237

Hay DW, Giardina GA, Griswold DE, Underwood DC, Kotzer CJ, Bush B, Potts W, Sandhu P, Lundberg D, Foley JJ, Schmidt DB, Martin LD, Kilian D, Legos JJ, Barone FC, Luttmann MA, Grugni M, Raveglia LF, Sarau HM (2002) Nonpeptide tachykinin receptor antagonists. III. SB 235375, a low central nervous system-penetrant, potent and selective neurokinin-3 receptor antagonist, inhibits citric acid-induced cough and airways hyper-reactivity in guinea pigs. J Pharmacol Exp Ther 300:314–323

Herbert ME, Brewster GS (2000) Myth: Codeine is an effective cough suppressant for upper respiratory tract infections. West J Med 173:283

Higenbottam TW (1987) Anticholinergics and cough. Postgrad Med J 63(Suppl 1):75–78

Hill R (2000) NK1 (substance P) receptor antagonists – why are they not analgesic in humans? Trends Pharmacol Sci 21:244–246

Holmes PW, Barter CE, Pierce RJ (1992) Chronic persistent cough: Use of ipratropium bromide in undiagnosed cases following upper respiratory tract infection. Respir Med 86:425–429

Hope-Gill BD, Hilldrup S, Davies C, Newton RP, Harrison NK (2003) A study of the cough reflex in idiopathic pulmonary fibrosis. Am J Respir Crit Care Med 168:995–1002

Huang SM, Bisogno T, Trevisani M, Al-Hayani A, De PL, Fezza F, Tognetto M, Petros TJ, Krey JF, Chu CJ, Miller JD, Davies SN, Geppetti P, Walker JM, Di Marzo V (2002) An endogenous capsaicin-like substance with high potency at recombinant and native vanilloid VR1 receptors. Proc Natl Acad Sci USA 99:8400–8405

Hunter DD, Undem BJ (1999) Identification and substance P content of vagal afferent neurons innervating the epithelium of the guinea pig trachea. Am J Respir Crit Care Med 159: 1943–1948

Hwang SW, Cho H, Kwak J, Lee SY, Kang CJ, Jung J, Cho S, Min KH, Suh YG, Kim D, Oh U (2000) Direct activation of capsaicin receptors by products of lipoxygenases: Endogenous capsaicin-like substances. Proc Natl Acad Sci USA 97:6155–6160

Irwin RS (2006) Assessing cough severity and efficacy of therapy in clinical research: ACCP evidence-based clinical practice guidelines. Chest 129:232S–237S

Ishiura Y, Fujimura M, Yamamori C, Nobata K, Myou S, Kurashima K, Michishita Y, Takegoshi T (2003) Effect of carbocysteine on cough reflex to capsaicin in asthmatic patients. Br J Clin Pharmacol 55:504–510

Ji RR, Woolf CJ (2001) Neuronal plasticity and signal transduction in nociceptive neurons: Implications for the initiation and maintenance of pathological pain. Neurobiol Dis 8:1–10

Ji RR, Baba H, Brenner GJ, Woolf CJ (1999) Nociceptive-specific activation of ERK in spinal neurons contributes to pain hypersensitivity. Nat Neurosci 2:1114–1119

Ji RR, Kohno T, Moore KA, Woolf CJ (2003) Central sensitization and LTP: Do pain and memory share similar mechanisms? Trends Neurosci 26:696–705

Jia Y, Wang X, Aponte SI, Rivelli MA, Yang R, Rizzo CA, Corboz MR, Priestley T, Hey JA (2002) Nociceptin/orphanin FQ inhibits capsaicin-induced guinea-pig airway contraction through an inward-rectifier potassium channel. Br J Pharmacol 135:764–770

Joos GF, De Swert KO, Schelfhout V, Pauwels RA (2003) The role of neural inflammation in asthma and chronic obstructive pulmonary disease. Ann N Y Acad Sci 992:218–230

Jordan D (2001) Central nervous pathways and control of the airways. Respir Physiol 125:67–81

Kagaya M, Lamb J, Robbins J, Page CP, Spina D (2002) Characterization of the anandamide induced depolarization of guinea-pig isolated vagus nerve. Br J Pharmacol 137:39–48

Kamei J (1996) Role of opioidergic and serotonergic mechanisms in cough and antitussives. Pulm Pharmacol 9:349–356

Kamei J (2002) Delta-opioid receptor antagonists as a new concept for central acting antitussive drugs. Pulm Pharmacol Ther 15:235–240

Karlsson JA (1996) The role of capsaicin-sensitive C-fibre afferent nerves in the cough reflex. Pulm Pharmacol 9:315–321

Karlsson JA, Fuller RW (1999) Pharmacological regulation of the cough reflex–from experimental models to antitussive effects in Man. Pulm Pharmacol Ther 12:215–228

Karlsson JA, Choudry NB, Zackrisson C, Fuller RW (1992) A comparison of the effect of inhaled diuretics on airway reflexes in humans and guinea pigs. J Appl Physiol 72:434–438

Karttunen P, Tukiainen H, Silvasti M, Kolonen S (1987) Antitussive effect of dextromethorphan and dextromethorphan-salbutamol combination in healthy volunteers with artificially induced cough. Respiration 52:49–53

Kato F, Shigetomi E (2001) Distinct modulation of evoked and spontaneous EPSCs by purinoceptors in the nucleus tractus solitarii of the rat. J Physiol 530:469–486

Kohrogi H, Nadel JA, Malfroy B, Gorman C, Bridenbaugh R, Patton JS, Borson DB (1989) Recombinant human enkephalinase (neutral endopeptidase) prevents cough induced by tachykinins in awake guinea pigs. J Clin Invest 84:781–786

Kollarik M, Ru F, Undem BJ (2007) Acid-sensitive vagal sensory pathways and cough. Pulm Pharmacol Ther 20:402–411

Kopriva F, Sobolova L, Szotkowska J, Zapalka M (2004) Treatment of chronic cough in children with montelukast,a leukotriene receptor antagonist. J Asthma 41:715–720
Kvale PA (2006) Chronic cough due to lung tumors: ACCP evidence-based clinical practice guidelines. Chest 129:147S–153S
Kwong K, Lee LY (2005) Prostaglandin E2 potentiates a TTX-resistant sodium current in rat capsaicin-sensitive vagal pulmonary sensory neurones. J Physiol 564:437–450
Lai J, Porreca F, Hunter JC, Gold MS (2004) Voltage-gated sodium channels and hyperalgesia. Annu Rev Pharmacol Toxicol 44:371–397
Lalloo UG, Fox AJ, Belvisi MG, Chung KF, Barnes PJ (1995) Capsazepine inhibits cough induced by capsaicin and citric acid but not by hypertonic saline in guinea pigs. J Appl Physiol 79: 1082–1087
Laude EA, Morice AH, Grattan TJ (1994) The antitussive effects of menthol, camphor and cineole in conscious guinea-pigs. Pulm Pharmacol 7:179–184
Laursen LC, Lindqvist A, Hepburn T, Lloyd J, Perrett J, Sanders N, Rocchiccioli K (2003) The role of the novel D2/beta2-agonist, Viozan (sibenadet HCl), in the treatment of symptoms of chronic obstructive pulmonary disease: Results of a large-scale clinical investigation. Respir Med 97(Suppl A):S23–S33
Lee PCL, Jawad MS, Eccles R (2000) Antitussive efficacy of dextromethorphan in cough associated with acute upper respiratory tract infection. J Pharm Pharmacol 52:1137–1142
Lee J, Lee J, Szabo T, Gonzalez AF, Welter JD, Blumberg PM (2001) N-(3-acyloxy-2-benzylpropyl)-N'-dihydroxytetrahydrobenzazepine and tetrahydroisoquinoline thiourea analogues as vanilloid receptor ligands. Bioorg Med Chem 9:1713–1720
Lee SY, Kim MK, Shin C, Shim JJ, Kim HK, Kang KH, Yoo SH, In KH (2003) Substance P-immunoreactive nerves in endobronchial biopsies in cough-variant asthma and classic asthma. Respiration 70:49–53
Lee J, Alrubaian J, Dores RM (2006) Are lungfish living fossils? Observation on the evolution of the opioid/orphanin gene family. Gen Comp Endocrinol 148:306–314
Lewis CA, Ambrose C, Banner K, Battram C, Butler K, Giddings J, Mok J, Nasra J, Winny C, Poll C (2007) Animal models of cough: Literature review and presentation of a novel cigarette smoke-enhanced cough model in the guinea-pig. Pulm Pharmacol Ther 20:325–333
Lilienfield LS, Rose JC, Princiotto JV (1976) Antitussive activity of diphenhydramine in chronic cough. Clin Pharmacol Ther 19:421–425
Lowry R, Higenbottam T, Johnson T, Godden D (1987) Inhibition of artificially induced cough in man by bronchodilators. Br J Clin Pharmacol 24:503–510
Lundberg JM, Hokfelt T, Martling CR, Saria A, Cuello C (1984) Substance P-immunoreactive sensory nerves in the lower respiratory tract of various mammals including man. Cell Tissue Res 235:251–261
Macpherson LJ, Dubin AE, Evans MJ, Marr F, Schultz PG, Cravatt BF, Patapoutian A (2007) Noxious compounds activate TRPA1 ion channels through covalent modification of cysteines. Nature 445:541–545
Mailleux P, Vanderhaeghen JJ (1992) Distribution of neuronal cannabinoid receptor in the adult rat brain: A comparative receptor binding radioautography and in situ hybridization histochemistry. Neuroscience 48:655–668
Malcangio M, Lessmann V (2003) A common thread for pain and memory synapses? Brain-derived neurotrophic factor and trkB receptors. Trends Pharmacol Sci 24:116–121
Mazzone SB (2005) An overview of the sensory receptors regulating cough. Cough 1:2–11
Mazzone SB, McGovern AE (2006) Na^+–K^+–$2Cl^-$ cotransporters and Cl^- channels regulate citric acid cough in guinea pigs. J Appl Physiol 101:635–643
Mazzone SB, McGovern AE (2007) Sensory neural targets for the treatment of cough. Clin Exp Pharmacol Physiol 34:955–962
Mazzone SB, Mori N, Canning BJ (2005) Synergistic interactions between airway afferent nerve subtypes regulating the cough reflex in guinea-pigs. J Physiol 569:559–573

McAlexander MA, Myers AC, Undem BJ (1998) Inhibition of 5-lipoxygenase diminishes neurally evoked tachykinergic contraction of guinea pig isolated airway. J Pharmacol Exp Ther 285:602–607

McGarvey LP, Morice AH (2006) Clinical cough and its mechanisms. Respir Physiol Neurobiol 152:363–371

McKemy DD, Neuhausser WM, Julius D (2002) Identification of a cold receptor reveals a general role for TRP channels in thermosensation. Nature 416:52–58

McLeod RL, Bolser DC, Jia Y, Parra LE, Mutter JC, Wang X, Tulshian DB, Egan RW, Hey JA (2002) Antitussive effect of nociceptin/orphanin FQ in experimental cough models. Pulm Pharmacol Ther 15:213–216

McLeod RL, Jia Y, Fernandez X, Parra LE, Wang X, Tulshian DB, Kiselgof EJ, Tan Z, Fawzi AB, Smith-Torhan A, Zhang H, Hey JA (2004) Antitussive profile of the NOP agonist Ro-64–6198 in the guinea pig. Pharmacology 71:143–149

McLeod RL, Fernandez X, Correll CC, Phelps TP, Jia Y, Wang X, Hey JA (2006) TRPV1 antagonists attenuate antigen-provoked cough in ovalbumin sensitized guinea pigs. Cough 2:10

Meis S, Pape HC (2001) Control of glutamate and GABA release by nociceptin/orphanin FQ in the rat lateral amygdala. J Physiol 532:701–712

Melis M, Pistis M, Perra S, Muntoni AL, Pillolla G, Gessa GL (2004) Endocannabinoids mediate presynaptic inhibition of glutamatergic transmission in rat ventral tegmental area dopamine neurons through activation of CB1 receptors. J Neurosci 24:53–62

Melzack R, Coderre TJ, Katz J, Vaccarino AL (2001) Central neuroplasticity and pathological pain. Ann N Y Acad Sci 933:157–174

Meunier JC, Mollereau C, Toll L, Suaudeau C, Moisand C, Alvinerie P, Butour JL, Guillemot JC, Ferrara P, Monsarrat B, Mazarguil H, Vassart G, Parmentier M, Costentin J (1995) Isolation and structure of the endogenous agonist of opioid receptor-like ORL_1 receptor. Nature 377:532–535

Midgren B, Hansson L, Karlsson JA, Simonsson BG, Persson CG (1992) Capsaicin-induced cough in humans. Am Rev Respir Dis 146:347–351

Mifflin SW, Felder RB (1990) Synaptic mechanisms regulating cardiovascular afferent inputs to solitary tract nucleus. Am J Physiol 259:H653–H661

Mochizuki H, Shimizu T, Morikawa A, Kuroume T (1995) Inhaled diuretics attenuate acid-induced cough in children with asthma. Chest 107:413–417

Monnet FP (2005) Sigma-1 receptor as regulator of neuronal intracellular Ca2+: Clinical and therapeutic relevance. Biol Cell 97:873–883

Monteillet-Agius G, Fein J, Anton B, Evans CJ (1998) ORL-1 and mu opioid receptor antisera label different fibers in areas involved in pain processing. J Comp Neurol 399:373–383

Mooraki A, Jenabi A, Jabbari M, Zolfaghari MI, Javanmardi SZ, Mahmoudian M, Bastani B (2005) Noscapine suppresses angiotensin converting enzyme inhibitors-induced cough. Nephrology (Carlton) 10:348–350

Moreaux B, Nemmar A, Vincke G, Halloy D, Beerens D, Advenier C, Gustin P (2000) Role of substance P and tachykinin receptor antagonists in citric acid-induced cough in pigs. Eur J Pharmacol 408:305–312

Morice AH, Geppetti P (2004) Cough. 5: The type 1 vanilloid receptor: a sensory receptor for cough. Thorax 59:257–258

Morice AH, Lowry R, Brown MJ, Higenbottam T (1987) Angiotensin-converting enzyme and the cough reflex. Lancet 2:1116–1118

Morice AH, Marshall AE, Higgins KS, Grattan TJ (1994) Effect of inhaled menthol on citric acid induced cough in normal subjects. Thorax 49:1024–1026

Morice AH, Fontana GA, Sovijarvi AR, Pistolesi M, Chung KF, Widdicombe JG, O'Connell F, Geppetti P, Gronke L, De JJ, Belvisi M, Dicpinigaitis P, Fischer A, McGarvey L, Fokkens WJ, Kastelik J (2004) The diagnosis and management of chronic cough. Eur Respir J 24:481–492

Morice AH, Menon MS, Mulrennan SA, Everett CF, Wright C, Jackson J, Thompson R (2007) Opiate therapy in chronic cough. Am J Respir Crit Care Med 175:312–315

Mulrennan S, Wright C, Thompson R, Goustas P, Morice A (2004) Effect of salbutamol on smoking related cough. Pulm Pharmacol Ther 17:127–131

Mutoh T, Bonham AC, Kott KS, Joad JP (1999) Chronic exposure to sidestream tobacco smoke augments lung C-fiber responsiveness in young guinea pigs. J Appl Physiol 87:757–768

Mutoh T, Joad JP, Bonham AC (2000) Chronic passive cigarette smoke exposure augments bronchopulmonary C-fibre inputs to nucleus tractus solitarii neurones and reflex output in young guinea-pigs. J Physiol 523(Part 1):223–233

Myers AC, Kajekar R, Undem BJ (2002) Allergic inflammation-induced neuropeptide production in rapidly adapting afferent nerves in guinea pig airways. Am J Physiol Lung Cell Mol Physiol 282:L775–L781

Nakajima T, Nishimura Y, Nishiuma T, Kotani Y, Nakata H, Yokoyama M (2006) Cough sensitivity in pure cough variant asthma elicited using continuous capsaicin inhalation. Allergol Int 55:149–155

Netzer R, Pflimlin P, Trube G (1993) Dextromethorphan blocks N-methyl-D-aspartate-induced currents and voltage-operated inward currents in cultured cortical neurons. Eur J Pharmacol 238:209–216

Nichol GM, Alton EW, Nix A, Geddes DM, Chung KF, Barnes PJ (1990) Effect of inhaled furosemide on metabisulfite- and methacholine-induced bronchoconstriction and nasal potential difference in asthmatic subjects. Am Rev Respir Dis 142:576–580

Nilius B, Owsianik G, Voets T, Peters JA (2007) Transient receptor potential cation channels in disease. Physiol Rev 87:165–217

O'Connell F, Thomas VE, Pride NB, Fuller RW (1994) Capsaicin cough sensitivity decreases with successful treatment of chronic cough. Am J Respir Crit Care Med 150:374–380

O'Connell F, Springall DR, Moradoghli-Haftvani A, Krausz T, Price D, Fuller RW, Polak JM, Pride NB (1995) Abnormal intraepithelial airway nerves in persistent unexplained cough? Am J Respir Crit Care Med 152:2068–2075

O'Connell F, Thomas VE, Studham JM, Pride NB, Fuller RW (1996) Capsaicin cough sensitivity increases during upper respiratory infection. Respir Med 90:279–286

Pacher P, Batkai S, Kunos G (2006) The endocannabinoid system as an emerging target of pharmacotherapy. Pharmacol Rev 58:389–462

Packman EW, Ciccone PE, Wilson J, Masurat T (1991) Antitussive effects of diphenhydramine on the citric acid aerosol-induced cough response in humans. Int J Clin Pharmacol Ther Toxicol 29:218–222

Pantaleo T, Bongianni F, Mutolo D (2002) Central nervous mechanisms of cough. Pulm Pharmacol Ther 15:227–233

Park KA, Vasko MR (2005) Lipid mediators of sensitivity in sensory neurons. Trends Pharmacol Sci 26:571–577

Patel HJ, Birrell MA, Crispino N, Hele DJ, Venkatesan P, Barnes PJ, Yacoub MH, Belvisi MG (2003) Inhibition of guinea-pig and human sensory nerve activity and the cough reflex in guinea-pigs by cannabinoid (CB2) receptor activation. Br J Pharmacol 140:261–268

Pavord I, Hall I, Wahedna I, Cooper S, Tattersfield A (1994) Effect of 443c81, an inhaled mu-opioid receptor agonist in asthma. Clin Exp Allergy 24:144–148

Pezet S, Spyropoulos A, Williams RJ, McMahon SB (2005) Activity-dependent phosphorylation of Akt/PKB in adult DRG neurons. Eur J Neurosci 21:1785–1797

Pillay V, Swingler G (2003) Symptomatic treatment of the cough in whooping cough. Cochrane Database Syst Rev:CD003257

Pounsford JC, Birch MJ, Saunders KB (1985) Effect of bronchodilators on the cough response to inhaled citric acid in normal and asthmatic subjects. Thorax 40:662–667

Pratter MR (2006) Cough and the common cold: ACCP evidence-based clinical practice guidelines. Chest 129:72S–74S

Pratter MR, Brightling CE, Boulet LP, Irwin RS (2006) An empiric integrative approach to the management of cough: ACCP evidence-based clinical practice guidelines. Chest 129: 222S–231S

Rees PJ, Clark TJ (1983) Assessment of antitussive effects by citric acid threshold. Br J Dis Chest 77:94–97

Reynolds SM, Mackenzie AJ, Spina D, Page CP (2004) The pharmacology of cough. Trends Pharmacol Sci 25:569–576
Riccio MM, Myers AC, Undem BJ (1996) Immunomodulation of afferent neurones in guinea-pig isolated airways. J Physiol 491:499–509
Ruhle KH, Criscuolo D, Dieterich HA, Kohler D, Riedel G (1984) Objective evaluation of dextromethorphan and glaucine as antitussive agents. Br J Clin Pharmacol 17:521–524
Sant'Ambrogio FB, Sant'Ambrogio G, Anderson JW (1993) Effect of furosemide on the response of laryngeal receptors to low-chloride solutions. Eur Respir J 6:1151–1155
Schroeder K, Fahey T (2002) Systematic review of randomised controlled trials of over the counter cough medicines for acute cough in adults. Br Med J 324:329–331
Schroeder K, Fahey T (2004) Over-the-counter medications for acute cough in children and adults in ambulatory settings. Cochrane Database Syst Rev:CD001831
Sekizawa K, Jia YX, Ebihara T, Hirose Y, Hirayama Y, Sasaki H (1996) Role of substance P in cough. Pulm Pharmacol 9:323–328
Sekizawa S, Joad JP, Bonham AC (2003) Substance P presynaptically depresses the transmission of sensory input to bronchopulmonary neurons in the guinea pig nucleus tractus solitarii. J Physiol 552:547–559
Shah S, Page CP, Spina D (1998) Nociceptin inhibits non-adrenergic non-cholinergic contraction in guinea-pig airway. Br J Pharmacol 125:510–516
Shannon R, Baekey DM, Morris KF, Li Z, Lindsey BG (2000) Functional connectivity among ventrolateral medullary respiratory neurones and responses during fictive cough in the cat. J Physiol 525(Part 1):207–224
Sheppard D, Rizk NW, Boushey HA, Bethel RA (1983) Mechanism of cough and bronchoconstriction induced by distilled water aerosol. Am Rev Respir Dis 127:691–694
Shin J, Cho H, Hwang SW, Jung J, Shin CY, Lee SY, Kim SH, Lee MG, Choi YH, Kim J, Haber NA, Reichling DB, Khasar S, Levine JD, Oh U (2002) Bradykinin-12-lipoxygenase-VR1 signaling pathway for inflammatory hyperalgesia. Proc Natl Acad Sci USA 99:10150–10155
Smart D, Gunthorpe MJ, Jerman JC, Nasir S, Gray J, Muir AI, Chambers JK, Randall AD, Davis JB (2000) The endogenous lipid anandamide is a full agonist at the human vanilloid receptor (hVR1). Br J Pharmacol 129:227–230
Smith CA, Adamson DL, Choudry NB, Fuller RW (1991) The effect of altering airway tone on the sensitivity of the cough reflex in normal volunteers. Eur Respir J 4:1078–1079
Smith BN, Dou P, Barber WD, Dudek FE (1998) Vagally evoked synaptic currents in the immature rat nucleus tractus solitarii in an intact in vitro preparation. J Physiol 512(Part 1):149–162
Soya A, Serino R, Fujihara H, Onaka T, Ozaki Y, Saito T, Nakamura J, Ueta Y (2005) Cannabinoids modulate synaptic activity in the rat supraoptic nucleus. J Neuroendocrinol 17:609–615
Spina D, Page CP, Morley J. (1998). Sensory neuropeptides and bronchial hyperresponsiveness.In: Said SI (ed) ProInflammatory and antiinflammatory peptides. Marcel Deckker, New York, pp 89–146
Stone RA, Barnes PJ, Chung KF (1993a) Effect of frusemide on cough responses to chloride-deficient solution in normal and mild asthmatic subjects. Eur Respir J 6:862–867
Stone RA, Worsdell YM, Fuller RW, Barnes PJ (1993b) Effects of 5-hydroxytryptamine and 5-hydroxytryptophan infusion on the human cough reflex. J Appl Physiol 74:396–401
Studham J, Fuller RW (1992) The effect of oral terfenadine on the sensitivity of the cough reflex in normal volunteers. Pulm Pharmacol 5:51–52
Suzuki K, Sato M, Morishima Y, Nakanishi S (2005) Neuronal depolarization controls brain-derived neurotrophic factor-induced upregulation of NR2C NMDA receptor via calcineurin signaling. J Neurosci 25:9535–9543
Szallasi A, Blumberg PM (1999) Vanilloid (Capsaicin) receptors and mechanisms. Pharmacol Rev 51:159–211
Tashkin DP, Reiss S, Shapiro BJ, Calvarese B, Olsen JL, Lodge JW (1977) Bronchial effects of aerosolized delta 9-tetrahydrocannabinol in healthy and asthmatic subjects. Am Rev Respir Dis 115:57–65

Tatar M, Webber SE, Widdicombe JG (1988) Lung C-fibre receptor activation and defensive reflexes in anaesthetized cats. J Physiol 402:411–420
Tatar M, Sant'Ambrogio G, Sant'Ambrogio FB (1994) Laryngeal and tracheobronchial cough in anesthetized dogs. J Appl Physiol 76:2672–2679
Trevisani M, Milan A, Gatti R, Zanasi A, Harrison S, Fontana G, Morice AH, Geppetti P (2004) Antitussive activity of iodo-resiniferatoxin in guinea pigs. Thorax 59:769–772
Trochtenberg S (1994) Nebulized lidocaine in the treatment of refractory cough. Chest 105: 1592–1593
Tucker RC, Kagaya M, Page CP, Spina D (2001) The endogenous cannabinoid agonist, anandamide stimulates sensory nerves in guinea-pig airways. Br J Pharmacol 132:1127–1135
Undem BJ, Weinreich D (1993) Electrophysiological properties and chemosensitivity of guinea pig nodose ganglion neurons in vitro. J Auton Nerv Syst 44:17–33
Undem BJ, Carr MJ, Kollarik M (2002) Physiology and plasticity of putative cough fibres in the Guinea pig. Pulm Pharmacol Ther 15:193–198
Ventresca PG, Nichol GM, Barnes PJ, Chung KF (1990) Inhaled furosemide inhibits cough induced by low chloride content solutions but not by capsaicin. Am Rev Respir Dis 142:143–146
Wang L, Bradley RM (1995) In vitro study of afferent synaptic transmission in the rostral gustatory zone of the rat nucleus of the solitary tract. Brain Res 702:188–198
Watanabe N, Horie S, Michael GJ, Spina D, Page CP, Priestley JV (2005) Immunohistochemical localization of vanilloid receptor subtype 1 (TRPV1) in the guinea pig respiratory system. Pulm Pharmacol Ther 18:187–197
Watanabe N, Horie S, Michael GJ, Keir S, Spina D, Page CP, Priestley JV (2006) Immunohistochemical co-localization of transient receptor potential vanilloid (TRPV)1 and sensory neuropeptides in the guinea-pig respiratory system. Neuroscience 141:1533–1543
Weinfeld D, Ternesten-Hasseus E, Lowhagen O, Millqvist E (2002) Capsaicin cough sensitivity in allergic asthmatic patients increases during the birch pollen season. Ann Allergy Asthma Immunol 89:419–424
Widdicombe JG (1995) Neurophysiology of the cough reflex. Eur Respir J 8:1193–1202
Widdicombe JG (2003) Overview of neural pathways in allergy and asthma. Pulm Pharmacol Ther 16:23–30
Woolf CJ, King AE (1989) Subthreshold components of the cutaneous mechanoreceptive fields of dorsal horn neurons in the rat lumbar spinal cord. J Neurophysiol 62:907–916
Woolf CJ, King AE (1990) Dynamic alterations in the cutaneous mechanoreceptive fields of dorsal horn neurons in the rat spinal cord. J Neurosci 10:2717–2726
Woolf CJ, Salter MW (2000) Neuronal plasticity: Increasing the gain in pain. Science 288: 1765–1769
Woolf CJ, Wall PD (1986) Relative effectiveness of C primary afferent fibers of different origins in evoking a prolonged facilitation of the flexor reflex in the rat. J Neurosci 6:1433–1442
Yang L, Zhang FX, Huang F, Lu YJ, Li GD, Bao L, Xiao HS, Zhang X (2004) Peripheral nerve injury induces trans-synaptic modification of channels, receptors and signal pathways in rat dorsal spinal cord. Eur J Neurosci 19:871–883
Yasumitsu R, Hirayama Y, Imai T, Miyayasu K, Hiroi J (1996) Effects of specific tachykinin receptor antagonists on citric acid-induced cough and bronchoconstriction in unanesthetized guinea pigs. Eur J Pharmacol 300:215–219
Yukioka H, Yoshimoto N, Nishimura K, Fujimori M (1985) Intravenous lidocaine as a suppressant of coughing during tracheal intubation. Anesth Analg 64:1189–1192
Zou X, Lin Q, Willis WD (2000) Enhanced phosphorylation of NMDA receptor 1 subunits in spinal cord dorsal horn and spinothalamic tract neurons after intradermal injection of capsaicin in rats. J Neurosci 20:6989–6997
Zygmunt PM, Petersson J, Andersson DA, Chuang HH, Sorgard M, DiMarzo M, Julius V, Hogestatt ED (1999) Vanilloid receptors on sensory nerves mediate the vasodilator action of anandamide. Nature 400:452–457

Central Mechanisms I: Plasticity of Central Pathways

C.-Y. Chen, J.P. Joad, J. Bric, and A.C. Bonham(✉)

Contents

Abstract Cough is the most common symptom for which individuals seek medical attention and spend health-care dollars. Despite the burden induced by cough, the current treatments for cough are only partially effective. Delineating the sites and mechanisms in the cough central network for changes in the cough reflex could lead to new therapeutic strategies and drug target sites for more effective treatments. The first synaptic target in the CNS for the cough-related sensory input is the second-order neurons in the nucleus tractus solitarius (NTS); these neurons reorganize the primary sensory information into a coherent output. The NTS neurons have been shown to undergo neuroplasticity under a variety of conditions, such as respiratory disorders, stress, and exposures to environmental pollutants. The NTS contains a rich innervation of substance P immunoreactive nerve terminals, suggesting that substance P might be important in altered cough reflex response. This chapter summarizes our current findings on the role of substance P in enhanced cough reflex as well as the potential NTS targets for the action of substance P.

A.C. Bonham
Department of Pharmacology, University of California, Davis School of Medicine, 4150 V Street, 1104 PSSB Sacramento, CA 95817, USA
ann.bonham@ucdmc.ucdavis.edu

K.F. Chung, J.G. Widdicombe (eds.), *Pharmacology and Therapeutics of Cough*, Handbook of Experimental Pharmacology 187,

1 Cough as a Symptom

Cough is a critical airway defense mechanism to clear foreign materials or accumulated secretions from the airways. However, too frequent or severe a cough can be annoying and thus individuals seek cough suppression. During vigorous cough, the intrathoracic pressure can reach up to 300 mmHg and systolic blood pressure can approach 140 mmHg. The pressure generated during cough can cause a variety of cardiovascular, CNS, gastrointestinal, genitourinary, musculoskeletal, neurological, ophthalmologic, psychosocial, skin, and respiratory complications as well as affecting the quality of life (Irwin 2006; Irwin et al. 2002). Cough is the most common symptom for which individuals seek medical attention and spend health-care dollars (Irwin et al. 1998). Of the estimated 963.6 million visits made to physician offices in the USA, cough (3.4%) was the most frequently mentioned reason regarding an illness or injury (Cherry et al. 2005). In the USA, the cough/cold market represents several billion dollars' worth of sales for over-the-counter products alone, not including the cost of prescription drugs or the treatment of chronic cough (Morice 2002). The prevalence and impact of cough in children is even greater. In a survey of 7,251 households with at least one child aged 1–5 years, 85% of the physician visits were for asthmalike symptoms including cough, with 35% of children making three or more visits during a 6-month period (Bisgaard and Szefler 2007). Over half of the children (67%) received cough treatment.

A major challenge is in treating the cough of unknown cause, which could make up as many as 40% of the patients with chronic cough (Chung 2007). Despite the burden induced by cough, the current treatments for cough are only partially effective at best (Morice 2002), even in some patients with identified associated causes (Chung 2007). The available cough-suppressant therapy is limited by the lack of effective agents and/or their unacceptable side effects. In 2007, the US Food and Drug Administration (FDA) re-examined the safety of the over-the-counter cough and cold medicine and whether the benefits justify any potential risks from the use of these products, especially in children under 2 years of age. These products can be harmful if more than the recommended amount is used, if it is given too often, or if more than one cough and cold medicine containing the same active ingredient is being used. The FDA has recommended that over-the-counter cough medications not be given to children under 2 years of two because of their lack of proven benefit and their risk of harm.

Delineating the sites and mechanisms in the cough central network for changes in the cough reflex could lead to new therapeutic strategies and drug target sites for more effective treatments.

2 Cough Reflex

The involuntary cough reflex can be triggered by a variety of conditions: common cold is the main cause for acute cough, while chronic cough is often due to more than one condition, including postnasal drip syndrome, asthma, gastroesophageal reflux disease, bronchitis, angiotensin-converting-enzyme inhibitor, nonbronchiectatic suppurative airway diseases, aspiration secondary to oral/pharyngeal dysphagia, and exposure to environmental or occupational pollutants (Irwin et al. 1998; Pratter 2006).

The tussive stimulus activates vagal sensory nerves in the lungs and airways. The first synaptic target in the CNS for the cough-related sensory input is the second-order neurons in the nucleus tractus solitarius (NTS); these neurons reorganize the primary sensory information into a coherent output and then transmit the information to a group of synaptically connected neurons in the ventrolateral medulla that form the central respiratory pattern generator (Lindsey et al. 2000; Shannon et al. 1998). The neuronal activity in the central respiratory pattern generator is transiently interrupted, perhaps by additional synaptic interactions transforming it into a central cough generator (Shannon et al. 1998; Morris et al. 2003; Bolser and Davenport 2002), to produce cough to engage what has been hypothesized as an endogenous cough network involving neurons located in the brainstem (Poliacek et al. 2007). Changes in neuronal behavior – short-term or long-term, at any site in the central network – could result in an exaggeration or suppression of the cough reflex.

The neuronal processing of the cough-related information by the NTS second-order neurons is complex, inasmuch as the neurons receive and monitor input from a number of sensory afferent fibers with varying effects on cough threshold, frequency, and intensity (Irwin et al. 1998; Pratter 2006; Widdicombe and Singh 2006). For example, activation of bronchopulmonary C-fiber, nasal, cardiac, splanchnic, esophageal, and nasal sensory afferents, while not directly mediating cough, regulates the intensity of the cough reflex (Widdicombe and Singh 2006). Acute stimulation of peripheral chemoreceptors and nasal sensory afferents upregulates the cough reflex, while stimulation of bronchopulmonary C-fiber, cardiac, and abdominal afferent nerve endings may downregulate it.

More recent studies have expanded on different phenotypes of sensory afferent fibers, including those associated with the cough reflex, on the basis of ganglionic location, ionic composition, specific chemosensitivity, mechanosensitivity, and neuropeptide content of the cell bodies of the afferent fibers and expression of receptors in their peripheral terminal endings (Lee and Undem 2004). It seems likely that the input signals monitored by the second-order neurons will vary still more with the differences in the phenotypes of the vagal synaptic partners. Presented with these wide ranges of sensory inputs, the NTS second-order neurons must (1) sort and process the inputs in such a way as to maintain the integrity of input relevant to the cough reflex, while at the same time allowing for the possibility that multiple inputs (i.e., regarding cardiovascular and respiratory status) might need to be coincidentally processed by the same neuron; (2) integrate the information spatially and temporally with other inputs from local neural networks or from higher brain regions (Chen and Bonham 1998; Aylwin et al. 1998); and (3) ultimately coordinate outputs to appropriate distal synapses to effect, modify, or suppress the cough reflex. Thus, any such changes in the NTS can change the activities/responsiveness of the downstream central sites and thus the nature of the reflex output into either exaggerating it or prolonging it or suppressing it or altering the output pattern.

2.1 NTS Plasticity and Relationship to Cough

Epidemiological studies have shown that, even in healthy subjects without pulmonary diseases, cough is one of the major symptoms related to exposure to air pollutants, including particulate matter, irritant gases, second-hand tobacco smoke, mixed pollutants, and molds (Joad et al. 2007). NTS neurons have been shown to undergo neuroplasticity under a variety of conditions, such as diseases, stress, and exposures to environmental pollutants (Joad et al. 2004; Chen et al. 2001, 2003). For example, the NTS neurons in primates exposed to either allergen and/or ozone displayed a more depolarized resting membrane potential that is associated with an increased input resistance, and an increased spiking activity in response to depolarization (Chen et al. 2001, 2003). The data provided early evidence for neuroplasticity in the NTS and suggest scenarios whereby subthreshold sensory inputs may evoke postsynaptic spikes and hence an increased NTS output to distal synapses in the cough reflex central network. In addition, suprathreshold inputs may evoke more action potentials resulting in the exposed animals, thereby amplifying the neuronal output at these synapses and thus a heightened responsiveness.

The plasticity in the NTS may not be uniform across exposure types, species, and subpopulations of neurons in NTS. The complexity of plasticity in the NTS is best demonstrated by the findings that even though both allergen and ozone depolarized the resting membrane potential and increased the spiking response in primates, only ozone exposure decreased the synaptic transmission between the primary afferent and the NTS second-order neurons (Bonham et al. 2006). In a preliminary finding in guinea pigs, exposures to second-hand tobacco smoke resulted in a more depolarized resting membrane potential while blunting the spiking response to depolarization in a subset of NTS neurons that displayed a fast spiking phenotype (Sekizawa et al. 2005). The data suggested that even though the exaggerated cough reflex under different conditions may share similar underlying mechanisms, there may be some changes that are unique to each pollutant and each species.

2.2 Substance P Mechanisms: Potential Role in NTS Plasticity

Interest in substance P in modulating autonomic reflexes in the NTS has waxed and waned over the past several years (Helke and Seagard 2004). The neuropeptide is widely distributed throughout the peripheral nervous system and the CNS. There are considerable data on the role of the neuropeptide as a neuromodulator; it has been proposed to play a role in asthma, cough and other respiratory reflexes, emesis, psoriasis, and a number of neuropsychiatric disorders. Attempts to develop therapeutic agents have been somewhat disappointing – which may reflect the complexity of the effects of substance P in the CNS.

Immunohistochemistry studies have shown that the NTS contains a rich innervation of substance P immunoreactive nerve terminals arising from cardiovascular, baroreceptor, and chemoreceptor afferent nerves, bronchopulmonary afferent nerves, and other brain regions (Davis and Smith 1999). Substance P in the NTS has been shown to provide a general mechanism for modulating autonomic function, including the respiratory reflex. We previously showed that near-threshold changes in expiratory time evoked by stimulation of the bronchopulmonary C-fibers were augmented tenfold by prior injections of substance P in the caudomedial NTS (Mutoh et al. 2000). In the same study, threshold bronchoconstriction and depressor as well as bradycardic response were also augmented. The substance P effects were abolished by blockade of the substance P neurokinin 1 (NK1) receptors. Similarly, in anesthetized guinea pigs, Mazzone and colleagues showed that direct microinjections of substance P into NTS decreased the electrical threshold for cough evoked from the tracheal mucosa by electrical stimulation or by citric acid (Mazzone et al. 2005). The data suggest that, just as in other autonomic reflexes, substance P acting on NK1 receptors modulates the cough reflex. A recent study by Mutolo and colleagues confirmed that, in anesthetized rabbits, bilateral microinjections of substance P into the NTS significantly increased the cough motor responses: the cough number, peak abdominal activity, and expiratory rise rate in response to stimulus (Mutolo et al. 2007).

Although *exogenous* application of substance P enhances the cough reflex, studies using various substance P antagonists suggest that substance P may not be functionally important under physiological conditions. For example, microinjections of substance P NK1 receptor antagonists, at doses that blocked the exogenous substance P effect on C-fiber mediated reflexes, failed to attenuate the tidal breathing pattern or cough evoked from the trachea by electrical stimulation or citric acid in control animals (Mazzone et al. 2005). In a conscious animal model, we previously showed that microinjections of a substance P NK1 receptor antagonist into the NTS had no significant effects on citric acid induced cough in freely moving control guinea pigs (Bonham et al. 2004). The data suggest that endogenous substance P plays a minor role, if any, in the cough reflex under physiological conditions.

The question is, then, does substance P play a role in the cough reflex in pathological conditions? The first data suggesting that this might be the case came from studies on the peripheral sensory afferent neurons. Vagal sensory afferent neurons, including those from the lungs, undergo plasticity in response to a number of challenges, including allergens, inflammation, and air pollutants (Carr and Lee 2006; Undem et al. 1999, 2002; Carr and Undem 2001). The plasticity has been shown to be associated with increases in the messenger RNA encoding substance P and de novo substance P expression in vagal afferent cell bodies (Fischer et al. 1996; Chuaychoo et al. 2005). The increased substance P expression suggests that substance P could be released both locally in the lungs and airways and also in the central terminals in the NTS to modify synaptic transmission of the sensory information to the NTS. For example, in conscious guinea pigs, exposure to second-hand tobacco smoke significantly increases the cough response to citric acid aerosol (Joad et al. 2007; Bonham et al. 2004). This second-hand tobacco smoke induced enhanced cough is attenuated by microinjection of a substance P NK1 receptor antagonist into the NTS (Joad et al. 2007), providing evidence that plasticity in the NTS in the cough pathway may involve substance P NK1 receptor mechanisms.

Similarly, repeatedly exposing infant primates to ozone resulted in an increased spiking response to intracellular injections of depolarizing currents that is mediated by endogenous substance P acting at NK1 receptors (Chen et al. 2003). Taken together, the data indicate that substance P NK1 receptors in the NTS play an important role in conditions associated with an enhanced cough reflex.

3 What Might the Role for Substance P Be at a Cellular Level?

The in vivo studies provide evidence for the role of substance P for neuroplasticity – an enhanced cough reflex via upregulation of the substance P NK1 receptor mechanisms. However, the underlying synaptic and cellular mechanisms are still not fully understood. Substance P may change the NTS outputs, by directly changing neuronal excitability and/or excitatory/inhibitory synaptic transmission.

In general, substance P has been shown to increase neuronal excitability, depolarize the resting membrane potential, decrease the action potential threshold and the magnitude of afterhypolarization, and increase the action potential duration and spiking response to depolarization (Sculptoreanu et al. 2007; Suwabe and Bradley 2007; Jun et al. 2004; Sun et al. 2004; Lee et al. 1995; Gilbert et al. 1998; Dray and Pinnock 1982; Murase et al. 1982; Davies and Dray 1980; Zieglgansberger and Tulloch 1979). The change in resting membrane potential sometimes is associated with a change in membrane resistance, a rough measurement of the membrane conductance, suggesting the involvement of ion channels (Suwabe and Bradley 2007; Dray and Pinnock 1982; Murase et al. 1982; Jafri and Weinreich 1996; Lepre et al. 1996).

The depolarization is often associated with an inward current under voltage-clamp condition, consistent with the excitatory effect of substance P (Jun et al. 2004; Ishimatsu 1994; Drew et al. 2005; Yang et al. 2003; Ito et al. 2002; Akasu et al. 1996). This substance P induced inward current has been shown to be mediated by activation of a nonselective cation channel (Jun et al. 2004; Lee et al. 1995; Drew et al. 2005; Ito et al. 2002) and/or inhibition of potassium channels (Drew et al. 2005; Akasu et al. 1996), including the voltage-activated potassium channels (Sun et al. 2004; Lepre et al. 1996) and the calcium-activated potassium channels (Gilbert et al. 1998; Lepre et al. 1996). Interestingly, Yang and colleagues showed that substance P can induce three inward currents in bullfrog dorsal root ganglion neurons: slow, fast, and moderately activating currents (Yang et al. 2003). The fast activating current was abolished by replacement of NaCl in the external solution, suggesting that it is likely mediated by the opening of Na^+-preferring nonselective cation channels. The opening of the nonselective cation channels did not appear to be mediated by activation of G-protein (Jun et al. 2004; Yang et al. 2003). The slow activating current was not affected by replacement of NaCl, but was blocked by Ba^{2+}, suggesting the involvement of G-protein-mediated closure of K^+ channels (Yang et al. 2003; Akasu et al. 1996). Not surprisingly, the substance P induced moderately activating inward current was mediated by both mechanisms. In the

NTS, Bailey and colleagues showed that activation of substance P NK1 receptors resulted in depolarization, evident as a slow inward current that is mediated by protein kinase C (Bailey et al. 2004).

Substance P can also increase the intracellular calcium concentration, partially due to the release of calcium from intracellular calcium stores (Jun et al. 2004; Ito et al. 2002). The release of intracellular stores of calcium has been shown to be mediated by the phospholipase C pathway (Ito et al. 2002). The effects of substance P on voltage-dependent calcium channels vary significantly between networks; for example, substance P induced calcium influx in spiral ganglion neurons (Sun et al. 2004) while having no effect on interstitial cells of Cajal (Jun et al. 2004). Different calcium channels may respond differently to substance P. In NTS neurons, the selective NK1 receptor agonist $[Sar^9,Met(O_2)^{11}]$-substance P caused facilitation of the L-type voltage-dependent calcium currents in about half of the neurons (Endoh 2006). The effect is mediated by Gαq/11-protein involving the protein kinase C pathway (Endoh 2006). Since calcium channels are important for neurotransmitter release at presynaptic terminals, enzyme activity, and gene expression, the data suggest that the presynaptic or postsynaptic location of the NK1 receptors may have different responses to substance P. In addition, it is still unclear if the difference in the response reflects the difference in the nature of the neurons: neurons in specific reflex pathways, input or output neurons, or the neuronal type (e.g., glutaminergic or GABAergic, etc.).

3.1 Effect of Substance P on Excitatory Synaptic Transmission

The fast excitatory synaptic transmission at the NTS second-order neurons is mediated by glutamate acting on ionotropic glutamate receptors, namely, the non-*N*-methyl-D-aspartate (non-NMDA) and NMDA receptors. The evidence presented by Mutolo and colleagues showing that blockade of non-NMDA receptors in NTS abolished cough responses suggests that glutamate acting on non-NMDA receptors is primarily responsible for signal transmission in the NTS in the cough pathway (Mutolo et al. 2007).

The effect of substance P on excitatory synaptic transmission may depend on the receptor loci. Acting on presynaptic terminals, substance P can reduce glutamate release from the central terminals of the vagal sensory neurons onto anatomically identified second-order lung afferent neurons in the NTS (Sekizawa et al. 2003). We showed that substance P decreased the amplitude of excitatory postsynaptic currents evoked by stimulation of the afferent fibers in the solitary tract and increased the paired-pulse ratio of two consecutive evoked responses. The effect was blocked by a NK1 receptor antagonist, suggesting a mechanism by which substance P might regulate release at the first central synapses between lung afferent fibers and the second-order neurons in the lung reflex pathways. The presynaptic inhibition of glutamate release against the background of an overall postsynaptic excitation may improve signal transmission by changing the incoming signals with a wide dynamic range (more subject to noise) to a smaller range (less subject to noise).

Postsynaptically, substance P has been shown to enhance currents mediated by the ionotropic glutamate receptors, both NMDA and non-NMDA receptors, in other networks (Pieri et al. 2005; Randic et al. 1990; Budai et al. 1992; Lieberman and Mody 1998; Liu et al. 1998; Wu et al. 2004). While enhancing the non-NMDA receptor mediated synaptic current has obvious effects on an enhanced rapid signal transmission, the effect of substance P on NMDA receptors may be more important in plasticity. For example, an enhanced NMDA current could result in the opening of the NMDA channel at less depolarizing membrane potentials, producing a slower-developing, longer-lasting component of glutamate signaling. The enhanced NMDA-mediated slow depolarization could prolong the time during which otherwise ineffective inputs could be integrated and enable the neurons to transduce afferent input, which has no apparent pattern, into more efficient spiking patterns (Tell and Jean 1991, 1993; Yen and Chan 1997). In addition, activation of the NMDA receptor could lead to a long-lasting increase in the strength of synaptic transmission, including long-term potentiation of synaptic transmission and pain hypersensitivity (Artola and Singer 1987; Thompson et al. 1990; Woolf and Thompson 1991). Of relevance to NMDA receptors in the NTS is the finding that even though the non-NMDA receptors in the NTS are primarily involved in the cough reflex evoked by mechanical stimulation of the tracheobrohchial tree, blockade of the NMDA receptors in the NTS significantly reduced the cough reflex (Mutolo et al. 2007). The data suggest a potential site for central plasticity in modulating the cough reflex and a potential new drug target.

3.2 *Effect of Substance P on Inhibitory Synaptic Transmission*

The major inhibitory synaptic transmission in the NTS is provided by γ-aminobutyric acid (GABA) acting on $GABA_A$ receptors. GABA is released spontaneously and through stimulation and has profound inhibitory effects on synaptic transmission in the NTS (Len and Chan 2001; Zhang and Mifflin 1998). Data from microinjection studies in an upregulated cough reflex model in guinea pigs suggest a complex interaction between substance P and GABA in neuroplasticity in the NTS.

Microinjection of a substance P NK1 receptor antagonist into the NTS attenuated the second-hand tobacco smoke induced enhanced cough (Joad et al. 2007) and had no effect in the filtered-air control group. Microinjection of the $GABA_A$ receptor antagonist bicuculline had no effect on the citric acid evoked cough in either the filtered-air or the second-hand smoke exposed groups. Interestingly, bicuculline blocked the substance P NK1 antagonist effect in the second-hand-smoke group (Fig. 1). The question is how might GABA and substance P interact to explain these outcomes. Presynaptically, we know from other studies that substance P has no effect on miniature GABAergic inhibitory postsynaptic currents (Ogier and Raggenbass 2003; Stacey et al. 2002), suggesting that the neuropeptide does not an amount as small as 1 μg was sufficient to potently inhibit the cough responses, various receptors such as 5-HT, acetylcholine, and *N*-methyl-D-aspartate (NMDA) receptors might be involved in cough responses in the NTS and adjacent regions. Surprisingly, direct administration of 1 μg of glycine into the cough center potentiated cough responses (Honda et al 1990), as indicated by an increase in the amplitude of the respiration curves of the cough responses. This finding is of particular interest, because enhancing cough responses is reported to prevent aspiration pneumonia in elderly people (Yamaya et al. 2001).

4.2 Glycine and Its Receptor

4.2.1 Action on Glycine-Induced Currents in Single Neurons

In the course of our study on the pharmacological actions of antitussives, we noticed that antitussives at toxic doses often produced convulsant activity in experimental animals, possibly due to blockade of receptors of inhibitory neurotransmitters such as glycine and γ-aminobutyric acid (GABA). This phenomenon reminded us that antitussives may have some affinity for inhibitory neurotransmitter receptors. Glycine and GABA are recognized as the main inhibitory neurotransmitters in the brain. In particular, glycine is mainly distributed in the lower brain stem and spinal cord, where the signaling pathway for cough responses is localized. Therefore, we first attempted to determine whether antitussives affect glycine-mediated responses in the CNS (Takahama et al. 1997a, b). In the experiment, we used acutely dissociated single neurons of the NTS. Neurons were freshly dissociated from 7–10-day-old Hartley guinea pigs, using standard methods. In these neurons, all of which were voltage-clamped to a holding potential (V_H) of –50 mV, glycine induced an inward current (I_{gly}) in a concentration-dependent manner at 3 μM to 3 mM. Electrophysiological and pharmacological analysis of the current showed that the current is caused by activation of glycine receptors.

Dextromethorphan did not induce membrane currents in NTS neurons at 0.1 mM; however, it depressed I_{gly} induced by 30 μM glycine in a concentration-dependent manner (Takahama et al. 1997a, b). The half-maximal inhibitory concentration (IC_{50}) of dextromethorphan for I_{gly} was 3.3 μM, 85 times the IC_{50} of strychnine. It has been reported that dextromethorphan inhibits voltage-dependent Ca^{2+} and Na^+ channels (Netzer et al. 1993). However, it is unlikely that the effect of dextromethorphan on I_{gly} is due to its action on these channels, because I_{gly} is carried by Cl^- in neurons of the CNS, including NTS neurons of guinea pigs. On the basis of electrophysiological and pharmacological analyses of the action of dextromethorphan on I_{gly}, we speculate that low concentrations of dextromethorphan act on the glycine receptor–ionophore complex, but not on the Cl^- channel of the complex. However, a relatively high concentration of dextromethorphan may affect the Cl^- channel of the complex. Interestingly, 100 μM dextromethorphan had no effect on the current induced by 30 μM GABA. The fact that GABA-induced current (I_{GABA}) in NTS neurons was depressed by bicuculline indicates that it was mediated by the $GABA_A$ receptor, further indicating that centrally acting antitussives have a negligible effect on $GABA_A$ receptors, although it has been reported that $GABA_B$ receptor agonists have antitussive action in experimental animals (Bolser et al. 1994). We also found that other antitussives, both narcotic and nonnarcotic, have inhibitory action on I_{gly} but not on I_{GABA}, but their action was less potent than that of dextromethorphan (Fukushima et al. 1998). These were the first documented examples indicating that I_{gly} inhibition is common to several centrally acting antitussives at the brain neuronal level, and the findings raise the question as to whether I_{gly} inhibition by antitussives is part of their mechanism of action.

4.2.2 Strychnine-Sensitive Glycinergic Transmission

The results of direct administration of antitussives into the cough center suggest the possibility of glycinergic transmission in the cough center; glycine receptors are distributed in the NTS (Kubo and Kihara 1987). To confirm this possibility, we performed a patch-clamp experiment using brain-slice preparations. Coronal sections of the lower brain stem, including the NTS, 200–250 μm in thickness, were prepared from 6–9-day-old male Hartley guinea pigs. Prior to preparing the slices, we first confirmed that the region from where the slices were to be taken included sites innervated by afferent neurons of the cough reflex, using a neuronal tracing technique with the fluorescent dye 1,1′-dioctadecyl-3,3,3′,3′-tetramethylindocarbocyanine perchlorate as the probe.

As shown in Fig. 2, three types of synaptic currents were recorded in the slices including the NTS. Of 70 synaptic currents evoked by electrical stimulation of the solitary tract, 13 were excitatory, 28 were inhibitory, and the remaining 29 were excitatory currents followed by inhibitory currents. Of the 28 inhibitory postsynaptic currents, 21 were blocked by strychnine. The inhibitory currents preceded by the excitatory ones were effectively reduced by bicuculline, a $GABA_A$ receptor antagonist. The excitatory currents were all blocked by combination of 2-amino-5-phosphonovalic acid and 6-cyano-7-nitroquinoxaline-2,3-dione, suggesting that glutamatergic transmission exist in the NTS. Bicuculline-sensitive inhibitory currents were found to be carried by Cl^- on the basis of the results of the electrophysiological analysis. A higher concentration of glycine reduced or abolished the evoked inhibitory currents, suggesting glycinergic transmission in the slices examined including the NTS. The strychnine-sensitive glycinergic transmission was effectively blocked by dextromethorphan (Fig. 2).

4.2.3 Microdialysis Experiment

By using a new device we had developed for floating a microdialysis probe, we studied whether the glycine level in the NTS of guinea pigs increases during coughing. This device stably maintains the tip of a microdialysis probe within the selected

NTS, Bailey and colleagues showed that activation of substance P NK1 receptors resulted in depolarization, evident as a slow inward current that is mediated by protein kinase C (Bailey et al. 2004).

Substance P can also increase the intracellular calcium concentration, partially due to the release of calcium from intracellular calcium stores (Jun et al. 2004; Ito et al. 2002). The release of intracellular stores of calcium has been shown to be mediated by the phospholipase C pathway (Ito et al. 2002). The effects of substance P on voltage-dependent calcium channels vary significantly between networks; for example, substance P induced calcium influx in spiral ganglion neurons (Sun et al. 2004) while having no effect on interstitial cells of Cajal (Jun et al. 2004). Different calcium channels may respond differently to substance P. In NTS neurons, the selective NK1 receptor agonist [Sar^9,$Met(O_2)^{11}$]-substance P caused facilitation of the L-type voltage-dependent calcium currents in about half of the neurons (Endoh 2006). The effect is mediated by Gαq/11-protein involving the protein kinase C pathway (Endoh 2006). Since calcium channels are important for neurotransmitter release at presynaptic terminals, enzyme activity, and gene expression, the data suggest that the presynaptic or postsynaptic location of the NK1 receptors may have different responses to substance P. In addition, it is still unclear if the difference in the response reflects the difference in the nature of the neurons: neurons in specific reflex pathways, input or output neurons, or the neuronal type (e.g., glutaminergic or GABAergic, etc.).

3.1 Effect of Substance P on Excitatory Synaptic Transmission

The fast excitatory synaptic transmission at the NTS second-order neurons is mediated by glutamate acting on ionotropic glutamate receptors, namely, the non-*N*-methyl-D-aspartate (non-NMDA) and NMDA receptors. The evidence presented by Mutolo and colleagues showing that blockade of non-NMDA receptors in NTS abolished cough responses suggests that glutamate acting on non-NMDA receptors is primarily responsible for signal transmission in the NTS in the cough pathway (Mutolo et al. 2007).

The effect of substance P on excitatory synaptic transmission may depend on the receptor loci. Acting on presynaptic terminals, substance P can reduce glutamate release from the central terminals of the vagal sensory neurons onto anatomically identified second-order lung afferent neurons in the NTS (Sekizawa et al. 2003). We showed that substance P decreased the amplitude of excitatory postsynaptic currents evoked by stimulation of the afferent fibers in the solitary tract and increased the paired-pulse ratio of two consecutive evoked responses. The effect was blocked by a NK1 receptor antagonist, suggesting a mechanism by which substance P might regulate release at the first central synapses between lung afferent fibers and the second-order neurons in the lung reflex pathways. The presynaptic inhibition of glutamate release against the background of an overall postsynaptic excitation may improve signal transmission by changing the incoming signals with a wide dynamic range (more subject to noise) to a smaller range (less subject to noise).

Postsynaptically, substance P has been shown to enhance currents mediated by the ionotropic glutamate receptors, both NMDA and non-NMDA receptors, in other networks (Pieri et al. 2005; Randic et al. 1990; Budai et al. 1992; Lieberman and Mody 1998; Liu et al. 1998; Wu et al. 2004). While enhancing the non-NMDA receptor mediated synaptic current has obvious effects on an enhanced rapid signal transmission, the effect of substance P on NMDA receptors may be more important in plasticity. For example, an enhanced NMDA current could result in the opening of the NMDA channel at less depolarizing membrane potentials, producing a slower-developing, longer-lasting component of glutamate signaling. The enhanced NMDA-mediated slow depolarization could prolong the time during which otherwise ineffective inputs could be integrated and enable the neurons to transduce afferent input, which has no apparent pattern, into more efficient spiking patterns (Tell and Jean 1991, 1993; Yen and Chan 1997). In addition, activation of the NMDA receptor could lead to a long-lasting increase in the strength of synaptic transmission, including long-term potentiation of synaptic transmission and pain hypersensitivity (Artola and Singer 1987; Thompson et al. 1990; Woolf and Thompson 1991). Of relevance to NMDA receptors in the NTS is the finding that even though the non-NMDA receptors in the NTS are primarily involved in the cough reflex evoked by mechanical stimulation of the tracheobrohchial tree, blockade of the NMDA receptors in the NTS significantly reduced the cough reflex (Mutolo et al. 2007). The data suggest a potential site for central plasticity in modulating the cough reflex and a potential new drug target.

3.2 *Effect of Substance P on Inhibitory Synaptic Transmission*

The major inhibitory synaptic transmission in the NTS is provided by γ-aminobutyric acid (GABA) acting on $GABA_A$ receptors. GABA is released spontaneously and through stimulation and has profound inhibitory effects on synaptic transmission in the NTS (Len and Chan 2001; Zhang and Mifflin 1998). Data from microinjection studies in an upregulated cough reflex model in guinea pigs suggest a complex interaction between substance P and GABA in neuroplasticity in the NTS.

Microinjection of a substance P NK1 receptor antagonist into the NTS attenuated the second-hand tobacco smoke induced enhanced cough (Joad et al. 2007) and had no effect in the filtered-air control group. Microinjection of the $GABA_A$ receptor antagonist bicuculline had no effect on the citric acid evoked cough in either the filtered-air or the second-hand smoke exposed groups. Interestingly, bicuculline blocked the substance P NK1 antagonist effect in the second-hand-smoke group (Fig. 1). The question is how might GABA and substance P interact to explain these outcomes. Presynaptically, we know from other studies that substance P has no effect on miniature GABAergic inhibitory postsynaptic currents (Ogier and Raggenbass 2003; Stacey et al. 2002), suggesting that the neuropeptide does not

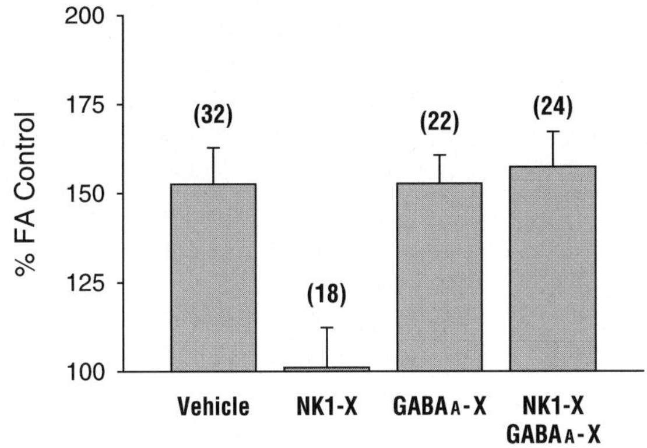

Fig. 1 Effect of a 5-week exposure to second-hand smoke on citric acid induced cough. Guinea pigs exposed to second-hand smoke coughed more than the filtered-air control. Substance P neurokinin 1 (NK1) receptor antagonist injected into the nucleus tractus solitarius (NTS) (NK1-X) normalized the second hand smoke induced cough. $GABA_A$ receptor antagonist alone ($GABA_A$-X) had no effect on the cough but significantly blocked the NK1 antagonism effect on the cough. The data are for citric acid induced cough after exposure to second-hand smoke, expressed as a percentage of the averaged filtered air control (n = 18–32 animals). *FA* filtered air. The number in parentheses indicates the number of animals exposed to second-hand smoke

play a significant role in modifying the release probability of GABA at the terminal site. However, substance P can enhance the general excitability of the GABA neurons via NK1 receptors located on the soma of the GABAergic interneurons, which could result in an enhanced GABA release (Ogier and Raggenbass 2003; Stacey et al. 2002). Postsynaptically, it has been shown that activation of substance P receptors significantly decreased the postsynaptic $GABA_A$ receptor function, an inhibition that is mediated by the protein kinase C but not the protein kinase A pathway (Si et al. 2004; Hu et al. 2004; Yamada and Akasu 1996). Thus, the enhanced GABA release could be compensated with a reduced postsynaptic modification of the $GABA_A$ receptors. These two actions of substance P could explain the lack of effect by the $GABA_A$ receptor blockade alone. However, since substance P also directly excites postsynaptic neurons, blocking $GABA_A$ would not be expected to modify the substance P effect. One possibility is that the GABA neurons tonically inhibit other non-GABAergic inhibitory inputs such that blocking $GABA_A$ receptor would disinhibit the non-GABAergic inhibitory inputs to offset the direct effect of substance P NK1 receptors on the general excitability (Fig. 2). A glycinergic inhibitory input is one of the likely candidates for such interaction because it is the second major inhibitory system throughout the CNS. In this regard, it has been shown that substance P may enhance the postsynaptic glycine receptor function via a protein kinase C or a Ca/calmodulin dependent protein kinase II pathway (Wang et al. 1999).

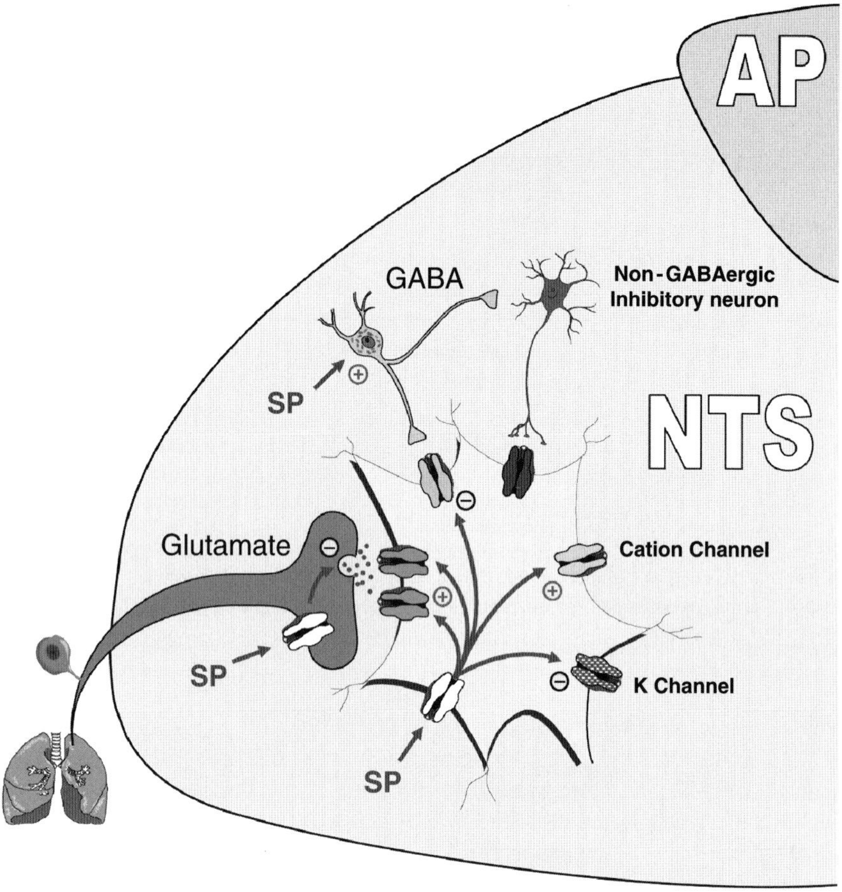

Fig. 2 Substance P induced NTS plasticity targets in the cough reflex. Plasticity targets include (1) increased intrinsic excitability via closure of potassium channels, activation of nonselective cation channels, opening of L-type calcium channel, and intracellular calcium release. (2) increased synaptic excitability via an increased ionotropic glutamate receptor function, a decreased $GABA_A$ receptor function. Other compensatory mechanisms may also occur, including an increased GABA release and decreased glutamate release from the afferent terminals. *AP* area postrema

4 Summary: The Model

The source and conditions under which substance P is released may determine the extent to which the presynaptic and postsynaptic mechanisms are integrated to shape the neuronal output from the NTS to distal synapses in the reflex pathways. Multiple central mechanisms are likely to contribute to the plasticity of cough reflex. In upregulated cough reflex, an increased substance P expression in the vagal neurons may be released at the central terminal in the NTS (Fischer et al. 1996; Chuaychoo et al. 2005). This enhanced substance P release in the NTS could contribute to the enhanced substance P–NK1 receptor mechanism mediating a heightened cough reflex response (Joad et al. 2007; Bonham et al. 2004). The increased substance P released from the primary sensory central terminals is likely to have more focused effect on the NTS second-order neurons, including the inhibitory interneurons. Upon binding to the NK1 receptors, the released substance P can increase the general output of the NTS second-order neurons by (1) closing potassium channels via the protein kinase C pathway, (2) opening nonselective cation channels, (3) increasing intracellular calcium concentration via opening L-type calcium channels, (4) increasing intracellular calcium concentration via the phospholipid C pathway, (5) increasing the glutamate synaptic transmission via enhanced NMDA and non-NMDA receptor function, (6) inhibiting postsynaptic $GABA_A$ receptor function, and (7) inhibiting other non-GABAergic inhibitory neurons via an enhanced GABA release (Fig. 2).

At the same time, several other mechanisms may be activated to compensate for some of the excitatory effect, perhaps to keep the system from being hyperreactive. These mechanisms include (1) presynaptic inhibition of glutamate release from the primary afferent terminals, (2) activation of the NK1 receptors on the soma/dendrite of GABAergic interneurons to enhance GABA release, and (3) an enhanced glycine synaptic transmission via an enhanced glycine receptor function through a protein kinase C or a Ca/calmodulin dependent protein kinase II pathway (Fig. 2).

5 Future Challenges

Another important consideration in examining the neuroplasticity in the NTS is that although the gustatory, gastrointestinal, and cardiorespiratory sensory inputs are organized somewhat viscerotopically in the NTS, neurons receiving different afferent inputs and serving different functions are intermingled throughout the nucleus (Loewy 1990). Functionally and anatomically identifying the second-order neuron by the use of fluorescent tracers is helpful in identifying the neurons receiving lung afferent inputs. However, challenges still exist in isolating the neurons specifically in the cough pathway, whether they are input neurons, interneurons, or output neurons. Given the limitation of identifying the specific NTS neurons in the cough pathway, there are more unknown than known cellular mechanisms by which substance P enhances the cough pathway. More studies are needed to sorting out the proposed potential sites for plasticity in different model of enhanced cough, whether acute or chronic.

References

Akasu T, Ishimatsu M, Yamada K (1996) Tachykinins cause inward current through NK1 receptors in bullfrog sensory neurons. Brain Res 713:160–167

Artola A, Singer W (1987) Long-term potentiation and NMDA receptors in rat visual cortex. Nature 330:649–652

Aylwin ML, Horowitz JM, Bonham AC (1998) Non-NMDA and NMDA receptors in the synaptic pathway between area postrema and nucleus tractus solitarius. Am J Physiol 275:H1236–H1246

Bailey CP, Maubach KA, Jones RS (2004) Neurokinin-1 receptors in the rat nucleus tractus solitarius: Pre- and postsynaptic modulation of glutamate and GABA release. Neuroscience 127: 467–479

Bisgaard H, Szefler S (2007) Prevalence of asthma-like symptoms in young children. Pediatr Pulmonol 42:723–728

Bolser DC, Davenport PW (2002) Functional organization of the central cough generation mechanism. Pulm Pharmacol Ther 15:221–225

Bonham AC, Sekizawa SI, Joad JP (2004) Plasticity of central mechanisms for cough. Pulm Pharmacol Ther 17:453–457

Bonham AC, Chen CY, Sekizawa S, Joad JP (2006) Plasticity in the nucleus tractus solitarius and its influence on lung and airway reflexes. J Appl Physiol 101:322–327

Budai D, Wilcox GL, Larson AA (1992) Modulation of *N*-methyl-D-aspartate and (R,S)-alpha-amino-3-hydroxy-5-methylisoxazole-4-propionate (AMPA) responses of spinal nociceptive neurons by a N-terminal fragment of substance P. Eur J Pharmacol 216:441–444

Carr MJ, Lee LY(2006) Plasticity of peripheral mechanisms of cough. Respir Physiol Neurobiol 152:298–311

Carr MJ, Undem BJ (2001) Inflammation-induced plasticity of the afferent innervation of the airways. Environ Health Perspect 109(Suppl 4):567–571

Chen CY, Bonham AC (1998) Non-NMDA and NMDA receptors transmit area postrema input to aortic baroreceptor neurons in NTS. Am J Physiol 275:H1695–H1706

Chen CY, Bonham AC, Schelegle ES, Gershwin LJ, Plopper CG, Joad JP (2001) Extended allergen exposure in asthmatic monkeys induces neuroplasticity in nucleus tractus solitarius. J Allergy Clin Immunol 108:557–562

Chen CY, Bonham AC, Plopper CG, Joad JP (2003) Neuroplasticity in nucleus tractus solitarius neurons after episodic ozone exposure in infant primates. J Appl Physiol 94:819–827

Cherry DK, Woodwell DA, Rechtsteiner EA (2007) National ambulatory medical care survey: 2005 summary. In: Advance data from vital and health statistics, vol 387. US Department of health and human services, USA, pp 1–40

Chuaychoo B, Hunter DD, Myers AC, Kollarik M, Undem BJ (2005) Allergen-induced substance P synthesis in large-diameter sensory neurons innervating the lungs. J Allergy Clin Immunol 116:325–331

Chung KF (2007) Chronic cough: Future directions in chronic cough: Mechanisms and antitussives. Chron Respir Dis 4:159–165

Davies J, Dray A (1980) Depression and facilitation of synaptic responses in cat dorsal horn by substance P administered into substantia gelatinosa. Life Sci 27:2037–2042

Davis BJ, Smith HM (1999) Neurokinin-1 receptor immunoreactivity in the nucleus of the solitary tract in the hamster. Neuroreport 10:1003–1006

Dray A, Pinnock RD (1982) Effects of substance P on adult rat sensory ganglion neurones in vitro. Neurosci Lett 33:61–66

Drew GM, Mitchell VA, Vaughan CW (2005) Postsynaptic actions of substance P on rat periaqueductal grey neurons in vitro. Neuropharmacology 49:587–595

Endoh T (2006) Dual effects of neurokinin on calcium channel currents and signal pathways in neonatal rat nucleus tractus solitarius. Brain Res 1110:116–127

Fischer A, McGregor GP, Saria A, Philippin B, Kummer W (1996) Induction of tachykinin gene and peptide expression in guinea pig nodose primary afferent neurons by allergic airway inflammation. J Clin Invest 98:2284–2291

Gilbert R, Ryan JS, Horackova M, Smith FM, Kelly ME (1998) Actions of substance P on membrane potential and ionic currents in guinea pig stellate ganglion neurons. Am J Physiol 274:C892–C903

Helke CJ, Seagard JL (2004) Substance P in the baroreceptor reflex: 25 years. Peptides 25:413–423

Hu WP, Li ZW, Ru LQ, Fan YZ (2004) Opposite modulatory effects of substance P on GABA-and 5-HT-activated currents in the same sensory neurons. Sheng Li Xue Bao 56:703–707

Irwin RS (2006) Complications of cough: ACCP evidence-based clinical practice guidelines. Chest 129:54S–58S

Irwin RS, Boulet LP, Cloutier MM, Fuller R, Gold PM, Hoffstein V, Ing AJ, McCool FD, O'Byrne P, Poe RH, Prakash UB, Pratter MR, Rubin BK (1998) Managing cough as a defense mechanism and as a symptom. A consensus panel report of the American college of chest physicians. Chest 114:133S–181S

Irwin RS, French CT, Fletcher KE (2002) Quality of life in coughers. Pulm Pharmacol Ther 15:283–286

Ishimatsu M (1994) Substance P produces an inward current by suppressing voltage-dependent and -independent K^+ currents in bullfrog primary afferent neurons. Neurosci Res 19:9–20

Ito K, Rome C, Bouleau Y, Dulon D (2002) Substance P mobilizes intracellular calcium and activates a nonselective cation conductance in rat spiral ganglion neurons. Eur J Neurosci 16: 2095–2102

Jafri MS, Weinreich D (1996) Substance P hyperpolarizes vagal sensory neurones of the ferret. J Physiol 493(Part 1):157–166

Joad JP, Munch PA, Bric JM, Evans SJ, Pinkerton KE, Chen CY, Bonham AC (2004) Passive smoke effects on cough and airways in young guinea pigs: Role of brainstem substance P. Am J Respir Crit Care Med 169:499–504

Joad JP, Sekizawa S, Chen CY, Bonham AC (2007) Air pollutants and cough. Pulm Pharmacol Ther 20:347–354

Jun JY, Choi S, Yeum CH, Chang IY, You HJ, Park CK, Kim MY, Kong ID, Kim MJ, Lee KP, So I, Kim KW (2004) Substance P induces inward current and regulates pacemaker currents through tachykinin NK1 receptor in cultured interstitial cells of Cajal of murine small intestine. Eur J Pharmacol 495:35–42

Len WB, Chan JY (2001) GABAergic neurotransmission at the nucleus tractus solitarii in the suppression of reflex bradycardia by parabrachial nucleus. Synapse 42:27–39

Lee LY, Undem BJ (2004) Mechanisms of chronic cough. Pulm Pharmacol Ther 17:463–464

Lee HK, Shuttleworth CW, Sanders KM (1995) Tachykinins activate nonselective cation currents in canine colonic myocytes. Am J Physiol 269:C1394–C1401

Lepre M, Olpe HR, Brugger F (1996) The effects of neurokinin-1 receptor agonists on spinal motoneurones of the neonatal rat. Neuropharmacology 35:511–522

Lieberman DN, Mody I (1998) Substance P enhances NMDA channel function in hippocampal dentate gyrus granule cells. J Neurophysiol 80:113–119

Lindsey BG, Morris KF, Segers LS, Shannon R (2000) Respiratory neuronal assemblies. Respir Physiol 122:183–196

Liu X, Andre D, Puizillout JJ (1998) Substance P post-synaptically potentiates glutamate-induced currents in dorsal vagal neurons. Brain Res 804:95–104

Loewy AD (1990) Central autonomic pathways. In: Loewy AD, Spyer KM (eds) Central regulation of autonomic functions. Oxford University Press, New York, pp 88–103

Mazzone SB, Mori N, Canning BJ (2005) Synergistic interactions between airway afferent nerve subtypes regulating the cough reflex in guinea-pigs. J Physiol 569:559–573

Morice AH (2002) Epidemiology of cough. Pulm Pharmacol Ther 15:253–259

Morris KF, Baekey DM, Nuding SC, Dick TE, Shannon R, Lindsey BG (2003) Invited review: Neural network plasticity in respiratory control. J Appl Physiol 94:1242–1252

Murase K, Nedeljkov V, Randic M (1982) The actions of neuropeptides on dorsal horn neurons in the rat spinal cord slice preparation: An intracellular study. Brain Res 234:170–176

Mutoh T, Bonham AC, Joad JP (2000) Substance P in the nucleus of the solitary tract augments bronchopulmonary C fiber reflex output. Am J Physiol Regul Integr Comp Physiol 279:R1215–R1223

Mutolo D, Bongianni F, Fontana GA, Pantaleo T (2007) The role of excitatory amino acids and substance P in the mediation of the cough reflex within the nucleus tractus solitarii of the rabbit. Brain Res Bull 74:284–293

Ogier R, Raggenbass M (2003) Action of tachykinins in the rat hippocampus: Modulation of inhibitory synaptic transmission. Eur J Neurosci 17:2639–2647

Pieri M, Severini C, Amadoro G, Carunchio I, Barbato C, Ciotti MT, Zona C (2005) AMPA receptors are modulated by tachykinins in rat cerebellum neurons. J Neurophysiol 94:2484–2490

Poliacek I, Corrie LW, Wang C, Rose MJ, Bolser DC (2007) Microinjection of DLH into the region of the caudal ventral respiratory column in the cat: Evidence for an endogenous cough-suppressant mechanism. J Appl Physiol 102:1014–1021

Pratter MR (2006) Overview of common causes of chronic cough: ACCP evidence-based clinical practice guidelines. Chest 129:59S–62S

Randic M, Hecimovic H, Ryu PD (1990) Substance P modulates glutamate-induced currents in acutely isolated rat spinal dorsal horn neurones. Neurosci Lett 117:74–80

Sculptoreanu A, de Groat WC (2007) Neurokinins enhance excitability in capsaicin-responsive DRG neurons. Exp Neurol 205:92–100

Sekizawa S, Joad JP, Bonham AC (2003) Substance P presynaptically depresses the transmission of sensory input to bronchopulmonary neurons in the guinea pig nucleus tractus solitarii. J Physiol 552:547–559

Sekizawa S, Bramham CR, Joad JP (2005) Second hand smoke (SHS) exposure decreases transient outward currents of nucleus tractus solitarius (NTS) in a subset of 2nd-order lung afferent neurons in young guinea pigs. ATS Annual Meeting C97, Poster 809

Shannon R, Baekey DM, Morris KF, Lindsey BG (1998) Ventrolateral medullary respiratory network and a model of cough motor pattern generation. J Appl Physiol 84:2020–2035

Si JQ, Zhang ZQ, Li CX, Wang LF, Yang YL, Li ZW (2004) Modulatory effect of substance P on GABA-activated currents from rat dorsal root ganglion. Acta Pharmacol Sin 25:623–629

Stacey AE, Woodhall GL, Jones RS (2002) Activation of neurokinin-1 receptors promotes GABA release at synapses in the rat entorhinal cortex. Neuroscience 115:575–586

Sun W, Ding DL, Wang P, Sun J, Jin X, Salvi RJ (2004) Substance P inhibits potassium and calcium currents in inner ear spiral ganglion neurons. Brain Res 1012:82–92

Suwabe T, Bradley RM (2007) Effects of 5-hydroxytryptamine and substance P on neurons of the inferior salivatory nucleus. J Neurophysiol 97:2605–2611

Tell F, Jean A (1991) Bursting discharges evoked in vitro, by solitary tract stimulation or application of *N*-methyl-D-aspartate, in neurons of the rat nucleus tractus solitarii. Neurosci Lett 124:221–224

Tell F, Jean A (1993) Ionic basis for endogenous rhythmic patterns induced by activation of *N*-methyl-D-aspartate receptors in neurons of the rat nucleus tractus solitarii. J Neurophysiol 70:2379–2390

Thompson SW, King AE, Woolf CJ (1990) Activity-dependent changes in rat ventral horn neurons in vitro; summation of prolonged afferent evoked postsynaptic depolarizations produce a d-2-amino-5-phosphonovaleric acid sensitive windup. Eur J Neurosci 2:638–649

Undem BJ, Hunter DD, Liu M, Haak-Frendscho M, Oakragly A, Fischer A (1999) Allergen-induced sensory neuroplasticity in airways. Int Arch Allergy Immunol 118:150–153

Undem BJ, Carr MJ, Kollarik M (2002) Physiology and plasticity of putative cough fibres in the Guinea pig. Pulm Pharmacol Ther 15:193–198

Wang DS, Xu TL, Li JS (1999) Modulation of glycine-activated chloride currents by substance P in rat sacral dorsal commissural neurons. Sheng Li Xue Bao 51:361–370

Widdicombe JG, Singh V (2006) Physiological and pathophysiological down-regulation of cough. Respir Physiol Neurobiol 150:105–117

Woolf CJ, Thompson SW (1991) The induction and maintenance of central sensitization is dependent on *N*-methyl-D-aspartic acid receptor activation; implications for the treatment of post-injury pain hypersensitivity states. Pain 44:293–299

Wu ZZ, Guan BC, Li ZW, Yang Q, Liu CJ, Chen JG (2004) Sustained potentiation by substance P of NMDA-activated current in rat primary sensory neurons. Brain Res 1010:117–126

Yamada K, Akasu T (1996) Substance P suppresses GABAA receptor function via protein kinase C in primary sensory neurones of bullfrogs. J Physiol 496(Part 2):439–449

Yang YL, Yao KH, Gu YZ, Guan BC, Li ZW (2003) Three kinds of current in response to substance P in bullfrog DRG neurons. Brain Res 981:70–77

Yen JC, Chan SH (1997) Interchangeable discharge patterns of neurons in caudal nucleus tractus solitarii in rat slices: Role of GABA and NMDA. J Physiol 504(Part 3):611–627

Zhang J, Mifflin SW (1998) Receptor subtype specific effects of GABA agonists on neurons receiving aortic depressor nerve inputs within the nucleus of the solitary tract. J Auton Nerv Syst 73:170–181

Zieglgansberger W, Tulloch IF (1979) Effects of substance P on neurones in the dorsal horn of the spinal cord of the cat. Brain Res 166:273–282

Central Mechanisms II: Pharmacology of Brainstem Pathways

D.C. Bolser

Contents

Abstract Following systemic administration, centrally acting antitussive drugs are generally assumed to act in the brainstem to inhibit cough. However, recent work in humans has raised the possibility of suprapontine sites of action for cough suppressants. For drugs that may act in the brainstem, the specific locations, types of neurones affected, and receptor specificities of the compounds represent important issues regarding their cough-suppressant actions. Two medullary areas that have received the most attention regarding the actions of antitussive drugs are the nucleus of the tractus solitarius (NTS) and the caudal ventrolateral respiratory column. Studies that have implicated these two medullary areas have employed both microinjection and in vitro recording methods to control the location of action of the antitussive drugs. Other brainstem regions contain neurones that participate in the production of cough and could represent potential sites of action of antitussive drugs. These regions include the raphe nuclei, pontine nuclei, and rostral ventrolateral medulla. Specific receptor subtypes have been associated with the suppression of cough at central sites, including 5-HT1A, opioid (μ, κ, and δ), GABA-B, tachykinin neurokinin-1 (NK-1) and neurokinin-2, non-opioid (NOP-1), cannabinoid, dopaminergic, and sigma receptors. Aside from tachykinin NK-1 receptors in the NTS, relatively little is known regarding the receptor specificity of putative

---

D.C. Bolser
Department of Physiological Sciences, University of Florida, College of Veterinary Medicine, Gainesville, FL 32610-0144, USA
bolser@ufl.edu

K.F. Chung, J.G. Widdicombe (eds.), *Pharmacology and Therapeutics of Cough*, Handbook of Experimental Pharmacology 187,

antitussive drugs in particular brainstem regions. Our understanding of the mechanisms of action of antitussive drugs would be significantly advanced by further work in this area.

1 Introduction

The pharmacology of centrally active antitussive drugs is a multifactorial topic that involves not only pharmacological and pharmacokinetic issues but neurophysiology as well. This review will focus on three primary matters related to the brainstem actions of these drugs: (1) location of action, (2) identity of neurones affected by the drugs, (3) receptor specificity. There are other informative reviews available (Reynolds et al. 2004).

2 Location of Action of Antitussive Drugs

It is widely accepted that several prominent drugs act in the central nervous system to inhibit cough, primarily by an action in the brainstem. The evidence supporting this concept is strong and is based largely on studies showing that decerebrate animals can cough and that antitussives will suppress coughing under these circumstances (Chou 1975; Wang et al. 1977; Domino et al. 1985; Lal et al. 1986; Bolser 1991; Bolser and DeGennaro 1994; Gestreau et al. 1997; Ohi et al. 2004, 2005).

The central control of cough is complex and there may be many potential sites in the brainstem at which a given drug may act to suppress this behavior. In this context, an understanding of the brainstem regions that may be involved in the production of cough is an important component of any approach to the investigation of the actions of antitussive drugs. It is critical to know "where to look" to design studies investigating the mechanisms of action of these agents. In this control system there may be many areas where antitussives could work, but only a few that are responsible for the cough-suppressant effects that result from systemic administration of these agents. It should be noted that the results of studies showing a brainstem action of antitussives do not preclude an effect of these drugs on suprapontine or spinal pathways in animals that have an intact neuraxis.

Spinal motoneurones (and their antecedant interneuronal pathways) are an often-overlooked component of the cough-generation system, but represent an important site at which regulation of the behavior can occur. Several classes of compounds that have antitussive activity also suppress spinal motor activity in other systems. Baclofen is a well-known muscle relaxant and inhibits spinal motor activity in low doses after intrathecal administration (Penn 1992). Opioids also inhibit motor activity after topical administration to the spinal cord in spinal cats (Schomburg and Steffens 1995). Central nervous system penetrant drugs gain access to the entire neuraxis within 5 min after vascular administration and compounds that are delivered to the cerebrospinal fluid (CSF) of the brain are rapidly transported to the spinal CSF (Xie and Hammarlund-Udenaes 1998). Therefore, centrally acting antitussive drugs probably reach the spinal cord after systemic administration. Preliminary results (Rose et al. 2004) have shown that intrathecal administration of baclofen has no effect on expiratory muscle electromyographic activity during tracheobronchial cough. However, the same dose of baclofen almost completely inhibits cough when administered via the vertebral artery. Similar results were obtained in preliminary studies with intrathecal administration of codeine. These preliminary findings are consistent with disfacilitation of expiratory spinal motor pathways by antitussive drugs acting in the brainstem.

The role of suprapontine pathways in the generation of cough and the effects of antitussive drugs is not well understood. It is likely that the potential role of these areas in the generation of cough may be much greater in conscious humans (and perhaps animals as well), given that humans can both initiate and suppress cough by voluntary means (Hutchings et al. 1993; Hutchings and Eccles 1994). Significant sensations also are associated with irritant-induced cough, indicating the involvement of suprapontine sensory systems during coughing. A model incorporating the potential influence of suprapontine pathways in the production of cough has recently been published (Bolser 2006). However, codeine has no effect on sensations during irritant-induced cough in humans, but in that study this opioid did not alter the musculomechanical aspects of cough (Davenport et al. 2007). In the absence of an effective antitussive agent in humans, the role of suprapontine pathways and sensations in the mechanism of action of these drugs will remain obscure.

3 Identity of Neurones Affected by Antitussive Drugs

A great deal of information is available regarding the locations of neurones that participate in the production of cough in the brainstem. Shannon and coworkers (Shannon et al. 1996, 1998, 2000; Baekey et al. 2001) have developed a model of the neurogenic mechanism for coughing that includes columns of neurones in both the ventrolateral and the dorsomedial regions of the medulla. Different classes of neurones in these regions interact with one another to control inspiratory- and expiratory-phase durations during cough, the magnitude of motor drive to spinal motoneurons, and the activation of laryngeal muscle motoneurones that determine the caliber of the larynx (Shannon et al. 1996, 1998, 2000; Baekey et al. 2001). These neurones also participate in the control of breathing. Sensory input to this cough pattern generation system is mediated by relay neurones in the nucleus of the tractus solitarius (NTS), located in the dorsomedial medulla.

Recent studies utilizing microinjection of a neurotoxin to induce chemical lesions have implicated the pons (Poliacek et al. 2004), raphe nuclei (Jakus et al. 1998), and lateral tegmental field (Jakus et al. 2000) in the production of coughing. The microinjection method is best suited to the study of regions or nuclei of the brain

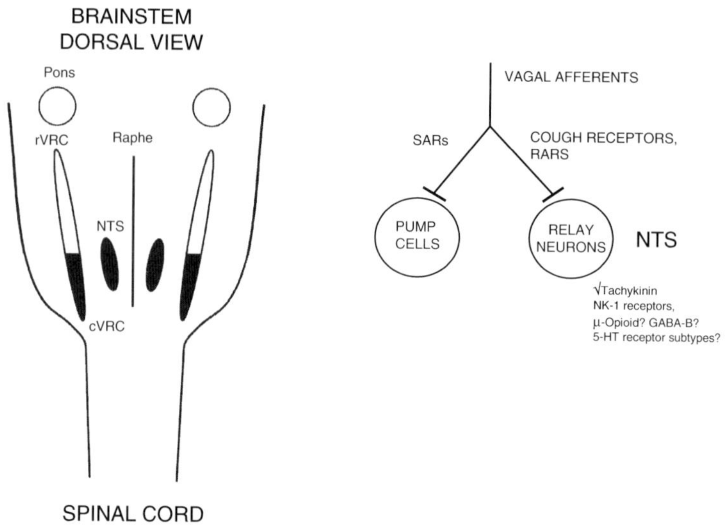

Fig. 1 Brainstem regions that have been specifically associated with the actions of antitussive drugs. *Outlined areas* on the *left* indicate regions of the brainstem in which cough-related neurones have been recorded. *Shaded areas* represent regions in which application of cough-suppressant drugs has inhibited cough. The model on the *right* shows second-order interneurones in pulmonary vagal afferent pathways. Pump cells mediate sensory information from pulmonary slowly adapting receptors and neurones designated "relay" mediate information from cough receptors and rapidly adapting receptors. The notation shows that although a number of different drugs may act on this population of neurones, receptor specificity only for tachykinin neurokinin-1 receptors has been demonstrated (Chen et al. 2003; Sekizawa et al. 2003). NTS nucleus of the tractus solitarius, rVRC rostral ventral respiratory column, cVRC caudal ventral respiratory column

that are roughly spherical because the diffusion pattern of microinjected chemicals approximates a sphere (Lipski et al. 1988). Use of this method in the study of the actions of the raphe nuclei presents unique issues because these nuclei are restricted to the midline in a "sheet" extending dorsoventrally. The lesions produced by Jakus and coworkers (Jakus et al. 1998) affected the raphe nuclei but extended into the nearby medial reticular formation as well, potentially implicating this region in the production of cough (Fig. 1).

Xu et al. (1997) have shown that one of the deep cerebellar nuclei, the nucleus interpositus, has an important modulatory role in the production of cough. Furthermore, the inferior olive, a brainstem nucleus that interacts with the deep cerebellar nuclei, was labeled in coughing animals in a *cFos* study (Gestreau et al. 1997).

From a theoretical standpoint, any of these areas could represent targets for the action of antitussive drugs. Important considerations are the route and dose of the cough-suppressant drug in question. Intracerebroventricular or vascular administration of antitussive agents lacks anatomic specificity (Bolser 1996). Vascular administration can be coupled with electrophysiological recordings of single neurones to provide anatomic specificity, but the drug will still reach all the tissue supplied by the brainstem circulation (Jakus et al. 1987). Intravenous or intra-arterial administration of antitussive drugs does have the advantage that the effects can be directly related to concentrations of the agents that are restricted to the therapeutic range. Effective concentrations of cough suppressants administered by the intracerebroventricular route also can be related to therapeutic doses obtained by vascular administration but only by the use of ratios (Bolser 1996). However, a drawback of intracerebroventricular studies is that doses are often not normalized to body weight.

The specific location at which an antitussive drug acts to suppress cough can be studied with the microinjection method. Microinjection has the advantage of localizing the drug to a relatively small region of the brain. However, the drug can still diffuse $500-1,000\,\mu m$ away from the injection site depending on factors such as volume injected, concentration, and the characteristics of the extracellular space (Lipski et al. 1988). An additional limitation of this method is that it is difficult to relate the dose of a drug that is delivered to therapeutic concentrations associated with systemic administration. Microinjection studies run the risk of delivering doses of cough suppressants that result in tissue concentrations that are higher than those that would be achieved after systemic administration. As such, the risk of type I errors in these types of studies may be high.

3.1 Nucleus of the Tractus Solitarius

Antitussive drugs have been microinjected into the region of the NTS. Codeine and dextromethorphan have been injected into the NTS in relatively large quantities ($0.5-10\mu g$) in large volumes ($0.5-10\mu L$), resulting in suppression of cough in the cat and guinea pig. As stated already it is difficult to know whether these amounts of antitussive drugs are consistent with local tissue concentrations that occur after systemic administration. Concentrations of microinjected drugs are usually restricted to the nanomolar range with volumes of 100 nL or less to limit the local concentration of the drug (Lipski et al. 1988). Gestreau and coworkers (Gestreau et al. 1997) have shown that *cFos* labeling of NTS neurones is reduced after systemic administration of a large dose of codeine ($17\,mg\,kg^{-1}$), which they suggested occurred because codeine acted on this population of neurones. Recent work suggests that codeine presynaptically inhibits glutaminergic excitatory neurotransmission from primary afferent fibers to putative second-order interneurones in the NTS (Ohi et al. 2007). Furthermore, glycinergic currents in acutely dissociated guinea pig NTS neurones were inhibited by dextromethorphan (Takahama et al. 1997). Similar findings were obtained for NTS neurones in slices of the guinea pig medulla for the effects of substance P on excitatory postsynaptic potentials induced from electrical stimulation of the tractus solitarius (Sekizawa et al. 2003). In this study, the authors specifically identified second-order interneurones by labeling of primary afferents with retrograde tracer. Another study by this group on NTS neurones from brainstem slices of nonhuman primates showed that substance P in the micromolar range decreased input resistance of these cells, indicative of inhibition (Chen et al. 2003). This effect was observed in control as well as ozone-challenged animals. Interestingly, tachykinin neurokinin-1 (NK-1) receptor antagonists also depressed the responses of these neurones to current injection, but not the excitatory effect of electrical stimulation of the tractus solitarius (Chen et al. 2003). Mutolo et al. (2007) have shown in the rabbit that microinjection of picomolar quantities of substance P into the commissural subnucleus of the NTS results in enhancement of coughing. Another study showed central suppression of cough by tachykinin NK-1 antagonists (Bolser et al. 1997) delivered to the brainstem circulation. The apparent conflicting findings of some of these studies may simply be a manifestation of the inherent complexity of the neural network for cough in this region of the medulla. Antitussive drugs may act presynatically, postsynaptically, and/or on multiple subpopulations of interneurones in this area, presenting significant challenges to the interpretation of microinjection studies.

3.2 Ventrolateral Respiratory Column

Jakus and coworkers (Jakus et al. 1987) have shown that the tracheobronchial cough-related discharge of caudal ventral respiratory group expiratory neurones is reduced after systemic administration of codeine. However, this observation does not prove that antitussive drugs act directly on any of these neurone groups. The findings of Jakus and coworkers Jakus et al. (1987) could be explained either by an inhibitory action of codeine on medullary expiratory neurones or by an action of this drug on other neurones that provide excitatory input to medullary expiratory neurones. Jakus et al. (1987) observed that previously silent caudal expiratory neurones were recruited during cough in the cat. These neurones have expiratory discharge patterns during cough that are very similar to abdominal motor discharge patterns. Merrill (1970, 1972) investigated spinal projections of both spontaneously active caudal expiratory neurones and nearby silent neurones. Both groups of neurones had contralateral spinal axons and Merrill (1974) noted that antidromic latencies of the silent neurones fluctuated with the respiratory cycle, indicating that these neurones had respiratory-modulated membrane potentials. Indeed, he suggested that this group of silent neurones was recruited during expulsive movements, such as cough.

Microinjection of the excitatory amino acid agonist D,L-homocysteic acid (DLH) into the caudal portion of the ventral respiratory group elicited prolonged suppression of coughing (more than 10min) but only brief (less than 1min) alterations in breathing in the cat (Poliacek et al. 2007). Spontaneously active neurones in this region of the medulla are almost exclusively premotor expiratory with few, if any, axon collaterals to the rest of the brainstem. These neurones are unlikely to have mediated the cough-suppressant effects that were observed. Given that DLH is an excitatory neurochemical, we have proposed that either silent or nonbreathing modulated neurones in or near the caudal ventral respiratory group are responsible for this cough suppression. We also have preliminary data showing that microinjection of codeine into this area in doses as low as 100 pmol suppresses cough. Given that codeine and the well-known excitatory neurochemical DLH have similar effects, it is possible that codeine also is acting as an excitatory agent under these conditions. This proposal represents a departure from the long-held concept that centrally acting antitussives *inhibit* the activity of neural elements involved in the production of cough. An alternative hypothesis is that these drugs may excite neurones in the central nervous system that in turn synaptically inhibit control elements in the cough network.

4 Receptor Specificity

The presence of receptors for a drug in a given brain region does not indicate that the drug will exert its effects by an action in that location. The receptor density and/or affinity for the drug may be low in that region, requiring a dose of the drug to alter the behavior of these neurones that is out of the therapeutic range. Furthermore, many of the brain regions that contain neurones associated with the production of cough also mediate other functions, such as breathing and cardiovascular control. The concept of receptor specificity is relevant not only to the receptor subtype actuated by a given drug, but also to the physiological system that is affected. For example, both morphine (a specific μ-opioid receptor agonist) and baclofen (a specific GABA-B receptor agonist) are known to be respiratory depressant agents (Adcock et al. 1988; Hey et al. 1995). However, in the guinea pig the doses of these drugs necessary for respiratory depression are greater than those required for inhibition of cough (Adcock et al. 1988; Bolser et al. 1994; Hey et al. 1995). Presumably, each drug acts specifically at its respective receptor subtype to inhibit both cough and breathing, but the potency of the drugs to inhibit these behaviors is different.

Studies to identify the receptor specificity of antitussive drugs have been restricted to animal models. In humans, investigations of the effects of antitussive drugs are usually focused on demonstrating efficacy. Studies in animal models usually involve the use of specific agonists and/or antagonists to demonstrate receptor specificity.

Opioids, most notably codeine, are effective cough suppressants in animal models (May and Widdicombe 1954; Chou 1975; Wang et al. 1977; Adcock et al. 1988; Bolser et al. 1993, 1999; Bolser and DeGennaro 1994; Reynolds et al. 2004), although the efficacy of codeine in humans has recently been questioned (Freestone and Eccles 1997; Smith et al. 2006). In guinea pigs, the action of codeine as a cough suppressant has been attributed to μ-opioid receptors (Kotzer et al. 2000). However, codeine does not appear to act through opioid receptors in the cat (Chau et al. 1983), because its antitussive action is not blocked by the opioid antagonist naloxone in this species. The particular receptor that codeine acts through to inhibit cough in the cat is unknown. The μ-opioid receptor agonist morphine does inhibit cough in the guinea pig and cat by a naloxone-sensitive (or naltrexone-sensitive) mechanism (Chau et al. 1983; Adcock et al. 1988). Morphine has been shown to be active as an antitussive in humans (Fuller et al. 1988; Morice et al. 2007). Furthermore, both δ- and κ-opioid receptor agonists inhibit cough in the guinea pig. Opioid antagonists do not influence the cough reflex in either animals or humans (Chau et al. 1983; Freestone and Eccles 1997; Kotzer et al. 2000), suggesting that endogenous opioids are not critical for the production of cough. These observations are consistent with the existence of a regulatory mechanism of endogenous opioids on cough that is normally quiescent. The conditions under which this putative mechanism could be active are unknown. Moreover, the role(s) that the different opioid receptor subtypes may have in the regulation of the expression of coughing in the human is not clear.

GABA-B receptor agonists can inhibit cough in animals and humans (Bolser et al. 1993, 1994, 1999; Dicpinigaitis 1996; Dicpinigaitis and Dobkin 1997; Dicpinigaitis et al. 1998). These effects are central in both guinea pigs and cats and the efficacy of baclofen, the prototypical centrally active GABA-B receptor agonist, is similar to that of opioids in animal models (Bolser et al. 1993). A specific receptor antagonist for GABA-B receptors, SCH 50911, had no effect on its own to alter cough but did block the antitussive effects of baclofen in the cat and guinea pig (Bolser et al. 1995). This observation suggests that γ-aminobutyric acid (GABA), acting through the GABA-B receptor subtype, does not modulate the production of cough under experimental conditions. The actions of other GABA receptor subtypes on the expression of cough are not well understood. The GABA-A receptor antagonist bicuculline induces seizures (Wood 1975). This limits the usefulness of bicuculline in the investigation of the effects of GABA-A receptors on cough as seizure activity itself may modify the excitability of coughing. The effects of GABA-C receptors on cough are unknown.

Serotonin receptor ligands can alter cough in animal models. Most of our current information indicates that 5-HT1A receptor agonists inhibit cough in small animals (Kamei et al. 1991a; Stone et al. 1997). Serotonin may modify the effects of other antitussive drugs, suggesting more complex actions of these drugs than simple suppression of a single type of neurone in the central nervous system. For example, in the rat depletion of serotonin attenuated the antitussive effect of morphine, dihydrocodeine, and dextromethorphan (Kamei et al. 1987b). The serotonin precursor L-tryptophan potentiated the antitussive action of dihydrocodeine (Kamei et al. 1990) and the serotonin antagonist methysergide reduced the antitussive effects of dextromethorphan and dihydrocodeine in rats (Kamei et al. 1986a, 1996). L-Tryptophan treatment also prevented the development of tolerance in rats to the antitussive effects of dihydrocodeine (Kamei et al. 1991b). Morphine dependence elicits a reduced sensitivity to opioid and non-opioid antitussives and these effects were associated with a reduction on serotonin receptors in the brainstems of morphine-dependent rats (Kamei et al. 1989). In cats, the serotonin precursor 5-hydroxytryptophan inhibited cough (Kamei et al. 1986b) and methysergide increased cough number and reduced the antitussive effects of dextromethorphan (Kamei et al. 1986a). In humans, serotonin and 5-hydroxytryptamine both inhibited cough due to inhalation of low-chloride solutions, but not coughing elicited by capsaicin aerosols (Stone et al. 1993). Our understanding of the effects serotonin receptor ligands on cough would be enhanced by further work in animal models other than the rat as well as in the human.

The data from the cat on the effects of serotonin receptor agonists and antagonists on cough appear to be consistent with the data from experiments conducted in rats. However, the rat model has been used by relatively few investigators owing to the comparative difficulty to induce cough in this species. Furthermore, a detailed analysis of the motor patterns induced by mechanical stimulation of the tracheobronchial region as well as inhalation of capsaicin in the rat revealed that the behaviors produced by these interventions did not conform to the well-accepted criteria for cough (Ohi et al. 2004). There also is a considerable amount of published information on antitussives and cough in the mouse (Kamei et al. 1993a–f, 1994a–d, 1995a, b, 1999, 2003, 2006, 2007; Morita and Kamei, 2003), but there is no corresponding analysis of the motor patterns of putative coughing in this species to confirm that the behaviors that were measured were coughs. Indeed, in their monograph, Korpas and Tomori (Korpas and Tomori 1979) indicate that mice do not cough in response to mechanical or chemical stimulation of airway afferents, but they do readily express another airway defensive behavior, the expiration reflex, in response to mechanical stimulation of the larynx. The expiration reflex is regulated differently than cough (Korpas and Tomori 1979; Tatar et al. 2008), is insensitive to antitussive drugs (first shown by May and Widdicombe 1954 and later confirmed by Korpas and Tomori 1979), and is generated by a brainstem neural network that is not organized in the same manner as that for coughing (Baekey et al. 2004).

Tachykinin receptor antagonists have been shown to be effective cough suppressants in the cat, dog, and guinea pig (Advenier et al. 1993; Girard et al. 1995; Kudlacz et al. 1996; Bolser et al. 1997; Hay et al. 2002; Chapman et al. 2004; Joad et al. 2004). Both tachykinin NK-1 and tachykinin neurokinin-2 receptor antagonists can inhibit cough by a central mechanism (Bolser et al. 1997; Joad et al. 2004; Mazzone et al. 2005). Although a tachykinin neurokinin-3 receptor antagonist can inhibit cough (Daoui et al. 1998), it is not known whether this effect is by a central site of action. There is evidence that substance P can enhance cough when microinjected into the region of the NTS in the rabbit (Mazzone et al. 2005; Mutolo et al. 2007). However, substance P inhibited excitatory postsynaptic potentials in guinea pig NTS putative relay neurones that were produced by electrical stimulation of the tractus solitarius (Sekizawa et al. 2003). This effect of substance P was presynaptic and blocked by a tachykinin NK-1 receptor antagonist (Sekizawa et al. 2003). These observations suggest that tachykinin NK-1 receptors are important in regulating the excitability of cough in the region of the NTS. The exact role that tachykinin receptors have in this complex area is not clear at this time. The extent to which tachykinin receptor antagonists act to regulate cough in the NTS and/or other brainstem areas after systemic administration in animal models is unknown. Only one study on the effects of tachykinin receptor antagonists on cough in humans has been published. Fahy et al. (1995) observed no effect of the tachykinin NK-1 receptor antagonist on hypertonic saline-induced cough in humans with mild asthma. Given the limited amount of information in the human, further studies should be performed to fully evaluate the antitussive potential of tachykinin receptor antagonists.

Sigma receptor agonists have been associated with cough suppression. The most prominent of these is the cough-suppressant drug dextromethorphan. Given that dextromethorphan binds to *N*-methyl-D-aspartate and sigma-1 receptors, the specificity of its action to suppress cough has not been clear until recently. Brown et al. (2004) recently showed that this drug acts at sigma-1 receptors to inhibit cough in the guinea pig. Other sigma-1 receptor agonists can inhibit cough as well (Chau et al. 1983; Brown et al. 2004). Carbetapentane, a sigma-1 receptor agonist that is approved for use in humans as an antitussive, inhibits cough in the cat (Talbott et al. 1975) and guinea pig (Brown et al. 2004). The role of sigma-2 receptors in the suppression of cough is less clear.

Various other receptors have been associated with cough suppression in animal models, such as non-opioid receptors (NOP-1) (Bolser et al. 2001; McLeod et al. 2001, 2004; Lee et al. 2006), cannabinoid receptors (Gordon et al. 1976; Jia et al. 2002; Morita and Kamei 2003; Patel et al. 2003), and dopaminergic receptors (Kamei et al. 1987a; Li et al. 2002). In humans, dopaminergic receptors appear to have no effect on irritant-induced cough in normal subjects (O'Connell 2002).

References

Adcock JJ, Schneider C, Smith TW (1988) Effects of codeine, morphine and a novel opioid pentapeptide BW443C, on cough, nociception and ventilation in the unanaesthetized guinea-pig. Br J Pharmacol 93:93–100

Advenier C, Girard V, Naline E, Vilain P, Emonds-Alt X (1993) Antitussive effect of SR 48968, a non-peptide tachykinin NK2 receptor antagonist. Eur J Pharmacol 250:169–171

Baekey DM, Morris KF, Gestreau C, Li Z, Lindsey BG, Shannon R (2001) Medullary respiratory neurones and control of laryngeal motoneurones during fictive eupnoea and cough in the cat. J Physiol 534:565–581

Baekey DM, Morris KF, Nuding SC, Segers LS, Lindsey BG, Shannon R (2004) Ventrolateral medullary respiratory network participation in the expiration reflex in the cat. J Appl Physiol 96:2057–2072

Bolser DC (1991) Fictive cough in the cat. J Appl Physiol 71:2325–2331

Bolser DC (1996) Mechanisms of action of central and peripheral antitussive drugs. Pulm Pharmacol 9: 357–364

Bolser DC (2006) Current and future centrally acting antitussives. Respir Physiol Neurobiol 152:349–355

Bolser DC, DeGennaro FC (1994) Effect of codeine on the inspiratory and expiratory burst pattern during fictive cough in cats. Brain Res 662:25–30

Bolser DC, Aziz SM, DeGennaro FC, Kreutner W, Egan RW, Siegel MI, Chapman RW (1993) Antitussive effects of GABAB agonists in the cat and guinea-pig. Br J Pharmacol 110:491–495

Bolser DC, DeGennaro FC, O'Reilly S, Chapman RW, Kreutner W, Egan RW, Hey JA (1994) Peripheral and central sites of action of GABA-B agonists to inhibit the cough reflex in the cat and guinea pig. Br J Pharmacol 113:1344–1348

Bolser DC, Blythin DJ, Chapman RW, Egan RW, Hey JA, Rizzo C, Kuo SC, Kreutner W (1995) The pharmacology of SCH 50911: A novel, orally-active GABA-beta receptor antagonist. J Pharmacol Exp Ther 274:1393–1398

Bolser DC, DeGennaro FC, O'Reilly S, McLeod RL, Hey JA (1997) Central antitussive activity of the NK1 and NK2 tachykinin receptor antagonists, CP-99,994 and SR 48968, in the guinea-pig and cat. Br J Pharmacol 121:165–170

Bolser DC, Hey JA, Chapman RW (1999) Influence of central antitussive drugs on the cough motor pattern. J Appl Physiol 86:1017–1024

Bolser DC, McLeod RL, Tulshian DB, Hey JA (2001) Antitussive action of nociceptin in the cat. Eur J Pharmacol 430:107–111
Brown C, Fezoui M, Selig WM, Schwartz CE, Ellis JL (2004) Antitussive activity of sigma-1 receptor agonists in the guinea-pig. Br J Pharmacol 141:233–240
Chapman RW, House A, Liu F, Celly C, Mei H, Hey JA (2004) Antitussive activity of the tachykinin NK1 receptor antagonist, CP-99994, in dogs. Eur J Pharmacol 485:329–332
Chau TT, Carter FE, Harris LS (1983) Antitussive effect of the optical isomers of mu, kappa and sigma opiate agonists/antagonists in the cat. J Pharmacol Exp Ther 226:108–113
Chen CY, Bonham AC, Plopper CG, Joad JP (2003) Neuroplasticity in nucleus tractus solitarius neurons after episodic ozone exposure in infant primates. J Appl Physiol 94:819–827
Chou D, Wang, SC (1975) Studies on the localization of the central cough mechanism: Site of action of antitussive drugs. J Pharmacol Exp Ther 194:499–505
Daoui S, Cognon C, Naline E, Emonds-Alt X, Advenier C (1998) Involvement of tachykinin NK3 receptors in citric acid-induced cough and bronchial responses in guinea pigs. Am J Respir Crit Care Med 158:42–48
Davenport PW, Bolser DC, Vickroy T, Berry RB, Martin AD, Hey JA, Danzig M (2007) The effect of codeine on the Urge-to-Cough response to inhaled capsaicin. Pulm Pharmacol Ther 20: 338–346
Dicpinigaitis PV (1996) Use of baclofen to suppress cough induced by angiotensin-converting enzyme inhibitors. Ann Pharmacother 30:1242–1245
Dicpinigaitis PV, Dobkin JB (1997) Antitussive effect of the GABA-agonist baclofen. Chest 111:996–999
Dicpinigaitis PV, Dobkin JB, Rauf K, Aldrich TK (1998) Inhibition of capsaicin-induced cough by the gamma-aminobutyric acid agonist baclofen. J Clin Pharmacol 38:364–367
Domino EF, Krutak-Krol H, Lal J (1985) Evidence for a central site of action for the antitussive effects of caramiphen. J Pharmacol Exp Ther 233:249–253
Fahy JV, Wong HH, Geppetti P, Reis JM, Harris SC, Maclean DB, Nadel JA, Boushey HA (1995) Effect of an NK1 receptor antagonist (CP-99,994) on hypertonic saline-induced bronchoconstriction and cough in male asthmatic subjects. Am J Respir Crit Care Med 152:879–884
Freestone C, Eccles R (1997) Assessment of the antitussive efficacy of codeine in cough associated with common cold. J Pharm Pharmacol 49:1045–1049
Fuller RW, Karlsson JA, Choudry NB, Pride NB (1988) Effect of inhaled and systemic opiates on responses to inhaled capsaicin in humans. J Appl Physiol 65:1125–1130
Gestreau C, Bianchi AL, Grelot L (1997) Differential brainstem Fos-like immunoreactivity after laryngeal-induced coughing and its reduction by codeine. J Neurosci 17:9340–9352
Girard V, Naline E, Vilain P, Emonds-Alt X, Advenier C (1995) Effect of the two tachykinin antagonists, SR 48968 and SR 140333, on cough induced by citric acid in the unanaesthetized guinea pig. Eur Respir J 8:1110–1114
Gordon R, Gordon RJ, Sofia D (1976) Antitussive activity of some naturally occurring cannabinoids in anesthetized cats. Eur J Pharmacol 35:309–313
Hay DW, Giardina GA, Griswold DE, Underwood DC, Kotzer CJ, Bush B, Potts W, Sandhu P, Lundberg D, Foley JJ, Schmidt DB, Martin LD, Kilian D, Legos JJ, Barone FC, Luttmann MA, Grugni M, Raveglia LF, Sarau HM (2002) Nonpeptide tachykinin receptor antagonists. III. SB 235375, a low central nervous system-penetrant, potent and selective neurokinin-3 receptor antagonist, inhibits citric acid-induced cough and airways hyper-reactivity in guinea pigs. J Pharmacol Exp Ther 300:314–323
Hey JA, Mingo G, Bolser DC, Kreutner W, Krobatsch D, Chapman RW (1995) Respiratory effects of baclofen and 3-aminopropylphosphinic acid in guinea-pigs. Br J Pharmacol 114:735–738
Hutchings HA, Eccles R (1994) The opioid agonist codeine and antagonist naltrexone do not affect voluntary suppression of capsaicin induced cough in healthy subjects. Eur Respir J 7:715–719
Hutchings HA, Morris S, Eccles R, Jawad MS (1993) Voluntary suppression of cough induced by inhalation of capsaicin in healthy volunteers. Respir Med 87:379–382
Jakus J, Tomori Z, Stransky A, Boselova L (1987) Bulbar respiratory activity during defensive airways reflexes in cats. Acta Physiol Hung 70:245–254

Jakus J, Stransky A, Poliacek I, Barani H, Bosel'ova L (1998) Effects of medullary midline lesions on cough and other airway reflexes in anaesthetized cats. Physiol Res 47:203–213
Jakus J, Stransky A, Poliacek I, Barani H, Bosel'ova L (2000) Kainic acid lesions to the lateral tegmental field of medulla: Effects on cough, expiration and aspiration reflexes in anesthetized cats. Physiol Res 49:387–398
Jia Y, McLeod RL, Wang X, Parra LE, Egan RW, Hey JA (2002) Anandamide induces cough in conscious guinea-pigs through VR1 receptors. Br J Pharmacol 137:831–836
Joad JP, Munch PA, Bric JM, Evans SJ, Pinkerton KE, Chen CY, Bonham AC (2004) Passive smoke effects on cough and airways in young guinea pigs: Role of brainstem substance P. Am J Respir Crit Care Med 169:499–504
Kamei J (1996) Role of opioidergic and serotonergic mechanisms in cough and antitussives. Pulm Pharmacol 9:349–356
Kamei J, Hosokawa T, Yanaura S, Hukuhara T (1986a) Effects of methysergide on the cough reflex. Jpn J Pharmacol 42:450–452
Kamei J, Hosokawa T, Yanaura S, Hukuhara T (1986b) Involvement of central serotonergic mechanisms in the cough reflex. Jpn J Pharmacol 42:531–538
Kamei J, Hukuhara T, Kasuya Y (1987a) Dopaminergic control of the cough reflex as demonstrated by the effects of apomorphine. Eur J Pharmacol 141:511–513
Kamei J, Ogawa M, Kasuya Y (1987b) Monoamines and the mechanisms of action of antitussive drugs in rats. Arch Int Pharmacodyn Ther 290:117–127
Kamei J, Mori T, Ogawa M, Kasuya Y (1989) Subsensitivity to the cough-depressant effects of opioid and nonopioid antitussives in morphine-dependent rats: Relationship to central serotonin function. Pharmacol Biochem Behav 34:595–598
Kamei J, Mori T, Kasuya Y (1990) Effects of L-tryptophan on the effects of antitussives. J Pharmacobiodyn 13:231–237
Kamei J, Mori T, Igarashi H, Kasuya Y (1991a) Effects of 8-hydroxy-2-(di-n-propylamino)tetralin, a selective agonist of 5-HT1A receptors, on the cough reflex in rats. Eur J Pharmacol 203: 253–258
Kamei J, Mori T, Kasuya Y (1991b) Effects of L-tryptophan on the development of tolerance to the antitussive effects of dihydrocodeine. Jpn J Pharmacol 55:403–406
Kamei J, Iwamoto Y, Kawashima N, Suzuki T, Nagase H, Misawa M, Kasuya Y (1993a) Possible involvement of mu 2-mediated mechanisms in mu-mediated antitussive activity in the mouse. Neurosci Lett 149:169–172
Kamei J, Iwamoto Y, Misawa M, Kasuya Y (1993b) Effects of rimcazole, a specific antagonist of sigma sites, on the antitussive effects of non-narcotic antitussive drugs. Eur J Pharmacol 242:209–211
Kamei J, Iwamoto Y, Misawa M, Nagase H, Kasuya Y (1993c) Antitussive effect of beta-endorphin is mediated by mu-opioid receptors, but not by kappa- or epsilon-opioid receptors. Eur J Pharmacol 233:251–254
Kamei J, Iwamoto Y, Suzuki T, Misawa M, Nagase H, Kasuya Y (1993d) Antitussive effects of naltrindole, a selective delta-opioid receptor antagonist, in mice and rats. Eur J Pharmacol 249:161–165
Kamei J, Iwamoto Y, Suzuki T, Misawa M, Nagase H, Kasuya Y (1993e) The role of the mu 2-opioid receptor in the antitussive effect of morphine in mu 1-opioid receptor-deficient CXBK mice. Eur J Pharmacol 240:99–101
Kamei J, Iwamoto Y, Suzuki T, Nagase H, Misawa M, Kasuya Y (1993f) Differential modulation of mu-opioid receptor-mediated antitussive activity by delta-opioid receptor agonists in mice. Eur J Pharmacol 234:117–120
Kamei J, Iwamoto Y, Misawa M, Nagase H, Kasuya Y (1994a) Antitussive effect of [Met5] enkephalin-Arg6-Phe7 in mice. Eur J Pharmacol 253:293–296
Kamei J, Iwamoto Y, Misawa M, Nagase H, Kasuya Y (1994b) Involvement of adenosine A1 receptors in antitussive effect in mice. Life Sci 55:PL383–388
Kamei J, Iwamoto Y, Suzuki T, Misawa M, Nagase H, Kasuya Y (1994c) Antitussive effect of dihydroetorphine in mice. Eur J Pharmacol 260:257–259

Kamei J, Iwamoto Y, Suzuki T, Misawa M, Nagase H, Kasuya Y (1994d) Involvement of delta 1-opioid receptor antagonism in the antitussive effect of delta-opioid receptor antagonists. Eur J Pharmacol 251:291–294
Kamei J, Saitoh A, Morita K, Nagase H (1995a) Antagonistic effect of buprenorphine on the antitussive effect of morphine is mediated via the activation of mu 1-opioid receptors. Life Sci 57:PL231–235
Kamei J, Saitoh A, Suzuki T, Misawa M, Nagase H, Kasuya Y (1995b) Buprenorphine exerts its antinociceptive activity via mu 1-opioid receptors. Life Sci 56:PL285–290
Kamei J, Morita K, Ohsawa M, Onodera K (1999) Effects of epinastine on the antitussive and rewarding effects of dihydrocodeine in mice. Methods Find Exp Clin Pharmacol 21:663–668
Kamei J, Morita K, Saitoh A, Nagase H (2003) The antitussive effects of endomorphin-1 and endomorphin-2 in mice. Eur J Pharmacol 467:219–222
Kamei J, Yoshikawa Y, Saitoh A (2006) Effect of *N*-arachidonoyl-(2-methyl-4-hydroxyphenyl) amine (VDM11), an anandamide transporter inhibitor, on capsaicin-induced cough in mice. Cough 2:2
Kamei J, Hayashi SS, Takahashi Y, Nozaki C (2007) Role of cyclin-dependent kinase 5 in capsaicin-induced cough. Eur J Pharmacol 566:181–184
Korpas J, Tomori Z (1979) Cough and other respiratory reflexes. S. Karger, Basel, New York
Kotzer CJ, Hay DW, Dondio G, Giardina G, Petrillo P, Underwood DC (2000) The antitussive activity of delta-opioid receptor stimulation in guinea pigs. J Pharmacol Exp Ther 292: 803–809
Kudlacz EM, Knippenberg RW, Logan DE, Burkholder TP (1996) Effect of MDL 105,212, a non-peptide NK-1/NK-2 receptor antagonist in an allergic guinea pig model. J Pharmacol Exp Ther 279:732–739
Lal J, Krutak-Krol H, Domino EF (1986) Comparative antitussive effects of dextrorphan, dextromethorphan and phencyclidine. Arzneimittelforschung 36:1075–1078
Lee MG, Undem BJ, Brown C, Carr MJ (2006) Effect of nociceptin in acid-evoked cough and airway sensory nerve activation in guinea pigs. Am J Respir Crit Care Med 173:271–275
Li JQ, Jia YX, Yamaya M, Arai H, Ohrui T, Sekizawa K, Sasaki H (2002) Neurochemical regulation of cough response to capsaicin in guinea-pigs. Auton Autacoid Pharmacol 22:57–63
Lipski J, Bellingham MC, West MJ, Pilowsky P (1988) Limitations of the technique of pressure microinjection of excitatory amino acids for evoking responses from localized regions of the CNS. J Neurosci Meth 26:169–179
May AJ, Widdicombe JG (1954) Depression of the cough reflex by pentobarbitone and some opium derivatives. Br J Pharmacol Chemother 9:335–340
Mazzone SB, Mori N, Canning BJ (2005) Synergistic interactions between airway afferent nerve subtypes regulating the cough reflex in guinea-pigs. J Physiol 569:559–573
McLeod RL, Parra LE, Mutter JC, Erickson CH, Carey GJ, Tulshian DB, Fawzi AB, Smith-Torhan A, Egan RW, Cuss FM, Hey JA (2001) Nociceptin inhibits cough in the guinea-pig by activation of ORL(1) receptors. Br J Pharmacol 132:1175–1178
McLeod RL, Jia Y, Fernandez X, Parra LE, Wang X, Tulshian DB, Kiselgof EJ, Tan Z, Fawzi AB, Smith-Torhan A, Zhang H, Hey JA (2004) Antitussive profile of the NOP agonist Ro-64–6198 in the guinea pig. Pharmacology 71:143–149
Merrill EG (1970) The lateral respiratory neurones of the medulla: Their associations with nucleus ambiguus, nucleus retroambigualis, the spinal accessory nucleus and the spinal cord. Brain Res 24:11–28
Merrill EG (1972) Temporal patterns of antidromic invasion latencies for the respiratory neurones of nucleus retroambigualis in cats. J Physiol 223:18P–20P
Merrill EG (1974) Proceedings: Antidromic activation of lateral respiratory neurones during their silent periods. J Physiol 241:118P–119P
Morice AH, Menon MS, Mulrennan SA, Everett CF, Wright C, Jackson J, Thompson R (2007) Opiate therapy in chronic cough. Am J Respir Crit Care Med 175:312–315
Morita K, Kamei J (2003) Antitussive effect of WIN 55212–2, a cannabinoid receptor agonist. Eur J Pharmacol 474:269–272

Mutolo D, Bongianni F, Fontana GA, Pantaleo T (2007) The role of excitatory amino acids and substance P in the mediation of the cough reflex within the nucleus tractus solitarii of the rabbit. Brain Res Bull 74:284–293
O'Connell F (2002) Central pathways for cough in man–unanswered questions. Pulm Pharmacol Ther 15:295–301
Ohi Y, Yamazaki H, Takeda R, Haji A (2004) Phrenic and iliohypogastric nerve discharges during tussigenic stimulation in paralyzed and decerebrate guinea pigs and rats. Brain Res 1021: 119–127
Ohi Y, Yamazaki H, Takeda R, Haji A (2005) Functional and morphological organization of the nucleus tractus solitarius in the fictive cough reflex of guinea pigs. Neurosci Res 53:201–209
Ohi Y, Kato F, Haji A (2007) Codeine presynaptically inhibits the glutamatergic synaptic transmission in the nucleus tractus solitarius of the guinea pig. Neuroscience 146:1425–1433
Patel HJ, Birrell MA, Crispino N, Hele DJ, Venkatesan P, Barnes PJ, Yacoub MH, Belvisi MG (2003) Inhibition of guinea-pig and human sensory nerve activity and the cough reflex in guinea-pigs by cannabinoid (CB2) receptor activation. Br J Pharmacol 140:261–268
Penn RD (1992) Intrathecal baclofen for spasticity of spinal origin: Seven years of experience. J Neurosurg 77:236–240
Poliacek I, Jakus J, Stransky A, Barani H, Halasova E, Tomori Z (2004) Cough, expiration and aspiration reflexes following kainic acid lesions to the pontine respiratory group in anesthetized cats. Physiol Res 53:155–163
Poliacek I, Corrie LW, Wang C, Rose MJ, Bolser DC (2007) Microinjection of DLH into the region of the caudal ventral respiratory column in the cat: Evidence for an endogenous cough-suppressant mechanism. J Appl Physiol 102:1014–1021
Reynolds SM, Mackenzie AJ, Spina D, Page CP (2004) The pharmacology of cough. Trends Pharmacol Sci 25:569–576
Rose MJ, Wang C, Golder FJ, Hammond JA, Bolser DC (2004) Influence of intrathecal and intra-arterial baclofen on laryngeal and tracheobronchial cough in the cat. FASEB J 18:A718
Schomburg ED, Steffens H (1995) Influence of opioids and naloxone on rhythmic motor activity in spinal cats. Exp Brain Res 103:333–343
Sekizawa S, Joad JP, Bonham AC (2003) Substance P presynaptically depresses the transmission of sensory input to bronchopulmonary neurons in the guinea pig nucleus tractus solitarii. J Physiol 552:547–559
Shannon R, Baekey DM, Morris KF, Lindsey BG (1996) Brainstem respiratory networks and cough. Pulm Pharmacol 9:343–347
Shannon R, Baekey DM, Morris KF, Lindsey BG (1998) Ventrolateral medullary respiratory network and a model of cough motor pattern generation. J Appl Physiol 84:2020–2035
Shannon R, Baekey DM, Morris KF, Li Z, Lindsey BG (2000) Functional connectivity among ventrolateral medullary respiratory neurones and responses during fictive cough in the cat. J Physiol 525 Pt 1:207–224
Smith J, Owen E, Earis J, Woodcock A (2006) Effect of codeine on objective measurement of cough in chronic obstructive pulmonary disease. J Allergy Clin Immunol 117:831–835
Stone RA, Worsdell YM, Fuller RW, Barnes PJ (1993) Effects of 5-hydroxytryptamine and 5-hydroxytryptophan infusion on the human cough reflex. J Appl Physiol 74:396–401
Stone RA, Barnes PJ, Chung KF (1997) Effect of 5-HT1A receptor agonist, 8-OH-DPAT, on cough responses in the conscious guinea pig. Eur J Pharmacol 332:201–207
Takahama K, Fukushima H, Isohama Y, Kai H, Miyata T (1997) Inhibition of glycine currents by dextromethorphan in neurones dissociated from the guinea-pig nucleus tractus solitarii. Br J Pharmacol 120:690–694
Talbott MW, Barch GK, Gabel LP (1975) A new method for evaluating antitussives in cats using an electrode-cannula. Eur J Pharmacol 34:59–63
Tatar M, Hanacek J, Widdicombe JG (2008) The expiration reflex from the trachea and bronchi. Eur Respir J 31:385–390
Wang SC, Chou DT, Wallenstein MC (1977) Studies on the potency of various antitussive agents. Agents Actions 7:337–340

Wood JD (1975) The role of gamma-aminobutyric acid in the mechanism of seizures. Prog Neurobiol 5:77–95
Xie R, Hammarlund-Udenaes M (1998) Blood–brain barrier equilibration of codeine in rats studied with microdialysis. Pharm Res 15:570–575
Xu F, Frazier DT, Zhang Z, Baekey DM, Shannon R (1997) Cerebellar modulation of cough motor pattern in cats. J Appl Physiol 83:391–397

Central Mechanisms III: Neuronal Mechanisms of Action of Centrally Acting Antitussives Using Electrophysiological and Neurochemical Study Approaches

K. Takahama(✉), T. Shirasaki, and F. Soeda

Contents

1 Introduction

Much remains to be discussed about the mechanisms underlying cough production. Furthermore, there are various types of cough in humans with different susceptibilities to various antitussive drugs. Although codeine is reportedly one of the most effective narcotic antitussives, it has recently been demonstrated not to be completely effective in suppressing cough, as shown from experimental results in guinea pigs. Also, chronic coughs are often resistant to codeine treatment. Pharmacological diversity is also found in cough responses produced in experimental animals. Although many antitussives have been developed and used in clinics and as over-the-counter drugs, it remains difficult to describe specific mechanisms of action of centrally acting antitussives. Therefore, in this review, we first describe codeine-sensitive and codeine-insensitive coughs in experimental animals. According to our

K. Takahama
Department of Environmental and Molecular Health Sciences, Graduate School of Pharmaceutical Sciences, Kumamoto University, 5–1 Oe-honmachi, Kumamoto 862–0973, Japan
takahama@gpo.kumamoto-u.ac.jp

K.F. Chung, J.G. Widdicombe (eds.), *Pharmacology and Therapeutics of Cough*,
Handbook of Experimental Pharmacology 187,

own data, codeine-sensitive cough is caused by activation of the afferent neurons, probably Aδ neurons, arising mainly from the upper trachea. This neuron is not desensitized by capsaicin treatment. On the other hand, codeine-insensitive cough may be produced by activation of the C-fibers arising mainly from the bifurcation of the trachea and the adjacent region. In addition, we review mechanisms of action of centrally acting antitussives on codeine-sensitive cough in experimental animals, addressing their actions at the neuronal level.

2 Codeine-Sensitive and Codeine-Insensitive Coughs

Although coughing is a very simple phenomenon ending in explosive expiration preceded by rapid inspiration, it is diverse in terms of not only clinical features but also pharmacological properties. For example, in guinea pigs with airway inflammation such as in the case of bronchitis caused by SO_2 gas exposure and cigarette smoking, coughing caused by various stimuli is often hard to inhibit using centrally acting antitussives. This is also the case of coughing augmented by treatment with an angiotensin-converting-enzyme inhibitor (ACEI).

In the course of our studies on antitussive effects of codeine, we observed that even at very high doses the drug failed to completely suppress coughing induced by inhalation of citric acid solution in guinea pigs (Takahama and Shirasaki 2007), although codeine is reportedly one of the most potent antitussives and is generally very effective against coughing caused by various stimuli. The result suggests that there is a type of cough that cannot be inhibited by codeine. Furthermore, we found that coughing caused by mechanical stimulation of the tracheal bifurcation was not responsive to codeine in guinea pigs, compared with that induced by mechanical stimulation of the trachea close to the larynx (Takahama et al. 1997a, b).

Among sensory receptors in airway vagal afferents, rapidly adapting receptors (RARs), Aδ-nociceptors, and bronchial C-fiber receptors appear to be involved in cough responses. RARs are myelinated Aδ-fibers and have a low threshold for mechanical stimuli, but are resistant to chemical stimuli. Aδ-nociceptors and unmyelinated C-fiber receptors have a high threshold for mechanical stimuli, but a low threshold for chemical stimuli such as bradykinin and capsaicin.

A classic study carried out by Widdicombe (1954) showed that RARs are distributed mainly to the larynx and tracheal bifurcation. On the other hand, chemoreceptors involved in cough responses are mainly distributed in the lower trachea, particularly around the tracheal bifurcation. In addition to these receptors, Canning et al. have recently found a "cough receptor" which is a subpopulation of myelinated, capsaicin-sensitive polymodal afferent neurons in guinea pigs (Canning et al. 2004). These "cough receptors" have properties similar to those of RARs, but have a slower conduction velocity and do not respond to stretching. These receptors are distributed in the larynx and upper trachea, and appear to play a primary role in regulating the cough response (Canning et al. 2004). Taken together, it is suggested that chemoreceptors distributed in the lower trachea, particularly around

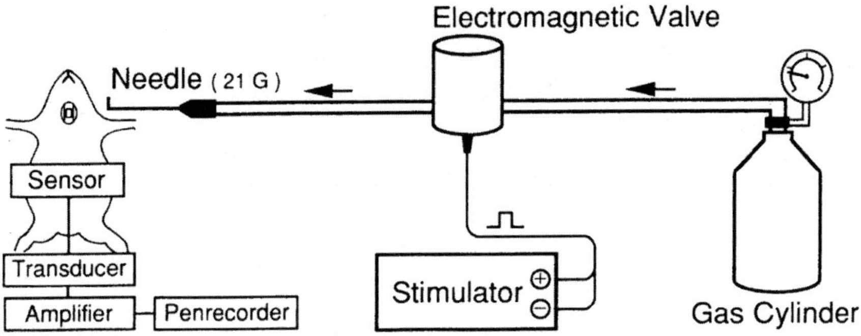

Fig. 1 The method of topical chemical stimulation for producing cough and cough-like responses in guinea pigs. Guinea pigs were lightly anesthetized with pentobarbital. A 10-μl aliquot of stimulant solution was ejected through a 21-gauge needle by applying $1.5\,\mathrm{kg\,cm^{-2}}$ air pressure for 30 ms. Switching of applying air pressure was done by an electromagnetic valve controlled by a stimulator. (From Takahama et al. 1995)

the tracheal bifurcation, may be involved in the production of cough resistant to codeine treatment. This suggestion is further supported by the following findings. ACEIs, which sensitize nociceptive fibers, induce codeine-resistant chronic cough in conscious guinea pigs (Fox et al. 1996), indicating that coughs mediated by nociceptive fibers are resistant to codeine. In our own preliminary study using a newly developed method for topical chemical stimulation of the lower airway (Fig. 1), topical administration of capsaicin to the tracheal bifurcation induced a cough-like response resistant to codeine, whereas when it was added to the laryngeal side of the trachea it did not induce a cough response (Takahama et al. 1995; Wakuda et al. 1994). Furthermore, our own preliminary study showed that substance P (SP)-like immunoreactivity is higher in the tracheal bifurcation than in the larynx side in guinea pigs. A similar finding has been obtained in human subjects. The density of SP-immunoreactive nerves is significantly higher in patients with cough-variant asthma than in normal subjects and patients with classic asthma (Lee et al. 2003). The abovementioned findings support the hypothesis that coughs mediated by nociceptive fibers may be resistant to codeine treatment.

In a guinea pig model of subacute bronchitis produced by SO_2 gas exposure, cough responses induced by mechanical stimulation of the trachea were more difficult to inhibit using codeine than those induced experimentally in normal guinea pigs (Miyata et al. 1999). It has been reported that SP content is elevated in the trachea of rats with bronchitis induced by SO_2 gas exposure (Killingsworth et al. 1996), probably owing to reduction in the level of neutral endopeptidase (NEP), which degrades various peptides, including bradykinin, SP, and other tachykinins (Matsas et al. 1984). In bronchitic guinea pigs, NEP levels and activity in the trachea and bronchus were significantly lower than those in normal guinea pigs, although no epithelial shedding was found in the lower airway. Administration of high doses of a NEP inhibitor has been shown to induce a cough response in normal guinea pigs (Takahama et al. 1995). Based on findings that bradykinin and tachykinins are potent inflammatory mediators, and that neurokinins such as SP are released from C-fiber (eNANC nerve) terminals, one hypothesis is that coughing induced by inflammatory peptides might be resistant to centrally acting antitussives such as codeine.

Codeine has, however, been found to significantly suppress the cough response induced or enhanced by NEP inhibitors (Takahama et al. 1995). In addition, opioids peripherally inhibit tachykinergic transmission in the guinea pig bronchus (Kamikawa and Shimo 1990; Frossard and Barnes 1987; Belvisi et al. 1989). Thus, it appears that NEP inhibition or tachykinin release from peripheral C-fiber terminals is not always sufficient to explain mechanisms underlying induction of codeine-resistant cough.

As described above, citric acid induced coughing was difficult to completely inhibit even with high doses of codeine alone. Interestingly, coadministration of codeine and an antagonist of the neurokinin-2 receptor almost suppressed citric acid induced coughing in conscious guinea pigs (Takahama and Shirasaki 2007), suggesting that neurokinins may induce codeine-resistant coughs in guinea pigs. Therefore, it seems likely that RAR or "cough-receptor"-mediated coughs are sensitive to codeine, but coughs triggered by nocieceptive fibers containing tachykinins such as neurokinin are resistant to the drug.

Recently, it has been reported that the expression of transient receptor potential vanilloid-1 (TRPV-1) is enhanced in airway nerves of patients with chronic cough (Groneberg et al. 2004). In addition to TRPV-1, acid-sensing ion channels are localized in Aδ-fibers of guinea pigs (Gu 2006). Therefore, differences in codeine sensitivity of acid-induced coughs may depend on the pH of the cough induction site. Further studies are needed to clarify such differences.

3 Mechanism of Induction of Codeine-Resistant Coughs and Opioid Receptors

Two putative mechanisms underlying the production of coughs resistant to centrally acting antitussives, including narcotic antitussives such as codeine, are proposed. One possible mechanism is that μ-opioid receptors may not be expressed in neurons of the cough reflex arc involved in the production of codeine-resistant coughs. Another possible mechanism is that neurotransmission in the cough reflex arc may be facilitated when codeine-resistant cough is produced.

As described already, several lines of evidence suggest that tachykinin-containing vagal afferent fibers contribute to the production of codeine-resistant cough. Such fibers spread from the airway and innervate the nucleus tractus solitarii (NTS) (Helke et al. 1981; Saria et al. 1988). Injection of a neurokinin-1 (NK_1) receptor antagonist into the NTS produced an antitussive effect in animals exposed to cigarette smoke, but not in animals exposed to filtered air (Joad et al. 2004). It has also been shown that an excitatory action of iontophoretically applied SP on NTS neurons is not inhibited by μ-agonists (Morin-Surun et al. 1984). These findings seem to suggest that μ-opioid receptors are not expressed in C-fibers such as tachykinin-containing fibers involved in the production of codeine-resistant cough.

μ-Opioid receptors are found in both postsynaptic and presynaptic sites in the NTS (Rhim et al. 1993; Cheng et al. 1996) as the predominant type of opioid receptor (Hassen et al. 1982; Xia and Haddad 1991). μ-Agonists activate the G-protein-coupled inwardly rectifying potassium (GIRK) channel current and hyperpolarize postsynaptic membrane potential in about 60% of NTS neurons, whereas they inhibit glutamatergic excitatory postsynaptic currents in almost all NTS neurons via a presynaptic mechanism, which is much more sensitive to μ-agonists than postsynaptic mechanisms (Rhim et al. 1993; Glatzer and Smith 2005). μ-Agonists also inhibit high-voltage-activated Ca^{2+} currents in the nodose ganglion (Rusin and Moises 1998), the origin of RARs (Ricco et al. 1996), and glutamate is the principal neurotransmitter in Aδ-fibers (Lawrence 1995; Meeley et al. 1989; Wilson et al. 1996). Excision of the nodose ganglion causes marked depletion of μ-opioid receptors in the dorsal and medial regions of the ipsilateral caudal NTS (Dashwood et al. 1988). Electron microscopy has shown that μ-opioid receptors localize in the plasma membrane of the terminals of vagal afferents derived from nodose ganglia (Aicher et al. 2000), but not in nociceptive fibers (Undem et al. 2004). In relation to these findings, our own preliminary study using a c-fos-like immunoreactivity expression technique revealed that c-fos-like proteins are expressed in the medial regions of the caudal NTS following mechanical stimulation of the larynx side of the airway.

Certainly, codeine only weakly inhibited coughs in animal models of allergic responses (Kamei et al. 1998; Winter and Flataker 1955; Myou et al. 2001), as well as of chronic bronchitis produced by SO_2 gas exposure (Killingsworth et al. 1996). It has been reported that infection and inflammation of the airway lead to the production of neurokinins in nonnociceptive RAR nerve terminals in the cell bodies of vagal sensory ganglia (Undem et al. 1999; Carr et al. 2002; Myers et al. 2002; Dinh et al. 2004, 2005). Therefore, it is intriguing that changes in the phenotype of vagal nerves and tachykinin release from RAR fibers might facilitate glutamatergic transmission in the nucleus involved in the cough reflex under pathological conditions such as airway inflammation, infections, and cigarette smoking. These changes might explain codeine-resistant cough.

4 Mechanisms of Action of Nonnarcotic Centrally Acting Antitussives

One strategy to determine the effects of centrally acting antitussives on single brain neurons forming the cough reflex arc seems to have an advantage for studying the mechanism of action of the drugs, because evidence from in vivo pharmacological and electrophysiological studies indicates that the primary sites of action of centrally acting antitussives are brain neurons. In particular, evidence from comparison of equieffective doses administered via different routes, such as the vein, common carotid artery, vertebral artery and cerebellomedullary cistern (Kase 1980), suggests that the primary site of action of centrally acting antitussives is the so-called cough center located in the region that includes part of the NTS and adjacent regions. This hypothesis is supported by data from analysis of actions of antitussives on evoked discharges in afferent and efferent nerves of the cough reflex. In addition, there have been studies of neuronal circuitry involved in integration of the cough response in the brain (Widdicombe 2002; Beakey et al. 2001). However, the precise mechanisms of action of centrally acting antitussives have yet to be clarified, particularly at the neuron level.

Various medicinal drugs show a particular characteristic based on their chemical structure. In other words, the pharmacological actions of such drugs are based on their chemical structure, as exemplified by various centrally acting drugs such as antipsychotic drugs, antianxiety drugs, and antidepressants. There is usually a relationship between pharmacological action and chemical structure. However, the chemical structures of centrally acting antitussives are diverse, because these drugs are derived from various parent compounds possessing various pharmacological actions. These facts bring us to the following issues: whether centrally acting antitussives have a common mechanism of antitussive action, and whether these drugs have a common effect on single brain neurons. To further clarify these issues, we must prepare single neurons from the cough center and related regions of the brain. Although there have been some reports about the location of the cough center, these studies have been carried out in cats. However, it seemed difficult to obtain a fresh single neuron from the cough center and related nuclei in cats. Therefore, prior to the study on the actions of antitussives at the neuron level, we first tried to clarify the location and pharmacological properties of the cough center in guinea pigs.

4.1 Cough Center and its Pharmacological Properties

In lightly anesthetized guinea pigs, we found an area responsible for producing cough-like responses when it was electrically stimulated. It is located within the region from 500 μm rostal to 100 μm caudal to the obex, from 800 to 400 μm from the midline, and from 1,000 to 600 μm below the dorsal surface of the medulla oblongata. This area corresponds to a region that includes the rostal part of the NTS and adjacent areas, and is nearly the same as that in cats reported by Kase (1980). The responses produced by electrical stimulation of this area were considered to be coughing because (1) the responses were depressed by an antitussive dose of codeine and (2) the responses consisted of a brief inspiration followed by an explosive expiration with the sound of coughing.

By using a microinjection technique in guinea pigs, we also determined the pharmacological properties of an area producing cough-like responses when it was electrically stimulated. The injection of a small amount of 5-hydroxytryptamine (5-HT), atropine, and DL-2-amino-5-phosphonopentanoic acid inhibited cough responses caused by mechanical stimulation of the mucosa of the tracheal bifurcation. Since

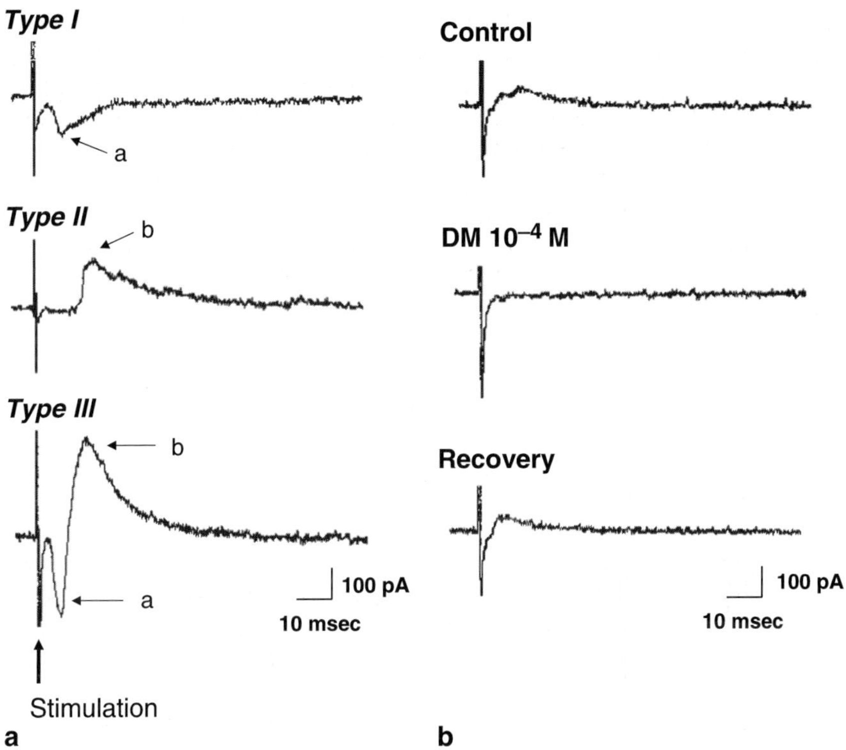

Fig. 2 Three types of waveforms of postsynaptic currents recorded from the slice including the nucleus tractus solitarii (NTS) (**A**). The postsynaptic currents were evoked by stimulation of the solitary tract or the adjacent region. The neurons of the NTS were voltage-clamped at −65-mV holding potential. **a**: excitatory postsynaptic current, **b**: inhibitory postsynaptic current

region of the lower brain stem, preventing the probe from moving as a result of respiration- and cardiac-pulse-induced fluctuations of the brain stem. In the experiments performed under stereotaxic fixation using lightly anesthetized guinea pigs, the glycine level in the NTS was markedly elevated in the microdialysis samples collected during a period of coughing caused by inhalation of an aerosol of 2% citric acid solution. The responses produced by citric acid under the condition described above were typical cough responses when estimated by the cough response curve and the sound of coughing. However, the levels of GABA, alanine, glutamic acid, and glutamine remained unchanged. The taurine level also significantly increased during coughing. Dimethylphenylpiperazinium (DMPP) produces a potent respiratory excitation similar to a cough response. Coughing is associated with transient increase in blood pressure, which is similar to that caused by bolus injection of vasopressors such as phenylephrine. Therefore, we examined the effects of DMPP and phenylephrine on the levels of glycine and taurine in the NTS. Treatment with both drugs did not cause any changes in the levels of glycine or taurine, suggesting that the abovementioned changes associated with coughing in the levels of glycine and taurine may not be caused by respiratory excitation and increase in blood pressure observed during coughing.

Taken together, it is suggested that glycine and its receptor may be involved in the cough response, although it remains unknown whether an increase in the glycine level causes coughing or is a result of coughing. However, it appears that inhibition of I_{gly} by centrally acting antitussives other than dextromethorphan does not necessarily contribute to antitussive action, because relatively high concentrations were needed to inhibit I_{gly}. Further studies are needed to elucidate these actions.

4.3 5-HT and Its Receptors

There have been findings that 5-HT may be involved in the regulation of respiration. For example, morphine-induced respiratory depression was reduced by treatment with reserpine, a monoamine depleter, and *p*-chlorophenylalanine (PCPA), an inhibitor of 5-HT synthesis. It is probable that antitussive activity of some drugs may be reduced by treatment with these pharmacological tools. Kamei et al. have reported that the 50% antitussive effective dose (AtD_{50}) of some antitussives in reserpinized rats is twofold to fourfold higher than that in normal rats (Kamei et al. 1987). This increase in AtD_{50} has also been found in rats treated with PCPA, which depletes 5-HT content in the brain, compared with normal rats. In contrast, treatment with α-methyl-*p*-tyrosine, which depletes catecholamines such as norepinephrine and dopamine in the brain, has little effect on antitussive activity, suggesting that catecholamines are not involved in antitussive activity. That is, reserpine and PCPA, which both deplete 5-HT content in the brain, reduce antitussive activity of centrally acting antitussives in rats. On the basis of these findings, Kamei et al. suggested that 5-HT but not catecholamines in the brain plays an important role in manifestation of antitussive activity (Kamei et al. 1987); however, the suggestion should be carefully verified because it is based on experimental data in anesthetized rats, in which it is often hard to produce cough. It is also well known that activation of serotonergic neurons in the brain causes respiratory depression.

An important question is which subtype of 5-HT receptors is involved in antitussive activity. Of the 5-HT receptor subtypes, 5-HT_{1A} receptors are considered as receptors involved in cough responses because these receptors are distributed throughout the medulla oblongata, and their receptors are involved in respiration regulation. Interestingly, systemic administration of 8-hydroxy-2-(di-*n*-propylamino)tetralin (8-OH-DPAT) suppresses coughing in guinea pigs (Stone et al. 1997) and rats (Kamei et al. 1991). Because 8-OH-DPAT is a 5-HT_{1A} receptor agonist possessing high selectivity to 5-HT_{1A} receptors, this finding suggests that direct activation of 5-HT_{1A} receptors may suppress cough responses, which is supported by the finding that the antitussive effects of dextromethorphan, a representative nonnarcotic antitussive in rats, are inhibited by 5-HT_{1A} receptor antagonists such as spiperone and methysergide, but not by the 5-HT_2 receptor antagonist ketanserine (Kamei 1986). The same result was obtained from an experiment on dihydrocodeine, a narcotic antitussive. Although it remains unclear whether these findings can be extrapolated to coughing in humans, such findings strongly suggest that activation of 5-HT$_{1A}$ receptors may induce an antitussive effect in experimental animals such as rodents.

From the abovementioned findings, the question remains whether antitussive drugs directly or indirectly activate 5-HT$_{1A}$ receptors. Therefore, we first examined the effects of dextromethorphan and other antitussives on single brain neurons acutely dissociated from the dorsal raphe, which expresses 5-HT$_{1A}$ receptors, using nystatin-perforated and conventional whole-cell recording techniques. In dorsal raphe neurons, voltage-clamped at $V_H = -80\,\text{mV}$, 1 nM to 1 μM 5-HT induced an inwardly rectifying K^+ current ($I_{5\text{-HT}}$) in a concentration-dependent manner. This $I_{5\text{-HT}}$ was blocked by spiperone, a 5-HT$_{1A}$ receptor antagonist, but not by ketanserine, a 5-HT$_2$ receptor antagonist. We also confirmed that $I_{5\text{-HT}}$ was carried by K^+, on the basis of results of electrophysiological analysis. In these neurons, 8-OH-DPAT, a 5-HT$_{1A}$ receptor agonist, similar to 5-HT, induced $I_{5\text{-HT}}$ in dorsal raphe neurons. However, 10 μM dextromethorphan did not cause any changes in the membrane current when administered alone. This suggests that dextromethorphan may not act agonistically on the 5-HT$_{1A}$ receptor. Surprisingly, dextromethorphan potently inhibited $I_{5\text{-HT}}$ (Ishibashi et al. 2000), with an IC_{50} of 14.3 μM and a Hill coefficient of 1.0. The effect of dextromethorphan on the maximum 5-HT response and the results obtained from the Lineweaver–Burk plot indicate that dextromethorphan inhibition was noncompetitive (Ishibashi et al. 2000).

4.4 Inhibitory Action on GIRK Channel

When 5-HT$_{1A}$ receptor agonists were applied to the recorded dorsal raphe neurons in the brain slice and acutely dissociated cell preparation, they caused hyperpolarization and cessation of spontaneous firing. 5-HT$_{1A}$ receptors are one of the G-protein-coupled receptors (GPCRs) and their activation induces inwardly rectifying K^+ channels through the direct action of G-protein subunits, Gβγ. This effect is caused by the activation of GIRK channels. It has been shown that intracellular perfusion with GTPγS, an irreversible activator of the G-protein subunit Gα, causes irreversible 5-HT$_{1A}$-receptor-mediated activation of GIRK channels, in the conventional whole-cell recording mode. The irreversible activation of K^+ current by intracellular perfusion with GTPγS reminded us of the direct activation of K^+ channels by G-protein, and enabled us to determine whether GIRK channels could be blocked by dextromethorphan even in the absence of an agonist. To carry out this protocol, we used a recording electrode filled with pipette solution containing GTPγS. Under this condition, a brief application of 5-HT induced a continuous and almost irreversible inward K^+ current. Dextromethorphan markedly inhibited the current activated by intracellular GTPγS in the absence of 5-HT (Ishibashi et al. 2000), whereas spiperone, a 5-HT$_{1A}$ receptor antagonist, had no effect on this current. This result confirmed that dextromethorphan had no effect on the 5-HT_{1A} receptor, making it different from receptor antagonists such as spiperone. Here, there appeared to be two possible explanations for the inhibition of GTPγS-activated currents by dextromethorphan, because it has been reported that the activation of GIRK channels is mediated by the G-protein βγ subunit. The first possible mechanism is blockade of the GIRK channel and the other is blockade of the G-protein-mediated activation of the channel by inhibiting the action of the G-protein βγ subunit. Although the site of action of dextromethorphan is not clearly indicated by available evidence, the fact that the onset and the offset of the dextromethorphan response were relatively rapid suggests that dextromethorphan may act at the level of the GIRK channel (Ishibashi et al. 2000). To clarify the site of inhibitory action of antitussives such as dextromethorphan on GIRK-channel-activated currents at the molecular level, we further performed additional studies. Recently, our own preliminary study has suggested that the site of action of dextromethorphan, one of the antitussives inhibiting the GIRK-channel-activated current, might be in the intracellular region of the GIRK channel, because such a drug had more potent action in low-pH solution, where it exists in the molecular form and not in the ionic form. Furthermore, in *Xenopus* oocytes expressing GIRK channels formed by GIRK1 and GIRK2 subunits, in which GPCRs such as the 5-HT_{1A} receptor are not expressed, the antitussive cloperastin inhibited the current activated by ethanol, which is known to activate the GIRK channel directly, suggesting that cloperastine may directly inhibit GIRK channel activity.

At this point, we felt it could be instructive to determine whether dextromethorphan also inhibits currents caused by activation of other neurotransmitter receptors coupled to G-protein, because GIRK channels are coupled to various GPCRs and other receptors, including adrenergic α_2 and $GABA_B$ receptors. Moreover, it is known that GIRK channels are heterotetrameric channels with two or more subunit isoforms, although little is known about such isoforms coupled to GPCRs such as 5-HT_{1A} receptors and α_2 adrenoceptors. In this situation, we felt that it was worth determining whether dextromethorphan also inhibits currents resulting from activation of other neurotransmitter receptors coupled to G-protein. Therefore, we examined the effect of dextromethorphan on a norepinephrine-induced current, which is mediated by α_2 adrenoceptors (Arima et al. 1998), in locus coeruleus neurons. We found that dextromethorphan reversibly inhibits the norepinephrine-induced current, with rapid onset and offset. The IC_{50} of dextromethorphan for the norepinephrine-induced current was very similar to that for 5-HT-induced currents in dorsal raphe neurons. Although it is known that GIRK channels are heterotetrameric channels with two or more subunit isoforms, little is known about GIRK channels coupled to 5-HT_{1A} receptors and α_2 adrenoceptors. The abovementioned results indicate that dextromethorphan has no specific effect on GIRK channels coupled to 5-HT_{1A} receptors in dorsal raphe neurons.

An important question at this point is whether blockade of GIRK channels by antitussives causes inhibition of coughing. Interestingly, all centrally acting antitussives studied inhibited $I_{5\text{-HT}}$ caused by 5-HT_{1A} receptor activation. IC_{50} values of various antitussives for the current caused by 0.1 μM 5-HT ranged from 1 to 10 μM.

Also, currents irreversibly activated by intracellular perfusion with GTPγS were also inhibited. Interestingly, inhibitory ratios of antitussives studied for $I_{5\text{-HT}}$ were comparable to those for the current caused by 5-HT under intracellular perfusion with GTPγS. This finding seems to suggest that inhibitory action of these antitussives on 5-HT-induced currents may be caused mostly by inhibition of GIRK channel activity. Furthermore, it may be a significant finding that antitussives such as dextromethorphan inhibit GIRK channels at relatively low concentrations, because only a few GIRK channel blockers are currently available.

Many antitussives have respiratory depressant action in experimental animals. In addition, it is well known that activation of serotonergic neurons in the brain depresses respiration. Therefore, we thought it would be helpful to determine whether general anesthetics such as pentobarbital sodium, which has potent respiratory depressant action, have inhibitory action on GIRK-channel-activated currents. Pentobarbital sodium had little effect on 5-HT-induced current, suggesting that it had a negligible effect on GIRK channels. Thus, it is likely that inhibition of GIRK channels by antitussive drugs contributes to antitussive action, although the mechanism by which inhibition of GIRK channels affects coughing remains unclear.

As described above, the results obtained from pharmacological studies using 5-HT_{1A} receptor agonists and antagonists suggest that direct activation of the 5-HT_{1A} receptor may inhibit coughing. However, a patch-clamp study using acutely dissociated single neurons revealed that centrally acting antitussives, including dextromethorphan, do not act directly on the 5-HT1A receptor but rather block GIRK channels coupled to the 5-HT_{1A} receptor and α_2 adrenoceptor via G-protein. This finding strongly indicates that centrally acting antitussives inhibit manifestation of the function of neurons induced by 5-HT_{1A} receptor activation. Because activation of 5-HT_{1A} receptors causes inhibition of neurons through activation of coupled GIRK channels, dextromethorphan should cause activation of neurons possessing 5-HT_{1A} receptors through inhibition of GIRK channels. Although these findings of pharmacological and patch-clamp studies appear contradictory, the following discussion and findings may resolve this dilemma.

4.4.1 Significance of Inhibitory Action on GIRK Channels

The results of a competitive binding assay indicate that dextromethorphan has no effect on 5-HT_{1A} receptors (Craviso and Musacchio 1983), and this finding is consistent with the results of our own patch-clamp study. It is well known that serotonergic neurons often have autoreceptors that may have an inhibitory effect on neurons. It has also been reported that the subtypes of autoreceptors in serotonergic neurons are 5-HT_{1A} receptors. The agonist sensitivity of autoreceptors is also known to be higher than that of postsynaptic receptors. Furthermore, activation of 5-HT_{1A} receptors causes hyperpolarization of these neurons. As shown in Fig. 3, we also demonstrated 5-HT-induced hyperpolarization in single dorsal raphe neurons. Therefore, a reasonable working hypothesis is that the inhibitory effect of dextromethorphan on GIRK channels coupled to 5-HT_{1A} receptors blocks feedback

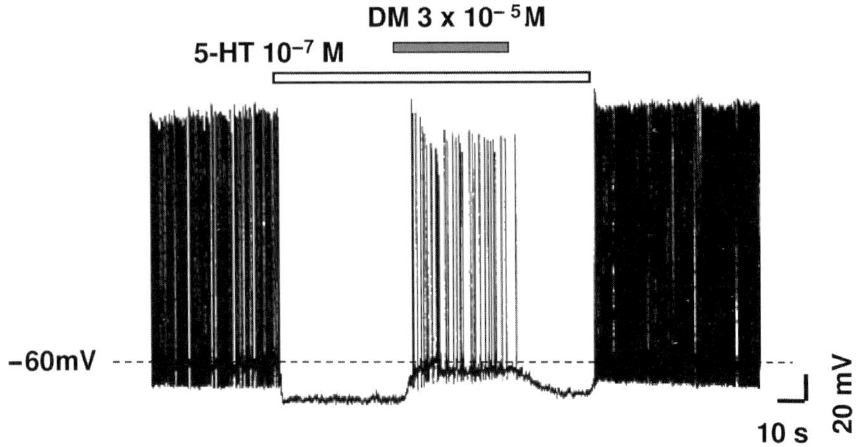

Fig. 3 5-Hydroxytryptamine (5-HT)-induced inhibition and its reactivation by dextromethorphan in a single neuron acutely dissociated from the raphe nucleus. Note that 10^{-7} M 5-HT induced hyperpolarization of the membrane potential followed by complete blocking of the occurrence of action potentials. Dextromethorphan at 3×10^{-5} M reactivated 5-HT-induced inhibition of the neuron

inhibition in serotonergic neurons, in turn augmenting the release of 5-HT. Certainly, dextromethorphan reactivated 5-HT-induced inhibition in dorsal raphe neurons containing 5-HT (Fig. 3). In fact, it has been reported that dextromethorphan and dihydrocodeine enhance the release of 5-HT from slices prepared from the NTS, an important relay center for modulating cough (Kamei et al. 1992b). Our preliminary study directly demonstrated that cloperastin, one of the centrally acting nonnarcotic antitussives, increases the 5-HT level in the cortex of rats as determined using a microdialysis probe.

In addition to 5-HT_{1A} receptor agonists, a few agonists for various pharmacological receptors have been reported to have antitussive effect in experimental animals. At this point, we will discuss the involvement of GIRK channels on the antitussive effect of these receptor agonists.

Bolser et al. have reported that $GABA_B$ receptor agonists such as baclofen and 3-aminopropylphosphinic acid have antitussive activity in cats and guinea pigs. On the basis of their study of these antagonists and comparison of equieffective doses administered via veins and arteries, they concluded that baclofen inhibits coughing by activating CNS $GABA_B$ receptors, whereas 3-aminopropylphosphinic acid inhibits coughing by acting at peripheral sites (Bolser et al. 1994). $GABA_B$ agonists acting on presynaptic receptors are known to inhibit glutamate exocytosis from synaptosomes, primary neuronal cultures and brain slices (Nicholls and Sanchez-Prieto 1998). It is likely that the inhibition of glutamate exocytosis results in the antitussive effect of baclofen. Also, it is important to remember that NMDA receptor antagonists have antitussive action in experimental animals (Kamei et al. 1989), although not all centrally acting antitussives block NMDA receptors. On the other hand, it has been reported that $GABA_B$ receptors couple to GIRK channels. If antitussives inhibit GIRK channels coupled to $GABA_B$ receptors, they should affect the GABA level in the brain, because $GABA_B$ receptors are also presynaptic autoreceptors. In turn, cough responses may be modulated via changes in the GABA level in the brain. We have already found that antitussives inhibit the current activated by baclofen, a $GABA_B$ receptor agonist, probably by inhibiting GIRK channels.

Recently, Kamei et al. reported that a δ-receptor antagonist exerts antitussive action via a CNS mechanism (Kamei et al. 1993a, b). However, Kotzer et al. recently reported that the selective δ-receptor antagonist SB 244525 has little antitussive effect in guinea pigs when administered alone (Kotzer et al. 2000). Surprisingly, Kotzer et al. have reported that the selective δ-receptor agonist SB 227122 has antitussive effect in guinea pigs (Kotzer et al. 2000), which is not compatible with results reported by Kamei et al. (1993a, b). As mentioned above, they have also reported that the selective SB 244525 has little antitussive effect in guinea pigs when administered alone. It has also been reported that the δ1-receptor agonist DPDPE has a negligible effect on experimentally induced cough in animals, but has an inhibitory action on the antitussive effect of the μ-receptor agonist DAMGO (Kamei et al. 1993a, b). In contrast, the δ_2-receptor antagonist DELT-II potentiates the antitussive effect of DAMGO (Kamei et al. 1993a, b). Studies of the involvement of δ-receptors in antitussive activity have produced some conflicting results. On the basis of our own findings regarding the relationship between antitussive drugs and GIRK channel activities, we hypothesized that the abovementioned conflicting results concerning agonists and antagonists of δ-receptors might be resolved by determining the effect of these substances on GIRK channel activities. As we expected, in our own study, the δ-receptor antagonists naltrindole and naltriben, known to have antitussive actions, both inhibited GIRK channels in single brain neurons (Shirasaki et al. 2004), although relatively large concentrations were needed to inhibit neuronal currents.

There are also several reports suggesting the involvement of σ-receptors in the mechanisms of action of antitussive drugs (Kamei et al. 1992a, b). This suggestion is supported by the finding that there is a high-affinity binding site for dextromethorphan on σ-receptors (Klein et al. 1989). Other studies suggest that σ-receptors are associated with K^+ channels, because some σ-ligands modify K^+ conductance in some tissues. Aside from these findings, it has also been reported that σ-ligands inhibit ligand-induced hyperpolarization mediated by α_2 adrenoceptors and μ-opioid receptors in locus coeruleus neurons (Bobker et al. 1989). As described above, dextromethorphan inhibits not only $I_{5\text{-HT}}$ in dorsal raphe neurons but also α_2 adrenoceptor-mediated currents in locus coeruleus neurons. α_2-Adrenoceptors in locus coeruleus neurons are coupled to GIRK channels (Arima et al. 1998); therefore, it is possible that σ-ligands exert antitussive action at least partly through inhibition of GIRK channels. There is a report that some σ-ligands modify K^+ conductance in some tissues. Moreover, a patch-clamp study by Nguyen et al. showed no correlation between binding affinities of antitussives at σ_1- or σ_2-binding sites and potency of inhibition of K^+ currents (Neuyen et al. 1998).

Coughing is regulated by higher brain areas, and is under voluntary control to some extent. It may be reasonable to speculate that the action of antitussive drugs is facilitated, at least in part, by a placebo effect (Eccles 2002), since low doses of some antitussives suppress cough with potency slightly greater than that of a placebo.

Recently, it has been reported that antitussives may act, at least in part, at the cortical level. Our microelectrophoretic study showed that all the antitussives studied inhibit spontaneous unit activities recorded from the pyramidal layer of the cerebral cortex of guinea pigs in vivo. To the best of our knowledge, the mechanism of inhibition of unit activities in the cerebral cortex has not yet been studied. However, one of its possible effects is the inhibition of GIRK channels, because GIRK channels are distributed in various regions of the brain, including the cerebral cortex (Ponce et al. 1996).

4.5 NMDA and NK_1 Receptors

4.5.1 NMDA Receptor

Glutamate often acts as an excitatory neurotransmitter in primary afferent neurons, including Aδ-fibers. Netzer et al. have reported that dextromethorphan inhibits NMDA-induced currents in brain neurons (Netzer et al. 1993). We have confirmed this finding, and in addition we found that dextromethorphan had no effect on kainate-induced current in single brain neurons. Codeine did not inhibit NMDA-induced current, although dextromethorphan and codeine both have a morphinan structure. For other antitussives, eprazinone also had no effect on the NMDA-induced current. These findings indicate that blockade of NMDA receptors is not essential for inducing antitussive effects. However, it has been reported that MK-801, an NMDA receptor antagonist, has an antitussive effect in experimental animals. Dextromethorphan showed a relatively potent inhibitory action on the NMDA-induced current. Sometimes it is said that among the centrally acting nonnarcotic antitussives dextromethorphan might be the most effective in inhibiting cough in the clinical setting. If this is true, the effect may be caused by multiplex actions, that is, inhibitory actions on GIRK-channel-activated, NMDA-induced, and glycine-induced currents.

4.5.2 SP and Its Receptor

SP has been reported to stimulate coughing through activation of NK_1 receptors in the airway, since inhalation of aerosol of SP solution induced coughing in guinea pigs. However, it is likely that SP may also be involved in signal transduction in the central pathway of the cough reflex. SP-containing nerve terminals are found in the NTS at high density (Kawai et al. 1989). SP receptors localize in postsynaptic neurons of these nerve terminals in the NTS. Studies using guinea pigs have found that SP in the NTS augments bronchopulmonary C-fiber reflex output, causing bronchoconstriction and changes in respiration, including coughing (Mutoh et al. 2000).

Although it is unknown whether clinically available centrally acting antitussives act on NK_1 receptors in the NTS, it seems likely that drugs which block NK_1 receptors in the NTS have antitussive action.

An interesting finding has been reported by Sasaki's group (Yamaya et al. 2001) regarding the relationship between SP and coughing. They first observed a high risk of pneumonia due to aspiration of oropharyngeal bacterial pathogens in older adults. Interestingly, an increase in the SP level significantly lowered the risk of pneumonia, apparently via recovery (or strengthening) of coughing and swallowing reflexes. SP is synthesized in nodose and jugular ganglia, and is transported peripherally and released in the airway. Increased release of SP is involved in the development of cough and neurogenic inflammation of the airway, and production of SP is regulated by dopaminergic neurons in the brain stem. It is reported that L-dopa, a precursor of dopamine, also inhibits manifestation of aspiration pneumonia, suggesting that L-dopa also strengthens the coughing reflex, although treatment with α-methyl-*p*-tyrosine, which depletes catecholamine in neurons, had little effect on the activity of antitussive drugs.

There is considerable morphological and physiological evidence that SP is also released at central synapses in the NTS (Mutoh et al. 2000). This suggests that NK_1 receptors in the NTS can serve as target receptors for new cough-remedy drugs. Moreover, cough stimulants such as capsaicin may lower the incidence of pneumonia in older people through improvement in upper protective respiratory reflexes such as the swallowing reflex and cough reflex in older people with a high risk for aspiration (Ebihara et al. 2005). Finally, studies of NK_1 receptors and the glycine receptor may provide useful data for developing a "cough recovery (or strengthening)"drug for older adults with decreased coughing and swallowing reflexes, which increase susceptibility to aspiration pneumonia.

5 Conclusion

At present, mechanisms of action of centrally acting antitussives still remain largely unknown. As mentioned earlier, various substances (5-HT_{1A} receptor agonist, δ-receptor agonist and antagonist, $GABA_B$ receptor agonist, and NMDA receptor antagonists, among others) show antitussive action in experimental animals. In addition, the chemical structures of centrally acting antitussives are very diverse, compared with those of other centrally acting drugs such as antianxiety drugs, antipsychotic drugs, and others. This seems to recall an idea that mechanisms of action of centrally acting antitussives are multiple and very complex. However, it should be noted that nonnarcotic antitussives used in the clinic and various test substances possessing antitussive action have a common pharmacological feature, that is, inhibitory action on GIRK-channel-activated currents in single brain neurons (Table 1). Kase proposed a "piperidino group theory" for structure–activity relationship of antitussive drugs (Kase and Yuizono 1959; Kase et al. 1963). In the theory, introduction of a piperidino group to a chemical substance with existing antitussive

Table 1 Effects of antitussive drugs on receptor-mediated or ion-channel-activated currents in single brain neurons

	Narcotic	Nonnarcotic		Others[a]	δ-Antagonist/
	Codeine	Dextromethorphan	Cloperastine		σ-agonist
I_{Gly}	↓↓	↓↓	→	↓	ND
I_{GABA}	→	→	→	→	ND
I_{NMDA}	→	↓↓	→	→	ND
$I_{Kainate}$	ND	→	ND	ND	ND
I_{5-HT}	↓	↓↓	↓↓↓	↓↓	↓↓
I_{NE}	ND	↓↓	↓↓↓	ND	↓↓
I_{Na}	ND	ND	↓	ND	ND
I_A	ND	ND	↓	ND	ND
I_{KD}	ND	ND	↓	ND	ND
GIRK channel	↓	↓↓	↓↓↓	↓↓	↓↓

Downward arrows indicate the magnitude of the inhibitory effect: the greater the number of arrows, the greater the inhibitory action on the currents. I_{Gly}, I_{GABA}, I_{NMDA}, $I_{Kainate}$, I_{5-HT}, and I_{NE} are the current induced by glycine, γ-aminobutyric acid, *N*-methyl-D-aspartate, kainate, 5-hydroxytryptamine, and norepinephrine, respectively. I_{Na}, I_A, and I_{KD} are Na^+ current, fast transient K^+ current, and delayed rectified K^+ current, respectively.
GIRK G-protein-coupled inwardly rectifying potassium, *ND* not determined
[a]For example, eprazinone (Takahama 2003 modified)

activity strengthens the activity, whereas introduction of the group to a substance without antitussive activity induces the activity. In our preliminary study, we found that introduction of a piperidino group to a chemical substance possessing antitussive activity strengthens its inhibitory action on GIRK-channel-activated currents. However, it seems insufficient to conclude that the mechanism of the action of centrally acting antitussives is based on its inhibitory action on GIRK-channel-activated currents. In the course of the pharmacological study of antitussive drugs, we found very interesting findings that centrally acting antitussives have multiple pharmacological effects in animal models of various diseases of the CNS. On the basis of findings from our own studies, we hypothesize that centrally acting antitussives with inhibitory action on GIRK-channel-activated currents may have a *stabilizing action* on brain function disturbed by various factors. We considered coughing as a state in which the function of neurons involved in cough responses may be "disturbed" (or activated) by neuronal inputs from the airway. In this context, it is considered that centrally acting antitussives would be drugs that stabilize the "disturbed" (or activated) function of neurons involved in cough responses, resulting in their antitussive effect. Therefore, possibilities remain with regard to definitive molecular targets for centrally acting antitussive drugs, although this idea is very speculative. If a drug has multi-inhibitory actions, one on GIRK-channel-activated currents as a basic action and another on a specific molecular target, it might produce a very effective antitussive activity. Additional studies are needed to further clarify essential molecules involved in cough responses.

Acknowledgements The authors thank T. Miyata, H. Kai, Y. Isohama, and H. Ishibashi, Graduate School of Pharmaceutical Sciences, Kumamoto University, for their collaboration. The authors also thank the following graduate students for their assistance: H. Honda, H. Fukushima, M. Otsuka, S. Nagayama, and Y. Terasako, Department of Pharmacology, and K. Kuwano, K. Abe, K. Yamasaki, K. Hashitani, and M. Kinoshita, Department of Environmental and Molecular Health Sciences, Graduate School of Pharmaceutical Sciences, Kumamoto University. The authors also sincerely thank the late Dr. Y. Kase, Emeritus Professor of Kumamoto University, for his continuous encouragement during his lifetime.

References

Aicher SA, Goldberg A, Sharma S, Pickel VM (2000) Mu-opioid receptors are present in vagal afferents and their dendritic targets in the medial nucleus tractus solitarius. J Comp Neurol 422:181–190

Arima J, Kubo C, Ishibashi H, Akaike N (1998) α_2-Adrenoceptor-mediated potassium currents in acutely dissociated rat locus coeuleus neurones. J Physiol (Lond) 508:57–66

Beakey DM, Morris K, Gestreau C, Li Z, Lindsey B, Shannon R (2001) Medullary respiratory neurones and control of laryngeal motoneurones during fictive eupnoea and cough in the cat. J Physiol 534:565–581

Belvisi MG, Ichinose M, Barnes PJ (1989) Modulation of non-adrenergic, non-cholinergic neural bronchoconstriction in guinea-pig airways via $GABA_B$-receptors. Br J Pharmacol 97: 1225–1231

Bobker DH, Shen K-Z, Surprenant A, William JT (1989) DTG and (+)-3PPP inhibit a ligand-activated hyperpolarization in mammalian neurons. J Pharmacol Exp Ther 251:840–845

Bolser DC, DeGennaro FC, O'Reilly S, Chapman RW, Kreutner W, Egan RW Hey JA (1994) Peripheral and central sites of action of GABA-B agonists to inhibit the cough reflex in the cat and guinea pig. Br J Pharmacol 113:1344–1348

Canning BJ, Mazzone SB, Meeker SN, Mori N, Reynolds SM, Undem BJ (2004) Identification of the tracheal and laryngeal afferent neurones mediating cough in anaesthetized guinea-pigs. J Physiol 557:543–558

Carr MJ, Hunter DD, Jacoby DB, Undem BJ (2002) Expression of tachykinins in nonnociceptive vagal afferent neurons during respiratory viral infection in guinea pigs. Am J Respir Crit Care Med 165:1071–1075

Cheng PY, Liu-Chen LY, Chen C, Pickel VM (1996) Immunolabeling of Mu opioid receptors in the rat nucleus of the solitary tract: Extrasynaptic plasmalemmal localization and association with Leu5-enkephalin. J Comp Neurol 371:522–536

Craviso G, Musacchio JM (1983) High affinity dextromethorphan binding sites in guinea pig brain, competition experiments. Mol Pharmacol 23:629–640

Dashwood MR, Muddle JR, Spyer KM (1988) Opiate receptor subtypes in the nucleus tractus solitarii of the cat: The effect of vagal section. Eur J Pharmacol 155:85–92

Dinh QT, Groneberg DA, Peiser C, Springer J, Joachim RA, Arck PC, Klapp BF, Fischer A (2004) Nerve growth factor-induced substance P in capsaicin-insensitive vagal neurons innervating the lower mouse airway. Clin Exp Allergy 34:1474–1479

Dinh QT, Mingomataj E, Quarcoo D, Groneberg DA, Witt C, Klapp BF, Braun A, Fischer A (2005) Allergic airway inflammation induces tachykinin peptides expression in vagal sensory neurons innervating mouse airways. Clin Exp Allergy 35:820–825

Ebihara T, Takahashi T, Ebihara S, Okazaki T, Sasaki T, Watando A, Nemoto M, Sasaki H (2005) Capsaicin troche for swallowing dysfunction in older people. J Am Geriatr Soc 53:824–828

Eccles R (2002) The powerful placebo. Do we need antitussives? Pulm Pharmacol Ther 15: 123–131

Fox AJ, Lalloo UG, Belvisi MG, Bernareggi M, Chung KF, Barnes PJ (1996) Bradykinin-evoked sensitization of airway sensory nerves: A mechanism for ACE-inhibitor cough. Nat Med 2: 814–817

Frossard N, Barnes PJ (1987) Mu-opioid receptors modulate non-cholinergic constrictor nerves in guinea-pig airways. Eur J Pharmacol 141:519–522

Fukushima K, Nagayama S, Otsuka M, Takahama K, Isohama Y, Kai H, Miyata T (1998) Inhibition of glycine-induced current by morphine in nucleus tractus solitarii neurones of guinea pigs. Meth Find Exp Clin Pharmacol 20:125–132

Glatzer NR, Smith BN (2005) Modulation of synaptic transmission in the rat nucleus of the solitary tract by endomorphin-1. J Neurophysiol 93:2530–2540

Groneberg DA, Niimi A, Dinh QT, Cosio B, Hew M, Fischer A, Chung KF (2004) Increased expression of transient receptor potential vanilloid-1 in airway nerves of chronic cough. Am J Respir Crit Care Med 170:1276–1280

Gu Q, Lee LY (2006) Characterization of acid signaling in rat vagal pulmonary sensory neurons. Am J Physiol Lung Cell Mol Physiol 291:L58–L65

Hassen AH, Feuerstein G, Faden AI (1982) μ Receptors and opioid cardiovascular effects in the NTS of rat. Peptides 3:1031–1037

Helke CJ, Jacobowitz DM, Thoa NB (1981) Capsaicin and potassium evoked substance P release from the nucleus tractus solitarius and spinal trigeminal nucleus in vitro. Life Sci 29:1779–1785

Honda H, Takahama K, Kawaguchi T, Fuchikami J, Kai H, Miyata T (1990) Involvement of *N*-methyl-D-aspartic acid (NMDA) receptor in the central mechanism of cough reflex. Jpn J Pharmacol 52(Suppl. I):308

Ishibashi H, Kuwano K, Takahama K (2000) Inhibition of the 5-HT_{1A}receptor-mediated inwardly rectifying K^+current by dextromethorphan in rat dorsal raphe neurones. Neuropharmacology 39:2302–2308

Joad JP, Munch PA, Bric JM, Evans SJ, Pinkerton KE, Chen CY, Bonham AC (2004) Passive smoke effects on cough and airways in young guinea pigs: Role of brainstem substance P. Am J Respir Crit Care Med 169:499–504

Kamei J, Hosokawa T, Yanaura S, Hukuhara T (1986) Effects of methysergide on the cough reflex. Jpn J Pharmacol 42:450–452

Kamei J, Ogawa M, Kasuya Y (1987) Monoamine and the mechanisms of action of antitussive drugs in rats. Arch Int Pharmacodyn 290:117–127

Kamei J, Tanihara H, Igarashi H, Kasuya Y (1989) Effects of *N*-methyl-D-aspartate antagonists on the cough reflex. Eur J Pharmacol 168:153–158

Kamei J, Mori T, Igarashi H, Kasuya Y (1991) Effects of 8-hydroxy-2-(di-n-propylamino)tetralin, a selective agonist of 5-HT_{1A} receptors, on the cough reflex in rats. Eur J Pharamcol 203: 253–258

Kamei J, Iwamoto Y, Misawa M, Kasuya Y (1992a) Involvement of haloperidol-sensitive σ-sites in antitussive effects. Eur J Pharmacol 224:39–43

Kamei J, Mori T, Igarashi H, Kasuya Y (1992b) Serotonin release in the nucleus of the solitary tract and its modulation by antitussive drugs. Res Commun Chem Pathol Pharmacol 76:371–374

Kamei J, Iwamoto Y, Suzuki T, Misawa M, Nagase H, Kasuya Y (1993a) Antitussive effect of naltrindole, a selective δ-opioid receptor antagonist, in mice and rats. Eur J Pharmacol 249: 161–165

Kamei J, Iwamoto Y, Suzuki T, Nagase H, Misawa M, Kasuya Y (1993b) Differential modulation of μ-opioid receptor-mediated antitussive activity by δ-opioid receptor agonists in mice. Eur J Pharmacol 234:117–120

Kamei J, Morita K, Kashiwazaki T, Ohsawa M (1998) Antitussive effect of moguisteine on allergic coughs in the guinea pig. Eur J Pharmacol 347:253–255

Kamikawa Y, Shimo Y (1990) Morphine and opioid peptides selectively inhibit the non-cholinergically mediated neurogenic contraction of guinea-pig isolated bronchial muscle. J Pharm Pharmacol 42:214–216

Kase Y (1980) Antitussive agents and their sites of action. Trends Pharmacol Sci 1:237–239

Kase Y, Yuizono T (1959) A contribution of piperidino group to manifestation of antitussive activity. Chem Pharm Bull 7:378–382

Kase Y, Yuizono T, Muto M (1963) Piperidino groups in antitussive activity. J Med Chem 6: 118–122

Kawai Y, Mori S, Takagi H (1989) Vagal afferents interact with substance P -immunoreactive structures in the nucleus of the tractus solitarius: Immunoelectron microscopy combined with an anterograde degeneration study. Neurosci Lett 101:6–10

Killingsworth CR, Paulauskis JD, Shore SA (1996) Substance P content and preprotachykinin gene-I mRNA expression in a rat model of chronic bronchitis. Am J Respir Cell Mol Biol 14:334–340

Klein M, Paturzo JJ, Musacchio JM (1989) The effects of prototypic σ-ligands on the binding of [^{3}H]dextromethorphan to guinea pig brain. Neurosci Lett 97:175–180

Kotzer CJ, Hay DWP, Dondio G, Giardina G, Petrillo P, Underwood DC (2000) The antitussive activity of δ-opioid receptor stimulation in guinea pigs. J Pharmacol Exp Ther 292:803–809

Kubo T, Kihara M (1987) Evidence for the presence of GABAergic and glycine-like systems responsible for cardiovascular control in the nucleus tractus solitarii of the rat. Neurosci Lett 74:331–336

Lawrence AJ (1995) Neurotransmitter mechanisms of rat vagal afferent neurons. Clin Exp Pharmacol Physiol 22:869–873

Lee SY, Kim MK, Shin C, Shim JJ, Kim HK, Kang KH, Yoo SH, In KH (2003) Substance P-immunoreactive nerves in endobronchial biopsies in cough-variant asthma and classic asthma. Respiration 70:49–53

Matsas R, Kenny AJ, Turner AJ (1984) The metabolism of neuropeptides. The hydrolysis of peptides, including enkephalins, tachykinins and their analogues, by endopeptidase-24.11. Biochem J 223:433–440

Meeley MP, Underwood MD, Talman WT, Reis DJ (1989) Content and in vitro release of endogenous amino acids in the area of the nucleus of the solitary tract of the rat. J Neurochem 53:1807–1817

Miyata T, Isohama Y, Tai S, Kai H, Takahama K (1999) Pharmacological evaluation of Bakumondo-to (Mai-men-dong-tang) as a curative for chronic inflammatory airway disease. In: Watanabe H, Shibuya T (eds) Pharmacological research on traditional herbal medicine. Harwood academic Publishing, Amsterdam, The Netherlands, pp 121–147

Morin-Surun MP, Jordan D, Champagnat J, Spyer KM, Denavit-Saubie M (1984) Excitatory effects of iontophoretically applied substance P on neurons in the nucleus tractus solitarius of the cat: Lack of interaction with opiates and opioids. Brain Res 307:388–392

Mutoh T, Bonham AC, Joad JP (2000) Substance P in the nucleus of the solitary tract augments bronchopulmonary C fiber reflex output. Am J Physiol Regul Integr Comp Physiol 279:R1215–R1223

Myers AC, Kajekar R, Undem BJ (2002) Allergic inflammation-induced neuropeptide production in rapidly adapting afferent nerves in guinea pig airways. Am J Physiol Lung Cell Mol Physiol 282:L775–L781

Myou S, Fujimura M, Kurashima K, Kita T, Tachibana H, Ishiura Y, Abo M, Nakao S (2001) Effects of suplatast tosilate, a new type of anti-allergic agent, on airway cough hypersensitivity induced by airway allergy in guinea-pigs. Clin Exp Allergy 31:1939–1944

Netzer R, Pflimlin P, Trube G (1993) Dextromethorphan blocks N-methyl D-aspartate-induced currents and voltage-operated inward currents in cultured cortical neurons. Eur J Pharmacol 238:209–216

Neuyen VH, Ingram SL, Kassiou M, Christie MJ (1998) σ-Binding site ligands inhibit K^+ currents in rat locus coeruleus neurons in vitro. Eur J Pharmacol 361:157–163

Nicholls DG, Sanchez-Prieto J (1998) Neurotransmitter release mechanism. In: Anne Stephenson F, Turner AJ (eds) Frontiers in neurobiology 3: Amino acid neurotransmission. Portland Press, London, pp 1–24

Ponce A, Bueno E, Kentros C, Vega-Saenz de Miera E, Chow A, Hillman D, Chen S, Zhu L, Wu MB, Wu X, Rudy B, Thornhill WB (1996) G-protein-gated inward rectifier K^+ channel proteins (GIRK1) are present in the soma and dendrites as well as in nerve terminals of specific neurons in the brain. J Neurosci 16:1990–2001

Rhim H, Glaum SR, Miller RJ (1993) Selective opioid agonists modulate afferent transmission in the rat nucleus tractus solitarius. J Pharmacol Exp Ther 264:795–800

Ricco MM, Kummer W, Biglari B, Myers AC, Undem BJ (1996) Interganglionic segregation of distinct vagal afferent fibre phenotypes in guinea-pig airways. J Physiol 496:521–530

Rusin KI, Moises HC (1998) Mu-opioid and $GABA_B$ receptors modulate different types of Ca^{2+} currents in rat nodose ganglion neurons. Neuroscience 85:939–956

Saria A, Martling CR, Yan Z, Theodorsson-Norheim E, Gamse R, Lundberg JM (1988) Release of multiple tachykinins from capsaicin-sensitive sensory nerves in the lung by bradykinin, histamine, dimethylphenyl piperazinium, and vagal nerve stimulation. Am Rev Respir Dis 137:1330–1335

Shirasaki T, Abe K, Soeda F, Takahama K (2004) δ-Opiod receptor antagonists inhibit GIRK channel currents in acutely dissociated brainstem neurons of rat. Brain Res 1006:190–197

Stone RA, Barnes PJ, Chung KF (1997) Effect of 5-HT1A receptor agonist, 8-OH-DPAT, on cough responses in the conscious guinea pig. Eur J Pharmacol 332:201–207

Takahama K (2003) Mechanisms of actions of centrally acting antitussive – electrophysiological and neurochemical analysis. In: Chung KF, Widdicombe JG, Boushey HA (eds) Cough: Causes, mechanisms and therapy. Blackwell Publishing, Oxford, UK, pp 225–236

Takahama K, Miyata T (1995) Cough-diversity and the peripheral mechanisms of production. Folia Pharmacol Jpn 105:41–52

Takahama K, Shirasaki T (2007) Central and peripheral mechanisms of narcotic antitussives: Codeine-sensitive and -resistant coughs. Cough 3:8 (http://www.coughjournal.com/content/3/1/8)

Takahama K, Fuchikami J, Kai H, Isohama Y, Miyata T (1995) Inhalation of phosphoramidon, a neutral endopeptidase inhibitor, induces cough in awake guinea-pigs. Arch Int Pharmacodyn Ther 330:241–250

Takahama K, Wakuda I, Fukushima H, Isohama Y, Kai H, Miyata T (1997a) Differential effect of codeine on coughs caused by mechanical stimulation of two different sites in the airway of guinea pigs. Eur J Pharmacol 329:93–97

Takahama K, Fukushima H, Isohama Y, Kai H, Miyata T (1997b) Inhibition of glycine currents by dextromethorphan in neurones dissociated from the guinea-pig nucleus tractus solitarii. Br J Pharmacol 120:690–694

Undem BJ, Hunter DD, Liu M, Haak-Frendscho M, Oakragly A, Fischer A (1999) Allergen-induced sensory neuroplasticity in airways. Int Arch Allergy Immunol 118:150–153

Undem BJ, Chuaychoo B, Lee MG, Weinreich D, Myers AC, Kollarik M (2004) Subtypes of vagal afferent C-fibres in guinea-pig lungs. J Physiol 556:905–917

Wakuda I, Takahama K, Isohama Y, Miyata T (1994) Pharmacological discrimination of cough reflexes induced by topically applied chemical stimuli to the guinea pig airways. Jpn J Pharmacol 46:148

Widdicombe JG (1954) Respiratory reflexes from the trachea and bronchi of the cat. J Physiol 123:55–70

Widdicombe JG (2002) Neuroregulation of cough: Implications for drug therapy. Curr Opin Pharmacol 2:256–263

Wilson CG, Zhang Z, Bonham AC (1996) Non-NMDA receptors transmit cardiopulmonary C fibre input in nucleus tractus solitarii in rats. J Physiol 496:773–785

Winter CA, Flataker L (1955) The effects of drugs upon a graded cough response obtained in sensitized guinea pigs exposed to aerosol of specific antigen. J Exp Med 101:17–24

Xia Y, Haddad GG (1991) Ontogeny and distribution of opioid receptors in the rat brainstem. Brain Res 549:181–193

Yamaya M, Yanai M, Ohrui T, Arai H, Sasaki H (2001) Interventions to prevent pneumonia among older adults. J Am Geriatr Soc 49:85–90

Central Mechanisms IV: Conscious Control of Cough and the Placebo Effect

R. Eccles

Contents

Abstract Early animal experiments on cough developed the concept that cough was an involuntary reflex controlled from areas in the brainstem and that cough could be inhibited by centrally acting medicines such as codeine. Studies on the voluntary control of cough, the urge to cough and the placebo effect of cough medicines have demonstrated that human cough is more complex than a brainstem reflex. The efficacy and mechanism of action of centrally acting cough medicines such as codeine

R. Eccles
Common Cold Centre, Cardiff School of Biosciences, Cardiff University, Cardiff CF10 3US, UK
eccles@cardiff.ac.uk

K.F. Chung, J.G. Widdicombe (eds.), *Pharmacology and Therapeutics of Cough*,
Handbook of Experimental Pharmacology 187,

and dextromethorphan is now in dispute, and codeine is no longer accepted as a gold-standard antitussive. This review puts forward a cough model that includes three types of cough: (1) reflex cough, caused by the presence of food or fluid in the airway – this type of cough is not under conscious control and can occur in the unconscious subject during general anaesthesia; (2) voluntary cough – under conscious control that is abolished with general anaesthesia; (3) cough in response to sensation of airway irritation – this type of cough causes an urge to cough that initiates voluntary cough and may only be present in the conscious subject. The review proposes that human cough associated with respiratory disease is under conscious control and is mainly related to a sensation of airway irritation and an urge to cough (type 3). The review discusses the summation of sensory input from the airway in a brainstem integrator that reaches a threshold to cause reflex cough. Subthreshold conditions in the cough integrator may be perceived as an urge to cough that is under voluntary control. The cough model presented in the review has implications for the development of cough medicines as it indicates that the older view of cough medicines acting in the brainstem area to inhibit the cough reflex may need to be revised to include conscious control of cough as an important mechanism of cough in man.

1 Introduction

Until relatively recently cough has been viewed as an involuntary reflex controlled from the medulla, with little interest in researching other areas of the brain that may influence cough. The view that cough is a reflex mediated from areas in the brainstem has developed from the early animal studies on cough using anaesthetised animals to study cough physiology and pharmacology (May and Widdicombe 1954; Widdicombe 1954). The antitussive activity of codeine, was first established by studies on anaesthetised animals, and these early classical studies in physiology and pharmacology have had a dominant effect on our understanding of cough as a brainstem reflex (Eddy et al. 1969), right up to the present day.

The early animal experiments focussed on the control of cough from the so called cough centre in the medulla, and the control of human cough was also explained as originating from the medulla, with antitussives such as codeine having a depressant action on the cough-control areas in the medulla (May and Widdicombe 1954). Some of the early reviews on the efficacy of codeine as an antitussive (Eddy et al. 1969) did briefly acknowledge that the cough 'centre' must be subject to influences from other areas of the brain apart from the brainstem, but the focus was clearly on cough as a reflex controlled from the medulla.

In the last decade there has been a gradual change in opinion on the control of cough as the research focus has moved more towards the study of human cough. Studies on the voluntary control of cough (Lee et al. 2002), placebo effects (Eccles 2006) and the sensation of the urge to cough (Davenport et al. 2007) have helped to develop the idea that cough is much more than a reflex, and that higher centres in the brain play a major role in controlling cough.

This article will look at the conscious control of cough and the placebo effect, and will focus mainly on human cough. The article will conclude with a model of cough control that attempts to incorporate recent developments in these areas.

2 Conscious and Unconscious Control of Cough

It is self-evident that cough is under conscious control as cough can be initiated voluntarily and can be suppressed (Lee et al. 2002) when it is inappropriate or dangerous to cough. The conscious control of cough may have a selective advantage for survival as "the perception of the urge to cough, which precedes cough itself, may allow the animal time to activate inhibitory descending pathways to suppress cough in situations ranging from basic survival to social embarrassment in a concert hall" (Gracely et al. 2007). Our understanding of conscious cough can be increased by looking at those conditions when conscious control may be inhibited or abolished, such as in general anaesthesia or sleep.

2.1 Cough in Unconscious Conditions

Cough can occur in the unconscious human, and most of the early animal studies on cough used anaesthetised or decerebrate animals to study cough. When discussing cough in the unconscious condition it is important to distinguish those conditions where there is a general depression of the brain such as unconsciousness induced by pharmacological means, e.g. general anaesthesia and alcohol intoxication, from those conditions where unconsciousness may be due to trauma or some specific damage to the cerebral cortex and higher centres.

In a study on the relationship between the depth of coma and the cough reflex in 76 comatose patients undergoing intubation, the integrity of the cough reflex was not related to the depth of coma (Moulton and Pennycook 1994). The cough reflex was depressed only in the group of patients with deep coma due to pharmacological causes such as drug or alcohol intoxication, and those patients with coma due to head injury often had a normal cough reflex. The results of this study may be explained on the basis that coma due to pharmacological depression of the brain with drug overdose would include depression of the cough-control areas in the brainstem, but in cases of coma due to trauma, the brainstem area may not have been depressed, and reflex cough initiated on intubation would have been relatively normal. The presence of the cough reflex in the unconscious patient nicely demonstrates that consciousness is not necessary for cough to occur in response to airway stimulation.

2.2 Cough and General Anaesthesia

General anaesthesia was classically divided by Guedel in 1920 into four stages of increasing depth of anaesthesia: I analgesia, II delirium, III surgical anaesthesia and IV medullary depression (Kennedy and Longnecker 1996). This classical division of anaesthesia was believed to be related to a descending depression of the brain, with initial depression of the cerebral cortex and eventually depression of the medulla. The cough reflex is still present under light surgical general anaesthesia. Local anaesthesia of the airway is necessary in order to abolish the cough reflex rather than increase the depth of anaesthesia to a level that may also affect respiratory and cardiovascular control (Calvert et al. 1966). General anaesthetics have a very small margin of safety, and increasing concentrations of anaesthetic cause medullary depression and loss of cardiovascular and respiratory control (Kennedy and Longnecker 1996). The depression of the medulla with increasing depth of general anaesthesia causes apnoea due to depression of respiratory control centres and it is at this level of anaesthesia that abolition of the cough reflex usually occurs. However, the depth of anaesthesia associated with inhibition of the cough reflex may vary with the anaesthetic agent, as the cough reflex was one of the most sensitive respiratory reflexes to general anaesthesia with enflurane (Nishino et al. 1988). The sensitivity of the cough reflex to general anaesthesia may also depend on other properties of the general anaesthetic, such as depression of neuromuscular transmission causing relaxation of respiratory muscles (Nishino et al. 1988).

Light general anaesthesia causes analgesia before loss of consciousness (stage I anaesthesia) (Kennedy and Longnecker 1996) but it is not known if the sensation of airway irritation associated with cough is also lost with light general anaesthesia.

2.3 Cough in Sleep

Cough in patients with asthma and other causes of chronic cough is reduced during sleep (Power et al. 1984; Hsu et al. 1994). Like general anaesthesia, sleep is subject to different levels or stages and there has been relatively little work on relating the control of cough to the stage of sleep. Patients with habit cough (cf. Tourette's syndrome) do not cough when asleep and, since their daytime cough is psychogenic, the cerebral cortex is almost certainly its site of origin (Widdicombe et al. 2006). The absence of habit cough during sleep indicates that this type of cough may not involve any component of reflex cough and that the urge to cough in these patients may not be related to any sensations derived from the airway.

There have been a few studies on induced cough during sleep. Jamal et al. (1983) showed that the citric acid threshold for cough was increased during human sleep. This confirmed the classical work by Sullivan et al. (1978, 1979) with sleeping dogs; they showed that during sleep, distilled water in the larynx or trachea was a far weaker stimulus to cough, and that the cough only occurred after arousal (i.e. in a sense only after the dogs were awakened). The studies on sleep and induced cough indicate that cough due to airway irritation is not abolished during sleep and that the stimulus of airway irritation may be perceived as a noxious stimulus that causes arousal from sleep so that the higher centres can initiate conscious control of the cough.

2.4 Voluntary Initiation and Inhibition of Cough

As mentioned already, it is self-evident that cough can be initiated voluntarily and can be suppressed (Lee et al. 2002), yet this voluntary and conscious control of cough has only recently been subject to experimental studies in natural cough due to common cold (Hutchings et al. 1993a) and chemically induced cough (Hutchings et al. 1993b; Hutchings and Eccles 1994).

It is a routine part of many cough studies to ask volunteers and patients to make several voluntary coughs in order to test or calibrate the equipment used to record cough (Hutchings et al. 1993b). These voluntary coughs cannot be distinguished from any of the other coughs recorded during the experiments and this indicates that the cough pathways involved in conscious cough link up with those involved in reflex cough to give a final common pathway for activation of respiratory muscles and cough. The voluntary initiation of cough is probably controlled from the cerebral cortex and recent studies in man support this view (Simonyan et al. 2007).

In a study on 15 healthy volunteers Simonyan et al. (2007) used functional magnetic resonance imaging techniques to determine which areas of the brain were activated during voluntary cough and sniff. The study reported that a widespread pattern of sensorimotor activation occurred along the Sylvian fissure of the cerebral cortex during voluntary cough. The study reported that voluntary cough was associated with activation of brain regions similar to those activated in voluntary breathing. The largest region of cerebral activation during voluntary coughing and breathing involved the ventrolateral sensorimotor cortex of the orofacial representation. Coughing was also associated with distinct activity in subcortical structures, with activation of areas of the midbrain–pons transition and parabrachial region. A limitation of the design of the study is that it did not allow the investigators to distinguish between motor activity during coughing and sensory feedback from airway receptors and respiratory muscle proprioceptors. Despite the limitations of this type of study it provides some interesting support for the involvement of the cerebral cortex in the voluntary control of cough.

Voluntary inhibition of cough is possible when it would be embarrassing or dangerous to cough as mentioned before. Perhaps the most surprising example of voluntary inhibition of cough is the voluntary suppression of capsaicin-induced cough described by Hutchings et al. (1993b). Capsaicin is potent irritant of sensory nerves in the airway and cough induced on inhalation of capsaicin was presumed to be reflex in origin (Fuller 1991), similar to cough reflex induced by inhalation of citric acid (Bickerman and Barach 1954). Cough challenge with citric acid or with capsaicin was presumed to initiate a cough reflex by stimulation of sensory nerves in the

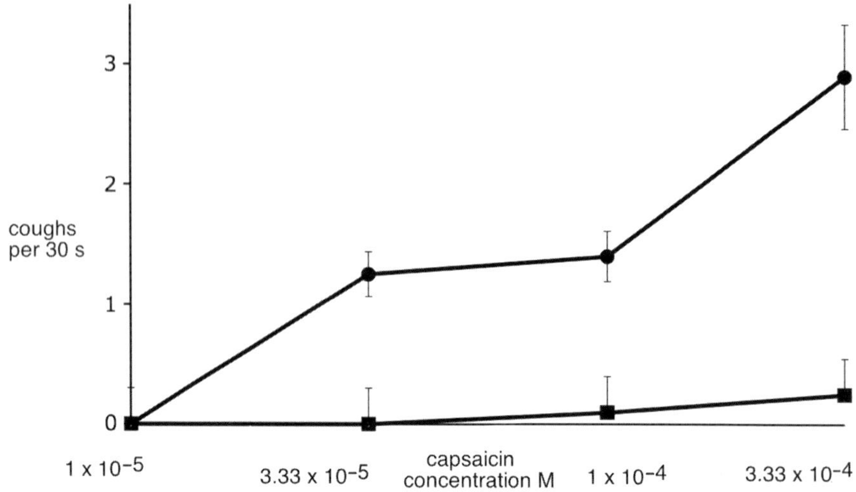

Fig. 1 Mean number of coughs per 30 s from 24 subjects produced on inhalation of capsaicin, with standard error bars. *Round symbols* indicate coughs without any voluntary suppression. *Square symbols* indicate cough during voluntary suppression. (Results redrawn from Hutchings et al. 1993b)

airway. Most studies on cough challenge refer to a cough 'reflex' (Fuller 1991) and this implies that the subjects have little or no voluntary control over the cough. In the study by Hutchings et al. (1993b) healthy subjects were challenged by increasing concentrations of inhaled capsaicin whilst sitting in front of a monitor screen. The screen could be illuminated to read either 'Please do not cough' or 'Just relax and cough if you wish' and the results of the study are illustrated in Fig. 1. In the non-suppressed challenge 23 of 24 subjects coughed, whereas in the suppressed challenge only three of 24 subjects coughed. This study clearly demonstrates that cough challenge with capsaicin induces cough that can be almost completely suppressed by voluntary control and it puts into question what is meant by the term 'reflex' that was applied previously to cough induced by challenge with citric acid or capsaicin. The conclusion that may be drawn from the study by Hutchings et al. (1993b) is that cough challenge creates a sensation of airway irritation and that subjects have considerable control over whether or not they decide to cough in response to the sensation of airway irritation. But of course any voluntary suppression of cough will depend on the strength of the stimulus and with strong stimuli it may not be possible to suppress cough.

3 Airway Sensations and Cough

The lower airway from the larynx downwards is supplied with sensory nerves that are supplied from branches of the vagus nerve (Canning et al. 2006). Cough has been considered to be solely triggered by stimulation of vagal nerve endings in the lower airways as stimulation of other cranial nerves in the upper airways triggers gagging and sneezing rather than cough (Widdicombe 2002; Eccles 2005). The hypothesis has been proposed that cough can be initiated in response to a sensation of airway irritation and that the vagal sensory nerves may provide this sensory input that is eventually processed at the level of the cerebral cortex to initiate cough under voluntary control (Eccles 2003). There has been little research on the airway sensations that initiate voluntary cough, and apart from 'a sensation of irritation' it is difficult to describe the qualities of the sensation. Much of the recent research has concentrated on the sensory transduction in cough-associated nerves (Kollarik and Undem 2006) rather than the sensations that these sensory pathways may mediate. The sensation of airway irritation may have an important role in the initiation of cough and it has been proposed that cough induced during general anaesthesia may be quite different from cough in the conscious state, as anaesthesia "may interfere with the conscious perception of airway irritation and the resultant urge to cough" (Canning et al. 2006).

Visceral sensations such as those related to stimulation of vagal airway receptors are difficult to describe and locate compared with somatic sensations such as touch. The sensation of airway irritation may initiate a vague sensation that is better described as an 'urge to cough' and this type of sensation can be readily measured using standard scales such as the Borg scale (Davenport et al. 2007). In a recent study on the urge to cough induced by inhalation of capsaicin, Davenport et al. (2007) state "that there is a sensory process, which initiates a cognitive awareness of the need to cough; this cognitive awareness then can modulate the brainstem motor cough pattern". An interesting aspect of research on the urge to cough is that the sensation was unaffected by administration of codeine in doses of 30 and 60 mg and this supports the previously reported lack of effect of codeine on voluntary control of cough (Hutchings and Eccles 1994) and the lack of effect of codeine on natural cough (Freestone and Eccles 1997).

The urge to cough can be initiated by inhalation of capsaicin and a recent study on human volunteers using functional magnetic resonance imaging indicates that the sensation may involve a variety of brain regions and especially areas of the cerebral cortex (Mazzone et al. 2007). A major issue with this type of study is that other sensations such as eye irritation or oral sensations may contribute to the brain activation and this makes it difficult to be specific about the areas of the brain involved in the sensation of the urge to cough.

4 Placebo Effects on Cough

Cough medicines are presently undergoing a revision similar to that seen in the twentieth century when dozens of cough remedies were lost owing to lack of any evidence of efficacy when the introduction of placebo-controlled trials was introduced as a standard way of assessing the efficacy of medicines. The 1899 edition of Merck's *Manual of the Materia Medica*, which lists all treatments for the practising physician, gives 61 treatments for cough, including carbolic acid, alcohol, cannabis indica, cod-liver oil, creosote, morphine, potassium bromide, sandalwood oil and zinc sulphate. In contrast the current British National Formulary (2007) for British physicians lists only three cough suppressants, dextromethorphan, codeine and pholcodine, and briefly mentions sedating antihistamines. The efficacy of these three remaining antitussives has been questioned and as discussed above the efficacy of codeine as an antitussive is debatable, and its previous position as a gold-standard antitussive is now doubted (Bolser and Davenport 2007).

Despite the rather pessimistic opinion on the pharmacological activity of current antitussives there is no doubt that the cough syrups available from the pharmacy and supermarket do have a significant effect on cough and this can be shown in clinical trials. However, the issue is that almost all of the effect of the medicines is due to a placebo effect.

In most placebo-controlled clinical trials on cough medicines the effect of placebo treatment on cough is almost as great as that caused by the active treatment. A literature survey on placebo-controlled clinical trials for cough associated with common cold has reported that the placebo response varied from 56% up to a maximum of 105% of the active-treatment response, with a mean of 85% (Eccles 2002). The results from one of the studies (Lee et al. 2000) are illustrated in Fig. 2. In this study the reduction in cough frequency following treatment with a single dose of cough medicine (30 mg dextromethorphan in capsule form) appears impressive, until it is compared with the placebo response, which is almost identical in magnitude and time course. This type of placebo response is typical of clinical trials on cough and it is often very difficult to demonstrate any benefit of a cough medicine above

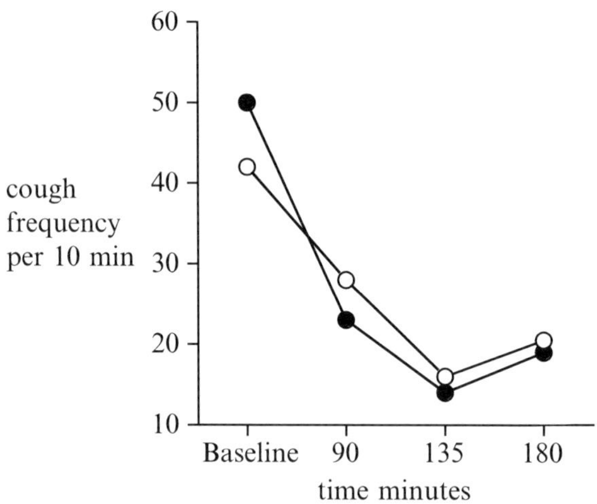

Fig. 2 Median cough frequency (per 10 min) for patients with cough associated with common cold. Immediately after the baseline measurement (0 min) patients were treated with either a single dose of 30 mg dextromethorphan powder in a hard gelatin capsule (*round symbols*, $n = 21$) or a matched placebo capsule containing lactose powder (*square symbols*, $n = 22$). (Lee et al. 2000)

that of the placebo treatment. Cough frequency is only one parameter of the impact of cough, and cough can also be measured in terms of cough effort or intensity, and in one study using cough intensity as a measure of acute cough the placebo response was only 60% of the active-treatment response (Parvez et al. 1996).

5 How Do Cough Medicines Work?

Several factors contribute to the overall change in cough associated with treatment with a cough medicine. The changes in cough frequency and intensity associated with treatment with an antitussive medicine, such as a sweet cough syrup containing dextromethorphan, can be attributed to at least five different factors: pharmacological, physiological, true placebo, psychological and non-specific (Fig. 3). These five factors are discussed in the following sections. This classification of the different components of the placebo response is an extension of the ideas put forward by Ernst and Resch (1995). The classification was first described by the author in 2003 (Eccles 2003). Ernst and Resch (1995) refer to the overall placebo effect measured in clinical trials as the 'perceived placebo effect' and propose that the perceived placebo effect may be considered as two components: a 'true placebo effect' and other 'non-specific effects' such as natural recovery from the disease and regression

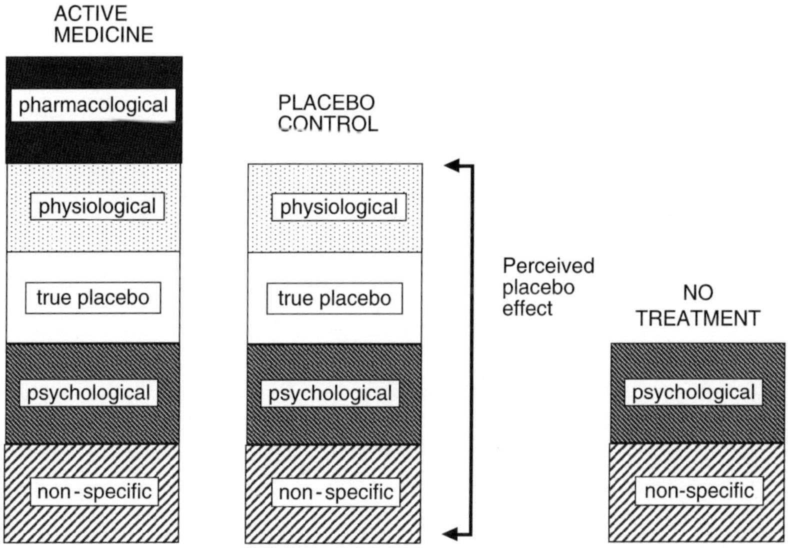

Fig. 3 Components of a cough medicine. The efficacy of a cough medicine can be attributed to at least five factors: pharmacological, physiological, true placebo, psychological and non-specific effects. The overall placebo response measured in clinical trials can be considered as a 'perceived' placebo effect that is made up of four factors: physiological, true placebo, psychological and non-specific effects

of symptom measurements towards the mean. This way of classifying the placebo effect has been reaffirmed by Ernst (2007). In this review the perceived placebo effect is classified as consisting of four components and includes a 'physiological' effect. The term 'physiological effect' is put forward for discussion. The term has been described previously (Eccles 2003), but it is not generally used in the literature on the placebo effect.

5.1 Pharmacological Effect

The pharmacological effect of treatment with a cough medicine is related to the active ingredient of the medicine, such as codeine or dextromethorphan. The pharmacologically active ingredient has a high affinity for a specific pharmacological receptor, such as the interaction of codeine with opioid receptors. Slight changes in the molecular structure of the active ingredient may have marked effects on its affinity with the receptor and its biological activity. The pharmacological effects of the opioids morphine and codeine are reported to be due to stimulation of μ-opioid receptors in the cough control areas of the brainstem (Reynolds et al. 2004). Opioid receptors have also been located in the periphery but their ability to inhibit cough by a peripheral mechanism is debatable (Reynolds et al. 2004).

In clinical trials on cough medicines, it is the pharmacological effect of the medicine that is under investigation, and any other effects of the treatment are controlled by comparison with the effects of a placebo medication that is identical in appearance, colour taste, etc. with the active medication but does not contain the pharmacologically active ingredient. The pharmacological effect of a medicine is measured by subtracting the effects of the placebo control from those of the active medicine as illustrated in Fig. 3.

Sedating antihistamines such as diphenhydramine are used in many cough medicines and the efficacy of these medicines as antitussives may be due to their general sedative effects rather than any specific interaction with pharmacological receptors.

5.2 Physiological Effects

The physiological effects of a cough medicine are the effects of the treatment that are initiated by the physical properties of the medicine that are perceived by the patient, such as colour, taste, smell, viscosity, acidity, temperature, texture, etc. These properties are related to the physical and chemical properties of the medicine and they may influence the magnitude of the placebo effect in at least three ways: (1) by triggering effects such as salivation and increased airway secretions, (2) by specific effects of sweet substances on cough and (3) by making the patient aware of the treatment by its sensory effects such as taste, smell, etc. The active pharmacological ingredient of the cough medicine may also contribute to the physiological effects of the medicine by its physical and chemical properties such as having a bitter taste or a distinctive smell.

A physiological effect on cough may be initiated by sensory stimuli such as the sweet or bitter taste of the medicine triggering reflex salivation and the secretion of mucus in the airways. Cough syrups that contain sapid substances such as sugar, honey, spicy substances such as capsicum and bitter-tasting substances such as lemon and citric acid will readily cause reflex salivation and may also promote secretion of airway mucus. In cases of dry unproductive cough the demulcent effects of a cough medicine may lubricate the pharynx and larynx and help to reduce coughing. In cases of productive cough the increase in airway secretions caused by a sapid cough syrup may increase mucociliary clearance from the airway by an expectorant effect. Gustatory rhinorroeah has been shown to occur after eating spicy foods and this observation demonstrates a link between gustation and airway secretion of mucus (Choudry et al. 1992). Some cough medicines contain capiscum, which is a potent gustatory stimulus and which may also promote airway secretions. The sweet taste of cough syrups may have been traditionally used to mask the taste of bitter-tasting plant extracts such as opium, but the fact that almost all modern cough medicines are formulated as sweet sapid syrups indicates that the physiological actions of the sweet syrup may contribute to the antitussive (cough suppressant) and expectorant activity of the treatment.

The sweet taste of a cough medicine may contribute to the placebo effect by triggering reward areas in the brain and this topic has recently been reviewed (Eccles 2006). Most cough medicines are formulated as sweet viscous syrups and the sweet taste may influence cough in two ways: firstly by stimulation of airway secretions as described above, and secondly by the generation of endogenous opioids (Jain et al. 2004; Eccles 2006). The hypothesis has been put forward that the sweet taste of cough medicines may cause the generation of endogenous opioids in the nucleus of the tractus solitarius and that since this nucleus is also involved in the control of cough there may be some inhibition of cough due to generation of endogenous opioids (Eccles 2006).

Cooling and warming agents are often added to give extra sensations to cough medicines and these agents influence the activity of cold and warm receptors. Cooling agents such as menthol are sometimes included as flavouring agents in cough medicines, although menthol may also have pharmacological activity as a local anaesthetic (Eccles 1994). The cooling properties of menthol and other cooling agents could also be considered as a pharmacological component of treatment, as there is some evidence that cooling properties are determined by interaction with a menthol type of pharmacological receptor on sensory nerves (Eccles 1994) and with interactions with transient receptor potential ion channels (McKemy et al. 2002). Although menthol is usually declared as a flavouring agent in cough medicines, there is some evidence that it may have specific antitussive activity (Laude et al. 1994; Morice et al. 1994; Pavesi et al. 2001).

The third component of the physiological effect involves the 'sensory impact' of the cough medicine. If the medicine could be administered without any sensory impact, i.e. without the patient perceiving that any treatment had been administered, then it is doubtful that there could be any true placebo effect as this is dependent on the conscious perception that a treatment has been administered. Consciousness

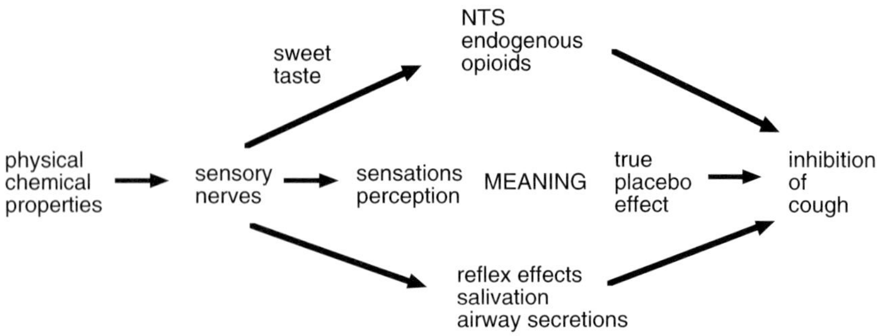

Fig. 4 Components of the physiological effect. The sensory impact of the cough medicine may cause three different effects that lead to an inhibition of cough as described in the text

and perception of treatment must be necessary for any true placebo effect as it is generally accepted that the treatment must have not only have been perceived but that it should also have some 'meaning' to the patient (Moerman 2002) and this level of cognition must involve consciousness.

The physiological effect of a cough syrup may exhibit similar characteristics to a pharmacological effect, with a time course of action, peak effect, cumulative effect and carryover effect, but at present there is no information on the pharmacodynamics of any physiological effect of treatment on cough. In the case of cough medicines, there is likely to be a large physiological effect with a cough syrup, but little if any physiological effect with a tablet or capsule formulation.

The three components of the physiological effect are illustrated in Fig. 4. The physical and chemical properties of the cough medicine stimulate sensory nerves and are responsible for the sensory impact of the treatment as discussed above. The sensory impact is interpreted in terms of its 'meaning' to the subject and if it is in the context of a positive belief about the treatment then this 'meaning' may be cause a true placebo effect. In this respect, some components of the physiological effects may be responsible for the true placebo effect.

5.3 *True Placebo Effect*

The perceived placebo effect is defined as the total effect of the placebo medicine, which includes the true placebo effect and other effects, such as any physiological effect, and non-specific effects such as natural recovery from the disease. The perceived placebo effect is normally measured in a placebo-controlled clinical trial, but it is not possible to estimate the contribution of the true placebo effect to any changes in cough severity from this parameter, as the perceived placebo effect also includes the physiological effect and non-specific effect of treatment as shown in Fig. 3.

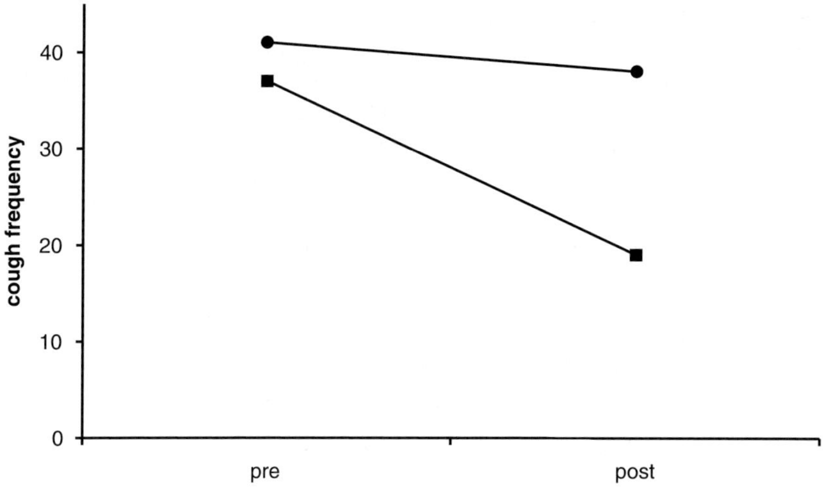

Fig. 5 Median cough frequency (per 15 min) before treatment and after treatment. *Round symbols* represent the 'no-treatment' group and *filled symbols* the 'placebo-treatment' group. (Redrawn from Lee et al. 2005)

In clinical trials where a 'no-treatment' group is included in the design, it is possible to control for any non-specific effects of treatment by subtracting any changes in the no-treatment group from those changes observed in the placebo-treatment group (Lee et al. 2005). This leaves us with a measure of any true placebo effect plus any physiological effect. In a study on patients with cough associated with upper respiratory tract infection, comparing the antitussive effects of 'no treatment' and placebo treatment, the no-treatment group had a 7% decrease in cough frequency compared with a 50% decrease in the placebo-treatment group as shown in Fig. 5 (Lee et al. 2005). In this study the placebo medicine was a capsule, rather than a sweet syrup, so the placebo effect cannot be explained by any physiological effect of sugar, or by rest, and it may be reasonably defined as a true placebo effect.

The true placebo effect of treatment with a cough medicine is related to the patient's belief about the efficacy of the medicine (Evans 2003) and the meaning that the patient relates to the treatment (Moerman 2002). The degree of belief in the treatment will depend on many factors, such as the healer–patient interaction, cultural beliefs about traditional treatments, the environment in which the medicine is administered, the properties of the medicine, such as taste, colour and smell, advertising and claims made about the efficacy of the medicine, the brand name of the medicine, and side effects associated with treatment that may reinforce the belief of efficacy. This list of factors that may influence the true placebo effect is not exhaustive and it illustrates how difficult it is to properly control and standardise studies on the true placebo effect.

The true placebo effect may be explained in terms of psychoneuropharmacology (Eccles 2002), which means that the belief in the effects of the medicine triggers a distinct nervous pathway with its own neurotransmitters that can be influenced by pharmacological intervention. The true placebo effect of cough medicines may be related to the generation of endogenous opioids, as a similar explanation for placebo analgesia has been proposed (Fields and Price 1999). If the true placebo effect on cough is mediated by the generation of endogenous opioids then this effect should be inhibited by pharmacological intervention with opioid antagonists. The analgesic effect of placebo treatment has been reported to be inhibited by administration of opioid antagonists such as naltrexone (Benedetti 1997) but at present there are no similar reports on the effects of opioid antagonists on natural cough, although there is one report of negative findings for induced cough (Hutchings and Eccles 1994).

5.4 Non-specific Factors

"Sick people often get better". In an acute illness natural recovery may occur and this is not due to any effect of the treatment (Ernst and Resch 1995).

Patients recruited to a clinical trial to determine the efficacy of a cough medicine are screened to determine the severity of cough, and only those patients with a high subjective score and/or objective measure of cough are recruited for the study. By recruiting only those patients with a severe or troublesome cough and excluding those patients with a mild cough, one skews the population of patients in the trial towards those with a severe cough. In these circumstances the cough severity of the patients in the trial is unlikely to increase during the course of the study and it is more likely that the cough severity will decrease owing to the process of natural recovery. The mean measure of cough severity is likely to decline during the course of the clinical trial and this statistical effect is often referred to as 'regression to mean' (Kienle and Kiene 1996).

It is not possible to control for the effects of rest and spontaneous recovery in controlled clinical trials that involve only placebo- and active-treatment groups, as both these treatment groups will be affected by rest and recovery. However, if a no-treatment group is included in the trial design then this will allow direct comparison with the placebo-treatment group and any effect of placebo above the effect of no treatment can be deemed to be caused by a 'true placebo effect' less any physiological effect (Ernst and Resch 1995).

5.5 Psychological Factors

Reflexes such as the flexion reflex and the eye blink reflex are affected by psychological factors such as fear and arousal (Koh and Drummond 2006), and it is reasonable to assume that cough may also be influenced by these types of psychological factors.

In a study on the voluntary suppression of acute cough it was found that subjects with high scores for the psychology parameter of obsessional symptoms had less control of their cough than those with lower scores, and the study concluded that it was important to consider psychological characteristics in situations that involve voluntary forms of symptomatic behaviour such as cough (Hutchings et al. 1993a).

6 Cough Model

A model of cough is illustrated in Fig. 6. Cough is proposed to be initiated in three ways:

1. Reflex cough, caused by the presence of food or fluid in the airway. This type of cough is not under conscious control and can occur in the unconscious subject during general anaesthesia.

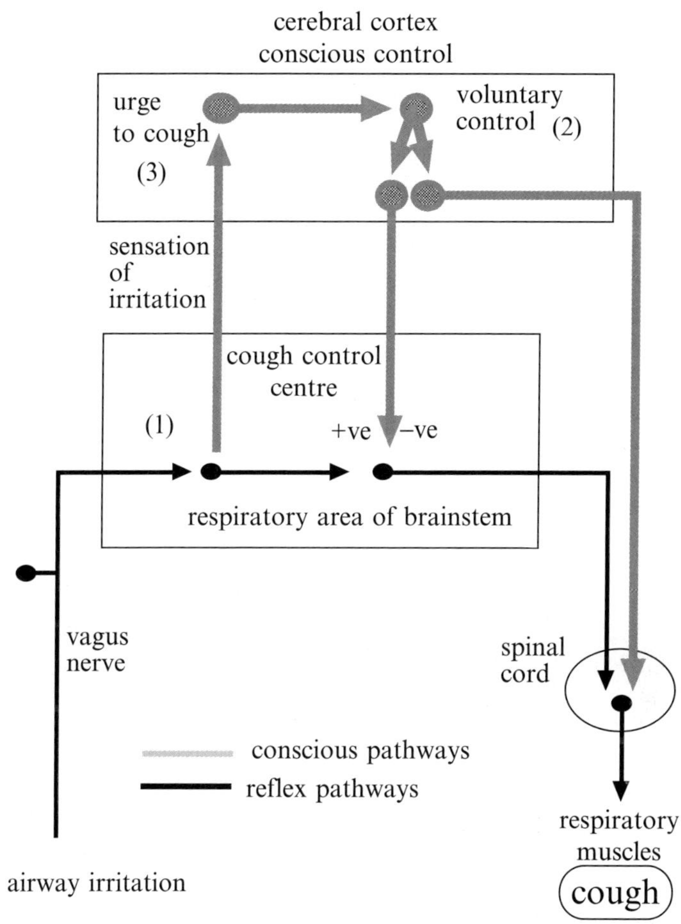

Fig. 6 A model of cough. Cough may be initiated in three ways: (1) reflex cough, caused by the presence of food or fluid in the airway – this type of cough is not under conscious control and can occur in the unconscious subject during general anaesthesia; (2) voluntary cough, under conscious control that is abolished with general anaesthesia; (3) cough in response to sensation of airway irritation – this type of cough causes an urge to cough that initiates voluntary cough and may only be present in the conscious subject. See the text for further explanation of the model

2. Voluntary cough, under conscious control that is abolished with general anaesthesia.
3. Cough in response to sensation of airway irritation. This type of cough causes an urge to cough that initiates voluntary cough. This urge to cough may only be present in the conscious subject.

6.1 Reflex Cough (Type 1)

The type of cough that was discussed in the early animal studies on cough is best described by the term ‘reflex cough’ and this is illustrated as type 1 in the model. This type of cough could also be described as a laryngeal cough reflex, that is “a brainstem-mediated involuntary reflex that is protective to the upper airway and is essential to the survival of the individual” (Stephens et al. 2003). This type of cough occurs owing to the sudden presentation of food or fluid to the airway, and results in an explosive and uncontrollable bout of coughing to clear the airway. The presence of food or fluid in the airway causes stimulation of vagal sensory nerves that relay directly in the brainstem to initiate a rapid and forceful type of cough. It is probable that this type of cough is initiated either before or simultaneously with a sensation of airway irritation.

In some respects there is some similarity between cough and pain reflexes. A powerful painful stimulus such as touching a very hot surface with the hand will induce a withdrawal reflex that is not under conscious control, and this is similar to the type 1 cough reflex.

Is reflex cough a component of human cough associated with respiratory disease? Are some coughs associated with disease entirely reflex with no opportunity for any voluntary inhibition? Is there a breaking point during suppression of cough when the urge to cough becomes uncontrollable and cough occurs despite a will to prevent cough? These types of questions were partly addressed in a study conducted on the ability of patients to suppress cough associated with common cold (Hutchings et al. 1993a). Patients were sat in front of a screen giving the instruction ‘Please do not cough’ and 27 out of a group of 79 patients with cough associated with common cold could suppress their cough for over 20 min. Of the remaining 52 patients, 32 could suppress cough for between 2.5 and 12.5 min and only 20 subjects coughed in the first 2.5 min of cough suppression. The study found a significant correlation between the cough suppression time and the severity of cough. The study demonstrates that cough can occur even though the patient is attempting to suppress the cough and therefore it provides some support for a type of reflex cough. The cough model proposed by the investigators to explain the results was a central integrator of cough sensory information located in the brainstem. The integrator acted like a leaky capacitor so that if the sensory input from airway irritant receptors was of a low level then it was possible to suppress cough indefinitely as the cough threshold was never reached. However, if the sensory input was greater, then a threshold for cough would be reached within a set time and an involuntary or reflex-type cough would occur. With aspiration of food and fluid, the sensory input to the cough integrator is so great that the cough threshold is reached almost immediately and there is no opportunity to suppress the cough. With airway irritation associated with respiratory disease, such as common cold, the coughs may be suppressed at will, perhaps by raising the threshold for cough in the cough integrator (Hutchings et al. 1993a).

6.2 Voluntary Cough (Type 2)

Cough can be initiated without any sensory input from the airway. If subjects are instructed to cough at intervals they can easily follow any protocol of voluntary cough. This type of cough occurs in the absence of any input from the airway and in the absence of any sensation of urge to cough. Voluntary cough may be related to psychological illness such as habit cough in Tourette's syndrome (Widdicombe et al. 2006) or psychogenic cough disorders such as conversion syndrome (Bhatia et al. 2002; Irwin et al. 2006). The cough model in Fig. 6 provides two pathways for the initiation of voluntary cough, one from the cerebral cortex straight to the spinal control areas for respiratory muscles, and a second pathway via the brainstem cough-control centre. At present it is not clear which of these pathways may be the most important for voluntary control of cough. Cough initiated via the brainstem may be similar to a reflex cough as it is likely that the cough is initiated from some sort of pattern generator. Cough initiated via the direct cortical pathway to the spinal cord may be variable in pattern similar to any voluntary contraction of skeletal muscles. Voluntary control of cough probably evolved with the voluntary control of the respiratory muscles associated with the use of the respiratory system for vocalisation and communication.

6.3 Urge To Cough (Type 3)

Probably the most common type of cough is cough initiated by a sensation of airway irritation (type 3 in the model). This type of cough may be involved in all types of human respiratory disease. The sensation of airway irritation is detected by vagal sensory fibres in the airway, which are sensitive to mechanical and chemical stimuli. In cough associated with common cold or chronic cough the sensory nerves may detect the presence of excess mucus in the airway or they may be in a condition of hyperreactivity. With airway hyperreactivity, a sensation of irritation may result from the presence of inflammatory mediators or in response to stimuli that would not normally be sensed as irritation such as cold air or airway inflation. The sensation of airway irritation causes an 'urge to cough' as described by Davenport (2007) and this can then initiate voluntary cough under conscious control. The urge to cough is discussed by Davenport in this volume. This urge to cough must be perceived by the conscious subject in order to initiate cough and it is likely to be abolished by loss of consciousness such as with general anaesthesia or sleep. In sleep the airway irritation may reach a level that causes arousal of the subject and then the subject will perceive the urge to cough and may initiate voluntary coughs. The type 3 cough may also stimulate brainstem pathways as with type 1 reflex cough, but it can be assumed that the level of stimulation is subthreshold as reflex cough does not occur and the cough is under voluntary control.

Type 3 cough may be processed in a cough integrator as described by Hutchings et al. (1993a). The integration of cough sensory input to the brainstem is similar to the urge to cough. As described above, the summation of cough sensory input to the brainstem areas may reach a threshold that initiates a cough that can no longer be suppressed. Similarly one can envisage that on occasions the urge to cough may reach a similar threshold and cough cannot be suppressed.

Type 1 and type 3 cough may be initiated by the same type of stimuli. For example, a little food or fluid in the airway may not cause a sensory response of sufficient magnitude to reach the threshold of a reflex cough. In this case the subject may have the control to clear the airway voluntarily in response to an urge to cough. However, if more food or fluid enters the airway the stimulus may reach the threshold for a reflex type 1 cough that is not under voluntary control. The type 3 cough reflex under control of the cerebral cortex may have evolved with the emergence of speech in order to clear the vocal cords prior to speech (Stephens et al. 2003).

7 Upper- and Lower-Airway Cough

Cough is often referred to as either a ‘dry’ cough that is unproductive (in the sense that coughing does not result in expectoration of airway mucus) and may be thought of as originating primarily at the level of the larynx and trachea, or as a ‘chesty’ productive cough that originates in the bronchi and lower airways. The cough model illustrated in Fig. 6 does not distinguish between these two types of cough as the sensory pathway for initiation of cough is shown merely as the vagus nerve with no differentiation into upper and lower airways. Both dry and chesty types of cough may have the same mechanism and both may involve cough initiated by a sensation of airway irritation and an urge to cough as with type 3 cough. In the past there has been a tendency to separate these two types of cough, and the chesty cough has been thought to fit in the category of chronic cough where there is some chronic disease of the lower airways that results in airway inflammation and the production of airway mucus. This mucus may combine with the inflammatory plasma exudates in the airway to form a viscous plug of mucus that causes some mechanical stimulation and physical stimulation of the airway to cause productive cough. However, there is no reason to believe that the central pathways associated with chesty cough or chronic cough are different from those associated with dry cough. It is apparent that both types of cough result from an urge to cough and although the peripheral mechanisms of cough may be different, the central pathways and mechanisms are likely to be the same.

When the first studies that reported the lack of efficacy of codeine and dextromethorphan in cough associated with common cold were published (Freestone and Eccles 1997; Lee et al. 2000) it was assumed that these results were not transferable to chronic cough owing to differences in the mechanisms of acute and chronic cough. However, if the central mechanisms of acute and chronic cough are the same, one would not expect any difference in the effects of centrally acting cough medicines on acute and chronic cough.

8 Discussion

8.1 *Unconscious Reflex Cough Versus Conscious Control of Cough*

This review puts forward the thesis that cough is not just a simple brainstem reflex but that cough is also under conscious control. The early animal studies on cough used decerebrate or anaesthetised animals to study the cough reflex and the effects of antitussive medicines, and up until quite recently cough was only discussed in terms of reflex cough. These early animal studies established morphine and codeine as a gold-standard antitussive that inhibited the cough reflex before abolition of spontaneous breathing (May and Widdicombe 1954). Later studies on man have adopted the concept that cough is a reflex controlled from the brainstem and that antitussives such as codeine and dextromethorphan inhibit cough by inhibiting the activity of the cough reflex at the level of the brainstem (Eddy et al. 1969; Braga and Allegra 1989).

Some studies on cough associated with common cold have failed to show any effect of codeine (Eccles et al. 1992; Freestone and Eccles 1997) or dextromethor phan (Lee et al. 2000) on cough. These results have been explained on the thesis that cough associated with common cold consists of two types of cough, reflex cough and voluntary cough, and that most cough associated with common cold is under voluntary control. The lack of effect of codeine on cough associated with common cold may be explained on the basis that codeine only acts on reflex cough and has little effect on cough that is initiated by voluntary control (Freestone and Eccles 1997). In relation to the present review, this would mean that codeine and other centrally acting antitussives would have an inhibitory effect on reflex cough (type 1) but would have no effect on voluntary cough (type 2) or the urge to cough (type 3). This concept is supported by a recent study on the urge to cough using capsaicin challenge to induce cough that reported that codeine in single doses of 30 and 60 mg had no effect on the urge to cough as perceived by the subjects (Davenport et al. 2007).

8.2 Placebo Effect and Cough Control

Placebo treatments do cause a great reduction in cough frequency and severity (Eccles 2002) but it is not understood how the placebo effect influences the three types of cough illustrated in the cough model shown in Fig. 6. It has been proposed that the placebo effect inhibits cough by the generation of endogenous opioids in the brainstem area (Eccles 2006). This placebo effect could be mediated by the sweet taste of a cough syrup causing the generation of endogenous opioids in the brainstem areas that control cough, or by the belief in the cough treatment activating cortical areas that have descending inhibitory pathways to the cough-control area in the brainstem (Eccles 2006).

If most human cough associated with respiratory disease is under conscious control, then the placebo effect of cough medicines may be due to some change in the intensity of the urge to cough. The urge to cough is a drive that must be sensed at higher centres in the brain as the subject is aware of a sensation of airway irritation that may eventually summate to cause a cough that is not under conscious control. The coughs that are under voluntary control may be influenced by treatment with any medicine that the subject believes may have an inhibitory effect on cough. It is difficult to say whether this inhibition of cough is a cognitive process or whether it is due to release of endogenous opioids as discussed earlier.

9 Conclusions

Human cough is much more than a brainstem reflex that protects the airway from aspiration of food and fluid. Cough can be initiated and inhibited by voluntary control and is also subject to inhibition by placebo treatments. Cough may be classified into three types: reflex cough that protects the airway and which is present even in the comatose and unconscious subject; voluntary cough initiated at will; and voluntary cough initiated in response to airway irritation. Reflex cough has been reported to be inhibited by opioids such as codeine, in animal studies, but codeine is no longer accepted as a gold-standard treatment for human cough associated with respiratory disease. Much cough associated with respiratory disease is under conscious control and occurs in response to an urge to cough initiated by a sensation of airway irritation.

References

Benedetti FAM (1997) The neurobiology of placebo analgesia: From endogenous opioids to cholecystokinin. Prog Neurobiol 52:109

Bhatia MS, Chandra R et al (2002) Psychogenic cough: A profile of 32 cases. Int J Psychiatr Med 32:353–360

Bickerman HA, Barach AL (1954) The experimental production of cough in human subjects induced by citric acid aerosols. Preliminary studies on the evaluation of antitussive agents. Clin Lab Sci 228:156–163

Bolser DC, Davenport PW (2007) Codeine and cough: An ineffective gold standard. Curr Opin Allergy Clin Immunol 7:32–36

Braga PC, Allegra L (1989) Centrally acting opioid drugs. In: Braga PC, Allegra L (eds) Cough. Raven Press, New York, pp 109–143

British National Formulary (2007) British Medical Association and Royal Pharmaceutical Society of Great Britain, London

Calvert JR, Steinhaus JE et al (1966) Halothane as a depressant of cough reflex. Anesth Analg 45:76–81

Canning BJ, Mori N et al (2006) Vagal afferent nerves regulating the cough reflex. Respir Physiol Neurobiol 152:223–242

Choudry NB, Harrison AJ et al (1992) Inhibition of gustatory rhinorrhea by intranasal ipratropium bromide. Eur J Clin Pharmacol 42:561–562

Davenport PW, Bolser DC et al (2007) The effect of codeine on the Urge-to-Cough response to inhaled capsaicin. Pulm Pharmacol Ther 20:338–346

Eccles R (1994) Menthol and related cooling compounds. J Pharm Pharmacol 46:618–630

Eccles R (2002) The powerful placebo in cough studies. Pulm Pharmacol Ther 15:303–308

Eccles R (2003) Placebo effects of antitussive treatments on cough associated with acute upper respiratory tract infection. In: Chung KF, Widdicombe JG, Boushey HA (eds) Cough, causes, mechanisms and therapy. Blackwell, Massachusetts, pp 259–268

Eccles R (2005) Acute cough: Epidemiology, mechanisms, and treatment. In: Redington AE, Morice AH (eds) Acute and chronic cough, vol 205. Marcel Dekker, New York, pp 215–236

Eccles R (2006) Mechanisms of the placebo effect of sweet cough syrups. Respir Physiol Neurobiol 152:340–348

Eccles R, Morris S et al (1992) Lack of effect of codeine in the treatment of cough associated with acute upper respiratory tract infection J Clin Pharm Ther 17:175–180

Eddy NB, Friebel H et al (1969) Codeine and its alternates for pain and cough relief. Bull World Health Org 40:425–454

Ernst E (2007) Placebo: New insights into an old enigma. Drug Discov Today 12:413–418

Ernst E, Resch KL (1995) Concept of true and perceived placebo effects. Br Med J 311:551–553

Evans D (2003) Placebo. The belief effect. Harper Collins, London

Fields HL, Price DD (1999) Towards a neurobiology of placebo analgesia. In: Harrington A (ed) The placebo effect. An interdisciplinary approach. Harvard University Press, Cambridge, USA, pp 93–116

Freestone C, Eccles R (1997) Assessment of the antitussive efficacy of codeine in cough associated with common cold. J Pharm Pharmacol 49:1045–1049

Fuller RW (1991) Pharmacology of inhaled capsaicin in humans. Respir Med 85:31–34

Gracely RH, Undem BJ et al (2007) Cough, pain and dyspnoea: Similarities and differences. Pulm Pharmacol Ther 20:433–437

Hsu JY, Stone RA et al (1994) Coughing frequency in patients with persistent cough: Assessment using a 24 hour ambulatory recorder. Eur Respir J 7:1246–1253

Hutchings HA, Eccles R (1994) The opioid agonist codeine and antagonist naltrexone do not affect voluntary suppression of capsaicin induced cough in healthy subjects. Eur Respir J 7:715–719

Hutchings HA, Eccles R et al (1993a) Voluntary cough suppression as an indication of symptom severity in upper respiratory tract infections. Eur Respir J 6:1449–1454

Hutchings HA, Morris S et al (1993b) Voluntary suppression of cough induced by inhalation of capsaicin in healthy volunteers. Respir Med 87:379–382

Irwin RS, Glomb WB et al (2006) Habit cough, tic cough, and psychogenic cough in adult and pediatric populations: ACCP evidence-based clinical practice guidelines. Chest 129:174S–179S

Jain R, Mukherjee K et al (2004) Influence of sweet tasting solutions on opioid withdrawal. Brain Res Bull 64:319–322

Jamal K, McMahon G et al (1983) Cough and arousal responses to inhaled citric-acid in sleeping humans. Am Rev Respir Dis 127:237

Kennedy S, Longnecker D (1996) History and principles of anesthesiology. In: Hardman JG, Limbird LE, Molinoff PB, Ruddon RW, Goodman Gilman A (eds) Goodman and Gilman's the pharmacological basis of therapeutics. McGraw Hill, New York, pp 295–306

Kienle G, Kiene H (1996) Placebo effect and placebo concept: A critical methodological and conceptual analysis of reports on the magnitude of the placebo effect. Altern Ther Health Med 2:39–54

Koh CW, Drummond PD (2006) Dissociation between pain and the nociceptive blink reflex during psychological arousal. Clin Neurophysiol 117:851–854

Kollarik M, Undem BJ (2006) Sensory transduction in cough-associated nerves. Am Rev Respir Dis 152:243–254

Laude EA, Morice AH et al (1994) The antitussive effects of menthol, camphor and cineole in conscious guinea-pigs. Pulm Pharmacol 7:179–184

Lee PCL, Jawad MSM et al (2000) Antitussive efficacy of dextromethorphan in cough associated with acute upper respiratory tract infection. J Pharm Pharmacol 52:1137–1142

Lee P, Cotterill-Jones C et al (2002) Voluntary control of cough. Pulm Pharmacol Ther 15:317–320

Lee PC, Jawad MS et al (2005) The antitussive effect of placebo treatment on cough associated with acute upper respiratory infection. Psychosom Med 67:314–317

May AJ, Widdicombe JG (1954) Depression of the cough reflex by pentobarbitone and some opium derivatives. Br J Pharmacol Chemother 9:335–340

Mazzone SB, McLennan L et al (2007) Representation of capsaicin-evoked urge-to-cough in the human brain using functional magnetic resonance imaging. Am J Respir Crit Care Med 176:327–332

McKemy DD, Neuhausser WM et al (2002) Identification of a cold receptor reveals a general role for TRP channels in thermosensation. Nature 416:52–58

Moerman D (2002) Deconstructing the placebo effect and finding the meaning response. Ann Intern Med 136:471–476

Morice AH, Marshall AE et al (1994) Effect of inhaled menthol on citric acid induced cough in normal subjects. Thorax 49:1024–1026

Moulton C, Pennycook AG (1994) Relation between Glasgow coma score and cough reflex. Lancet 343:1261–1262

Nishino T, Hiraga K et al (1988) Respiratory reflex responses to stimulation of tracheal mucosa in Enflurane-anesthetized humans. J Appl Physiol 65:1069–1074

Parvez L, Vaidya M et al (1996) Evaluation of antitussive agents in man. Pulm Pharmacol Ther 9:299–308

Pavesi L, Subburaj S et al (2001) Application and validation of a computerized cough acquisition system for objective monitoring of acute cough A meta-analysis. Chest 120:1121–1128

Power JT, Stewart IC et al (1984) Nocturnal cough in patients with chronic-bronchitis and emphysema. Am Rev Respir Dis 130:999–1001

Reynolds SM, Mackenzie AJ et al (2004) The pharmacology of cough. Trends Pharmacol Sci 25:569–576

Simonyan K, Saad ZS et al (2007) Functional neuroanatomy of human voluntary cough and sniff production. NeuroImage 37:401–409

Stephens RE, Addington WR et al (2003) Effect of acute unilateral middle cerebral artery infarcts on voluntary cough and the laryngeal cough reflex. Am J Phys Med Rehabil 82:379–383

Sullivan CE, Murphy E et al (1978) Waking and ventilatory responses to laryngeal stimulation in sleeping dogs. J Appl Phys 4:681–689

Sullivan CE, Kozar LF et al (1979) Arousal, ventilatory, and airway responses to bronchopulmonary stimulation in sleeping dogs. J Appl Physiol.47:17–25

Widdicombe JG (1954) Respiratory reflexes from the trachea and bronchi of the cat. J Physiol 123:55–70

Widdicombe JG (2002) Neuroregulation of cough: Implications for drug therapy. Curr Curr Opin Pharmacol 2:256–263

Widdicombe JG, Eccles R et al (2006) Supramedullary influences on cough. Respir Physiol Neurobiol 152:320–328

Clinical Cough I: The Urge-To-Cough: A Respiratory Sensation

P.W. Davenport

Contents

Abstract Cough is generated by a brainstem neural network. Chemical and mechanical stimulation of the airway can elicit a reflex cough and can elicit a cognitive sensation, the urge-to-cough. The sensation of an urge-to-cough is a respiratory-related sensation. The role of the respiratory sensation of an urge-to-cough is to engage behavioral modulation of cough motor action. Respiratory sensations are elicited by a combination of modalities: central neural, chemical, and mechanical. Stimulation of respiratory afferents or changes in respiratory pattern resulting in a cognitive awareness of breathing are mediated by central neural processes that are the cognitive neural basis for respiratory sensations, including the urge-to-cough. It is proposed that the urge-to-cough is a component of the cough motivation-to-action system. The urge-to-cough is induced by stimuli that motivate subjects to protect their airway by coughing. Cough receptor stimulation is gated into suprapontine brain systems. In the proposed cough motivation system, the cough stimulus would produce an urge-to-cough which then matches with the cognitive desire for a response to the urge. If a cough is produced by the motor action system, the descending cognitive drive modulates the brainstem cough neural network. Receptors within the respiratory system provide sensory feedback indicating if the cough occurred, the motor pattern, and the magnitude. The limbic system uses that information to determine if the coughing behavior satisfied the urge. Cough is stopped if the urge-to-cough is satisfied; if the urge has not been satisfied then the urge-to-cough will continue to

---

P.W. Davenport
Department of Physiological Sciences, Box 100144, HSC, University of Florida, Gainesville, FL 32610, USA
davenportp@mail.vetmed.ufl.edu

K.F. Chung, J.G. Widdicombe (eds.), *Pharmacology and Therapeutics of Cough*, Handbook of Experimental Pharmacology 187,

motivate the central nervous system. The central component within this cough motivation system is the intrinsic brain mechanism which can be activated to start the cycle for motivating a cough, the urge-to-cough. Eliciting a cognitive urge-to-cough is dependent on the integration of respiratory afferent activity, respiratory motor drive, affective state, attention, experience, and learning.

1 Introduction

Cough is generated by a brainstem neural network (Shannon et al. 2000, 2004). Chemical and mechanical stimulation of the airway can elicit a cough (Canning et al. 2006; Dicpinigaitis 2003; Ho et al. 2001; Mazzone 2005). While cough does not require cognitive awareness or the cerebral cortex, cough can elicit a cognitive sensation. It has been reported that inhalation of aerosolized capsaicin stimulates cough and results in a sensation termed the "urge-to-cough" (Davenport et al. 2002). It was also reported that this cognitive sensation of an urge-to-cough increases with increasing capsaicin concentration. There are correlations between the urge-to-cough, the number of coughs, and the capsaicin concentration (Davenport et al. 2002, 2007a). When subjects were asked to pay attention to their breathing, the urge-to-cough occurred at a capsaicin concentration less than the capsaicin concentration eliciting the motor reflex cough (Davenport et al. 2002). This suggests that the cough cognitive sensory process can precede the cough motor event; hence, the stimulation of receptors that initiate irritant reflex cough projects to cognitive cortical centers eliciting the perception of a need to cough and the discriminative sensation of cough. The sensation of an urge-to-cough suggests that this cognitive awareness is a respiratory sensation. The role of reflex cough is to protect the airway. The role of respiratory sensations is to engage the cognitive motivation system to modulate respiratory motor behavior. The first component of the cognitive motivation system is initiation of a biological urge which then initiates a cascade of neural events leading to conscious behavioral modulation of motor behavior (Bradley 2000). Hence, the role of the respiratory sensation of an urge-to-cough is to engage behavioral modulation of cough motor action.

Respiratory sensations are not a single sensory modality but rather a combination of modalities: central neural, chemical, and mechanical (O'Donnell et al. 2007). Respiratory sensations usually include the various forms of dyspnea (air hunger, chest tightness, effort of breathing); however, other respiratory sensations are common, such as sensations of an urge-to-breathe, urge-to-cough, urge-to-sneeze, sense of suffocation, airway irritation, laryngeal burning, and similar cognitive sensations related to breathing. Disruption of breathing, stimulation of respiratory afferents, and changes in respiratory pattern result in a cognitive awareness of breathing which is mediated by neural processes. Often this cognitive awareness of breathing leads to distressing emotions and, as such, motivates and elicits behavioral breathing-related compensations. Respiratory sensations of sufficient magnitude can dominate cognitive awareness; hence, there has to be a cognitive neural basis for respiratory sensations. It follows that appropriate manipulation of these neural processes will provide insight into the mechanisms mediating the specific forms of respiratory sensations, including the urge-to-cough.

Research on respiratory sensations and the urge-to-cough use physiological changes to understand psychological processes and use psychological changes to understand physiological processes. In order to investigate the urge-to-cough and other respiratory sensations, several things must be considered: the modality mediating the sensation, threshold, magnitude of stimulation, neural processing mechanisms, and motor outcomes/compensations.

2 Respiratory Sensation

The significance of respiratory sensations is the fact that respiratory sensory information processing is fundamental to the perception of ventilatory status. When ventilation is obstructed, stimulated, challenged, or attended to, cognitive awareness of breathing occurs. The effects of respiratory sensations range from a simple awareness of breathing to highly distressing fear and anxiety in both humans and animals. This is the result of neural gating into the cerebral cortex of respiratory afferent input eliciting a somatosensory cognitive awareness of breathing and an affective response. It is hypothesized that (1) neural information from respiratory afferents must pass through a subcortical threshold gate before sensory information is transferred to the cerebral cortex, (2) respiratory sensory information simultaneously projects to the discriminative (i.e., somatosensory cortex) and affective (i.e., anterior cingulate, insular cortex, amygdala) neural centers, and (3) gating of cortical neural activity is modulated by convergent afferent input to the neural region that may be the gate mechanism for respiratory afferent information to the discriminative and the affective areas. Subsequent neural processing recruits neural centers in motor pathways to initiate respiratory compensatory behaviors.

Respiratory sensations can originate with the activation of multiple sensory receptors associated with respiration, such as arterial chemoreceptors and mechanoreceptors in the lung, and airway and respiratory muscles (O'Donnell et al. 2007). Respiratory stimuli known to elicit respiratory sensations are lung hyperinflation, lung deflation, inhaled irritants, hypercapnia, and respiratory mechanical loads (O'Donnell et al. 2007). These stimuli are produced by changes in lung volume, Pa_{CO_2}, Pa_{O_2}, bronchial smooth-muscle tone, airway pressure, pleural pressure, respiratory muscle force, respiratory muscle length, airflow, and lung mechanics.

It is known that respiratory stimuli elicit neural activity in suprapontine areas (O'Donnell et al. 2007). In human studies (Fig. 1) respiratory-related evoked potentials (RREP), positron emission tomography (PET), and functional magnetic resonance imaging (fMRI) have been used to study human brain activity induced by respiratory mechanical loads, reducing tidal volume, hypercapnia, and hypoxia (O'Donnell et al. 2007). Neural activation has been reported in the primary sensory cortex, thalamus, hypothalamus, anterior cingulate, amygdala, anterior insular

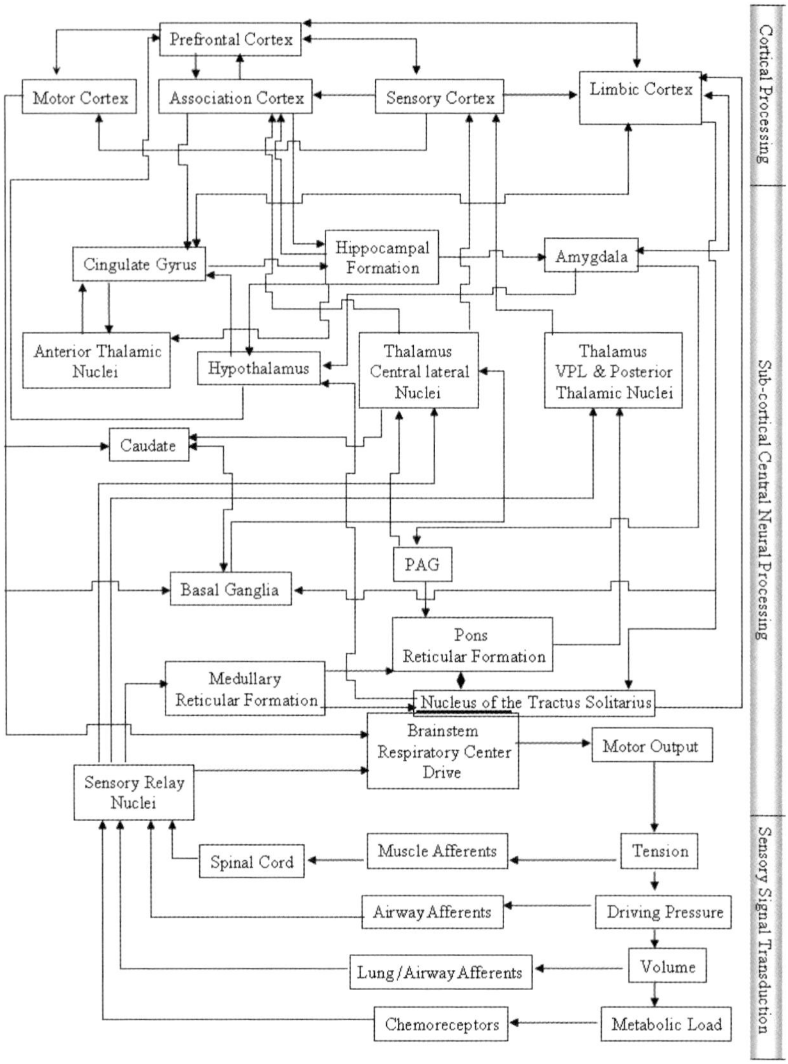

Fig. 1 Neural areas identified as active in response to respiratory stimuli that elicit respiratory sensations (Banzett et al. 2000; Brannan et al. 2001; Davenport et al. 1996; Evans et al. 2002; Gozal et al. 1995, 1996; Isaev et al. 2002; Liotti et al. 2001; Logie et al. 1998; Mazzone et al. 2007; Peiffer et al. 2001; Simonyan et al. 2007)

cortex, basal ganglia, caudate, frontal operculum (Banzett et al. 2000; Davenport et al. 1996; Evans et al. 2002; Logie et al. 1998), supplementary motor area, putamen, prefrontal cortex (Davenport et al. 1996; Gozal et al. 1995, 1996; Logie et al. 1998; Peiffer et al. 2001), amygdala, posterior insular cortex, and gyri in the occipital and temporal lobes (Brannan et al. 2001; Isaev et al. 2002; Liotti et al. 2001).

While PET and fMRI methods allow identification of neural areas activated by respiratory stimuli, these brain-imaging methods cannot discriminate between sensory and motor activation. The pattern of RREP peaks provides neural indicators of the temporal sequence of respiratory sensation. However, RREP studies provide limited information on the specific neural nuclei of the respiratory sensory pathways to the activated discriminative and affective cortices (Bloch-Salisbury and Harver 1994; Bloch-Salisbury et al. 1998; Davenport et al. 1986, 1996; Davenport and Hutchison 2002; Harver et al. 1995; Knafelc and Davenport 1997, 1999; Logie et al. 1998; Revelette and Davenport 1990; Webster et al. 2002; Webster and Colrain 2000a, b; Zhao et al. 2002).

The results of the RREP, PET, and fMRI studies indicate common areas are activated by multiple respiratory modalities, suggesting convergent neural pathways for respiratory perception that can lead to specific respiratory sensations. The differences in brain activations with the different respiratory sensory modalities are likely due to the afferent systems activated. Each respiratory sensory modality has a unique convergent and divergent afferent input to the central nervous system and a unique temporal pattern of afferent activity. Despite well-known afferent responses to respiratory stimuli, it remains unknown which of the fMRI- and PET-imaged activated central neural structures are essential to produce specific respiratory sensations.

Respiratory sensory information is continuously sent to the respiratory centers in the brainstem yet does not activate higher brain cognitive centers on a breath-by-breath basis. This suggests that respiratory afferent input is continually present in the central nervous system but is normally gated-out of higher brain cognitive centers, allowing individuals to breathe without conscious attention to their ventilation. However, changes in the respiratory system can occur that recruit or increase respiratory afferent input. If a sensory threshold is exceeded (the detection threshold), respiratory sensory information passes through the gate, activating higher brain centers and resulting in cognitive awareness of breathing. This means that one component of the neural mechanism for initiation of cognitive respiratory sensation is threshold gating (Fig. 2). The presence of a gated neural mechanism predicts that a stimulus magnitude must exceed a threshold for respiratory afferents to activate the cerebral cortex (Chou and Davenport 2007; Davenport et al. 2007b). It is also well known that the threshold for load detection can be modulated with background loads (Chou and Davenport 2007; Wiley and Zechman 1966). Loads that were detected with minimal background resistance become undetectable when a background load was increased. Increasing the magnitude of the background stimulus of the same respiratory sensory modality inhibits detection and cortical neural activity (Chou and Davenport 2007). Increasing nonrespiratory background stimuli may also modulate respiratory perception. These results suggest that gating of cortical neural activity is modulated by the background sensory state. It is likely that the sensation of the urge-to-cough uses similar neural mechanisms for the generation of this form of respiratory sensation.

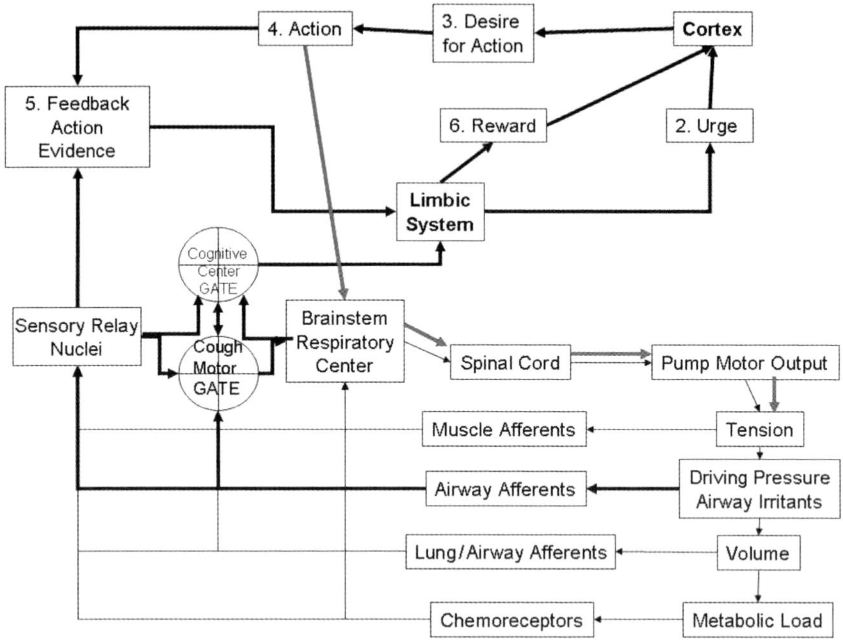

Fig. 2 Motivation model of urge-to-cough gated into the brainstem and suprapontine neural control systems for discriminative and affective control of cough

3 Urge-To-Cough

Reflex cough has been the most extensively investigated type of cough. Reflex cough is initiated by sensory stimulation of peripheral airway afferents using chemical irritants or mechanical stimulation. Chemical irritants, such as capsaicin, stimulate airway afferents, and the transduction of the cough-initiating stimuli has been well studied and characterized (Canning et al. 2006; Ho et al. 2001; Kollarik et al. 2007; Mazzone 2005). The brainstem mechanisms for generating respiratory cough motor behavior have also been described (Bolser and Davenport 2002; Shannon et al. 2000, 2004). It is well established that sensory afferents have neural connections into the brainstem cough network (Canning et al. 2006; Kollarik et al. 2007; Mazzone 2005). Reflex cough often occurs in the absence of a cognitive awareness of the cough. However, it has been reported that activation of cough afferents elicits cognitive cough sensations which have been defined as the urge-to-cough (Davenport et al. 2002). The cognitive awareness of an impending cough or urge-to-cough means afferents have been activated via cognitive neural pathways which have properties of detection, stimulus evaluation, and behavioral response (Davenport et al. 2007a). The magnitude of the sensory stimulation has been shown to be directly related to the stimulus evaluation (magnitude estimation), the number of coughs elicited, and the cough intensity (Davenport et al. 2007a; Vovk et al. 2007). The experimental methods used to investigate the urge-to-cough are critical for interpreting the cognitive results. Experiments need to be specifically designed to determine the relationship between the perceptual magnitude of the cough, the threshold for eliciting the cough, and the cough intensity. Threshold evaluation can be done by using the doubling dose paradigm (Dicpinigaitis 2003; Dicpinigaitis and Dobkin 1997; Dicpinigaitis et al. 1999; Dicpinigaitis and Rauf 1998). As the cough stimulus dose increases, the probability of eliciting a cough increases. At a specific magnitude of cough stimulus (such as capsaicin concentration), coughs are initiated, hence identifying the stimulus magnitude necessary for eliciting a motor cough (Dicpinigaitis 2003; Dicpinigaitis and Dobkin 1997; Dicpinigaitis et al. 1999; Dicpinigaitis and Rauf 1998). This is one measure of the motor threshold for a reflex cough. For cognitive perception of the cough stimulus, it is essential, however, that the subject cannot predict the stimulus magnitude that is to be presented. To determine the magnitude estimation of the cough stimulus it is, thus, necessary to blind the subject to the concentration of the cough stimulus. The stimulus range (capsaicin concentration) has to be randomized so the subject cannot anticipate a continuously increasing cough stimulus. It is also important to present the stimulus range multiple times to eliminate an order effect in the magnitude estimation for individual stimuli. The subject is blinded to the cough stimulus with a randomized-block method which has the individual increments for the stimulus range randomized into a block of stimulus presentations (Davenport et al. 2002, 2007a). Each stimulus increment is repeated in a second independently randomized block. The stimulus-range blocks can be repeated as often as required by the experimental protocol, with each block containing the full stimulus range increments independently randomized. Previous urge-to-cough experiments have used three randomized blocks, resulting in three presentations of each capsaicin concentration (Davenport et al. 2002, 2007a). It is critical, however, when performing repeated presentations of cough irritant stimuli such as capsaicin to allow sufficient time between each presentation to avoid tachyphylaxis. Previous urge-to-cough experiments have found a recovery period of a minimum of 1 min, and preferably 2 min, between each presentation to result in no significant difference in the magnitude estimation or cough motor response between three concentration blocks (Davenport et al. 2007a). It is also important to use a clearly defined cognitive scale for magnitude estimation of the cough stimulus (Davenport et al. 2002, 2007a). The question or statement provided to the subject of what is to be evaluated is of greatest importance because subjects will answer the specific question they are asked. The use of a modified Borg category scale has been successful in providing stable, reproducible, within- and between-subject magnitude estimations of capsaicin-elicited cough (Davenport et al. 2002, 2007a).

A linear relationship (using a log–log transform) between capsaicin dose and the magnitude estimation of the urge-to-cough has been reported (Davenport et al. 2002, 2007a). This is consistent with Steven's psychophysical law, i.e., the magnitude estimation of the stimulus is exponentially related to the magnitude of the physical stimulus. The slope of the line from the log–log relationship is a measure of the sensitivity of the subject to the cough stimulus. The capsaicin concentration that elicits an magnitude estimation greater than zero is a measure of the cognitive threshold for the urge-to-cough. Recording the motor cough response and determining the minimum capsaicin concentration to elicit a cough, as measured either by auditory or by respiratory mechanical responses, allows the determination of the cough motor threshold (Davenport et al. 2007a; Vovk et al. 2007). The relationship between the cognitive sensory and motor response to a cough stimulus allows the comparison of the threshold for the initial awareness of an urge-to-cough and the threshold for producing a cough. It has been reported that the cognitive threshold for eliciting an urge-to-cough occurs at a lower capsaicin concentration than the threshold for the cough motor response (Davenport et al. 2002, 2007a; Vovk et al. 2007). The functional significance of this difference in threshold is to allow the subject to include the decision-making "desire" step in the motor response of producing a cough. For example, if the subject desires to suppress the cough because of the environmental circumstances, then the cognitive awareness of an urge-to-cough preceding the motor action of a cough allows the subject to behaviorally modify the motor response as a function of the subject's motivational system. This reflex cough experimental paradigm also allows the determination of the relationship between the urge-to-cough and measures of cough intensity (Davenport et al. 2007a; Vovk et al. 2007). There is a direct relationship between the urge-to-cough, the number of coughs, and the expiratory muscle electromyographic response to increasing capsaicin concentrations (Vovk et al. 2007). It is clear from these results that increasing peripheral stimulation by a cough-eliciting irritant increases the perceived need to cough and in fact makes cough obligatory despite the motivational desire of the subject.

Voluntary cough uses a different perceptual paradigm for magnitude production (Hutchings et al. 1993; Lasserson et al. 2006). In each trial the subject is asked to produce a cough. The cough the subject is asked to generate is based on a cough magnitude that the experimenter wants the subject to produce. For example, a subject may be asked to produce a cough rated as strong, moderate, or weak. The subject will produce the cough on the basis of his/her internal cognitive perception of what constitutes a cough intensity corresponding to each requested magnitude. In this paradigm the subject's airway afferents are not stimulated and the sensory component is bypassed. This experiment was performed in a group for subjects asked to provide three cough magnitudes (Sapienza et al. 1997). The result was a predictable relationship for both inspiratory and expiratory airflows. There was a significantly increasing peak inspiratory airflow from weak to strong coughs, indicating that the subjects took a larger inspiratory priming breath before generating the cough expiratory airflow. There was also a significant increase in peak expiratory airflow that was directly proportional to the requested magnitude of the cough (weak to strong). Similarly it was reported that there was an increasing peak abdominal electromyographic response with increasing voluntary cough strength (Sapienza et al. 1997). This type of behavioral paradigm allows the investigation of cognitive cough motor behavior that is independent of an airway sensory activation (Hutchings et al. 1993; Lee et al. 2002; Widdicombe et al. 2006). This also shows that voluntary cough can be used to investigate the motor pattern related to anticipated cough strength in the absence of cough-eliciting-afferent stimulation.

fMRI studies of voluntary cough (Simonyan et al. 2007) and capsaicin inhalation (Mazzone et al. 2007) have shown activations in the primary sensory cortex, the primary motor cortex, superior temporal gyri, orbitofrontal cortex, anterior midcingulate cortex, insula, and cerebellum. Activations more specifically related to the magnitude estimation of the capsaicin stimulation were found in the anterior midcingulate cortex, right primary somatosensory cortex, and supplementary motor area. These areas of the brain are similar to other respiratory sensory modalities (Table 1), suggesting the urge-to-cough may be using neural mechanisms common to respiratory sensation (Fig. 1). The activation of the cingulate gyrus and insular cortex is consistent with subjects experiencing an urge-to-swallow (Kern et al. 2001). However, the neural mechanisms mediating the urge-to-cough remain unknown.

Table 1 Suprapontine neural areas activated by stimuli that elicit respiratory sensations (Banzett et al. 2000; Brannan et al. 2001; Davenport et al. 1996; Evans et al. 2002; Gozal et al. 1995, 1996; Isaev et al. 2002; Liotti et al. 2001; Logie et al. 1998; Mazzone et al. 2007; Peiffer et al. 2001; Simonyan et al. 2007)

	Respiratory sensation	Cough
	Thalamus	Lateral thalamus (VC)
	Hypothalamus	
	Amygdala	Amygdala (VC)
	Caudate nucleus	
	Red nucleus	
		Pontomesencephalic junction (VC)
Basal ganglia	Putamen	
	Putamen claustrum	
	Glubus pallidus	
Cerebellum	Cerebellum: dentate nucleus	Cerebellum (UtC)
	Cerebellum: anterior lobe culme	Cerebellar vermis (VC)
	Cerebellum: posterior lobe declive	
	Cerebellum: posterior lobe pyramis	
	Cerebellum: posterior lobe uvula	
	Cerebellum: posterior lobetonsil	
Cerebral cortex	Primary sensory cortex	Primary sensory cortex (UtC + VC)
	Supplementary motor area	Primary motor cortex (UtC + VC)
	Prefrontal cortex	Orbitofrontal cortex (UtC)
	Inferolateral precentral gurus	
	Anterior cingulate	Anterior midcingulate (UtC)
	Anterior insular cortex	Insular cortex (UtC)
	Posterior insular cortex	
	Gyri in the occipital lobe	
	Gyri in the temporal lobe	Superior temporal gyri (UtC)
	Parietal cortex: supramarginal gyrus	
	Parietal cortex: inferior parietal lobe	Lingual and inferior temporal gyri (VC)
	Frontal operculum	Parieto-opercular cortex (VC)

UtC urge-to-cough, *VC* voluntary cough

4 Model of Urge-To-Cough and Cognitive Respiratory Sensory Information Processing

It is proposed that the urge-to-cough is a respiratory sensation that is a brain component of the cough motivation-to-action system (Fig. 2). The significance of the urge-to-cough is based on the brain mechanisms that have been described for other biological urges (Bradley 2000; Cameron 2002). Extensive studies have been made on biological urges such as the urge-to-swallow and urge-to-eat and the motivational urges involved in drug dependencies (Bradley 2000; Cameron 2002; Kern et al. 2001). Cortical control of motor regulation and action has been attributed to activation of the primary sensorimotor cortex (Toogood et al. 2005). Higher-order motor processing, including attention, has been attributed to the nonprimary motor regions such as the anterior cingulate cortex (Toogood et al. 2005). Perceptual and cognitive processing of interoceptive stimuli has been attributed to the parieto-occipital cortex (Toogood et al. 2005). A highly simplified model has been produced which summarizes the brain pathways that are part of motivation-to-action systems. The motivation system includes the intrinsic biological urge as one of the fundamental components (Straker 2002). The urge-to-cough is a biological urge (Bradley 2000; Cameron 2002) which is induced by stimuli that motivate the subject to protect his/her airway by coughing. In this simplified view, after the cough receptor airway stimulation exceeds the threshold and is gated into suprapontine brain systems, there are six stages for motivation-to-action (similar to other biological urges) for the urge-to-cough (Fig. 2). The first stage is the stimulus. Cough sensory receptors are located throughout the upper and lower respiratory tract and are activated by chemical or mechanical stimuli. These afferents project to central neural structures and trigger a sequence of intrinsic neural events which can be gated into suprapontine neural networks and lead to activation of the "urge" neural system (Davenport et al. 2007a; Davenport et al. 2002; Widdicombe et al. 2006). The second stage, activation of the neural urge, converts the physical stimuli into a biological urge. For other sensory systems such as the urge-to-eat (Bradley 2000; Cameron 2002; Straker 2002), the thalamus has been implicated as the neural center that transfers the sensory initiated "urge" signal into the limbic system. This limbic-system-mediated urge-to-action then projects, in stage 3, to the cortex, where the urge is translated into a targeted desire, such as the desire to cough or desire to suppress cough. The urge–desire cortical activation produces a conscious motivation towards a particular goal which can be to either suppress or elicit a motor action (cough). The strength of the urge is critically important in this process because, regardless of the desire, if the urge is of sufficient strength, the cough may be obligatory despite a desire to suppress cough. The fourth stage is the descending neural drive for motor action. Eventually the urge–desire achieves its goal and the subject makes the physical response to satisfy the urge–desire, which can be a suppression of cough, motor production of a cough, or behaviorally modulated cough (such as covering their mouth). The internal motivational system can force an uncomfortable (affective component) action despite the desire to behaviorally modify or suppress that action. The cough motor action phase is hypothesized to result from a descending cortical motor drive activating the brainstem cough neural network (Bolser and Davenport 2002; Shannon et al. 2000, 2004). It is the brainstem cough neural network that generates the cough. The cough action then stimulates a feedback system, the fifth stage, which provides the central nervous system with evidence of the motor action. The brain receives evidence or feedback on the pattern of the action and if that action was completed (discriminative component). The feedback must come from cough-activated sensory receptors and central neural drive centers. The feedback must be specific for the actions that occurred, i.e., cough motor pattern. The feedback then projects via a limbic pathway which mediates the affective sense of reward for the action, the sixth and final stage of this hypothesized cough motivation-to-action system. The limbic-system-mediated sense of satisfaction of the urge (affective component) arises from the evidence provided by the feedback generated by the motor cough action. The brain rewards the urge by providing cognitive information to the subject that the right action occurred and rewards the continuance of the action. Alternatively the reward system signals that the action is sufficient to stop further action, producing a feeling of fulfillment, well-being, or relief. In the proposed cough motivation system, the airway stimulus would produce an urge-to-cough which then matches with the cognitive desire for a response to the urge-to-cough. If a cough is produced by the motor action system, the descending cognitive drive to cough activates the brainstem cough neural network. Receptors within the respiratory system provide sensory feedback to the brain indicating if the cough occurred, the cough motor pattern, and the magnitude of the cough. The limbic system uses that information to determine whether or not the coughing behavior satisfied the urge. Cough is stopped if the urge-to-cough is satisfied; if the urge has not been satisfied then the urge-to-cough will continue to motivate the central nervous system. The central component within this cough motivation system is the intrinsic brain mechanism (Fig. 1) which can be activated to start the cycle for motivating a cough, the urge-to-cough.

Respiratory sensations, including the urge-to-cough, thus, appear to be the result of two cortical processes (1) discriminative processing – awareness of the spatial, temporal, and intensity components of the respiratory input (i.e., what is sensed) and (2) affective processing – evaluative and emotional components of the respiratory input (i.e., how it feels). Discriminative processing involves somatosensory neural pathways resulting in somatosensory cortical activation. Affective processing includes the amygdala and associated structures such as the periaqueductal gray, anterior cingulate, and anterior insula (Fig. 1). The proposed model (Fig. 2) also suggests that respiratory sensation is a gated process. Cough motor drive is generated in the brainstem respiratory neural network. This cough drive produces the motor cough pattern. Cough is monitored by multiple sensory systems, which include only four major categories: muscle afferents, lung receptors, airway receptors, and chemoreceptors. These afferent systems provide sensory input to the brainstem respiratory network and it is also known that these afferents project to higher brain centers. It is further known that respiratory sensations are produced by respiratory-related changes that preferentially activate one or more of these groups of afferents.

However, these sensations do not occur with normal respiratory mechanics, ventilation and eupneic breathing patterns unless voluntary controlled attention is directed to breathing. This implies that a change of sufficient magnitude (threshold) in these respiratory sensory systems changes central neural information processing (gating), resulting in a cognitive awareness of breathing. It is also known that changes in breathing effort can be perceived. It has been reported that higher brain centers are activated when ventilatory drive is increased and some neurons show a respiratory rhythm during eupneic breathing. This suggests that respiratory motor drive is integrated with sensory input by gated comparator mechanisms into the cognitive sense that is normally gated-out of cognitive centers (Fig. 2). The urge-to-cough is further modulated by attention, experience/learning, and affective state. Attending to breathing results in cognitive awareness of breathing and can potentiate the sense of an urge-to-cough.

Experience and learning are also important components of behaviorally modulated cough. Respiratory perception and urge-to-cough studies begin with a familiarization or training session to instruct the subject on the sensation elicited by the specific ventilatory perturbation. This means the subject must experience the respiratory change and learn to associate that change with the sensation it produces. The association cortex is the brain region that mediates attention, experience, and learning; hence, it is likely that the urge-to-cough is modulated by the association cortex. Anxiety and distress elicit profound changes in ventilation and strong respiratory sensations. Thus, it is predicted that the urge-to-cough is modulated by the affective neural control system. Other sensory modalities can interact to change the sensory threshold. Eliciting a cognitive urge-to-cough is dependent on the integration of respiratory afferent activity, respiratory motor drive, affective state, attention, experience, and learning.

References

Banzett RB, Mulnier HE, Murphy K, Rosen SD, Wise RJ, Adams L (2000) Breathlessness in humans activates insular cortex. Neuroreport 11(10):2117–2120

Bloch-Salisbury E, Harver A (1994) Effects of detection and classification of resistive and elastic loads on endogenous event-related potentials. J Appl Physiol 77(3):1246–1255

Bloch-Salisbury E, Harver A, Squires NK (1998) Event-related potentials to inspiratory flow-resistive loads in young adults: Stimulus magnitude effects. Biol Psychol 49(1–2):165–186

Bolser DC, Davenport PW (2002) Functional organization of the central cough generation mechanism. Pulm Pharmacol Ther 15(3):221–225

Bradley MM (2000) Emotion and motivation. In: Cacioppo JT, Tassinary LG, Bernston GG (eds) Handbook of psychophysiology, 2nd edn. Cambridge University Press, New York, pp 602–642

Brannan S, Liotti M, Egan G, Shade R, Madden L, Robillard R, Abplanalp B, Stofer K, Denton D, Fox PT (2001) Neuroimaging of cerebral activations and deactivations associated with hypercapnia and hunger for air. Proc Natl Acad Sci USA 98(4):2029–2034

Cameron OG (2002) Visceral sensory neuroscience: Interoception. Oxford University Press, New York

Canning BJ, Mori N, Mazzone SB (2006) Vagal afferent nerves regulating the cough reflex. Respir Physiol Neurobiol 152(3):223–242

Chou YL, Davenport PW (2007) The effect of increased background resistance on the resistive load threshold for eliciting the respiratory related evoked potential. J Appl Physiol 103:2012–2017
Davenport PW, Hutchison AA (2002) Cerebral cortical respiratory-related evoked potentials elicited by inspiratory occlusion in lambs. J Appl Physiol 93(1):31–36
Davenport PW, Friedman WA, Thompson FJ, Franzen O (1986) Respiratory-related cortical potentials evoked by inspiratory occlusion in humans. J Appl Physiol 60(6):1843–1848
Davenport PW, Colrain IM, Hill PM (1996) Scalp topography of the short-latency components of the respiratory-related evoked potential in children. J Appl Physiol 80(5):1785–1791
Davenport PW, Sapeinza CM, Bolser DC (2002) Psychophysical assessment of the Urge-to-Cough. Eur Respir J 12(85):249–253
Davenport PW, Bolser DC, Vickroy T, Berry RB, Martin AD, Hey JA, Danzig M (2007a) The effect of codeine on the Urge-to-Cough response to inhaled capsaicin. Pulm Pharmacol Ther 20(4):338–346
Davenport PW, Chan PY, Zhang W, Chou YL (2007b) Detection threshold for inspiratory resistive loads and respiratory-related evoked potentials. J Appl Physiol 102(1):276–285
Dicpinigaitis PV (2003) Short- and long-term reproducibility of capsaicin cough challenge testing. Pulm Pharmacol Ther 16(1):61–65
Dicpinigaitis PV, Dobkin JB (1997) Antitussive effect of the GABA-agonist baclofen. Chest 111(4):996–999
Dicpinigaitis PV, Rauf K (1998) The influence of gender on cough reflex sensitivity. Chest 113(5):1319–1321
Dicpinigaitis PV, Grimm DR, Lesser M (1999) Cough reflex sensitivity in subjects with cervical spinal cord injury. Am J Respir Crit Care Med 159(5 Part 1):1660–1662
Evans KC, Banzett RB, Adams L, McKay L, Frackowiak RS, Corfield DR (2002) BOLD fMRI identifies limbic, paralimbic, and cerebellar activation during air hunger. J Neurophysiol 88(3):1500–1511
Gozal D, Omidvar O, Kirlew KA, Hathout GM, Hamilton R, Lufkin RB, Harper RM (1995) Identification of human brain regions underlying responses to resistive inspiratory loading with functional magnetic resonance imaging. Proc Natl Acad Sci USA 92(14):6607–6611
Gozal D, Omidvar O, Kirlew KA, Hathout GM, Lufkin RB, Harper RM (1996) Functional magnetic resonance imaging reveals brain regions mediating the response to resistive expiratory loads in humans. J Clin Invest 97(1):47–53
Harver A, Squires NK, Bloch-Salisbury E, Katkin ES (1995) Event-related potentials to airway occlusion in young and old subjects. Psychophysiology 32(2):121–129
Ho CY, Gu Q, Lin YS, Lee LY (2001) Sensitivity of vagal afferent endings to chemical irritants in the rat lung. Respir Physiol 127(2–3):113–124
Hutchings HA, Morris S, Eccles R, Jawad MS (1993) Voluntary suppression of cough induced by inhalation of capsaicin in healthy volunteers. Respir Med 87(5):379–382
Isaev G, Murphy K, Guz A, Adams L (2002) Areas of the brain concerned with ventilatory load compensation in awake man. J Physiol 539(Part 3):935–945
Kern MK, Jaradeh S, Arndorfer RC, Shaker R (2001) Cerebral cortical representation of reflexive and volitional swallowing in humans. Am J Physiol Gastrointest Liver Physiol 280(3): G354– G360
Knafelc M, Davenport PW (1997) Relationship between resistive loads and P1 peak of respiratory-related evoked potential. J Appl Physiol 83(3):918–926
Knafelc M, Davenport PW (1999) Relationship between magnitude estimation of resistive loads, inspiratory pressures, and the RREP P(1) peak. J Appl Physiol 87(2):516–522
Kollarik M, Ru F, Undem BJ (2007) Acid-sensitive vagal sensory pathways and cough. Pulm Pharmacol Ther 20(4):402–411
Lasserson D, Mills K, Arunachalam R, Polkey M, Moxham J, Kalra L (2006) Differences in motor activation of voluntary and reflex cough in humans. Thorax 61(8):699–705
Lee PC, Cotterill-Jones C, Eccles R (2002) Voluntary control of cough. Pulm Pharmacol Ther 15(3):317–320

Liotti M, Brannan S, Egan G, Shade R, Madden L, Abplanalp B, Robillard R, Lancaster J, Zamarripa FE, Fox PT, Denton D (2001) Brain responses associated with consciousness of breathlessness (air hunger). Proc Natl Acad Sci USA 98(4):2035–2040

Logie ST, Colrain IM, Webster KE (1998) Source dipole analysis of the early components of the RREP. Brain Topogr 11(2):153–164

Mazzone SB (2005) An overview of the sensory receptors regulating cough. Cough 1:2

Mazzone SB, McLennan L, McGovern AE, Egan GF, Farrell MJ (2007) Representation of capsaicin-evoked urge-to-cough in the human brain using functional magnetic resonance imaging. Am J Respir Crit Care Med 176(4):327–332

O'Donnell DE, Banzett RB, Carrieri-Kohlman V, Casaburi R, Davenport PW, Gandevia SC, Gelb AF, Mahler DA, Webb KA (2007) Pathophysiology of dyspnea in chronic obstructive pulmonary disease: A roundtable. Proc Am Thorac Soc 4(2):145–168

Peiffer C, Poline JB, Thivard L, Aubier M, Samson Y (2001) Neural substrates for the perception of acutely induced dyspnea. Am J Respir Crit Care Med 163(4):951–957

Revelette WR, Davenport PW (1990) Effects of timing of inspiratory occlusion on cerebral evoked potentials in humans. J Appl Physiol 68(1):282–288

Sapienza CM, Bolser DC, Schulz GM, Scott K, Davenport PW (1997) Respiratory muscle and airflow pattern during graded voluntary cough in normal subjects. Am J Respir Crit Care Med (Abstracts)155:A577

Shannon R, Baekey DM, Morris KF, Li Z, Lindsey BG (2000) Functional connectivity among ventrolateral medullary respiratory neurones and responses during fictive cough in the cat. J Physiol 525(Part 1):207–224

Shannon R, Baekey DM, Morris KF, Nuding SC, Segers LS, Lindsey BG (2004) Production of reflex cough by brainstem respiratory networks. Pulm Pharmacol Ther 17(6):369–376

Simonyan K, Saad ZS, Loucks TM, Poletto CJ, Ludlow CL (2007) Functional neuroanatomy of human voluntary cough and sniff production. Neuroimage 37(2):401–409

Straker D (2002) The Brain's Urge System. <http://changingminds.org/explanations/brain/urge_system.htm>. Accessed 15 March, 2007

Toogood JA, Barr AM, Stevens TK, Gati JS, Menon RS, Martin RE (2005) Discrete functional contributions of cerebral cortical foci in voluntary swallowing: A functional magnetic resonance imaging (fMRI) "Go, No-Go" study. Exp Brain Res 161(1):81–90

Vovk A, Bolser DC, Hey JA, Danzig M, Vickroy T, Berry R, Martin AD, Davenport PW (2007) Capsaicin exposure elicits complex airway defensive motor patterns in normal humans in a concentration-dependent manner. Pulm Pharmacol Ther 20(4):423–432

Webster KE, Colrain IM (2000a) The relationship between respiratory-related evoked potentials and the perception of inspiratory resistive loads. Psychophysiology 37(6):831–841

Webster KE, Colrain IM (2000b) The respiratory-related evoked potential: Effects of attention and occlusion duration. Psychophysiology 37(3):310–318

Webster KE, Adey SA, Colrain IM (2002) The effect of stimulus probability on P3 in the respiratory-related evoked potential. Psychophysiology 39(1):9–15

Widdicombe JG, Eccles R, Fontana G (2006) Supramedullary influences on cough. Respir Physiol Neurobiol 152(3):320–328

Wiley RL, Zechman FW Jr (1966) Perception of added airflow resistance in humans. Respir Physiol 2(1):73–87

Zhao W, Martin AD, Davenport PW (2002) Respiratory-related evoked potentials elicited by inspiratory occlusions in double-lung transplant recipients. J Appl Physiol 93(3):894–902

Clinical Cough II: Therapeutic Treatments and Management of Chronic Cough

A.H. Morice and L. McGarvey(✉)

Contents

Abstract Chronic cough is a common and frequently disruptive symptom which can be difficult to treat with currently available medicines. Asthma/eosinophilic airway disease and gastro-oesophageal reflux disease are most commonly associated with chronic cough but it may also trouble patients with chronic obstructive pulmonary disease, pulmonary fibrosis and lung cancer. Over the last three decades there have been a number of key advances in the clinical approach to cough and a number of international guidelines on the management of cough have been developed. Despite the undoubted benefit of such initiatives, more effective treatments for cough are urgently needed. The precise pathophysiological mechanisms of chronic cough are unknown but central to the process is sensitization (upregulation) of the cough reflex. One well-recognized clinical consequence of this hypersensitive state is bouts of

L. McGarvey
Department of Medicine, Institute of Clinical Science, The Queen's University of Belfast, Grosvenor Road, Belfast, BT12 6BJ, UK
l.mcgarvey@qub.ac.uk

K.F. Chung, J.G. Widdicombe (eds.), *Pharmacology and Therapeutics of Cough*,
Handbook of Experimental Pharmacology 187,

coughing triggered by apparently trivial provocation such as scents and odours and changes in air temperature. The main objective of new treatments for cough would be to identify ways to downregulate this heightened cough reflex but yet preserve its crucial role in protecting the airway. The combined efforts of clinicians, scientists and the pharmaceutical industry offer most hope for such a treatment breakthrough. The aim of this chapter is to provide some rationale for the current treatment recommendations and to offer some reflections on the management of patients with chronic cough.

1 Introduction

Coughing is the most widely recognized clinical sign and symptom but probably the least well understood. Despite notable advances over the last 50 years in our understanding of the basic science and clinical aspects of cough, major gaps in knowledge remain and few if any effective treatments exist. It is not surprising therefore that although patients with a troublesome cough frequently present to primary- and secondary-care physicians, managing this symptom is often undertaken with little hope for diagnostic or treatment success. However, there is now some consensus on the management of patients with cough based on both published evidence and the expert opinion of clinicians with a specialist interest in this area. In fact the last decade has seen a number of published cough management guidelines endorsed by international respiratory societies (Irwin et al. 2006a; Morice et al. 2004; 2006). Although some have encountered criticism for a reliance on low-level evidence, there has been general support for the principles behind such efforts and a call for increased endeavour in this area (Chung et al. 2006).

The purpose of this chapter is to provide some rationale for the current recommendations and to offer some personal reflections on the treatment of patients with chronic cough.

2 Definition of Chronic Cough

Coughing serves as a crucial protective reflex for the upper airway. A problem cough is one which appears to serve no useful physiological purpose. The term 'chronic cough' has been used to describe a cough that has persisted for more than 8 weeks (Morice et al. 2004).

The definition of 'cough' has been a subject of intense debate (Fontana and Widdicombe 2007). It is unlikely that any single definition exists which completely satisfies the physician, the scientist and the worried patient. For the purpose of this chapter, we prefer the most relevant and accurate definition of a clinical cough (Morice et al. 2006), namely "Cough is a forced expulsive manoeuvre, usually against a closed glottis and which is associated with a characteristic sound and this definition is accepted by the majority of those working in the field" (Morice et al. 2007).

3 How Common is Chronic Cough?

Chronic cough is one of the commonest syndromes to affect the general population. The prevalence of chronic cough in the general population reported in the literature varies and depends very much on the questionnaire used and the population studied. On the basis of North American, British and European surveys, estimates of between 11 and 20% of the general population have been reported (Barbee et al. 1991; Cullinan 1992; Lundback et al. 1991). Environmental factors including smoking prevalence and exposure (including passive) and environmental pollution all contribute to these figures (Cerveri et al. 2003; David et al. 2005; Viegi et al. 1999). A recent survey in Yorkshire, UK, showed that 12% of an unselected population complained of coughing bouts on a greater-than-weekly basis. Furthermore, 7% of the respondents felt that cough interfered with activities of daily living (Ford et al. 2006). A survey of patients who had requested information following a British radio programme on chronic cough reported that the average number of physicians consulted was six and that clearly inadequate treatment strategies had been applied by these physicians (Everett et al. 2007). Cough as an isolated symptom accounts for between 10 and 38% of all new referrals to respiratory specialists (Irwin et al. 1990, McGarvey et al. 1998b). The most troubled patients often seek specialist help and it is clear that the success rates in treating cough in specialist clinics contrasts markedly with those seen in non-specialist clinics (McGarvey et al. 1998b). Many patients in primary care, once trials of asthma therapy have failed, are told that they must live with their asthma. Yet chronic cough can be socially and physically debilitating, leading to major decrement in the patient's quality of life (Birring et al. 2003b; French et al. 1998).

4 Mechanisms of Chronic Cough

A detailed commentary of the physiological basis of cough and the pathophysiologal processes involved in clinical cough has been covered elsewhere in this book. Coughing is a reflex act and in its simplest form is served by afferent receptors located on airway sensory nerves which when stimulated input via the vagus nerve to the brain stem, from where efferent fibres relay to both thoracic and abdominal musculature and laryngeal structures responsible for coughing. The cerebral cortex also appears to exert higher voluntary control over these processes and may be involved in the sensation of the 'need' or 'urge' to cough (Davenport et al. 2007; Mazzone et al. 2007).

A number of receptors (nociceptors) located on vagal afferent sensory airway nerves are responsive to a variety of noxious stimuli (chemical and mechanical) known to induce cough. In particular, the rapidly adapting receptor present on myelinated Aδ-type nerves, which appears polymodal (responding to both chemical and mechanical stimuli) and the small unmyelinated C-fibres, sensitive to the chemical capsaicin, the tussive extract from hot chilli peppers, are widely believed to be important in the mediation of cough. However, this classification of nociceptors is based on the response of individual nerve fibres and their nerve conduction velocities rather than the pharmacological and molecular characteristics of the individual receptor molecules located on these neurones. More recently, evidence has emerged for the existence of a dedicated cough receptor which has no specific features typical of the rapidly adapting receptor or C-fibre receptors which is related to a specific sodium ATPase (Canning 2006). Establishing the precise molecular receptor(s) responsible for cough and their endogenous ligands or agonists is crucial not least because they provide the greatest potential for future therapeutic modulation. Most work currently evaluating this has come from experiments involving capsaicin, which also readily evokes cough in both animals and humans and is likely to do so by stimulation of the transient receptor potential vanilloid 1 (TRPV-1) receptor (Morice and Geppetti 2004), a member of the vanilloid family of ion channels located on C-fibres (Caterina et al. 1997). This receptor is also activated by heat and low pH. There is much interest in the effect of alterations in airway pH which may evoke or trigger events including cough and bronchospasm (Canning et al. 2006; Kollarik and Undem 2002). Increased expression of TRPV-1 receptors has been demonstrated in the airways of humans, including patients with chronic cough (Groneberg et al. 2004; Mitchell et al. 2005). Another receptor from the transient receptor potential family, the noxious cold receptor TRPA1, is exciting much interest and since it is colocalized in neurones with TRPV-1 it may well have a modulatory role in cough reflex sensitivity. Like TRPV-1, another receptor which is responsive to acidification is the acid-sensing ion channel. It appears to have an established role in nociceptive responses in particular pain (Krishtal 2003). Currently no information exists on its presence in the human airway and on any specific role in mediating cough.

A key mechanistic principle underlying chronic cough is upregulation of the cough reflex. This sensitization process may occur peripherally and/or centrally (Carr et al. 2006; Bonham et al. 2004). The clinical consequence of this hypersensitive state is a cough triggered at low threshold by stimuli such as cold air, scents and odours. Inflammation occurs within the airways in patients with chronic cough and is likely to be responsible for the sensitization of the cough reflex. The pathophysiological changes which typify chronic cough include disruption of the airway epithelial lining, basement membrane thickening and chronic inflammatory infiltrate (Boulet et al. 1994; Irwin et al. 2006b; Niimi et al. 2005). These changes are mimicked by the profile of inflammatory cytokines in induced sputum and airway lavage samples (Birring et al. 2003a, 2004a, b; Jatakanon et al. 1999; McGarvey et al. 1999). Airway inflammation is a feature of asthma and chronic obstructive pulmonary disease, both of which are frequently characterized by bothersome cough and which exhibit a heightened response to the tussive effects of capsaicin and citric acid (Doherty et al. 2000; Wong and Morice 1999). Bradykinin and prostaglandin E_2 (both elevated in the airways of these patients) seem particularly important in sensitizing the cough reflex (Choudry et al. 1989). Bradykinin is also said to have an important mechanistic role in the cough associated with angiotensin-converting-enzyme inhibitors (Fox et al. 1996). Others suggest that neuropeptides such as substance P (Morice et al. 1987) are the major precipitant. Neurokinin A and cacitonin gene related peptide, which are stored and released by C-fibres, are also considered to be important cough reflex sensitizers. Recently, a number of published findings have suggested that neuropeptides may have an important role in human cough (O'Connell et al. 1995; Patterson et al. 2007). Other potentially important inflammatory sensitizers include adenosine, or more specifically its metabolite ATP. Both inhalation and airway challenge with ATP induce vigorous coughing in healthy and asthmatic subjects (Basoglu et al. 2005; Crummy et al. 2005).

A number of the inflammatory mediators mentioned above have the ability to activate the TRPV-1 receptor indirectly by activation of the protease-activated receptors and through protein kinase dependent pathways(Dai et al. 2004; Gatti et al. 2006; Ramachandran et al. 2007).

5 General Principles on the Management of Cough

The management of chronic cough relies on the application of a simple basic principle. Where possible the cause of the cough should be identified and treated and cough suppression should be only considered where such a strategy fails or is impractical. Implicit in this idea is that there are clear-cut diagnostic categories for the causes of cough. In certain circumstances, such as an inhaled foreign body, this is clearly true, with removal of the offending item producing an immediate and sustained improvement in cough. Even in these circumstances the diagnosis may be obscure with radiolucent hollow foreign bodies, such as a ballpoint pen top, or in the example of a chicken vertebra illustrated in Fig. 1, the object may be lodged without obvious sign for many years.

In clinical practice it is rare that the patient with the syndrome of chronic cough presents with an obvious single diagnostic entity. Indeed, in a number of studies multiple causes of chronic cough are described within a single patient (Irwin et al. 1981, 1990). This could have two explanations; firstly that there are genuinely

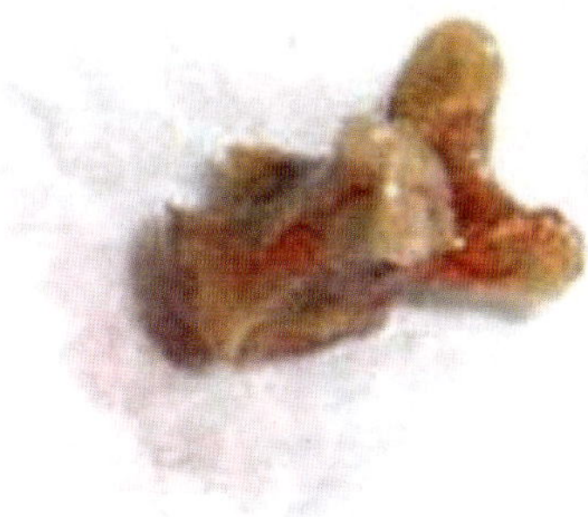

Fig. 1 A chicken vertebra removed from the right main bronchus 18 months after a chicken dinner followed by an intractable cough

multiple causes of chronic cough, which frequently coexist, or secondly chronic cough may have one cause which presents with protean manifestations.

In this chapter we deliberately avoid such philosophical issues and concentrate on the therapeutic options facing the clinician in a patient with chronic cough.

5.1 Management Protocols

The algorithms recommended for the management of a patient with chronic cough have recently been extensively reviewed in national and international guidelines. Despite differences in the methods of producing the guidelines, i.e. whether they were mainly evidence-based or opinion-based, the European Respiratory Society's guidelines (Morice et al. 2004), the British Thoracic Society's guidelines (Morice et al. 2006) and the American College of Chest Physicians' guidelines (Irwin et al. 2006) all endorse a single management philosophy based around the clinical picture with limited simple investigations followed by therapeutic trials based on the most likely cause. This approach has been tested and demonstrated to be the quickest and most cost effective strategy (Lin et al. 2001). It avoids unnecessary and frequently unpleasant investigations (which themselves are of dubious accuracy) (Patterson et al. 2004) and achieves the effect desired by the patient: the decrease or abolition of the cough in the shortest possible time.

A schematic overview of the management of adults with chronic cough as recommended by the British Thoracic Society is presented in Figs. 2 and 3. Figure 2 outlines the initial clinical assessment, including baseline investigations, and a suggested sequence of trials of disease-specific treatment that could be initiated by the treating physician in primary care. The persistence of symptoms and the need for more-specialist investigations such as bronchial challenge testing, oesophageal studies and induced-sputum sampling and analysis should prompt referral to secondary care and ideally to a unit with a specialist interest in the evaluation of cough.

In selected cases, a number of additional investigations, including bronchoscopy, high-resolution computed tomography of the chest, cardiac studies and psychological assessment, may be useful (see Fig. 3).

The approach described above is valid because in primary- and secondary-care settings, airway diseases (including upper-airway conditions) and gastro-oesophageal reflux disease are the conditions most frequently associated with chronic cough in adults. Each will be considered in more detail in the following sections.

5.2 Treatment of Asthma/Airway Eosinophilia Associated Cough Syndromes

Cough is sometimes the only symptom in asthma. In such circumstances and in the absence of airflow obstruction, the condition 'cough variant asthma' may be considered. Like classic asthma, the nature of airway inflammation in cough-variant

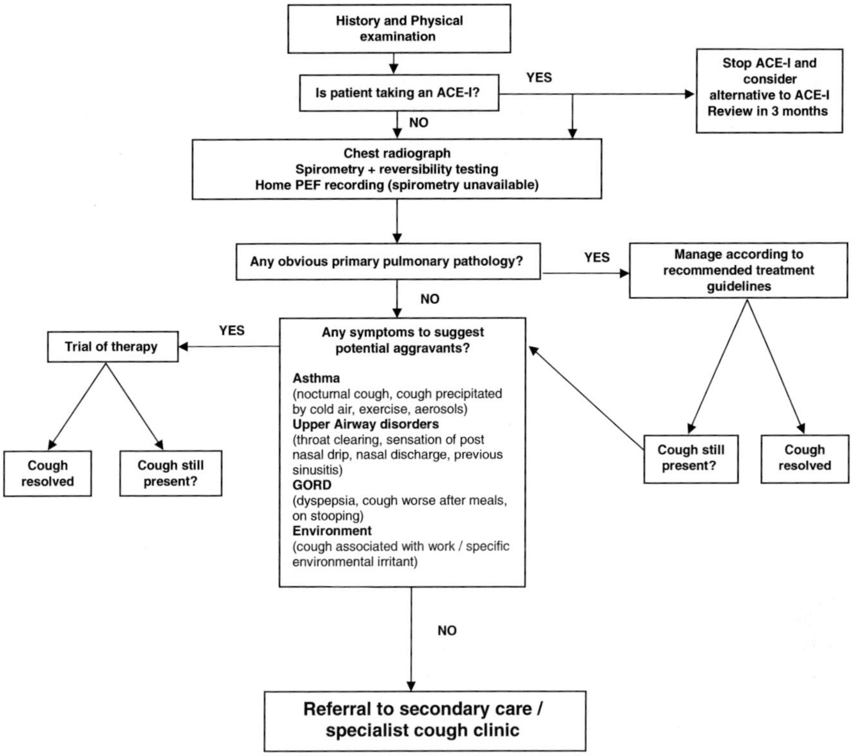

Fig. 2 Protocol for the evaluation of chronic cough in an adult (primary/secondary care) (modified with permission from Morice et al. 2006)

asthma is eosinophilic and typified with raised sputum and lavage eosinophil counts and increased exhaled nitric oxide concentrations (Brightling and Pavord 2000; Chatkin et al. 1999). However, 20 years ago Gibson et al. described a clinical syndrome of 'eosinophilic bronchitis' which also presents with isolated cough and is likewise associated with an airway eosinophilia (Gibson et al. 1989). What distinguishes this syndrome from cough-variant asthma is the absence of airway hyperresponsiveness (Pavord 2004). Finally, the condition of atopic cough described largely in the Japanese population is also typified by cough, normal lung function and airway eosinophila (Fujimura et al. 2003). One opinion has been that all of these conditions represent some part of the clinical spectrum of asthma (McGarvey and Morice 2003), although the view that they represent distinct clinical entities with some 'overlap' features has been suggested (see Fig. 4) (McGarvey et al. 2007). What is clinically more important is the feature distinguishing all three conditions from other forms of chronic cough, which is the treatment success with corticosteroids. In patients with chronic cough and evidence of airway eosinophila, inhaled corticosteroids may be dramatically effective. The time course to response has been claimed to be slower than that in classic asthma (Brightling et al. 2000). Patients

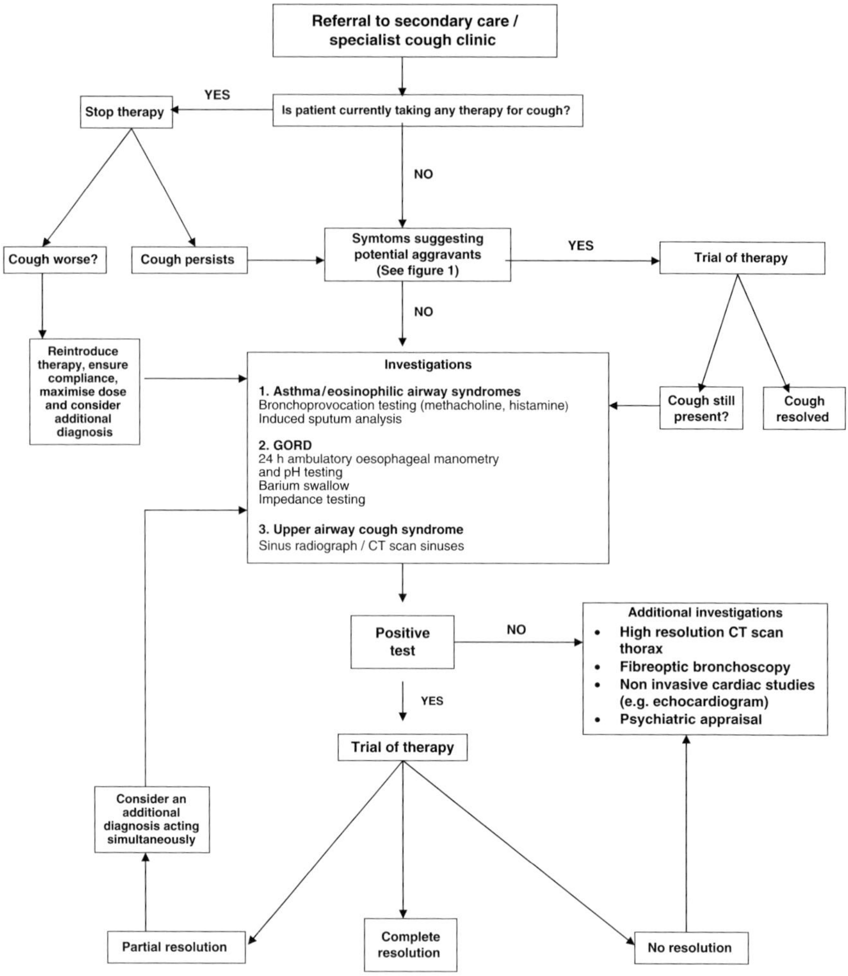

Fig. 3 Evaluation and management of an adult with chronic cough (secondary care/specialist clinic) (modified with permission from Morice et al. 2006)

with eosinophilic cough syndromes have a heightened cough reflex and with successful treatment cough reflex sensitivity returns to normal. This is illustrated in Fig. 5, which shows the change in cough reflex sensitivity to citric acid following administration of budesonide to a patient with cough-variant asthma. This normalization of cough reflex sensitivity is particularly well illustrated in inflammatory mechanisms of cough where reduction in the inflammatory milieu around the cough receptor returns sensitivity to normal.

On a practical note, for patients with cough and airway eosinophilia which fails to respond to inhaled corticosteroids, leukotriene antagonists are a much more effective second-line agent than long-acting β-agonists, as are conventionally used in

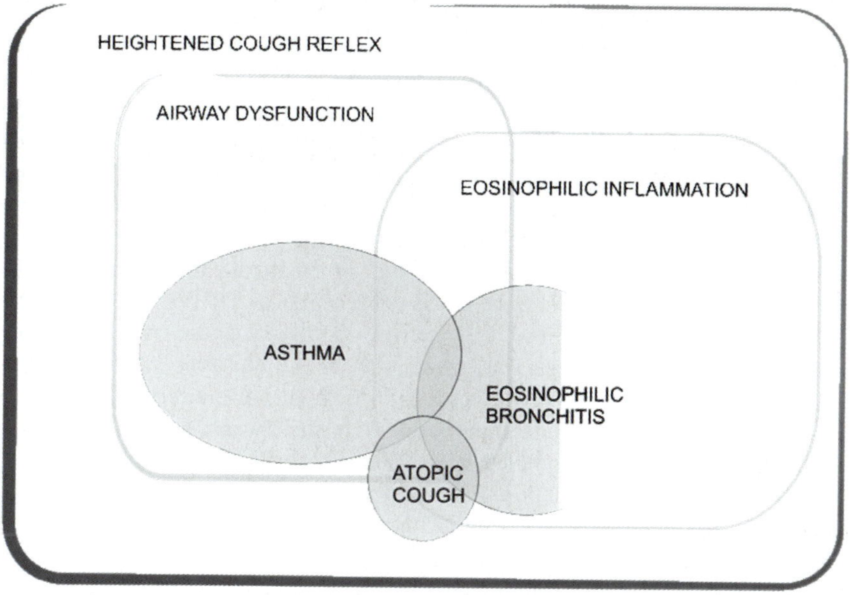

Fig. 4 Overlap between cough, airway dysfunction and eosinophilic inflammation (reproduced with permission from McGarvey et al. 2007)

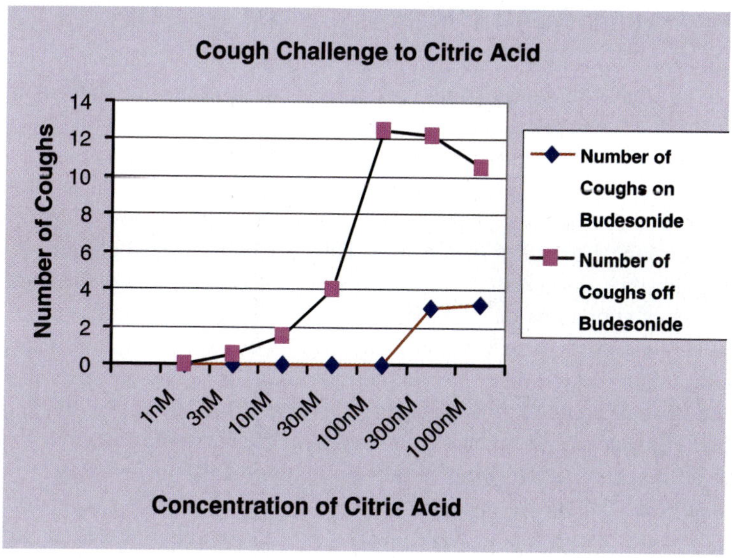

Fig. 5 Citric acid cough challenge in a patient with cough-variant asthma: the effect of inhaled corticosteroids

classic asthma (Dicpinigaitis et al. 2002). This difference in pharmacological action is logical in that patients with cough-variant asthma and eosinophilic bronchitis by definition have little alteration in airway tone, and giving a long-acting bronchodilator could only therefore alter bronchial hyperresponsiveness – a completely separate reflex from the cough reflex. Leukotriene antagonists have been demonstrated to ameliorate cough challenge in patients with cough-variant asthma but not in those with classic asthma (Dicpinigaitis and Dobkin 1999). This may be because the mediators released through the lipoxygenase pathway (1) are not responsive to corticosteroids and (2) may be directly involved in sensitizing the cough reflex. One of the putative cough receptors, the capsaicin receptor or TRPV-1, has intracellular binding sites which are known to be sensitive to lipoxygenase products (Hwang et al. 2000). Thus, leukotriene antagonists may have a dual role altering the profile of TH2-type inflammation within chronic cough and also directly affecting the sensitivity of one of the major cough receptors. In such instances a trial of steroids via the oral route may be indicted where administration is required to target alternative sources of inflammation. Failure of patients to respond to 30 mg per day of prednisolone administered orally for 2 weeks excludes a significant eosinophilc component to their chronic cough and all forms of steroid therapy should be withdrawn (Morice et al. 2006). Alternative diagnoses and therefore therapeutic avenues should be considered.

5.3 Treatment of Reflux Cough

Reflux is an important cause of chronic cough, but it differs fundamentally from the conventional textbook description of gastro-oesophageal reflux disease, and the treatment of its cardinal symptom, heartburn. Acid exposure is the major provocateur of the symptom of heartburn and the consequent oesophageal inflammation. In contrast, reflux cough is due to, at least in part, the volume effects of reflux, which may or may not be acidic in nature. Thus, the conventional wisdom in gastro-oesophageal reflux disease that adequate suppression of the acid alone should produce a relief of symptoms may be incorrect in treating reflux cough.

However, it is worth considering the evidence for a mechanistic role of acid reflux in contributing to cough. Firstly, close temporal relationships between acid reflux events and cough bouts have been observed (Irwin et al. 1993). Secondly, patients with acid reflux have a heightened cough reflex to capsaicin (Ferrari et al. 1995) and this can be heightened following direct stimulation of the lower oesophagus with acidic solutions (Javorkova et al. 2007). Finally, elevated airway levels of neuropeptides in patients with acid reflux and coexistent cough and asthma suggest a link between acid reflux and neural activation (Patterson et al. 2007). The precise relationship between acid and cough is likely to be complex as recent evidence suggests direct acid stimulation of the upper airways may actually reduce sensitivity to subsequent mechanical stimulation (Phua et al. 2005).

There is no doubt that some patients with reflux cough respond to acid suppression. Surprisingly, however this has been difficult to show in clinical studies (Chang et al. 2006). This is partly due to the poor design of the studies and is also due to a placebo effect of treatment in the early stages of the disease. In particular, reflux cough is subject to a number of psychological influences and relief of stress and dietary modifications may produce a pleasing non-pharmacological improvement. Fewer than 50% of patients with clinically diagnosed reflux cough suffer from heartburn, with the other classic symptom of reflux, regurgitation, being a prominent feature. It is therefore not surprising that acid suppression with once-daily proton-pump inhibitors is poorly efficacious in the treatment of reflux cough. Even twice-daily suppression, or even full acid suppression (Khoury et al. 1999) with twice-daily proton-pump inhibitor coupled with ranitidine at night (most nocturnal acid secretion is histaminergic) produces a response of approximately 50%. However, this regime of full acid suppression is endorsed by various guideline statements and appears to be effective in a number of individuals. Since it is relatively cheap and safe there is some justification for its recommendation even in the absence of strong evidence-based medicine support.

It is often recommended that acid suppression is given in chronic cough for a more prolonged period that it is in heartburn (Irwin et al. 2006a, b). This may represent a genuine delayed relief from this therapy. The cynic might comment that this gives a greater period of time for regression to the mean to take place!

If full acid suppression is ineffective or causes inadequate suppression of cough, then promotility agents are advocated (Poe et al. 2003). The main drugs which work on oesophageal motility are the dopamine agonists metoclopramide and domperidone. Both are used initially at the dose of 10 mg three times daily. Unfortunately, the side-effect profile sometimes prevents long-term use, with drowsiness, galactorrhoea and rarely dystonic movements a feature of their use. It is recommended that each of these agents should be tried in turn for 1 month to assess efficacy.

Baclofen is a $GABA_A$ agonist and mimics the inhibitory effect of the vagus nerve on transient opening of the lower oesophageal sphincter. It can reduce cough dramatically in selected patients particularly when vagal inhibition is reduced in an autonomic neuropathy (Menon et al. 2005; Dicpinigaitis et al. 1998). There may be a central component of activity in that baclofen inhibits capsaicin cough sensitivity in normal subjects (Dicpinigaitis et al. 2000). Again side effects are a limiting feature, particularly dysphoria.

Finally, two agents that have been relatively poorly researched can be useful in oesophageal dysmotility. Erythromycin at a low dose is frequently used in paediatric practice and acts as a motilin agonist. Our personal experience is that a number of adult patients also respond to this as single-agent therapy. Magnesium, either in liquid magnesium preparations or as magnesium glycerophosphate, can be dramatically effective in some patients. Since this agent is very poorly absorbed from the gastrointestinal tract it is presumably working as a topical modulator of oesophageal smooth-muscle function.

A considerable body of evidence supports the surgical intervention in patients with intractable reflux cough (Allen and Anvari 1998; Chen and Thomas 2000;

Ekstrom and Johansson 2000; Fitzgerald et al. 1989; Novitsky et al. 2002; Thoman et al. 2002). The usual procedure, Nissan fundoplication, effectively blocks significant reflux from the stomach into the oesophagus. A laparoscopic fundoplication is relatively atraumatic but requires considerable surgical skill in order to achieve the correct lower-oesophageal sphincter pressure. The procedure is almost universally successful in heartburn and therefore many surgeons are happy to operate on patients who are suffering from this symptom in addition to reflux cough (Farrell et al. 2001). In those patients without heartburn the success rate is approximately two thirds. It has recently been suggested that those patients who do not respond have abnormal oesophageal motility with bolus arrest at the region of the aortic arch (Fox and Bredenoord 2007). This produces oesophagopharyngeal reflux and spasm of the lower oesophagus. In this case fundoplication may actually be harmful. The long-term success rate of this procedure is somewhat debated with insufficient evidence as to its long-term 'cure' rate. In obese patients other antireflux procedures, such a Rou-en-Y, may have the dual effect of being bariatric as well as antitussive (Stanford et al. 2003).

5.3.1 Treatment of Laryngopharyngeal Reflux

Gastric contents can reflux as far as the larynx and hypopharynx and contribute to a number of symptoms, including dysphonia, dysphagia, cough and throat clearing. Symptoms do not always correlate well with laryngeal findings and most patients do not report heartburn (Franco 2006; Khan et al. 2006). Intuitively, patients should respond to antireflux therapy, but there is very little clinical trial evidence on which to base recommendations.

5.4 Treatment of Upper-Airway Cough Syndrome

This is a new term introduced by the American College of Chest Physicians Consensus Panel 2006 and includes what was previously termed postnasal drip syndrome (PNDS) (Irwin et al. 2006a, b). PNDS was previously considered the most common cause of chronic cough (Pratter et al. 1993). In the European Respiratory Society guidelines, PNDS is discounted as a common cause of cough and indeed its very existence as a syndrome rather than a symptom has been questioned (Morice 2004). Rhinitis is the preferred European terminology. However, the suggestion the PNDS/upper-airway cough syndrome was an important cause of cough was based on the assumption that improvement in cough following use of sedating antihistamines suggested PNDS as the aggravating factor (Irwin et al. 1990). This treatment response is suggested to be due to a non-specific action of the antihistamine on the central nervous system. Although an abundance of literature describes the relationship between PNDS and cough, there is evidence that PNDS often coexists without cough (O'Hara and Jones 2006). Any relationship between cough and upper-airway symptoms is not likely to be unified by mechanical stimulation of the pharynx by mucus. A more plausible explanation for this symptom association is inflammation of the entire airway (O'Hara and Jones 2005). Current recommendations on managing upper-airway cough syndrome include an empirical trial of sedating antihistamine/decongestant (Pratter 2006) or intranasal corticosteroids (Morice et al. 2006). Later we suggest that the response to antihistamines is not due to an effect on PNDS but on the TRPV-1 receptor and is therefore non specific in nature.

5.5 Idiopathic Cough

Despite a meticulous protocol involving diagnostic testing and trials of empirical therapy, there may be no obvious cause for a chronic cough in up to 42% of patients referred for specialist evaluation (Haque et al. 2005; McGarvey et al. 1998; Polley et al. 2005). In some cases, failure to consider causes including the asthma/eosinophilic airway syndromes such as eosinophilic bronchitis and atopic cough, or non-acid gastro-oesophageal reflux disease may explain diagnostic failure. However, a distinct group of patients may be considered to have true idiopathic cough (McGarvey and Ing 2004). Current published evidence suggests a certain patient phenotype, namely middle-aged women with prolonged non-productive cough and cough reflex hypersensitivity (Haque et al. 2005). Recent studies have suggested a relationship between idiopathic cough and organ-specific autoimmune disease (Birring et al. 2004b). Almost nothing else is known about this clinical entity and currently no therapy exists.

6 Cough Suppression

As implied earlier, for some patients 'disease-specific' cough therapy is ineffective. Strategies involving cough suppression can produce highly satisfactory amelioration of symptoms with long-term benefit. Two basic strategies can be invoked. Firstly, the use of opiates in cough has been described for many hundreds of years. However, until recently there were no randomized controlled trials demonstrating the efficacy of opiates in chronic cough. The opiate which is frequently used in animal studies and indeed is regarded as the archetypal opiate for the treatment of cough is codeine. Codeine however is a very variable metabolism in humans depending on the cytochrome P4502D6 (Kathiramalainathan et al. 2000). In some subjects very high blood levels of morphine are produced by oral codeine administration, whereas in others who are poor metabolizers low circulating levels lead to a low-side-effect profile but also very low efficacy. A study therefore was undertaken of use of low-dose morphine (5 mg twice daily) in a slow-release formulation in chronic cough (Morice et al. 2007). There was a highly significant reduction in diary record card cough and Leicester Cough Questionnaire score but an insignificant change in cough reflex sensitivity. Approximately one third of patients with intractable cough responded to this regime, with a further one third responding to 10 mg twice daily in an open-label extension study. The major side effect was constipation in approximately half the patients. Cough suppression with opiates appears to be a highly effective strategy when 'disease-specific' remedies have proven to be ineffective.

Until recently first-generation antihistamine drugs were only advocated within American College of Chest Physicians' cough guidelines (Irwin et al. 2006a, b). The European experience has been these drugs have relatively poor efficacy. However, a number of first-generation antihistamines are not licensed in Europe and from this dichotomy of experience the hypothesis was formed that certain specific antihistamines might have important additional effects in chronic cough. Dexbrompheniramine, an isomer of brompheniramine, is a first-generation antihistamine which has been demonstrated in a number of studies to suppress cough (Pratter 2006). Dexbrompheniramine was tested in an in vitro assay of cloned putative cough receptor, the capsaicin TRPV-1 receptor. In this assay capsaicin response was highly effectively blocked by dexbrompheniramine but not by several other antihistamines (Sadofsky et al. 2005). It appears, therefore, that this class of drug has specific actions on a putative cough receptor and our clinical experience is that it is not specific for a particular type of cough. Approximately two thirds of patients with intractable otherwise unresponsive cough respond to this agent and it is a valuable addition to the armamentarium of the cough specialist.

7 Non-pharmacological Treatments

Behavioural modification is a potential non-pharmacological treatment strategy. A psychological approach presumes some psychogenic component to the symptom and most evidence for its efficacy has come from case series reported in the paediatric literature (Gay et al. 1987).

Recently the role of speech disorder management in chronic cough has been tested in a randomized controlled trial (Vertigan et al. 2006). In this study, while both the behavioural strategy and the placebo programme reduced cough scores, the magnitude of improvement was significantly greater in the speech disorder intervention. The improvement observed was maintained throughout the 2-month follow-up although complete resolution of cough was not seen in any case. It is possible that such treatment may be a helpful adjunct to the management of some subjects with unresolved cough.

8 The Future

A number of agents are in development. As inferred earlier, suppression of the TRPV-1 capsaicin receptor is a very likely target for cough inhibition. A number of pharmaceutical companies have developed TRPV-1 antagonists which are undergoing clinical trial although predominately in clinical areas of migraine, neuropathic and dental pain. Clinical studies in cough are yet to be undertaken, but whatever the result it will be an important pharmacological tool to investigate the different forms of cough reflex stimulation.

Nocioceptin in the guise of its receptor NOP1 may be an important modulatory agent in sensory nerve activity. By decreasing the excitability of the afferent nerve, one reduces the stimulus to cough. Initial clinical studies are promising, but in general nerve inhibition, such as occurs with lignocaine, has proven to be a less promising route (Howard et al. 1977).

Finally, much neurogenic inflammation within the airways is a consequence of release of the neuropeptides substance P and other neurokinins. Antagonists of these agents have proven to be particularly poor in the treatment of asthma but have been developed in therapy of chemotherapy-induced nausea and vomiting. Their efficacy in chronic cough remains to be determined.

9 Conclusions

The clinical pharmacological treatment of chronic cough relies on two strategies, specific treatment of aetiological factors and non-specific cough suppression. Many patients have important psychological problems brought on by the severe decrement in the quality of life caused by chronic cough. Until recently, the treatment of chronic cough was viewed by physicians with pessimism. In the last decade in particular significant steps have been taken to focus scientific, clinical and pharmacological expertise to better understand how and why patients cough and what evaluation and treatment strategies are most appropriate. It is hoped that the momentum generated by these efforts will replace current nihilism with optimism for the future management of this important symptom.

References

Allen CJ, Anvari M (1998) Gastro-oesophageal reflux related cough and its response to laparoscopic fundoplication. Thorax 53:963–968

Barbee RA, Halonen M, Kaltenborn WT, Burrows B (1991) A longitudinal study of respiratory symptoms in a community population sample. Correlations with smoking, allergen skin-test reactivity, and serum IgE. Chest 99:20–26

Basoglu OK, Pelleg A, Essilfie-Quaye S, Brindicci C, Barnes PJ, Kharitonov SA (2005) Effects of aerosolized adenosine 5′-triphosphate vs adenosine 5′-monophosphate on dyspnea and airway caliber in healthy nonsmokers and patients with asthma. Chest 128:1905–1909

Birring SS, Brightling CE, Symon FA, Barlow SG, Wardlaw AJ, Pavord ID (2003a) Idiopathic chronic cough: Association with organ specific autoimmune disease and bronchoalveolar lymphocytosis. Thorax 58:1066–1070

Birring SS, Prudon B, Carr AJ, Singh SJ, Morgan MD, Pavord ID (2003b) Development of a symptom specific health status measure for patients with chronic cough: Leicester Cough Questionnaire (LCQ). Thorax 58:339–343

Birring SS, Parker D, Brightling CE, Bradding P, Wardlaw AJ, Pavord ID (2004a) Induced sputum inflammatory mediator concentrations in chronic cough. Am J Respir Crit Care Med 169:15–19

Birring SS, Murphy AC, Scullion JE, Brightling CE, Browning M, Pavord ID (2004b) Idiopathic cough and organ-specific autoimmune diseases: A case-control study. Respir Med 98:242–246

Bonham AC, Sekizawa SI, Joad JP (2004) Plasticity of central mechanisms for cough. Pulm Pharmacol Ther 17:453–457; discussion 469–470

Boulet LP, Milot J, Boutet M, St Georges F, Laviolette M (1994) Airway inflammation in nonasthmatic subjects with chronic cough. Am J Respir Crit Care Med 149:482–489

Brightling CE, Pavord ID (2000) Eosinophilic bronchitis: An important cause of prolonged cough. Ann Med 32:446–451

Brightling CE, Ward R, Wardlaw AJ, Pavord ID (2000) Airway inflammation, airway responsiveness and cough before and after inhaled budesonide in patients with eosinophilic bronchitis. Eur Respir J 15:682–686

Canning BJ (2006) Anatomy and neurophysiology of the cough reflex: ACCP evidence-based clinical practice guidelines. Chest 129:33S–47S

Canning BJ, Farmer DG, Mori N (2006) Mechanistic studies of acid-evoked coughing in anesthetized guinea pigs. Am J Physiol Regul Integr Comp Physiol 291:R454–463

Carr MJ, Lee LY (2006) Plasticity of peripheral mechanisms of cough. Respir Physiol Neurobiol 152:298–311

Caterina MJ, Schumacher MA, Tominaga M, Rosen TA, Levine JD, Julius D (1997) The capsaicin receptor: A heat-activated ion channel in the pain pathway. Nature 389:816–824

Cerveri I, Accordini S, Corsico A et al (2003) Chronic cough and phlegm in young adults. Eur Respir J 22:413–417

Chang AB, Lasserson TJ, Kiljander TO, Connor FL, Gaffney JT, Garske LA (2006) Systematic review and meta-analysis of randomised controlled trials of gastro-oesophageal reflux interventions for chronic cough associated with gastro-oesophageal reflux. Br Med J 332:11–17

Chatkin JM, Ansarin K, Silkoff PE et al (1999) Exhaled nitric oxide as a noninvasive assessment of chronic cough. Am J Respir Crit Care Med 159:1810–1813

Chen RY, Thomas RJ (2000) Results of laparoscopic fundoplication where atypical symptoms coexist with oesophageal reflux. Aust N Z J Surg 70:840–842

Choudry NB, Fuller RW, Pride NB (1989) Sensitivity of the human cough reflex: effect of inflammatory mediators prostaglandin E2, bradykinin, and histamine. Am Rev Respir Dis 140: 137–141

Chung KF, McGarvey L, Widdicombe JG (2006) American College of chest physicians' cough guidelines. Lancet 367:981–982

Crummy F, Livingston M, Ardill JE, Adamson C, Ennis M, Heaney LG (2005) Endobronchial adenosine monophosphate challenge causes tachykinin release in the human airway. J Allergy Clin Immunol 116:312–317

Cullinan P (1992) Persistent cough and sputum: Prevalence and clinical characteristics in south east England. Respir Med 86:143–149

Dai Y, Moriyama T, Higashi T et al (2004) Proteinase-activated receptor 2-mediated potentiation of transient receptor potential vanilloid subfamily 1 activity reveals a mechanism for proteinase-induced inflammatory pain. J Neurosci 24:4293–4299

Davenport PW, Bolser DC, Vickroy T et al (2007) The effect of codeine on the Urge-to-Cough response to inhaled capsaicin. Pulm Pharmacol Ther 20:338–346

David GL, Koh WP, Lee HP, Yu MC, London SJ (2005) Childhood exposure to environmental tobacco smoke and chronic respiratory symptoms in non-smoking adults: The Singapore Chinese health study. Thorax 60:1052–1058

Dicpinigaitis PV, Dobkin JB (1999) Effect of zafirlukast on cough reflex sensitivity in asthmatics. J Asthma 36:265–270

Dicpinigaitis PV, Rauf K (1998) Treatment of chronic, refractory cough with baclofen. Respiration 65:86–88

Dicpinigaitis PV, Grimm DR, Lesser M (2000) Baclofen-induced cough suppression in cervical spinal cord injury. Arch Phys Med Rehabil 81:921–923

Dicpinigaitis PV, Dobkin JB, Reichel J (2002) Antitussive effect of the leukotriene receptor antagonist zafirlukast in subjects with cough-variant asthma. J Asthma 39:291–297

Doherty MJ, Mister R, Pearson MG, Calverley PM (2000) Capsaicin responsiveness and cough in asthma and chronic obstructive pulmonary disease. Thorax 55:643–649

Ekstrom T, Johansson KE (2000) Effects of anti-reflux surgery on chronic cough and asthma in patients with gastro-oesophageal reflux disease. Respir Med 94:1166–1170

Everett CF, Kastelik JA, Thompson RH, Morice AH (2007) Chronic persistent cough in the community: A questionnaire survey. Cough 3:5

Farrell TM, Richardson WS, Trus TL, Smith CD, Hunter JG (2001) Response of atypical symptoms of gastro-oesophageal reflux to antireflux surgery. Br J Surg 88:1649–1652

Ferrari M, Olivieri M, Sembenini C et al (1995) Tussive effect of capsaicin in patients with gastroesophageal reflux without cough. Am J Respir Crit Care Med 151:557–561

Fitzgerald JM, Allen CJ, Craven MA, Newhouse MT (1989) Chronic cough and gastroesophageal reflux. Can Med Assoc J 140:520–524

Fontana GA, Widdicombe JG (2007) What is cough and what should be measured? Pulm Pharmacol Ther 20:307–312

Ford AC, Forman D, Moayyedi P, Morice AH (2006) Cough in the community: A cross sectional survey and the relationship to gastrointestinal symptoms. Thorax 61:975–979

Fox MR, Bredenoord AJ (2007) Oesophageal high-resolution manometry – moving from research into clinical practice. Gut 57(3):405–423

Fox AJ, Lalloo UG, Belvisi MG, Bernareggi M, Chung KF, Barnes PJ (1996) Bradykinin-evoked sensitization of airway sensory nerves: A mechanism for ACE-inhibitor cough. Nat Med 2: 814–817

Franco RA Jr (2006) Laryngopharyngeal reflux. Allergy Asthma Proc 27:21–25

French CL, Irwin RS, Curley FJ, Krikorian CJ (1998) Impact of chronic cough on quality of life. Arch Intern Med 158:1657–1661

Fujimura M, Ogawa H, Nishizawa Y, Nishi K (2003) Comparison of atopic cough with cough variant asthma: is atopic cough a precursor of asthma? Thorax 58:14–18

Gatti R, Andre E, Amadesi S et al (2006) Protease-activated receptor-2 activation exaggerates TRPV-1-mediated cough in guinea pigs. J Appl Physiol 101:506–511

Gay M, Blager F, Bartsch K, Emery CF, Rosenstiel-Gross AK, Spears J (1987) Psychogenic habit cough: Review and case reports. J Clin Psychiatr 48:483–486

Gibson PG, Dolovich J, Denburg J, Ramsdale EH, Hargreave FE (1989) Chronic cough: Eosinophilic bronchitis without asthma. Lancet 1:1346–1348

Groneberg DA, Niimi A, Dinh QT et al (2004) Increased expression of transient receptor potential vanilloid-1 in airway nerves of chronic cough. Am J Respir Crit Care Med 170:1276–1280

Haque RA, Usmani OS, Barnes PJ (2005) Chronic idiopathic cough: A discrete clinical entity? Chest 127:1710–1713

Howard P, Cayton RM, Brennan SR, Anderson PB (1977) Lignocaine aerosol and persistent cough. Br J Dis Chest 71:19–24

Hwang SW, Cho H, Kwak J et al (2000) Direct activation of capsaicin receptors by products of lipoxygenases: endogenous capsaicin-like substances. Proc Natl Acad Sci USA 97:6155–6160

Irwin RS, Corrao WM, Pratter MR (1981) Chronic persistent cough in the adult: the spectrum and frequency of causes and successful outcome of specific therapy. Am Rev Respir Dis 123: 413–417

Irwin RS, Curley FJ, French CL (1990) Chronic cough. The spectrum and frequency of causes, key components of the diagnostic evaluation, and outcome of specific therapy. Am Rev Respir Dis 141:640–647

Irwin RS, French CL, Curley FJ, Zawacki JK, Bennett FM (1993) Chronic cough due to gastroesophageal reflux. Clinical, diagnostic, and pathogenetic aspects. Chest 104:1511–1517

Irwin RS, Baumann MH, Bolser DC et al (2006a) Diagnosis and management of cough executive summary: ACCP evidence-based clinical practice guidelines. Chest 129:1S–23S

Irwin RS, Ownbey R, Cagle PT, Baker S, Fraire AE (2006b) Interpreting the histopathology of chronic cough: a prospective, controlled, comparative study. Chest 130:362–370

Jatakanon A, Lalloo UG, Lim S, Chung KF, Barnes PJ (1999) Increased neutrophils and cytokines, TNF-alpha and IL-8, in induced sputum of non-asthmatic patients with chronic dry cough. Thorax 54:234–237
Javorkova N, Varechova S, Pecova R et al (2007) Acidification of the oesophagus acutely increases the cough sensitivity in patients with gastro-oesophageal reflux and chronic cough. Neurogastroenterol Motil 20:119–124
Kathiramalainathan K, Kaplan HL, Romach MK et al (2000) Inhibition of cytochrome P450 2D6 modifies codeine abuse liability. J Clin Psychopharmacol 20:435–444
Khan AM, Hashmi SR, Elahi F, Tariq M, Ingrams DR (2006) Laryngopharyngeal reflux: A literature review. Surgeon 4:221–225
Khoury RM, Katz PO, Hammod R, Castell DO (1999) Bedtime ranitidine does not eliminate the need for a second daily dose of omeprazole to suppress nocturnal gastric pH. Aliment Pharmacol Ther 13:675–678
Kollarik M, Undem BJ (2002) Mechanisms of acid-induced activation of airway afferent nerve fibres in guinea-pig. J Physiol 543:591–600
Krishtal O (2003) The ASICs: Signaling molecules? Modulators? Trends Neurosci 26:477–483
Lin L, Poh KL, Lim TK (2001) Empirical treatment of chronic cough–a cost-effectiveness analysis. Proc AMIA Symp :383–387
Lundback B, Nystrom L, Rosenhall L, Stjernberg N (1991) Obstructive lung disease in northern Sweden: respiratory symptoms assessed in a postal survey. Eur Respir J 4:257–266
McGarvey LP, Ing AJ (2004) Idiopathic cough, prevalence and underlying mechanisms. Pulm Pharmacol Ther 17:435–439
Mazzone SB, McLennan L, McGovern AE, Egan GF, Farrell MJ (2007) Representation of capsaicin-evoked urge-to-cough in the human brain using functional magnetic resonance imaging. Am J Respir Crit Care Med 176:327–332
McGarvey L, Morice AH (2003) Atopic cough: little evidence to support a new clinical entity. Thorax 58:736–737; author reply 737–738
McGarvey LP, Heaney LG, Lawson JT et al (1998a) Evaluation and outcome of patients with chronic non-productive cough using a comprehensive diagnostic protocol. Thorax 53:738–743
McGarvey LP, Heaney LG, MacMahon J (1998b) A retrospective survey of diagnosis and management of patients presenting with chronic cough to a general chest clinic. Int J Clin Pract 52:158–161
McGarvey LP, Forsythe P, Heaney LG, MacMahon J, Ennis M (1999) Bronchoalveolar lavage findings in patients with chronic nonproductive cough. Eur Respir J 13:59–65
McGarvey LP, Polley L, Macmahon J (2007) Review Series: Chronic cough: Common causes and current guidelines. Chron Respir Dis 4:215–223
Menon MS, Mulrennan SA, Everett CF, Thompson R, Morice AH (2005) Experience with baclofen in cough secondary to gastro-oesophageal reflux disease. Proc Am Thorac Soc 2:A323
Mitchell JE, Campbell AP, New NE et al (2005) Expression and characterization of the intracellular vanilloid receptor (TRPV-1) in bronchi from patients with chronic cough. Exp Lung Res 31:295–306
Morice AH (2004) Post-nasal drip syndrome–a symptom to be sniffed at? Pulm Pharmacol Ther 17:343–345
Morice AH (2007) Rebuttal: Cough is an expiratory sound. Lung 186:S7–S9
Morice AH, Geppetti P (2004) Cough. 5: The type 1 vanilloid receptor: A sensory receptor for cough. Thorax 59:257–258
Morice AH, Lowry R, Brown MJ, Higenbottam T (1987) Angiotensin-converting enzyme and the cough reflex. Lancet 2:1116–1118
Morice AH, Fontana GA, Sovijarvi AR et al (2004) The diagnosis and management of chronic cough. Eur Respir J 24:481–492
Morice AH, McGarvey L, Pavord I, British Thoracic Society Cough Guideline Group (2006) Recommendations for the management of cough in adults. Thorax 61(Suppl 1):i1–i24
Morice AH, Menon MS, Mulrennan SA et al (2007) Opiate therapy in chronic cough. Am J Respir Crit Care Med 175:312–315

Niimi A, Torrego A, Nicholson AG, Cosio BG, Oates TB, Chung KF (2005) Nature of airway inflammation and remodeling in chronic cough. J Allergy Clin Immunol 116:565–570
Novitsky YW, Zawacki JK, Irwin RS, French CT, Hussey VM, Callery MP (2002) Chronic cough due to gastroesophageal reflux disease: Efficacy of antireflux surgery. Surg Endosc 16:567–571
O'Connell F, Springall DR, Moradoghli-Haftvani A et al (1995) Abnormal intraepithelial airway nerves in persistent unexplained cough? Am J Respir Crit Care Med 152:2068–2075
O'Hara J, Jones NS (2005) The aetiology of chronic cough: a review of current theories for the otorhinolaryngologist. J Laryngol Otol 119:507–514
O'Hara J, Jones NS (2006) "Post-nasal drip syndrome": most patients with purulent nasal secretions do not complain of chronic cough. Rhinology 44:270–273
Patterson RN, Johnston BT, MacMahon J, Heaney LG, McGarvey LP (2004) Oesophageal pH monitoring is of limited value in the diagnosis of "reflux-cough". Eur Respir J 24:724–727
Patterson RN, Johnston BT, Ardill JE, Heaney LG, McGarvey LP (2007) Increased tachykinin levels in induced sputum from asthmatic and cough patients with acid reflux. Thorax 62: 491–495
Pavord ID (2004) Cough and asthma. Pulm Pharmacol Ther 17:399–402
Phua SY, McGarvey LP, Ngu MC, Ing AJ (2005) Patients with gastro-oesophageal reflux disease and cough have impaired laryngopharyngeal mechanosensitivity. Thorax 60:488–491
Poe RH, Kallay MC (2003) Chronic cough and gastroesophageal reflux disease: experience with specific therapy for diagnosis and treatment. Chest 123:679–684
Polley L, Cardwell C, Costello R, McGarvey L (2005) Predictors of idiopathic cough: outcome of a retrospective analysis. Thorax 60:107 (Abstract)
Pratter MR (2006) Chronic upper airway cough syndrome secondary to rhinosinus diseases (previously referred to as postnasal drip syndrome): ACCP evidence-based clinical practice guidelines. Chest 129:63S–71S
Pratter MR, Bartter T, Akers S, DuBois J (1993) An algorithmic approach to chronic cough. Ann Intern Med 119:977–983
Ramachandran R, Sadofsky LR, Xiao Y et al (2007) Inflammatory mediators modulate thrombin and cathepsin-G signaling in human bronchial fibroblasts by inducing expression of proteinase-activated receptor-4. Am J Physiol Lung Cell Mol Physiol 292:L788–798
Sadofsky LR, Compton SJ, Morice AH (2005) Dexbrompheniramine maleate inhibits capsaicin and resinferatoxin activation of TRPV-1. Thorax 60:ii108
Stanford A, Glascock JM, Eid GM et al (2003) Laparoscopic Roux-en-Y gastric bypass in morbidly obese adolescents. J Pediatr Surg 38:430–433
Thoman DS, Hui TT, Spyrou M, Phillips EH (2002) Laparoscopic antireflux surgery and its effect on cough in patients with gastroesophageal reflux disease. J Gastrointest Surg 6:17–21
Vertigan AE, Theodoros DG, Gibson PG, Winkworth AL (2006) Efficacy of speech pathology management for chronic cough: A randomised placebo controlled trial of treatment efficacy. Thorax 61:1065–1069
Viegi G, Pedreschi M, Baldacci S et al (1999) Prevalence rates of respiratory symptoms and diseases in general population samples of North and Central Italy. Int J Tuberc Lung Dis 3: 1034–1042
Wong CH, Morice AH (1999) Cough threshold in patients with chronic obstructive pulmonary disease. Thorax 54:62–64

Clinical Cough III: Measuring the Cough Response in the Laboratory

P.V. Dicpinigaitis

Contents

Abstract As interest in clinical cough research grows, measurement of cough reflex sensitivity will assume an increasingly important role. With proper equipment and meticulous attention to methodological details, cough reflex sensitivity can be safely, accurately, and reproducibly determined. Such precise measurement allows the evaluation of the effect of pharmacological or nonpharmacological interventions on the sensitivity of the cough reflex, or the comparison of cough reflex sensitivity

---

P.V. Dicpinigaitis
Einstein Division/Montefiore Medical Center, 1825 Eastchester Road, New York, NY 10461, USA
pdicpinigaitis@pol.net

K.F. Chung, J.G. Widdicombe (eds.), *Pharmacology and Therapeutics of Cough*, Handbook of Experimental Pharmacology 187,

between different subject populations. In addition to the method, other vital components of optimal cough challenge testing include proper interpretation of data and appropriate selection of study populations.

1 Inhalation Cough Challenge

Inhalation cough challenge allows the measurement of cough reflex sensitivity, a vital component of clinical research studies in which differences between subject or patient groups are assessed, and effects of pharmacological or other interventions on cough reflex sensitivity are determined. Given the progressively increasing interest in cough-related research within the academic and pharmaceutical industrial communities, inhalation cough challenge will continue to assume an increasingly important role. Reflecting the growing interest in high-quality cough research, the European Respiratory Society (ERS) has recently published the first-ever guidelines on the assessment of cough, in which cough inhalation challenge plays a prominent role (Morice et al. 2007).

2 Tussive Agents used in Inhalation Cough Challenge

A relatively small number of provocative agents has been employed in clinical cough research (Table 1). Experimentally induced cough in humans was initially described over a half century ago, employing citric acid as the tussive agent (Bickerman et al. 1954). The past two decades have witnessed the emergence of capsaicin as the provocative agent of choice. Cough induction employing ultrasonically nebulized distilled water (UNDW), or fog, has a limited but growing clinical experience. Each of these tussive agents is discussed in more detail in the following sections.

2.1 Capsaicin

The use of capsaicin to induce cough in humans was first described in 1984 (Collier and Fuller 1984). Capsaicin induces cough through stimulation of the tran-

Table 1 Tussive agents used in cough inhalation challenge

Vanilloid compounds
Capsaicin
Acid tussive agents
Citric acid
Tartaric acid
Acetic acid
Ultrasonically nebulized distilled water (fog)

sient receptor potential vanilloid 1 (TRPV-1) receptor on cough sensory nerves (Geppetti et al. 2006). During the past two decades, capsaicin has earned the role of the experimental tussive agent of choice because of its ability to induce cough in a safe, reproducible, and dose-dependent manner (Midgren et al. 1992; Dicpinigaitis 2003a; Dicpinigaitis and Alva 2005). Further advantages of capsaicin cough challenge include the lack of pharyngeal discomfort and choking sensation that has been observed with citric acid challenge, and the absence of significant tachyphylaxis, especially if the interval between cough challenges is at least 4 h (Midgren et al. 1992; Morice et al. 1992; O'Connell et al. 1992).

2.2 Citric Acid and Other Acid Tussive Agents

The mechanism of acid-induced cough remains incompletely understood, but likely involves the effect of lowered pH on vagal bronchopulmonary sensory nerves on which TRPV-1 and acid sensitive ion channel membrane receptors have been identified (Kollarik et al. 2007). As mentioned already, citric acid was the first tussive agent described in clinical research (Bickerman et al. 1954), and continues to be used by investigators at present. The occurrence of pharyngeal discomfort and a choking sensation is a more common occurrence with this agent than with capsaicin. Marked tachyphylaxis has been demonstrated if citric acid challenge is repeated at short (10-min) intervals (Morice et al. 1992). Furthermore, cross-tachyphylaxis has been described between capsaicin and citric acid (Morice et al. 1992).

Other acid tussives have been employed in human cough research, but experience with these agents is limited. Tussive acids that have been used in clinical trials include tartaric acid (Fujimura et al. 2000) and acetic acid (Mochizuki et al. 1995).

2.3 Ultrasonically Nebulized Distilled Water

Ultrasonically nebulized distilled water (UNDW) was initially employed over four decades ago to induce bronchoconstriction in subjects with hyperreactive airways. The ability of UNDW or "fog" challenge to provoke severe coughing in both healthy subjects and patients with bronchial hyperresponsiveness was noted by early investigators (Cheney and Butler 1968). Since then, UNDW challenge has developed an independent role as a useful tussive agent in basic and clinical research. The mechanism by which UNDW induces cough involves the provision to the respiratory tract of an aqueous solution with reduced concentration of permeant anion (Eschenbacher et al. 1984).

As opposed to the delivery of capsaicin and citric acid during experimental cough challenge, which will be discussed in detail later, water aerosols must be provided by an ultrasonic nebulizer, which typically produces a much larger output per unit volume of air than a conventional nebulizer (Mercer 1981). As experience with fog challenge has increased, one limitation of this method that has been demonstrated is the inability of UNDW, at maximal attainable output, to induce cough in a significant percentage of subjects (15–20%) who nonetheless cough upon exposure to capsaicin and citric acid (Fontana et al. 1997). The topic of water aerosols and cough induction has recently been reviewed (Fontana et al. 2005).

3 Capsaicin Cough Challenge: Method

Recently, the need for standardization of cough challenge methods has been addressed by the publication of the ERS guidelines on the assessment of cough (Morice et al. 2007). The author's method of inhalation cough challenge testing using capsaicin, which is the recommended method in the ERS guidelines, will be reviewed here in detail. Many aspects of the following discussion will be relevant to cough challenge with citric acid as well; certain points of similarity and distinction between the two tussive agents will be mentioned. The topic of cough challenge testing employing citric acid has recently been reviewed (Kastelik 2005).

3.1 Preparation and Storage of Capsaicin Solutions

Historically, most investigators have used purified capsaicin (97–98% pure) supplied by the Sigma Chemical Company (St. Louis, MO, USA, and European branches). Recently, the US Food and Drug Administration has required American investigators to use a pharmaceutical-grade capsaicin product obtained from Formosa Laboratories (Taiwan, China). The author has verified this product to be approximately 99% pure, and to yield results in cough challenge studies similar to the Sigma product (unpublished data).

The author prepares capsaicin solutions as previously described (Dicpinigaitis 2003a). Capsaicin (30.5 mg) is dissolved in 1 ml of 100% ethanol and 1 ml polyoxyethylene sorbitan (Tween 80), and further dissolved in 8 ml physiological saline to yield 10 ml of 0.01 M stock solution. If the detergent Tween 80 is not used, a cloudy rather than crystal-clear solution is obtained. The stock solution is further diluted with physiological saline to produce serial doubling concentrations ranging from 0.49 to 1,000 μM. If healthy subjects rather than patients with cough are to be tested, the lowest concentration of capsaicin prepared is 0.98 μM since, in the author's experience, induction of cough at this concentration is rare.

Stock solution is maintained at approximately −10°C. Fresh dilutions are prepared on each day of testing. Although there is no consensus regarding how often fresh stock solution should be prepared, one study confirmed that capsaicin solutions of concentrations 4°μM and greater remain stable for 1 year if stored at 4°C protected from light (Kopec et al. 2002).

3.2 Administration of Capsaicin Aerosol

The two main methods of aerosol delivery in capsaicin (and citric acid) cough challenge testing are the single-dose and dose–response methods (Morice et al. 2001). In the single-dose method, one specific concentration of capsaicin is utilized. Typically, the number of coughs induced by a fixed time period of tidal breathing would be determined.

In the dose–response method, aerosol administration may occur either by single, vital-capacity inhalations of incremental concentrations of capsaicin via a dosimeter-controlled nebulizer, or by the tidal-breath inhalation of ascending concentrations of capsaicin, each over a fixed time period, typically 15–60 s. Typically, successive concentrations are administered at 1-min intervals (Dicpinigaitis 2003a).

The author prefers the single-breath dose–response method of capsaicin administration because of the accuracy and reproducibility of the delivered dose, and the ease with which a cough response can be determined. A potential concern with capsaicin inhalation occurring over a prolonged (15–60-s) time period is that variations in respiratory rate and tidal volume may cause significant variations in the amount and concentration of aerosol delivered to the airways, both from subject to subject, as well as from one concentration stage to another in an individual subject. This would be particularly relevant during administration of concentrations that induce substantial coughing, thereby preventing the subject from inhaling the tussive agent for a significant portion of the fixed-time period of aerosol delivery. A recent study, however, compared the single-breath and tidal-breathing (15 s) methods of capsaicin aerosol delivery and found both to be reproducible, with good agreement between the two methods (Nejla et al. 2000).

4 Optimizing Reproducibility of Cough Challenge Studies

A high degree of reproducibility is essential for the performance of quality research, especially when serial cough challenges are required. Several aspects of inhalation cough challenge methods must be recognized and controlled to yield maximally reproducible results.

4.1 Inspiratory Flow Rate

The rate of inspiratory flow will affect the deposition of aerosol within the respiratory tract, as well as its concentration at the site of action, depending on the method of delivery. Indeed, variations in inspiratory flow rate have been demonstrated to affect the results of capsaicin (Barros et al. 1991) and citric acid (Barros et al. 1990) cough challenge, with higher rates of inspiratory flow in those studies being inversely correlated with the number of coughs induced. Unless the inspiratory flow rate is specifically controlled, variable amounts of tussive agent will be delivered to different subjects, and even breath-to-breath variations may occur within the same challenge in a given subject. Consequently, such potential variability in aerosol delivery may significantly affect the results of studies in which reproducibility of cough challenge is essential, such as in epidemiological studies comparing different subject populations, and in pharmacological trials measuring cough reflex sensitivity before and after drug therapy.

To ensure a consistent inspiratory flow rate during cough challenge studies, the author employs a compressed-air-driven nebulizer (model 646; De Vilbiss Health Care, Somerset, PA, USA) controlled by a dosimeter (KoKo DigiDoser; nSpire Health, Louisville, CO, USA) that is modified by the addition of an inspiratory flow regulator valve (RIFR; nSpire Health). The valve limits inspiratory flow rate to $0.5\,\mathrm{l\,s^{-1}}$ regardless of excessive inspiratory force, thus guaranteeing a consistent and reproducible inspiratory effort with each breath (see Fig. 1 in Dicpinigaitis 2005). Thus, with appropriate instruction to subjects to inhale with sufficient force, an identical inspiratory flow rate is achieved with each inhalation of tussive agent.

4.2 Nebulizer Characteristics

With the use of a standard nebulizer, significant variation may occur in the amount of aerosol delivered per inhalation, even if inspiratory flow rate is controlled (as discussed earlier). A second important determinant of nebulizer output is related to the actual structure of the nebulizer. In the De Vilbiss 646 model, for example, the straw and baffle assembly is a detachable component of the nebulizer that is removed for washing. When this component is reattached, variable distances may result the straw and baffle assembly and the jet orifice, which is the source of pressurized air (see Fig. 2 in Dicpinigaitis 2005). This variation in distance, albeit minute, may result in variable nebulizer output. Thus, to optimize reproducibility, the author uses a nebulizer with an inspiratory flow regulator valve, as described earlier, and, with the straw and baffle assembly welded in place, thereby eliminating the variations in nebulizer output that may result when this component is detached and reattached, resulting in variable distances between the jet orifice and straw.

Once these modifications have been performed, the exact output (milliliters per minute) of the nebulizer is measured (characterized nebulizer; nSpire Health). With the exact nebulizer output known, modulation of the duration of aerosol delivery by the dosimeter allows the determination of aerosol output per inhalation. For example, the author currently uses a characterized nebulizer with an output of $1.007\,\mathrm{ml\,min^{-1}}$, with a dosimeter programmed to deliver pressurized air for 1.2 s, thus allowing the nebulizer to deliver exactly 0.02 ml aerosol per inhalation. Since investigators worldwide will likely continue to use different equipment for the performance of cough challenge studies, one way to achieve a measure of standardization would be to control nebulizer output per breath, with the aforementioned 0.02 ml being proposed here as a standard.

4.3 Instructions to Subjects

Proper instruction to subjects before performance of the cough challenge is essential to ensure optimal results. Subjects should specifically be instructed not to suppress any coughs, and not to speak for 15 s after inhalation of the tussive agent, as this may potentially suppress cough. The author prefaces each challenge study with the following instruction to the subject: "Allow yourself to cough if you need to, and as much as you need to." Subjects should not be told that the induction of a specific number of coughs determines the end of the study (see the discussion of end points in Sect. 5).

4.4 Placebo Inhalations During Cough Challenge

To enhance cough challenge blindness, placebo breaths (inhalations of physiological saline) should be randomly interspersed between incremental concentrations of capsaicin (Dicpinigaitis 2003a; Morice et al. 2001; O'Connell et al. 1996). This technique may reduce the effects of voluntary suppression or conditioned responses in subjects who otherwise would be anticipating progressively higher concentrations of provocative agent.

4.5 Determination of Tussive Response

In the single-breath method of inhalation cough challenge, the tussive response (if present) to each dose of aerosol is immediate and brief; therefore, it is suggested that only coughs occurring within 15 s of capsaicin (or citric acid) delivery be counted (Dicpinigaitis 2005). Coughs occurring beyond this time interval may not be the result of the tussive agent.

4.6 Published Data on Reproducibility of Inhalation Cough Challenge

Numerous investigators have documented the high degree of reproducibility of capsaicin cough challenge using the dose–response method with the single-breath technique (Dicpinigaitis 2003a; Nejla et al. 2000; O'Connell et al. 1996), as well as with the fixed time period of capsaicin inhalation (Collier and Fuller 1984; Fujimura et al. 1992; Midgren et al. 1992; Morice et al. 1992; Nejla et al. 2000). Most of these studies, performed mainly in healthy subjects, examined short-term (20 min to 14 days) reproducibility of capsaicin-induced cough. Two studies that employed the single-breath dose–response method have confirmed long-term reproducibility of capsaicin cough challenge data over periods of 3 months (Nieto et al. 2003) and 6 months or greater (Dicpinigaitis 2003a). The latter study evaluated 40 healthy subjects in whom between-study intervals ranged from 6 to 62 months (mean 16.7 $\pm$2.4 months).

Reproducibility of citric acid cough challenge has also been reported with both the single-breath technique (Barber et al. 2005; Bickerman et al. 1954, 1957; Grattan et al. 1995; Morice et al. 1994; Rostami-Hodjegan et al. 2001; Schmidt et al. 1997) and fixed-time period of inhalation (Bickerman et al. 1954, 1957; Morice et al. 1992). Recently, the reproducibility of citric acid cough challenge using the KoKo DigiDoser method described earlier has been reported (Wright et al. 2007).

The reproducibility of cough reflex sensitivity and cough intensity measurements with UNDW has also been documented in dose–response challenges (Fontana et al. 1997; 1999; Morice et al. 1992).

5 Interpretation of Cough Challenge Data

The standard end points of cough challenge testing, particularly with the dose–response method, are C_2 and C_5, the concentrations of tussive agent inducing two or more and five or more coughs, respectively (Figs. 1, 2). Typically C_2 and C_5 have been defined as the first administered dose inducing two or more and five or more coughs, respectively. It has been suggested that interpolation of log-transformed concentration–response data to determine C_2 and C_5 would be the optimal (though more complicated) analysis, but differences in the values arrived at by these two methods are small and unlikely to be of clinical significance (Prudon et al. 2005).

There is no current consensus on which is the more highly reproducible end point, C_2 or C_5. It appears from examination of the recent literature that most investigators, including the author, prefer C_5, though some studies have supported C_2 as the more reproducible end point (O'Connell et al. 1996).

A potential problem with the use of C_2 as an end point, especially when serial cough challenges are performed, is the "startle phenomenon" (Dicpinigaitis 2003a). Occasionally, subjects undergoing their first-ever capsaicin cough challenge may cough, typically one to three times, at a particular concentration of tussive agent, but then will not cough (or cough less) after one or two subsequent incremental doses. As the cough challenge proceeds, a normal dose–response curve usually results. Since the number of coughs representing the "startle phenomenon" is almost always less than five, the end point of C_5 is much less likely to succumb to this potential pitfall. The performance of a preliminary familiarization challenge may eliminate this concern, but would of course complicate study protocols by requiring an additional challenge.

Another potential issue relevant to cough challenge end points primarily affects measurement of C_5. In a small minority of subjects, mainly those with high cough

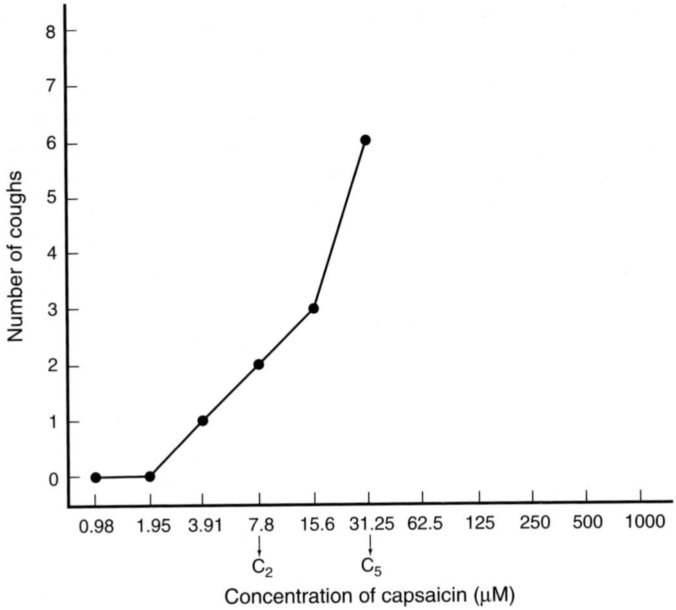

Fig. 1 Dose–response curve of capsaicin cough challenge. In this study, C_2, the concentration of capsaicin inducing two or more coughs, is 7.8 μM and C_5, the concentration of capsaicin inducing five or more coughs, is 31.25 μM

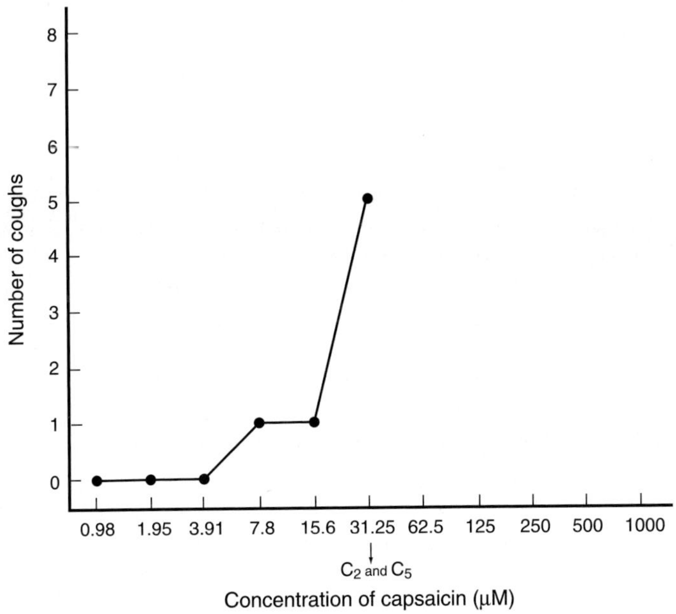

Fig. 2 Dose–response curve of capsaicin cough challenge. In this study, C_2 and C_5 are identical, 31.25 μM (see the text for the definitions of C_2 and C_5)

thresholds, the inhalation of high concentrations of capsaicin (almost exclusively concentrations greater than C_2) is precluded by upper-airway discomfort due to the pungent nature of the tussive agent. Consequently, the subject is unable to perform a full inhalation of capsaicin aerosol. If fewer than five coughs are induced by such a partial inhalation, C_5 cannot be determined. Such subjects would need to be excluded from studies in which the determination of C_5 is required. Given the potential issues surrounding both end points, the author suggests that both C_2 and C_5 be reported in clinical trial results (Morice et al. 2007).

6 Safety of Inhalation Cough Challenge

The most rigorous analysis of safety in cough challenge testing has been with capsaicin as the tussive agent. A review of the greater than 20 year human experience with inhaled capsaicin failed to document a single serious adverse event associated with cough challenge testing (Dicpinigaitis and Alva 2005). This review encompassed 122 published studies describing close to 5,000 subjects undergoing capsaicin cough challenge, including not only healthy adults and children, but also patients with pathological cough, asthma, chronic obstructive pulmonary disease, acute respiratory tract infection, heart–lung transplantation, cervical spinal cord injury, and other conditions. Side effects typically reported with capsaicin cough challenge consist almost exclusively of transient throat irritation in a minority of subjects. Capsaicin does not induce clinically significant bronchoconstriction in healthy volunteers or asthmatics (Dicpinigaitis and Alva 2005).

The safety of citric acid cough challenge was reported during the initial development of the method over a half century ago (Bickerman et al. 1954, 1957), but more recent data have not been published. Citric acid inhalation can induce a small decrease in the forced expiratory volume in 1 s (less than 5%) that is unlikely to be of clinical significance (Laude et al. 1993).

In contrast to inhalation of capsaicin and citric acid, inhalation of UNDW may be associated with clinically-significant bronchoconstriction in susceptible subjects (Fontana et al. 2005). Thus, a heightened level of vigilance is required during the performance of fog challenge studies.

7 Use of Inhalation Cough Challenge in Clinical Research

The performance of high-quality clinical cough research requires not only meticulous attention to the cough challenge procedure itself, but also an appreciation of issues relevant to study population, placebo effects, and proper interpretation of study data.

7.1 Subject Population

The significant gender difference in cough reflex sensitivity has been firmly established. Multiple studies have demonstrated that women have a more sensitive cough reflex than men, both in terms of healthy subjects (Dicpinigaitis and Rauf 1998; Fujimura et al. 1996) as well as in terms of patients with chronic cough (Kastelik et al. 2002). Thus, any study comparing cough reflex sensitivity between distinct populations must be matched for gender. Ethnic differences in cough reflex sensitivity, however, have not been demonstrated (Dicpinigaitis et al. 2001).

Studies in children have confirmed the experience in adult subjects that capsaicin cough challenge is safe, reproducible, affected by rate of inspiratory flow, and devoid of significant short-term tachyphylaxis (Chang et al. 1996). Further mirroring conclusions of adult studies, pediatric trials have demonstrated enhanced cough reflex sensitivity in subjects with pathological cough (Chang et al. 1997a) and asthmatic cough (Chang et al. 1997b).

Smoking status is an important consideration when designing clinical trials that will include cough challenge. Healthy cigarette smokers have a diminished cough reflex sensitivity relative to gender-matched nonsmokers, likely due to chronic cigarette smoke-induced desensitization of airway cough receptors (Dicpinigaitis 2003b). This desensitization appears to be dynamic and reversible, since studies have demonstrated enhancement of cough reflex sensitivity within 2 weeks of smoking cessation in former chronic smokers (Dicpinigaitis et al. 2006), and prompt inhibition of cough reflex sensitivity after resumption of smoking (Sitkauskiene et al. 2007). Given these findings, clinical trials should control for smoking status, and subjects having recently discontinued smoking should not be enrolled since there may be an unstable baseline cough reflex sensitivity.

A viral upper respiratory tract infection (URTI) induces transient cough receptor hyperresponsiveness to capsaicin (O'Connell et al. 1996); thus, subjects in whom baseline cough reflex sensitivity measurements will be required should be excluded from enrolment in clinical trials if symptoms consistent with URTI have occurred within the previous 4 weeks.

7.2 Placebo Effect

In subjects with acute cough due to viral URTI, clinical trials have shown that a majority of the reduction in subjectively or objectively measured cough is attributable to placebo, with only a minority of the effect being attributable to the active agent (Eccles 2002). Proposed explanations for this observation include an antitussive effect of endogenous opioids, a demulcent effect if the placebo is administered in the form of a syrup, and the expected transient nature of acute URTI (Eccles 2002). A placebo effect has also been described in cough challenge studies employing citric acid as the tussive agent (Rostami-Hodjegan et al. 2001).

7.3 Extrapolation of Subject Data to a Patient Population

Typically, a potential new pharmaceutical agent undergoes initial trials in a healthy population. In the setting of research with putative cough suppressants however, it must be appreciated that absence of an antitussive effect in healthy volunteers does not preclude an effect in subjects with pathological cough, whose cough receptors are presumably in a state of heightened sensitivity. For example, guaifenesin, commonly known as an expectorant, has been shown to inhibit capsaicin-induced cough in subjects with URTI, but not in healthy volunteers (Dicpinigaitis and Gayle 2003). The leukotriene receptor antagonist zafirlukast suppressed both subjective cough severity and cough reflex sensitivity in subjects with cough-variant asthma, but failed to inhibit induced cough in healthy volunteers and stable asthmatics without cough (Dicpinigaitis et al. 2002).

Conversely, the demonstration of an antitussive effect in the laboratory does not imply that a pharmacological agent will prove efficacious in a patient population. The opioid narcotic codeine, for example, has been shown in some clinical trials to inhibit capsaicin-induced cough (Dicpinigaitis et al. 1997; Fuller et al. 1988), yet was ineffective against cough due to URTI (Freestone and Eccles 1997).

References

Barber CM, Curran AD, Bradshaw LM, Morice AH, Rawbone R, Fishwick D (2005) Reproducibility and validity of a Yan-style portable citric acid cough challenge. Pulm Pharmacol Ther 18:177–180

Barros MJ, Zammattio SL, Rees PJ (1990) Importance of inspiratory flow rate in the cough response to citric acid inhalation in normal subjects. Clin Sci (London) 78:521–525

Barros MJ, Zammattio SL, Rees PJ (1991) Effect of changes in inspiratory flow rate on cough responses to inhaled capsaicin. Clin Sci (London) 81:539–542

Bickerman HA, Barach AL, Itkin SE, Drimmler F (1954) The experimental production of cough in human subjects induced by citric acid aerosols. Preliminary studies on the evaluation of antitussive agents. Am J Med Sci 228:156–163

Bickerman HA, German E, Cohen BM (1957) The cough response of healthy human subjects stimulated by citric acid aerosol. Am J Med Sci 234:191–205

Chang AB, Phelan PD, Roberts RGD, Robertson F (1996) Capsaicin cough receptor sensitivity in children. Eur Respir J 9:2220–2223

Chang AB, Phelan PD, Sawyer SM, Robertson CF (1997a) Airway hyperresponsiveness and cough receptor sensitivity in children with recurrent cough. Am J Respir Crit Care Med 155: 1935–1939

Chang AB, Phelan PD, Robertson CF (1997b) Cough receptor sensitivity in children with acute and non-acute asthma. Thorax 52:770–774

Cheney FW, Butler J (1968) The effects of ultrasonically produced aerosols on airway resistance in man. Anesthesiology 29:1099–1106

Collier JG, Fuller RW (1984) Capsaicin inhalation in man and the effects of sodium cromoglycate. Br J Pharmacol 81:113–117

Dicpinigaitis PV (2003a) Short- and long-term reproducibility of capsaicin cough challenge testing. Pulm Pharmacol Ther 16:61–65

Dicpinigaitis PV (2003b) Cough reflex sensitivity in cigarette smokers. Chest 123:685–688

Dicpinigaitis PV (2005) Capsaicin inhalation cough challenge. In: Redington AE, Morice AH (eds) Acute and chronic cough. Taylor & Francis, Boca Raton, USA, pp 161–175
Dicpinigaitis PV, Alva RV (2005) Safety of capsaicin cough challenge testing. Chest 128:196–202
Dicpinigaitis PV, Gayle YE (2003) Effect of guaifenesin on cough reflex sensitivity. Chest 124:2178–2181
Dicpinigaitis PV, Rauf K (1998) The influence of gender on cough reflex sensitivity. Chest 113:1319–1321
Dicpinigaitis PV, Dobkin JB, Rauf K (1997) Comparison of the antitussive effects of codeine and the GABA-agonist baclofen. Clin Drug Invest 14:326–329
Dicpinigaitis PV, Allusson VRC, Baldanti A, Nalamati JR (2001) Ethnic and gender differences in cough reflex sensitivity. Respiration 68:480–482
Dicpinigaitis PV, Dobkin JB, Reichel J (2002) Antitussive effect of the leukotriene receptor antagonist zafirlukast in subjects with cough-variant asthma. J Asthma 39:291–297
Dicpinigaitis PV, Sitkauskiene B, Stravinskaite K, Appel DW, Negassa A, Sakalauskas R (2006) Effect of smoking cessation on cough reflex sensitivity. Eur Respir J 28:786–790
Eccles R (2002) The powerful placebo in cough studies? Pulm Pharmacol Ther 15:303–308
Eschenbacher WL, Boushey HA, Sheppard D (1984) Alteration in osmolarity of inhaled aerosols causes bronchoconstriction and cough, but absence of a permeant anion causes cough alone. Am Rev Respir Dis 129:211–215
Fontana GA, Pantaleo T, Lavorini F, Boddi V, Panuccio P (1997) A noninvasive electromyographic study on threshold and cough intensity in humans. Eur Respir J 10:983–989
Fontana GA, Pantaleo T, Lavorini F, Maluccio NM, Mutolo D, Pistolesi M (1999) Repeatability of cough-related variables during fog challenges at threshold and suprathreshold stimulus intensity in humans. Eur Respir J 13:1447–1450
Fontana GA, Lavorini F, Pistolesi M (2005) water aerosols and cough. In: Redington AE, Morice AH (eds) Acute and chronic cough. Taylor & Francis, Boca Raton, USA, pp 195–214
Freestone C, Eccles R (1997) Assessment of the antitussive efficacy of codeine in cough associated with the common cold. J Pharm Pharmacol 49:1045–1049
Fujimura M, Sakamoto S, Kamio Y, Matsuda T (1992) Effects of methacholine-induced bronchoconstriction and procaterol-induced bronchodilation on cough receptor sensitivity to inhaled capsaicin and tartaric acid. Thorax 47:441–445
Fujimura M, Kasahara K, Kamio Y, Naruse M, Hashimoto T, Matsuda T (1996) Female gender as a determinant of cough threshold to inhaled capsaicin. Eur Respir J 9:1624–1626
Fujimura M, Kamio Y, Myou S, Hashimoto T (2000) Effect of oral mexiletine on the cough response to capsaicin and tartaric acid. Thorax 55:126–128
Fuller RW, Karlsson J-A, Choudry NB, Pride NB (1988) Effect of inhaled and systemic opiates on responses to inhaled capsaicin in humans. J Appl Physiol 65:1125–1130
Geppetti P, Materazzi S, Nicoletti P (2006) The transient receptor potential vanilloid 1: Role in airway inflammation and disease. Eur J Pharmacol 533:207–214
Grattan TJ, Marshall AE, Higgins KS, Morice AH (1995) The effect of inhaled and oral dextromethorphan on citric acid induced cough in man. Br J Clin Pharmacol 39:261–263
Kastelik JA (2005) Acid inhalation cough challenge. In: Redington AE, Morice AH (eds) Acute and chronic cough. Taylor & Francis, Boca Raton, USA, pp 177–194
Kastelik JA, Thompson RH, Aziz I, Ojoo JC, Redington AE, Morice AH (2002) Sex-related differences in cough reflex sensitivity in patients with chronic cough. Am J Respir Crit Care Med 166:961–964
Kollarik M, Ru F, Undem BJ (2007) Acid-sensitive vagal sensory pathways and cough. Pulm Pharmacol Ther 20:402–411
Kopec SE, DeBellis RJ, Irwin RS (2002) Chemical analysis of freshly prepared and stored capsaicin solutions: Implications for tussigenic challenges. Pulm Pharmacol Ther 15:529–534
Laude EA, Higgins KS, Morice AH (1993) A comparative study of the effects of citric acid, capsaicin and resiniferatoxin on the cough challenge in guinea-pig and man. Pulm Pharmacol 6:171–175

Mercer TT (1981) Production of therapeutic aerosols. Principles and techniques. Chest 80 (suppl. 6):813–818
Midgren B, Hansson L, Karlsson J-A, Simonsson BG, Persson CG (1992) Capsaicin-induced cough in humans. Am Rev Respir Dis 146:347–351
Mochizuki H, Shimizu T, Maeda S, Tokuyama K, Morikawa A, Kuruome T (1995) Relationship between ultrasonically nebulized distilled water-induced bronchoconstriction and acetic acid-induced cough in asthmatic children. J Allergy Clin Immunol 96:193–199
Morice AH, Higgins KS, Yeo WW (1992) Adaptation of the cough reflex with different types of stimulation. Eur Respir J 5:841–847
Morice AH, Marshall AE, Higgins KS, Grattan TJ (1994) Effect of inhaled menthol on citric acid induced cough in normal subjects. Thorax 49:1024–1026
Morice AH, Kastelik JA, Thompson R (2001) Cough challenge in the assessment of cough reflex. Br J Clin Pharmacol 52:365–375
Morice AH, Fontana GA, Belvisi MG, Birring SS, Chung KF, Dicpinigaitis PV, Kastelik JA, McGarvey LP, Smith JA, Tatar M, Widdicombe JG (2007) ERS guidelines on the assessment of cough. Eur Respir J 29:1256–1276
Nejla S, Fujimura M, Kamio Y (2000) Comparison between tidal breathing and dosimeter methods in assessing cough receptor sensitivity to capsaicin. Respirology 5:337–342
Nieto L, DeDiego A, Perpina M, Compte L, Garrigues V, Martinez E, Ponce J (2003) Cough reflex testing with inhaled capsaicin in the study of chronic cough. Respir Med 97:393–400
O'Connell F, Thomas VE, Pride NB (1992) Adaptation of the cough reflex with different types of stimulation(letter). Eur Respir J 5:1296–1297
O'Connell F, Thomas VE, Studham JM, Pride NB, Fuller RW (1996) Capsaicin cough sensitivity increases during upper respiratory tract infection. Respir Med 90:279–286
Prudon B, Vara DD, Pavord ID, Birring SS (2005) Analysis of cough reflex sensitivity data. Thorax 60(suppl 2):ii107
Rostami-Hodjegan A, Abdul-Manap R, Wright CE, Tucker GT, Morice AH (2001) The placebo response to citric acid-induced cough: Pharmacodynamics and gender differences. Pulm Pharmacol Ther 14:315–319
Schmidt D, Jorres RA, Magnussen H (1997) Citric acid-induced cough thresholds in normal subjects, patients with bronchial asthma, and smokers. Eur J Med Res 2:384–388
Sitkauskiene B, Stravinskaite K, Sakalauskas R, Dicpinigaitis PV (2007) Changes in cough reflex sensitivity after cessation and resumption of cigarette smoking. Pulm Pharmacol Ther 20: 240–243
Wright CE, Jackson J, Thompson RH, Morice AH (2007) Validation of the citric acid cough challenge using the KoKo DigiDoser system in healthy adult volunteers (abstract). Am J Respir Crit Care Med 175:A344

Clinical Cough IV: What is the Minimal Important Difference for the Leicester Cough Questionnaire?

A.A. Raj, D.I. Pavord, and S.S. Birring(✉)

Contents

Abstract *Background:* The Leicester Cough Questionnaire (LCQ) is a valid, reproducible, responsive self-reported cough-specific health status measure. It has been used to assess overall efficacy of treatments for cough, but its threshold for clinical significance, or patient importance, is unknown. The aim of this study was to determine the minimal important difference (MID) of the LCQ for patients with chronic cough; this is the smallest change in quality-of-life score considered to be clinically meaningful.

Methods: The LCQ MID was first estimated by a multidisciplinary panel of experts who reviewed two cases of chronic cough. It was subsequently determined using a standardized method. Fifty-two patients with chronic cough of more than 8 weeks' duration attending a respiratory outpatient clinic were recruited. Participants completed the LCQ at initial evaluation and repeated the LCQ with four Global

S.S. Birring
Department of Respiratory Medicine, King's College Hospital, London SE5 9RS, UK
surinder.birring@kch.nhs.uk

K.F. Chung, J.G. Widdicombe (eds.), *Pharmacology and Therapeutics of Cough*,
Handbook of Experimental Pharmacology 187,

Rating of Change Questionnaires (GRCQ) more than 2 months later. The LCQ total score ranges from 3 to 21 and from 1 to 7 for physical, psychological and social domains; a higher score indicates a better health-related quality of life. The GRCQ, a 15-point scale scored between +7 (a great deal better) and −7 (a great deal worse), was used to record patient ratings of change in cough symptoms. The MID was defined as the change in LCQ health status corresponding to a small change in the GRCQ score.

Results: The mean (standard deviation) LCQ MID corresponding to a small change in the GRCQ score was 1.3 (3.2); the MIDs for domains were as follows: physical 0.2 (0.8), social 0.2 (1.1) and psychological 0.8 (1.5). This MID for LCQ total score was similar to that determined by the expert panel. The global rating of change scores correlated significantly with the change in LCQ total and domain scores ($r = 0.4 - 0.5$; $p < 0.005$).

Conclusion: We have demonstrated that the LCQ MID is 1.3. The LCQ MID should aid clinicians and researchers to make meaningful interpretations of health-related quality-of-life data relating to chronic cough.

1 Introduction

Chronic cough is a common condition that often leads to considerable physical and psychological morbidity (McGarvey et al. 1998; French et al. 2002). We have previously reported the development of the Leicester Cough Questionnaire (LCQ) (Birring et al. 2003), a validated self-reported health-related quality-of-life (QOL) measure specifically for cough. The LCQ is increasingly being used as an outcome measure of cough in clinical trials to assess therapeutic response (Morice et al. 2007; Decalmer et al. 2007). Whilst statistically significant changes in health status are seen after specific therapy for cough (Birring et al. 2003; Morice et al. 2007), the clinical relevance and importance to the patient is not known. Ideally, changes in health status scores should be presented in a context that is meaningful to patients and healthcare professionals. This can be facilitated by determining the minimal important difference (MID) of health status questionnaires. The MID is defined as the smallest change in the health domain of interest which patients perceive as beneficial and which would mandate, in the absence of troublesome side effects or excessive cost, a change in the patient's management (Jaeschkle et al. 1989). A number of methods are available to determine the MID. These include MID ratings by an expert multidisciplinary panel, anchor-based methods that compare changes in QOL with other measures that assess change in health status, and distribution-based methods such as the standard error of measurement that assess the effect size and the variability of QOL measurements (Jaeschkle et al. 1989; Jones 2002; Wyrwich and Tierry 2002; Guyatt et al. 2002b; Beaton et al. 2002). The aim of this study was to determine the MID for the LCQ, so that clinicians can readily detect clinically significant changes in the health status of patients with chronic cough, both in clinical and in research settings.

2 Methods

2.1 Phase 1

The LCQ MID was determined by two methods. Phase 1 consisted of an estimation of the LCQ MID by a multidisciplinary panel ($n = 4$) consisting of a respiratory physician, a respiratory physiotherapist, a respiratory researcher and a technician, all experienced in administering the LCQ in routine clinical practice to patients with chronic cough. Three members of the panel had been involved in the development of the LCQ. The panel members independently reviewed two hypothetical case scenarios of chronic cough associated with moderate and severely impaired health-related QOL assessed by the LCQ and were asked to indicate the minimal change (positive or negative) in each LCQ item they considered clinically important, resulting from a 3-month trial of proton-pump-inhibitor therapy or inhaled corticosteroids. These changes were used to calculate new LCQ scores for each case scenario; the change in the LCQ score from the baseline was then calculated.

Case 1: A 35-year-old man with a 6-month history of chronic cough that disturbs sleep a little of the time, who has coughing bouts sometimes during the day, is embarrassed a little of the time, is concerned about what others think a little of the time and has cough that annoys his partner occasionally.

Case 2: A 56-year-old woman with a 6-year history of chronic cough that results in chest or stomach pains a good bit of the time, who is bothered by sputum production most of the time, has coughing bouts several times a day, has 'hardly any energy', is anxious most of the time, is worried about serious illness a good bit of the time, is concerned about what others think most of the time and has cough that interferes with her job and enjoyment most of the time.

2.2 Phase 2

Phase 2 consisted of a prospective study to determine the LCQ MID by validated methods described by other groups (Juniper et al. 1994; Jones 2002; Guyatt et al. 1987). (see Sects. 2.3 to 2.6)

2.3 Subjects

The subjects were 52 consecutive patients with chronic cough referred to a respiratory outpatient clinic. Chronic cough was defined as a cough lasting more than 8 weeks that remained unexplained after assessment by the primary-care physician. The patients underwent investigation and treatment according to a standardized diagnostic protocol described previously (Brightling et al. 1999). Each subject gave informed consent. The study was approved by the local ethics research committee.

2.4 Protocol

All patients completed the LCQ at the first clinic visit and a repeat LCQ and four Global Rating of Change Questionnaires (GRCQ; Appendix 1) more than 2 months later. The repeat questionnaires were either completed at a follow-up clinic visit or posted to patients for completion.

2.5 Leicester Cough Questionnaire

The LCQ is a well-validated, self-completed, cough-specific health status questionnaire that has been shown to be both repeatable and responsive in patients with chronic cough (Birring et al. 2003). The total score range for the LCQ is 3–21 and a higher score indicates a better QOL. A seven-point Likert response scale is used for all 19 LCQ items. The LCQ has three domains: physical, psychological and social (range 1–7). Domain scores are calculated by averaging scores from items in each domain; the total score is the sum of the domain scores. The LCQ can be downloaded from http://thorax.bmjjournals.com/cgi/content/full/58/4/339.

2.6 Global Rating of Change Questionnaire

The GRCQ is a 15-point scale widely used to determine the MID of health-related QOL questionnaires (Juniper et al. 1994). The GRCQ is used to ask patients to make global ratings of changes in their cough over a time interval. The response scale ranges from −7 (a great deal worse) to +7 (a great deal better). All subjects were asked to complete four GRCQs, each relating to one of the LCQ domains and overall health status. The score for each GRCQ was classified as unchanged (scores −1/0/1), a small change (−3, −2, 2, 3), a moderate change (−5, −4, 4, 5) or large change (−7, −6, 6, 7) (Juniper et al. 1994). The MID was defined as the change in LCQ health status corresponding to a small change in GRCQ score as defined by Juniper et al. (1994) and Jaeschkle et al. (1989). The final accepted LCQ MID was determined by phase 2 of this study.

2.7 Statistical Analysis

SPSS version 12 was used for data analysis. Data are presented as the mean (standard deviation) or median (range). The mean change in LCQ score was calculated for each domain. In accordance with previous studies we expressed change of the global rating score as an absolute number, i.e. when the change was negative, the sign was reversed as was the sign of the change in LCQ score (Jaeschkle et al. 1989; Juniper et al. 1994). Correlations between variables were analysed using Pearson's correlation coefficients (r). Paired t tests were used for group comparisons.

3 Results

3.1 Phase 1

A multidisciplinary panel of four individuals independently reviewed two case scenarios of chronic cough. The mean (standard deviation) LCQ MID determined by the expert multidisciplinary panel for total score was 1.49 (0.32) and for domain scores the MID were as follows: physical 0.52 (0.18), psychological 0.47 (0.14) and social 0.50 (0). There were no significant differences in MID scores between the two cases of chronic cough (Table 1).

3.2 Phase 2

Subject characteristics are shown in Table 2. The causes of cough were asthma (21.2%), eosinophilic bronchitis (9.6%), bronchiectasis (5.8%), gastro-oesophageal reflux disease (5.8%), chronic bronchitis (5.8%), chronic obstructive pulmonary disease (3.8%), chronic enlarged tonsils (3.8%), rhinitis (1.9%), sarcoidosis (1.9%), interstitial lung disease (1.9%) and idiopathic chronic cough (36.5%). The clinical course of patients between visits were investigations (12%), diagnostic trial of therapy (proton-pump inhibitors, inhaled corticosteroids and nasal corticosteroids) (31%), treatment of underlying condition causing cough (29%), chest physiotherapy and speech therapy (18%) and observation only (10%). The median time (range) from the first LCQ to completion of the second LCQ was 43 weeks (9–133 weeks).

Patients with chronic cough had reduced overall health status at baseline; the mean (standard deviation) LCQ total score was 13.6 (4.1); (normal QOL score is 21). Health status was reduced in all domains of QOL (Table 2). The mean change

Table 1 Mean minimal important differences estimated by a multidisciplinary panel for two cases of chronic cough

	Case 1	Case 2	Mean MID (SD) (cases 1 + 2)
MID physical domain	0.63	0.41	0.52 (0.18)
MID psychological domain	0.57	0.37	0.47 (0.14)
MID social domain	0.50	0.50	0.50 (0)
MID LCQ total score	1.70	1.25	1.49 (0.32)

MID minimal important difference, *SD* standard deviation, *LCQ* Leicester Cough Questionnaire

Table 2 Patient baseline characteristics ($n = 52$)

Age (years)	55 (13.9)
Female (n) (%)	29 (56)
FEV_1(% predicted)	95 (4)
Duration of cough (months)	67 (77)
Smoking history	
Ex-smokers (n) (%)	17 (38)
Non-smokers (n) (%)	27 (61)
LCQ scores	
Physical LCQ score	4.8 (1.1)
Psychological LCQ score	4.4 (1.6)
Social LCQ score	4.4 (1.6)
Overall QOL	13.6 (4.1)

Data are presented as the mean (standard deviation).
FEV_1 forced expiratory volume in 1 s, *QOL* quality of life

Table 3 Mean change in Leicester Cough Questionnaire score per global rating category

	Global rating of change category			
	Same (-1/0/1)	Minimal important difference (-3/–2/2/3)	Moderate (–5/–4/4/5)	Large (–7/–6/6/7)
LCQ: total	−1.3 (3.1), $n = 26$	1.3 (3.3), $n = 14$	1.7 (2.3), $n = 10$	2.7 (4.6), $n = 7$
LCQ: physical	−0.6 (0.9), $n = 25$	0.2 (0.8), $n = 14$	0.7 (1), $n = 14$	1.3 (1.2), $n = 4$
LCQ: social	−0.3(1.2), $n = 34$	0.2 (1.1), $n = 6$	0.7 (0.9), $n = 12$	1.3 (2.2), $n = 5$
LCQ: psychological	−0.1(1.1), $n = 28$	0.8 (1.5), $n = 11$	0.5 (0.8), $n = 12$	1.2 (2.1), $n = 6$

Data are presented as the mean (standard deviation).
n number of patients

in LCQ score for each GRCQ category at follow-up visit is given in Table 3. The LCQ MID, corresponding to a small change in the GRCQ score was 1.3 and the MIDs of the health domains were as follows: physical 0.2(0.8), social 0.2(1.1) and psychological 0.8(1.5) (Table 3). Patients reporting a positive change in global rating (+1 to +7) had improved health status (median change in LCQ score of 1.6; $n = 17$) compared with those reporting a negative change (median LCQ change of –1.3; $n = 22$; $p = 0.5$). The global rating of change scores correlated significantly with the change in the LCQ total and domain scores ($r = 0.4 - 0.5$; all $p < 0.005$). This was higher than correlation of global rating scores with post-LCQ scores ($r = -0.2$ to -0.3) and pre-LCQ scores ($r = 0$ to 0.2). The post-LCQ total and domain scores were significantly higher than the baseline LCQ scores in those patients who indicated an improvement on the global rating scale (mean difference in the LCQ total score of 2.7; $p = 0.01$; physical $p = 0.007$; psychological $p = 0.04$; social $p = 0.04$). There were no statistically significant differences in the change in LCQ scores when they were analysed per global rating category. The LCQ total score MID was 1.2 when patients with a global rating change of –1/+1 were included in the MID group.

4 Discussion

The LCQ is a brief, well-validated self-completed cough-specific health status questionnaire that has been shown to be repeatable and responsive. The purpose of this study was not to validate the LCQ, which has previously been done, but to determine the MID. We have demonstrated that the MID for the LCQ total health status score is 1.3 and the MID for LCQ health status domains ranges from 0.2 to 0.8. This study represents an advance in the clinical utility of cough severity outcome measures and should aid clinicians and researchers in making meaningful interpretations of health-related QOL data relating to chronic cough.

We determined the LCQ MID by two methods that assessed the patients' and healthcare professionals' perception of change in health status and found them to be very similar. The expert panel consensus method is simple and brief and provides an estimation of the MID from a clinicians' perspective (Jaeschkle et al. 1989; Jones et al. 1991). We chose the anchor-based method of Juniper et al. (Juniper et al. 1994) and Jaeschke et al. (Jaeschkle et al. 1989) as the main determinant of the LCQ MID, since this is the most widely used and published method and emphasizes the patients' opinion (Juniper et al. 1994). Distribution methods such as those based on the standard deviation are being used increasingly since they are easier to generate than anchor based methods (Guyatt et al. 2002b). Their limitations are the dependence on heterogeneity of the population being studied and the arbitrary nature of the units of measure (Guyatt et al. 2002b). These problems are minimized by using the standard error of measurement method, which has been found to correlate well with anchor-based methods.(Wyrwich et al. 1999) We chose multiple methods to determine the MID to enhance the interpretability of our QOL questionnaire and overcome the limitations of individual methods (Guyatt et al. 2002b; Beaton et al. 2002).

The GRCQ has been used to determine the MID for a wide range of QOL questionnaires, in which patients quantify the magnitude of the change in health status. We chose to include patients with a global rating score of –1/+1 in the unchanged category to be consistent with others (Juniper et al. 1994). It is possible that some patients in this group may have experienced a minimal important change in health status but there was little change in the LCQ MID when patients with GRCQ scores of –1/+1 were included in the MID group. The LCQ MID applies to patients whose health improved or deteriorated. We found a similar change in LCQ health status score in those scoring an improvement in their global rating of change to those with deterioration. It is possible that patients with deteriorating health may have a different LCQ MID from those experiencing an improvement. Studies with larger numbers of patients with deterioration in health status would be required to determine a separate MID value for deteriorating health status.

The GRCQ has limitations that may lead to some imprecision in determining the MID; it is a subjective instrument with arbitrary scales and categories, and individual patients may interpret the wording of GRCQ questions differently, leading to variability in GRCQ scoring. This may have contributed to the variability seen in the LCQ MID, but this is probably true for all QOL instruments (Decalmer et al. 2007). The GRCQ is subject to recall bias, particularly if patients are unable to recall their prior health status and may reflect current health status in some subjects (Guyatt et al. 2002b; Beaton et al. 2002). This may have been a particular problem for some of our patients who completed the repeat LCQ several months after the first. The long duration for repeat questionnaires for some patients reflects the prolonged treatment trials and follow-up that are often required for patients with chronic cough. The fact that the LCQ was administered only once before and patients were asked to recall this visit may have improved the performance of the GRCQ and minimized recall bias. Despite these limitations, the LCQ MID is similar to the MID of other QOL questionnaires and is consistent with the observations by Juniper et al. that QOL questionnaires utilizing a seven-point Likert response scale have a MID of approximately 0.5 points per item or question (Redelmeier et al. 1996; Guyatt et al. 1987; Guyatt et al. 1989). A change of LCQ score of 0.5 per question equates to a change in LCQ domain of 0.5 and LCQ overall score change of 1.5. We found significant correlations between global rating of change scores and the change in LCQ scores and weaker correlations with post-LCQ and pre-LCQ scores, in keeping with findings of Guyatt et al. (2002a), supporting the use of the global rating of change scale. We generally found a stepwise increase in the change in LCQ scores across GRCQ categories (with the exception of the psychological domain), which suggests that this instrument was able to discriminate between patients with small and large changes in health status related to chronic cough. Furthermore, the LCQ MID of 1.3 is consistent with our clinical experience of using the LCQ in routine clinical practice over the past 5 years.

We found an overall reduction in LCQ score in patients with chronic cough who indicated their cough remained unchanged on the GRCQ. This group included patients scoring +1/-1 on the GRCQ (hardly any change), so it is possible that some patients may have experienced more appreciable changes in health status in this category. Another unexpected finding was a higher LCQ psychological domain MID score than that of patients indicating moderate change in their psychological domain. Further review of our data revealed that this was due to three patients with large improvements in psychological domain scores who rated themselves as slightly changed in the global rating. This may reflect that for some patients perception of overall psychological status is not easily captured with a single global rating of change question and requires specific psychological questionnaires.

We were unable to perform a subanalysis to determine whether the LCQ MID varied according to age, gender or cause of cough owing to the small number of patients studied. The change in LCQ scores within individual global rating of change categories was not statistically significant, a finding also likely to have been influenced by the small number of subjects in the subgroup analysis. Larger studies are required to address this and it is hoped that the findings of this study will be used to determine the sample size of future studies to demonstrate both statistically and clinically significant differences. The findings of this study should be considered preliminary since the MID is likely to be refined by further studies and experience with the LCQ. Our study represents a useful starting point for interpreting QOL data and should stimulate further studies. The MID for acute cough may differ from that of chronic cough and needs to be determined. Objective cough assessment tools, such as 24-h ambulatory cough monitors, may allow further refinement of the MID for health status questionnaires.

Chronic cough frequently impacts on QOL and often profoundly. Health-related QOL questionnaires are increasingly being used as end points in clinical trials as they provide a standardized method to quantify the impact of cough on health status (Morice et al. 2007). It is of critical importance that health status data are interpreted in the correct context. The LCQ MID will facilitate this and aid interpretation of health status data when used in outpatient clinics as well as assessing risk–benefit profiles of new medicines and their cost-effectiveness. The MIDs for other measures of cough severity, such as cough reflex sensitivity, diary scores and cough frequency measurement, are not known and future studies should address this.

Acknowledgements We are very grateful to the outpatient, clinic preparation and secretarial staff for their assistance in this study.

References

Beaton DE, Boers M, Wells GA (2002) Many faces of the minimal clinically important difference (MCID): A literature review and directions for future research. Curr Opin Rheumatol 14: 109–114

Birring SS, Prudon B, Carr AJ et al (2003) Development of a symptom specific health status measure for patients with chronic cough: Leicester cough questionnaire (LCQ). Thorax 58:339–343

Brightling CE, Ward R, Goh VV et al (1999) Eosinophilic bronchitis is an important cause of cough. Am J Respir Crit Care 160:406–410

Decalmer S, Webster D, Kelsall A, McGuinness K, Woodcock A, Smith A (2007) Chronic cough: How do cough reflex sensitivity and subjective assessments correlate with objective cough counts during ambulatory monitoring? Thorax 62:329–334

French CT, Irwin RS, Fletcher KE et al (2002) Evaluation of a cough-specific quality-of-life questionnaire. Chest 121:1123–1131

Guyatt GH, Berman LB, Townsend M et al (1987) A measure of quality of life for clinical trials in chronic lung disease. Thorax 42:773–778

Guyatt GH, Nogradi S, Halcrow VV et al (1989) Development and testing of a new measure of health status for clinical trials in heart failure. J Gen Intern Med 4:101–107

Guyatt GH, Norman GR, Juniper EF, Griffith LE (2002a) A critical look at transition ratings. J Clin Epidem 55:900–908

Guyatt GH, Osoba D, Wu AW, Wyrwich KW, Norman GR (2002b) Methods to explain the clinical significance of health status measures. Mayo Clin Proc 77:371–383

Jaeschkle R, Singer J, Guyatt GH (1989) Measurements of health status: Ascertaining the minimally clinically important difference. Control Clin Trials 10:407–415

Jones PW (2002) Interpreting thresholds for a clinically significant change in health status in asthma and COPD. Eur Respir J 19:398–404

Jones PW, Quirk FH, Baveystock CM (1991) The St George's Respiratory Questionnaire. Respir Med 85:25–31

Juniper EF, Guyatt GH, Willan A et al (1994) Determining a minimal important change in a disease-specific quality of life questionnaire. J Clin Epidemiol 47:81–87

McGarvey LPA, Heaney LG, MacMahon J (1998) A retrospective survey of diagnosis and management of patients presenting with chronic cough to a general chest clinic. Int J Clin Pract 52:158–161

Morice AH, Menon MS, Mulrennan SA et al (2007) Opiate therapy in chronic cough. Am J Respir Crit Care Med 175:312–315

Redelmeier DA, Guyatt GH, Goldstein RS (1996) Assessing the minimal important difference in symptoms: A comparison of two different techniques. J Clin Epidemiol 49:1215–1219

Wyrwich KW, Tierry WM (2002) Using the standard error of measurement to identify important changes in the asthma quality of life questionnaire. Qual Life Res 11:1–7

Wyrwich KW, Nienaber NA, Tierney WM, Wolinsky FP (1999) Linking clinical relevance and statistical significance in evaluating intra-individual changes in health-related quality of life. Med Care 37:469–478

Appendix 1: Global Rating of Change Questionnaire

Question 1: Since your last clinic visit, has there been any change in the impact of your cough-related symptoms?
Question 2: Since your last clinic visit, has there been any change in your feelings (e.g. anxiety, embarrassment) as a consequence of your cough?
Question 3: Since your last clinic visit, has there been any change in the impact your cough has had on your work or social life?
Question 4: Since your last clinic visit, has there been any change in the impact your cough has had on your overall quality of life?

Please circle the response that best applies to you, for each question.
–7 A very great deal worse
–6 A great deal worse
–5 A good deal worse
–4 Moderately worse
–3 Somewhat worse
–2 A little worse
–1 Almost the same, hardly any worse at all
0 No change
1 Almost the same, hardly any better at all
2 A little better
3 Somewhat better
4 Moderately better
5 A good deal better
6 A great deal better
7 A very great deal better

Clinical Cough V: Complementary and Alternative Medicine: Therapy of Cough

J.G. Widdicombe(✉) and E. Ernst

Contents

Abstract We review the actions of complementary and alternative medicines (CAMs) in the treatment of cough and of the conditions associated with it; in particular asthma and upper respiratory tract infections. These therapies may work (1) peripherally, at the sites in the airways and lungs at which cough is being activated, (2) in the brainstem, where the neural "cough center" is situated, or (3) at the cerebral cortex, where cough can be initiated, suppressed or modified by conscious or unconscious controls. Of the large number of trials of CAMs against cough, most are inadequate in design. It may be difficult to randomize selection. Blinding is often impossible both for the patient and the therapist, and adequate placebo controls may be difficult to devise. The patient can usually identify the "active" treatment

J.G. Widdicombe
University of London, 116 Pepys Road, London SW20 8NY, UK
johnwiddicombej@aol.com

K.F. Chung, J.G. Widdicombe (eds.), *Pharmacology and Therapeutics of Cough*,
Handbook of Experimental Pharmacology 187,

by the taste or smell of a medicine, or from the approach and apparatus being used. Pure chemicals can be extracted from many of the herbs used as antitussives, and can be shown to be effective in randomized, blind, and controlled trials, but it does not follow that the herb itself, used in the recommended formula and shown to be antitussive, acts by this agency unless a placebo effect is ruled out. A few herbs are identified where the evidence points to a true antitussive action. Of nonherbal treatments, the few positive results are usually outweighed by the larger number of negative ones. Thus, in general, CAMs for cough are welcomed enthusiastically by the patient but lack sound evidence for their efficacy. Antitussive chemicals can be extracted from many herbs, but it is no more than a reasonable hypothesis that the herb itself acts through this pathway.

1 Introduction

Cough is one of the commonest symptoms for patients to consult a physician, only pain and physical disabilities surpassing it, and, with pain, it is the commonest reason to visit the pharmacist (Morice 2003). Over £300 million is spent each year by patients in the UK and over $2,000 million in the USA on over-the-counter (OTC) treatments (including herbal) for cough (Morice 2002). Acute and subacute coughs (by definition lasting less than 3 and between 3 and 8 weeks, respectively), which are predominantly caused by upper respiratory tract infections (URTIs), are usually viral (Higenbottam 2002; Morice et al. 2004; Irwin et al. 2006); the majority of patients with acute or subacute cough do not go to the community physician, local clinic, or hospital but instead go to the pharmacist and buy remedies on the basis of their experience or recommendations by others, or ask the pharmacist's advice. The treatment is usually symptomatic (nonspecific), directed at the cough rather than at its cause. Chronic cough, lasting over 8 weeks, often sends patients to community practices and hospital clinics. Here, if the cause of the cough is identified, specific (disease-specific) treatment for the condition is preferred and, if successful, should cure the cough (Higenbottam 2002; McGarvey 2003).

In idiopathic cough, where the cause cannot be identified, symptomatic treatment is the only option. In the USA, up to 20% of chronic coughs are idiopathic (Irwin et al. 2006), and in the UK 20–40% (Morice et al. 2004). In Japan, idiopathic chronic cough seems rather rare (Niimi 2007). These figures may point to the numbers of patients who potentially are attracted to complementary and alternative medicine (CAM): those with acute or subacute cough who do not go to a physician, and those with chronic undiagnosed cough whose physician cannot provide a satisfactory diagnosis and treatment. It should also be noted that each of the only three antitussive drugs currently available on prescription in the UK, codeine, pholcodine, and dextromethorphan, has adverse effects, so the physician may be wary of prescribing and the patient of taking them. Interestingly, in 1899, there were 61 antitussive drugs available on prescription (Eccles 2008); it is not clear whether the reduction in numbers is due to the efficiency of the three retained on the list or the inefficiency of the other 58. By comparison, for 2003/2004 there were 60 approved OTC treatments for cough, many of them herbal or herbal-based.

Conventional antitussive drugs and CAMs can be tested in five different ways: (1) in healthy (noncoughing) subjects using provocation of cough by an irritant aerosol such as citric acid, capsaicin or distilled water (fog) (Dicpinigaitis 2007, 2008; Morice et al. 2007); (2) in patients with cough due to airways' disease (Irwin et al. 2006; Morice et al. 2007); (3) in the latter group but with cough induced by an irritant aerosol (Dicpinigaitis 2007,2008); in animals with either (4) spontaneous cough due to disease (Korpas and Tomori 1979) or (5) induced by an irritant (Belvisi and Hele 2003; Mackenzie et al. 2004). The distinction between these methods is important in conventional medicine because the sensitivity to an antitussive drug depends on the condition of the subject and the method chosen (Fuller 2003). Such a consideration does not seem to have been applied to CAM and cough. Furthermore there are different patterns of cough, depending primarily on the cause. "Wet" and "dry" coughs are an obvious example, and the balance between inspiratory and expiratory efforts within a cough bout is another (Korpas and Tomori 1979; Fontana and Widdicombe 2007; Tatar et al. 2008). Conventional antitussive agents and general anesthetics have different potencies on different patterns of cough in animals and humans (May and Widdicombe 1954; Nishino et al. 1996,2004; Tatar et al. 2008), but this approach does not seem to have been extended to CAM.

1.1 Mechanisms of Action of Antitussive Agents

Most recent extensive research has identified four probable peripheral nervous "sensors" for cough stimuli in the larynx and lower respiratory tract (Canning et al. 2006; Canning 2008). These send nerve impulses up the vagus nerves to the brainstem, where they enter a complex neural network, the respiratory rhythm generator (previously called the "respiratory center"). The patterns of cough may depend on the nature of its stimulus and on the sensors activated. Different neural networks in the brainstem (in the medullary "cough center") have been delineated to explain these patterns (Shannon et al. 2004; Bolser 2008, Chen et al. 2008). The pharmacology of the cough sensors has been extensively studied, including their membrane receptors that can excite or inhibit the nerves (Geppetti 2008; Belvisi 2008; Lee and Gu 2008; Mazzone and Undem 2008). The application of these pharmacological results to the mode of action of antitussive agents is an emerging field of study, and is very relevant to understanding the modes of action of herbal and other CAM treatments.

Most conventional antitussives are thought to act at brainstem level, and several mediators and transmitters involved in cough have been identified here (Chen et al. 2008; Takahama et al. 2008), offering possible sites of action for antitussive agents. Other afferent inputs, for example, from the esophagus, nose, and ear, are known to modulate the cough reflex initiated from the lower airways (Hanacek et al. 2006), and some of these could be activated by herbal and other CAM treatments.

Finally, the role of the cerebral cortex in cough must be considered. We can voluntarily produce or inhibit cough, and there is growing evidence that the cortex may play a modulating role in the cough due to disease and its responses to conventional antitussive agents (Widdicombe et al. 2006; Davenport 2008; Eccles 2008). This is another possible site of action of CAM antitussive agents and treatments.

1.2 The Placebo Effect

The power of placebo is very relevant to antitussive treatments in CAM. Eccles (2002, 2006, 2008) and Parvesi et al. (2001) have shown that, with conventional antitussive treatments for acute cough due to URTI, the placebo effect accounts for about 80–85% of the suppression of cough caused by standard doses (e.g., 30 mg) of codeine or dextromethorphan. The meta-analysis by Parvesi et al. (2001) found that the residual nonplacebo effect (15–20% of total) was statistically significant. However, with cough due to other causes, e.g., bronchitis, asthma, or chronic lung diseases, the placebo effect may be much smaller or even absent (Sevelius et al. 1971; Alyward et al. 1984; Morice et al. 2006). Many of the early experiments with conventional treatments did not meet current standards; e.g., although they were usually placebo-controlled and double-blind, patient selection and treatments were not randomized or not stated as such. However, as will be described, most of the assessments of antitussive actions with herbal medicines and other CAM treatments were neither placebo-controlled nor double-blind nor randomized.

It has been proposed that at least part of the action of placebos is the release of endorphins in the cerebral cortex owing to, for example, sweet- or bitter-activated sensory inputs from the mouth (Eccles 2006, 2008); this has been described as a "physiological" component of the placebo effect. It may be a semantic quibble to suggest that this part of the placebo response to herbal agents is a "true" antitussive action. It might be said that for any drug that causes sensation, for example, taste or smell, a true placebo would only be possible if the antitussive ingredient could be removed from the preparation, without affecting the latter's taste and smell. This is probably impractical.

2 Herbal CAM Treatments

2.1 Introduction

CAM therapies are extremely popular. A Google search (4 February 2008) of "cough/alternative medicine" resulted in 421,000 hits and "common cold/alternative medicine" generated 325,000 hits. A scan of seven lay books on complementary medicine showed that, among them, they recommended 96 different treatments for a common cold with embarrassingly little agreement among them (Ernst et al. 2006). For another example, Herbmed (the herbal equivalent of PubMed) lists 966 references under "cough, garlic," while PubMed lists only eight. A similar search for "cough, onion" gives 385 and six references, respectively. These examples could be multiplied.

There are several problems is assessing the efficacy of herbal treatments for cough:

1. It is often difficult to fully eliminate the placebo effect (see earlier), since most herbal remedies have clear smells, tastes, and even appearances. This difficulty is well illustrated by a recent report (Paul et al. 2007), which received much newspaper publicity, concerning honey. (Is honey herbal – it is derived from plants!) This report stated that honey, administered orally by syringe, was more effective than dextromethorphan in treating children with night cough, and that dextromethorphan was little more effective then the "placebo" – an empty syringe. The second result is not surprising since it confirmed much published evidence (Eccles 2006, 2008). However the empty syringe was not an adequate placebo for honey, since it lacked the smell, taste, and physical nature of honey. Of course the latter qualities could have a "physiological action" as part of the placebo effect (Eccles 2006, 2008). To test the placebo effect, the therapies could have been given in the form of swallowed capsules, but this would not permit the claimed action of honey, which is said to be a demulcent or soothing action on the pharyngeal mucosa. A better, but possibly impractical, approach would have been to give the agents while the children were asleep, since the placebo effect, almost by definition, depends on consciousness. But interpretation of the results is even more complex. The honey, given while the children were awake, could exert a long-lasting "placebo effect" while the children were later asleep, for example, by the action of endorphins released in the cerebral cortex. Thus, we still cannot say whether honey, given as in this study, acted by a specific or by a placebo effect; the patient may not care, but the medical scientist should. We provide this example at some length, because it is rather typical of much research on CAM and cough.
2. There are surprisingly few herbal treatments for cough itself. Far more exist for conditions that cause cough, such as URTIs (including colds) and acute bronchitis and asthma. In other words, the treatments are disease-specific and, as a result, little evidence is usually presented in these trials about changes in cough itself. If the disease is alleviated or cured so, presumably, is the cough. Symptomatic herbal remedies for cough are uncommon.
3. Many herbal remedies that may be "antitussive" could act indirectly on cough: for example, by expectorant, demulcent, bronchodilator, or local anesthetic actions. Ziment (2002) lists 15 such indirect ways in which herbal "antitussives" may work (see also later) and lists over 120 herbs which are claimed, nearly always with little or no evidence, to act via these indirect mechanisms. With the exception of bronchodilation, there are no effective means of assessing these indirect actions, and the relationship between bronchodilation and cough is little understood.

4. Nearly all herbal medicines consist of mixtures of ingredients each potentially active against cough, and the relative role of the individual ingredients, if any, has seldom been assessed or discussed. Nor has the possibility that different ingredients may have synergistic (or even occlusive) actions. The problem is compounded because many herbal treatments for cough are mixtures of several or many different plant (and even animal) components that could have antitussive or protussive actions.
5. A single active ingredient of a herb may have more then one action that could relieve cough. For example, menthol, extracted from *Mentha* species, is said to be a local anesthetic, a Ca^{2+}-channel blocker, a smooth-muscle relaxant, and a cooling agent. Isolated purified menthol inhibits cough (see later), but which of its modes of action are important has not been determined.
6. Analysis of a number of herbal extracts has identified chemicals later shown to have "true" antitussive actions in experimental animals and sometimes in humans. To conclude that the herb itself is a remedy for cough is clearly unjustifiable on this evidence, although it is a reasonable hypothesis that could be tested. To give an example, pure opioids such as morphine extracted from poppy (*Papiver somniferum)* are established antitussives in animals and humans, but there seems to be no evidence that nonpurified extracts of the plant, as would be used in herbal medicine, are also effective; if they were, they could be working by a placebo effect.
7. The use of nonhuman animal experiments in studying treatment of cough is important. They can analyze mechanisms of action which may be relevant to treating patients (Belvisi and Hele 2003; Mackenzie et al. 2004); they allow research procedures which might be considered inappropriate or unethical in humans; they may provide the basis and background for future important human studies; and they may indicate the undesirability of proceeding with human studies. To cite one example: there have recently been extensive studies of the antitussive actions in cats of polysaccharides derived from a number of plant species, including *Althaea officinalis*, *Articum lappa*, *Prunus persica*, *Paederia foetida*, and *Rudbeckia fulgida* (Nosolova et al. 1989, 2000, 2007; Sutovska et al. 2007), some of which are used as antitussives in CAM; the positive results could lead to a valuable new approach to human antitussive therapy, either using the herbs or using their purified extracts.

But interpretation of animal experiments should be done with caution. For example, animal work with conventional antitussives such as codeine and dextromethorphan usually requires doses many times (often 10–20) greater in terms of body weight than those given to humans to be effective (Belvisi and Hele 2003; Usmani et al. 2004). And in most animal experiments placebo controls have not been included, although there is no reason to suppose that only humans respond to a placebo. This problem is sometimes circumvented when control tests, using some other agent and identical procedures, give negative or far smaller responses than those for the active agent. The absence of blinding may not matter, at least as far as the animal is concerned. (The experimenter seems never to be "blinded.") Randomization is seldom or never mentioned. There seems to be little or no discussion as to whether these are matters that should be of real concern for animal experimentalists.

2.2 Antitussive Herbs

In the description that follows, we will initially divide herbal antitussives into their proposed modes of action, with examples. With few exceptions, it is claimed that individual herbs have more than one type of action that will affect cough, but the relative strengths of these actions have seldom been assessed. For brevity, we will not discuss safety concerns, which are undoubtedly important for many herbal antitussives (Ziment 2002; Ernst et al. 2006). Only herbs used in treatment of cough and its causes, unlike those aimed at preventing a disease from starting, will be considered.

Many of the modes of action are not strictly "antitussive," and some may even enhance cough (e.g., expectorants). But since the patient with coughs benefits from the treatment, we will use the term "antitussive" in a broad, if not always semantically correct, sense.

2.2.1 Centrally Acting Antitussive Herbs

The most prescribed centrally acting antitussives of plant origin are codeine and morphine, while the derivative diamorphine is even more effective; all originate from *Papiver somniferum* (poppy). It is therefore surprising that poppy itself does not feature strongly as a herbal treatment for cough, although it is widely used for hallucinogenic effects. This is presumably because the latter can dominate the response. *Erythroxylum coca* (cocaine) and *Cannabis sativa* (marijuana), both used as cough suppressants, may also have a central nervous system as well as peripheral antitussive actions, in view of their well-known hallucinogenic actions.

2.2.2 Demulcent Herbs

These act by a "soothing" effect mainly on the pharynx and larynx if the solid or liquid preparation is given orally, or on all the airways if it is a vapor. The patient should feel better and experiences the cough as less irritating or painful. The problem is that demulcent activity has never been measured. The only way to do it would be by subjective patient scores (How soothed do you feel?), and even these do not seem to have been used. In addition, a placebo action is difficult to eliminate; indeed it seems inevitable in view of the relevant sensations aroused by the herb. To give the solid or liquid agents by capsule would bypass their site of action, and if given as a vapor an appropriate placebo would be hard to select.

Demulcent remedies for cough, such as guanefenesin and bromhexine, are popular in OTC medications. In herbal medicine, those given orally include *Althaea officinalis* (marshmallow), *Cerraria icelandica* (Iceland moss), *Glycyrrhiza glabra* (licorice), *Limon usitatissimum* (flaxseed), *Tilia cordata* (linden), and *Verbascum densiflorium* (mullein), as well as syrupy herb extracts and very many others. Although these treatments are undoubtedly popular, there seems to be no evidence that their action on cough, if any, is other than by a placebo effect. Those inhaled include *Mentha peperita* (menthol) and *Eucalyptus* species (eucalyptus oil) vapors (see later).

2.2.3 Local Anesthetics

Some herbal treatments for cough contain local anesthetics that might act on nervous sensors in the mucosa of the pharynx and larynx before the medicine reaches the stomach. These include *Erythroxylum coca* (cocaine), *Cannabis sativa* (marijuana), and possibly *Syzygium aromaticum* (clove oil). However, there seems to be no evidence that these herbs have antitussive actions by this mechanism. Most local anesthetics also taste bitter, which could also have an antitussive action (see later).

Researchers in the field of conventional medicine are now searching intensively for antitussive agents that act on the membrane receptors for cough sensors in the airways (peripherally acting antitussives) (Chung 2006, 2007,2008; Barnes 2007), but similar studies do not seem to have been conducted for herbal CAM therapies.

2.2.4 Surfactants, Saponins

Saponins and similar substances have surfactant-like properties and might be expected to ease clearance of mucus from the pharyngeal and laryngeal airway epithelia, thus providing a demulcent-like response. Such herbs used to treat cough or its causes include *Hedera helix* (ivy), *Primula* species (primrose), *Quillaja saponaria* (soap bark), *Grindelia camporum* (gumplant), and *Polygana senega* (senega); however, their surfactant activities have been demonstrated only in vitro, and there is no evidence that this effect is present in vivo or that it is "antitussive." In addition, many of them are bitter or even irritating and, if they are effective, this could be their mechanism of action (see later).

2.2.5 Expectorant, Mucolytic, and Mucokinetic Herbs

Expectorants increase the production and/or removal of phlegm from the lower airways. The only practical way to measure this is to make the patient spit out phlegm into a pot (while talking precautions to avoid contamination with saliva) (Rubin 2003). This is inaccurate and is only appropriate when the patient is producing copious amounts of phlegm, as in chronic bronchitis and bronchorrhea. Increased expectoration could be due to (1) increased production of mucus, (2) increased ease or rate of its clearance (e.g., by a surfactant action; see earlier), which can be measured by the rate of movement of inhaled technitium-99m, a technique used in laboratory studies of cough, but rarely in herbal medicine (Hasani et al. 2003), or (3) a change in its physical properties – mucolysis – which can be measured in vitro but not in vivo (Foster 2002; Rubin 2003). The terms "mucokinesis" and "mucokinetics" are sometimes used but are seldom or never defined, and they seem to mean no more that the processes just described. "Mucoregulation" and "mucoregulator" are terms also in use, and may be useful general expressions in view of the fact that the precise mode of action of drugs on mucus is usually unknown.

It is therefore not surprising that, in herbal medicine, an "expectorant" means a treatment that makes patients feel they can more easily "clear their chests," and the mechanisms are not defined and the response is not quantified. Herbs said to treat cough by improving "expectoration" include *Capsicum* species (paprika, peppers), *Adhatoda vasica* (malibar nut),*Ascarium sieboldi* (wild ginger), *Fritillaria verticillata* (fritillary), *Eriobotrya japonica* (loquat), *Mulberry* species (mulberry), *Citrus* species (orange peel), *Pinellia ternate*, *Pinellia ternate bulbil* (Zhi bau), *Boerhaavia diffuza* (hogweed), *Hypericum perforatum* (St. John's wort), and many others.

2.2.6 Antispasmodic Herbs

Many antitussive herbs are claimed to work by an antispasmodic or bronchodilator action (Ernst 1998). It is true that they cause either bronchial muscle relaxation in vitro, or decreases in airways' resistance and increases in the forced expiratory volume in 1 s (FEV_1)in vivo. The relationship between bronchomotor tone and cough and its sensitivity has been much studied. In general bronchoconstriction causes or enhances the sensitivity of cough, while bronchodilation does the opposite (Pavord 2004). Thus, if cough is associated with bronchoconstriction, as in asthma, one would expect a herbal treatment causing bronchodilation to be more effective against the associated cough, than with a cough not related to bronchospasm. In other words, bronchodilator herbs should be also antitussive, but not necessarily the opposite. This possibility does not seem to have been shown or tested.

Herbs claimed or shown to have antispasmodic effects, in vitro and/or in vivo, and to be antitussive, include *Adhartoda vasica* (malibar nut), *Petasites hybridus* (butterbur), *Ephedra sinica* (ephedra), *Mentha peperita* (peppermint), *Atropa* species (atropine), *Curcuma longa* (turmeric), *Ocimum sanctum* (basil), and *Piper longum* (pepper).

2.2.7 Anti-inflammatory, Antioxidant, and Antiallergic Herbs

Many antitussive herbs are claimed also to have anti-inflammatory, antioxidant, or antiallergic properties, or a combination, usually on the basis of in vitro studies. Presumably they would be especially effective against coughs due to URTI such as common colds, acute bronchitis, and allergic asthma. They include *Amni visnaga* (cromones), *Glycyrrhiza glabra* (licorice), *Aloysia triphylla* (lemon erbena), *Cymboconon citrates* (lemon grass), *Commiphora mukul* (myrrh), and *Allium cepa* (onion).

2.2.8 Strong Tasting Herbs

Many antitussive herbs taste bitter (including, paradoxically, local anesthetics) and it has been suggested that the afferent input from nerves responding to bitter-tasting drugs may inhibit cough (Eccles 2006, 2008; Ziment 2002). Oddly enough, many other antitussive drugs taste sweet or sugary, and the use of honey to treat cough has been described already; others taste spicy or peppery. The cynic might say that all you need for an effective herbal antitussive is a strong flavor, and that it would work by a placebo effect. For obvious reasons, it is difficult to test for a placebo effect with these agents. If the subject can identify the "active" treatment there can be no true placebo.

Strongly bitter herbal antitussives include *Berberis vulgais* (berberis), *Sanguinaria canadensis* (bloodroot), *Marrubium vulgare* (horehound), *Hedera helix* (ivy), *Schisandra chinensis* (schisandra), *Citrus* species (lemon juice), and *Polygala senega* (senega). Sugary herbal treatments are usually sugary owing to addition of sugar or honey to the preparation; of the 60 OTC antitussive preparations approved in 2003/2004, 58 had added sugar or syrup, but some herbs contain much natural sugar. Spicy/peppery herbs used as antitussives include *Capsicum* species (capsaicin) and *Piper longum* (pepper).

2.2.9 Decongestant Herbs

Some herbs used to treat cough and its causes are said to act as decongestants, mainly because they are claimed to clear a blocked nose. There is no evidence that their action against cough is based on decongestion, although if the cough is related to a congested nose or lower-airways' mucosa, this is a theoretical possibility. Examples include *Ephedra* species (ma huang) and *Magnolia* species (magnolia).

2.2.10 Conclusions

Thus there are at least nine ways in which herbal medicines may act to treat cough, and each may have several subdivisions (such as mucokinetics). Examples of herbs are given for each mechanism, but the total list is many times greater (Ziment 2001, 2002; Blumenthal 2000; Fetrow and Avila 2001; Skidmore-Roth 2001). Many herbs are thought to act by more than one mechanism. Many herbal treatments contain five to 12 different herbs (and even animal products). Even if it is shown that a particular herbal therapy successfully treats a particular type of cough, it is very difficult to define the exact mechanisms involved. Of course, the same is true of some conventional medical treatments of cough.

2.3 *Particular Herbal Cough Therapies*

In this section we select, inevitably rather arbitrarily, four herbal cough treatments for which the effectiveness and mechanisms have been studied. With many other herbs there are a few pointers in the literature that the herb may be objectively considered an effective antitussive in humans, but the evidence is so thin as to be unconvincing (Table 1). In some instances, research only on animals or animal preparations has been done.

2.3.1 Andrographis Paniculata (King of Bitters)

This ayurvedic herb contains andrographolides in its leaves and roots which, in addition to tasting very bitter, have anti-inflammatory, antioxidant, and antiallergic properties and act on the immune system of the body (Coon and Ernst 2004).

Extracts of this herb have been used to treat URTIs, although not specifically for cough (Poolsup et al. 2004; Spasov et al. 2004; Ernst et al. 2006). However, only subjective changes in cough were reported in most of the trials. Most of the seven clinical controlled trials showed significant superiority over placebo or antibiotic treatment, and the method of the trials was generally acceptable (Coon and Ernst 2004; Gabrielian et al. 2002; Caceres et al. 1999; Melchior et al. 2000). One trial showed improvement in most symptoms of the common cold, but not cough (Gabrielian et al. 2002). No results of experiments on animals or with cough-provocation tests in humans seem to have been published.

2.3.2 Ephedra Sinica (Ma Huang)

Ma huang is one of the most popular Chinese (and Western) herbs and is used to treat a variety of respiratory conditions, such as asthma and rhinitis, and also cough. It contains ephedrine and pseudoephedrine. It is a sympathomimetic, a bronchodilator, and a mucosal decongestant; it is presumably these properties that give it antitussive actions (Ziment 2002). It also has anorexiant and ergogenic properties, about which there is an extensive literature. It is officially banned in the USA because of its popularity as a dietary supplement to lose weight, elevate mood, and improve athletic performance. (It has been called "one of the most potent and dangerous herbal medications in common use" (Bent et al. 2003); one might add "and popular.")

Although ma huang is effective in preventing induced cough in guinea pigs and mice (Miyagoshi et al. 1986; Minamizawa et al. 2006), there seem to be no adequately conducted and controlled studies in humans (Ernst et al. 2006).

Table 1 Some herbal treatments for cough

Species	Common name	'Pure extract'	Action(s)	Refs
Adhatoda vasica	Malibar nut	Vascinine	Bronchodilator Expectorant Bitter	Dhuley (1999) Blumenthal (2000)
Allium sativum	Garlic	Allicin	Antibacterial	Adeleye and Opiah (2003) Bielowy (2004)
Carum copticum	Ajwain	Carvacrol	Bronchodilator	Boskabady et al. (1998, 2005)
		Thymol	Vasodilator Antihistamine	
Crocus sativus	Saffron	Safronal	Anti-inflammatory	Hosseinzadeh and Ghenaati (2006) Duke and Ayenau (1985)
Fritillaria sps	Beimu Snake's head	Alkaloids	Anti-inflammatory Expectorant Antibacterial	Saliba et al. (2004) Jiang et al. (2005) Lin et al. (2007)
Glaucum flavium		Glaucine	Antitussive Bronchodilator	Nosalova et al. (1989) Dierckx et al. (1981)
Glycyrrhizae glabra, radix	Licorice	Liquiritin Glycyrrhizinate	Anti-cAMP	Kamei et al. (2003) Bielowy (2004)
Panex quinquefolium	Ginseng	Ginsenoside	Antioxidant Antibacterial	Bielowy et al. (2004) McElhaney (2006)
Pinene sps	Pine oil	Pinene Sobrerol	Expectorant	Medici et al. (1985) Blumenthal (2002)
Piper longum	Pepper	Piperine	Spicy	Bielory et al. (2004) Ziment (2004)
Theobroma cocoa	Chocolate Cocoa	Theobromine *Epicatechin*	Bronchodilator	Usmani et al. (2004) Sanno et al. (1993)
Sesamus Indicum	Sesame oil	Omega 9 Sesamin	Bronchodilator Mucolytic Surfactant	Akoachere et al. (2002) Saab et al. (2006)
Zingiber Officinale	Ginger	Sesquit Penoids	Anti-inflammatory	Thomas et al. (2007) Hoffman (2007)

The list is selected, somewhat arbitrarily, to give examples of herbs popular in the treatment of cough, especially in western countries. Those detailed in the text are not included. For much longer lists see Blumenthal (2000), Fetrow and Avila (2001), Skidmore-Roth (2001), and Ziment (2001, 2002).

2.3.3 Echinacea Species (Sesame Oil)

The roots and leaves, especially species *Echinacea augustifolia*, *Echinacea pallida*, and *Echinacea purpurea*, of these plants have been extensively used to treat respiratory infections and many other medical conditions. Likewise, there is a large amount of literature on the subject. Extracts contain many potentially active constituents, including alkylamides, glycoproteins, and polysaccharides. Their main actions are anti-immune, anti-inflammatory, local anesthetic, antiviral, and antioxidant (Linde et al. 2007). Several recent reviews of the literature relating to *Echinacea* and URTIs have been published (Barratt et al. 1999; Giles et al. 2000; Ernst et al. 2006; Caruso and Gwaltney 2005; Linde et al. 2007; Melchart et al. 1999). These include several meta-analyses, not all of them with positive conclusions. The most recent one included 14 randomized controlled trials. Its results show that *Echinacea* decreased the odds of developing a cold by 58% and the duration of a cold (and therefore presumably the associated cough) by 1.4 days (Yale and Liu 2004). These findings were robust in several subgroup analyses (Shah et al. 2007).

2.3.4 Mentha Species (Spearmint, Peppermint, Pennyroyal)

Mint extracts are used in many antitussive treatments, either orally or by injunction (usually chest rub). They contain about 80 potentially active compounds, those thought to be most active including menthol (peppermint camphor), pulegone, cineole, and limonene. Menthol acts as a calcium-channel antagonist and by spasmolysis. Its vapor is an expectorant in rabbits (Boyd and Sheppard 1968). Whether it is a decongestant is disputed. It stimulates nerve sensors in the respiratory tract (but not those causing cough) and gives the sensation of coldness and may inhibit breathing (Eccles 1994). In guinea pigs it inhibits the cough caused by inhaled citric acid (Laude et al. 1994).

In healthy subjects, menthol vapor inhibits citric acid induced cough, unlike other aromatic oils which were used to act as placebos (Packman and London 1980; Morice et al. 1994). Unfortunately no properly designed clinical trials of vaporized menthol against cough seem to have been done, possibly for lack of an adequate placebo, but the basic studies suggest that it is potentially an active antitussive.

2.3.5 Conclusions

We have selected four herbs which seem to provide the most evidence for their antitussive effectiveness and mechanisms. Table 1 lists a further 13 antitussive herbs with some published evidence as to their effectiveness and modes of action. Very many more herbs, with virtually no good evidence for their efficacy, are listed elsewhere (Blumenthal 2000; Skidmore-Roth 2001; Fetrow and Avila 2001; Ziment 2001, 2002).

3 Nonherbal CAM Treatments

3.1 Introduction

Few of the many types of nonherbal CAMs have been tested against cough as a symptom, although many have been tried against common causes of cough such as asthma and URTI. Changes in cough associated with these conditions have been reported infrequently.

3.2 Acupuncture

Acupuncture has been tried in many studies as a treatment of asthma and other respiratory conditions associated with cough. The results are "highly inconsistent" (Ernst et al. 2006). Symptomatic treatment of cough itself has seldom been studied. One study showed that acupuncture worked significantly better than conventional medicine in asthma (Maa et al. 2003). However, none of the studies were blinded or adequately controlled to test for placebo effect.

3.3 Aromatherapy, Inunctions

Both aromatherapy and inunctions (rubs) have been used for centuries, but only recently have their possible methods of action been studied. L-Menthol (*Mentha piperita*), discussed earlier as an oral treatment for cough, is the most used and most studied, and is present in the popular Vick's Vaporub®. It is thought to act by several mechanisms: local anaesthetic, nerve stimulant (paradoxically), Ca^{2+}-channel antagonist, vasodilator, and bronchodilator (Eccles 1994; Ziment 2002). Inhaled as a vapor (with eucalyptus oil), menthol blocks citric acid induced cough in humans when compared, for example, with pine oil as a "placebo" (Morice et al. 1994). When applied as an injunction, it also blocks citric acid induced cough, as does camphor but not cineole (Packman and London 1980). Given as a vapor to unanesthetized guinea pigs, it blocks citric acid induced cough (Laude et al. 1994); here a placebo effect is less likely, although not impossible. Menthol inhibits the cough associated with endotracheal intubation (Haidl et al. 2001); such coughs due to mechanical stimulation of the larynx and trachea (unlike those with chemical irritants) are known to be resistant to antitussive agents (Korpas and Tomori 1979; Tatar et al. 2008). Hasani et al. (2003) showed that menthol significantly increased lung mucociliary clearance in patients with chronic airways obstruction; interestingly it also *increased* cough frequency, perhaps by an "expectorant" action. Apart from this study, there seem to be no acceptable studies on the effect of menthol on natural cough.

Although the abovementioned results may be suggestive, there seem to have been no rigorous clinical trials of the commercial and popular Vick's Vaporub® or Friars Balsam®, both of which contain many potentially antitussive ingredients.

3.4 Breathing Exercises, Yoga

There have been many reports of breathing exercises, such as Buteyko and pranayama and also yoga, as a treatment for asthma (Vedanthan et al. 1998; Manocha et al. 2002; Cooper et al. 2003) and one for pulmonary tuberculosis (Visweswaraiah and Telles 2004). Some have included "cough scores," usually by the patient, but cough itself is seldom specified. Many of the studies were controlled, using a different treatment procedure, and some randomized, but blinding was clearly impossible, at least by the patient. These studies have been reviewed (Ernst 2000; Huntley et al. 2002; Steurer-Stey et al. 2002; Holloway and Ram 2004; Ernst et al. 2006) and the general conclusion is that the treatments may improve the symptoms and need for medication by the patients, sometimes significantly, but tests such as FEV_1 and bronchial provocation show no significant change.

There seem to be no studies on URTI, or the cough related to it, and no studies of treatments directed specifically at cough.

3.5 Diet and Nutrition

There have been trials of dietary supplements, especially vitamin C, omega 3, zinc, and selenium, in preventing and treating asthma and URTI (Ernst et al. 2006). Nearly all the results have been inconclusive or at the best encouraging, except for those with vitamin C in the treatment of URTIs analyzed in a Cochrane review (Douglas et al. 2004); vitamin C significantly reduced the duration of the illness, and presumably therefore reduced the duration of the coughs. Other prospective studies on phlegm-producing cough have shown that its incidence is significantly reduced by diets rich in polysaccharides and isoflavones (Butler et al. 2004), and in mainly vegetarian diets compared with meat-rich diets (Butler et al. 2006). The mechanisms were not analyzed, but could involve the relative contents of antioxidants.

3.6 Exercise

There are several papers that suggest that regular exercise has a tendency to prevent URTIs (Nieman et al. 1998; Ernst et al. 2006). It does not seem to have been recommended as a preventative or therapeutic approach to asthma, or as a symptom-directed treatment for cough. This is not perhaps surprising since exercise may cause postexercise hyperreactivity, which usually includes coughing.

3.7 Homeopathy

Most homeopathic trials have been directed to treat conditions associated with cough, such as URTI including colds, rather than cough itself. Few studies have reached acceptable scientific standards and, of those that have, the results have been various.

A typical example is a study in 57 primary-care clinics in eight countries, with 1,577 patients with upper respiratory tract and ear complaints that usually included cough (Haidvogl et al. 2007). The conclusion was that "homeopathic treatment... was not inferior to conventional treatment." This is not surprising because the selection was not randomized and neither the patients nor the practitioners were "blinded." A similar study, neither randomized nor blinded, with 485 patients found that "homeopathic therapy was effective to a similar degree or greater than the conventional therapies" (Rabe et al. 2004).

A Cochrane Database meta-analysis (Vickers and Smith 2004) of homeopathy on influenza and influenza-like conditions treated by *Oscillococcinum* (extracted from the viscera of wild duck) included four studies, assessing symptoms such as cough, which were judged scientifically appropriate for inclusion (Casanova 1984; Ferley et al. 1989; Papp et al. 1998). "Most of the studies favor homeopathy, although not all reach statistical significance." A similar conclusion was reached in another analysis of homeopathy and URTI (Ernst et al. 2006). It must be stressed that all published studies seem to deal with treatment of URTI with cough as a symptom, and none with homeopathic treatment of cough per se.

3.8 Hypnosis

Hypnotherapy (self-hypnosis) has been shown to be effective in treating "habit cough" in children (Anbar and Hall 2004; Anbar and Stothower 2006). Obviously the trials were not blinded or controlled, and the positive result is hardly surprising. It has been used in several trials to treat asthma with encouraging but not fully convincing positive results (Ernst et al. 2006; Maher-Loughnan et al. 1962); the trials did not address cough specifically. Hypnosis does not seem to have been tried with URTIs.

3.9 Other Nonherbal CAMs

These include massage, mediation, osteopathy, and reflexology, and have been tried as treatments for asthma and URTIs, but seldom or not at all for treatments of cough itself. They provide no convincing evidence for their effectiveness.

4 Conclusions

Herbal treatments for cough and its underlying causes are very popular and, in the opinion of the users, often successful. It is difficult to rule out a placebo effect, a problem that also applies to conventional antitussive agents such as codeine. The antitussive effects are said to be via many possible pharmacological routes. Some of these can be measured ("true" antitussive, bronchodilator actions), while others cannot or are not (demulcent, "mucoregulation," expectorant, etc.). Many herbs claimed to be antitussive contain pure chemicals which, when extracted, can be shown to have antitussive actions. It is a reasonable hypothesis, but not evidence, that the intact herb or its concoctions may act via this pure ingredient. Evidence that other (nonherbal) forms of CAM might be effective against cough is less convincing, again largely because of the difficulty in elimination of placebo effects. They have usually been directed more to prevent cough-causing diseases such as asthma and URTI than to treat the cough per se.

This rather negative conclusion may be counterbalanced by the positive advice to consume the following herbal antitussive menu proposed by Ziment (2002):

Chicken soup with pepper, garlic, ginger and mint, to be followed by oysters flavored with Tabasco sauce. This should be succeeded with a salad containing onions, radish seedlings and guava leaves. The main course should consist simply of water, following which the next dish should be a syrupy, demulcent, mucilaginous, scrumptious dessert, to be capped by mint tea flavored with eucalyptus and licorice. Then, one may choose any of a number of herbs that offer central stimulation of the cerebral conversation center along with suppression of the cough cells of the medulla. As long as it will please, such a meal can do no harm and, provided conventional post-prandial cigars are not lit up, all coughing should be abated. At this point in the menu of choice cough agents, it would be appropriate to debate the antitussive potential of one of the most important herbal agents: the fermented fruit of *Vitis vinifera* as prepared in selected regions of the world in accordance with traditional oeniologic principles.

One might decide to add a spoonful of honey at some stage of the meal.

Acknowledgement We are grateful to Irwin Ziment for his expert assessment of the literature and for permission to quote his "ideal antitussive diet."

References

Adeleye IA, Opiah L (2003) Antimicrobial activity of extracts of local cough mixtures on upper respiratory tract bacterial pathogens. West Indian Med J 52:188–190

Akoachere JF, Ndip RN, Chenwl EB et al (2002) Antibacterial effect of Zingibar officinale and Garcinia kola on respiratory tract pathogens. East Afr Med J 79:88–92

Anbar RD, Hall HR (2004) Childhood habit cough treated with self-hypnosis. J Pediatr 144: 213–217

Anbar RD, Slothower MP (2006) Hypnosis for treatment of insomnia in school-age children: A retrospective chart survey. BMC Pediatr 6:23

Aylward M, Maddock J, Davies DE et al (1984) Dextromethorphan and codeine: Comparison of plasma kinetics and antitussive effects. Eur J Respir Dis 65:283–291

Barnes PJ (2007) The problem of cough and development of novel antitussives. Pulm Pharmacol Ther 20:416–422

Barratt B, Vohmann M, Calabrese C (1999) Echinacea for upper respiratory infection. J Fam Pract 48:628–637

Belvisi M (2008) Cough sensors II: Carrabinoid and opioid membrane receptors onvagal sensory nerves. In: Handbook of experimental pharmacology: Pharmacology and therapeutics of cough. Springer, Berlin, this volume

Belvisi MG, Hele DJ (2003) Animal models of cough. In: Chung KF, Widdicombe JG, Boushey HA (eds) Cough: Causes, mechanisms and therapy. Blackwell, Oxford, pp 217–224

Bent S, Tiedt T, Odden M, Shlipak M (2003) The relative safety of ephedra compared with other herbal products. Ann Intern Med 138:468–471

Bielory L (2004) Complementary and alternative interventions in asthma, allergy and immunology. Ann Allergy Asthma Immunol 93:S45–S54

Bielory L, Russin J, Zuckermen GB (2004) Clinical efficacy, mechanisms of action, and adverse effects of complementary and alternative medicine therapies for asthma. Allergy Asthma Proc 25:283–291

Blumenthal M (ed) (2000) Herbal medicine. Expanded commission E monographs. Integrative medicine Communications, Newton, MA

Bolser D (2008) Central mechanisms II: Pharmacology of brainstem pathways. In: Chung KF, Widdicombe JG (eds) Handbook of pharmacology: Pharmacology and therapeutics of cough. Springer, Berlin, this volume

Boskabady MH, Rakhshandah H, Moetamedshariati V (1998) Bronchodilatory and anticholinergic effects of *Carum copticum* on isolated guinea pig tracheal chains. Med J I R Iran 11:329–334

Boskabady MH, Jandaghi P, Kiani S et al (2005) Antitussive effect of Carum copticum in guinea pigs. J Ethnopharmacol 97:79–82

Boyd EM, Sheppard EP (1968) the effect of steam inhalation of volatile oils on the output and composition of respiratory tract fluid. J Pharmacol Exp Ther 163:250–256

Butler LM, Koh W-P, Lee MC et al (2004) Dietary fiber and reduced cough with phlegm. A cohort study in Singapore. Am J Respir Crit Care Med 170:279–287

Butler LM, Koh W-P, Lee H-P et al (2006) Prospective study of dietary patterns and persistent cough with phlegm among Chinese Singaporeans. Am J Respir Crit Care Med 173:264–270

Caceres DD, Hancke JL, Burgos RA et al (1999) Use of visual analogue measurements (VAS) to assess the effectiveness of standardized Andrographis paniculata extract SHA-10 in reducing the symptoms of common cold. A randomized double-blind placebo-controlled study. Phytomedicine 6:217–213

Canning B (2008) Mechanisms II: Sensors and afferent pathways for cough. In: Chung KF, Widdicombe JG (eds) Handbook of Experimental pharmacology. Pharmacology and therapeutics of cough. Springer, Berlin, this volume

Canning BJ, Mori N, Mazzone SB (2006) Vagal afferent nerves regulating the cough reflex. Respir Physiol Neurobiol 152:223–242

Caruso TJ, Gwaltney JM Jr (2005) Treatment of the common cold with Echinacea: A structured review. Clin Infect Dis 40:1237–1241

Casanova P (1984) Homeopathy, flu syndrome and double-blinding. Tonus 7:25–26

Chen C-Y, Joad JP, Bric J et al (2008) Central mechanisms I: Plasticity of central pathways. In: Chung KF, Widdicombe JG (eds) Handbook of pharmacology: Pharmacology and therapeutics of cough. Springer, Berlin, this volume

Chung KF (2006) Drugs to suppress cough. Expert Opin Investig Drugs 14:19–27

Chung KF (2007) Effective antitussives for the cough patient: An unmet need. Pulm Pharmacol Therap 20:438–445

Chung KF (2008) Clinical cough V: The need for new therapies for cough: Disease-specific and symptom-related antitussives. In: Chung KF, Widdicombe JG (eds) Handbook of pharmacology: Pharmacology and therapeutics of cough. Springer, Berlin, this volume
Coon JT, Ernst E (2004) *Andrographis paniculata* in the treatment of upper respiratory tract infections: A systematic review of safety and efficacy. Planta Med 70:293–298
Cooper S, Oborne J, Newton S et al (2003) Effect of two breathing exercises (Buteyko and pranayama) in asthma: A randomized controlled trial. Thorax 58:674–679
Davenport P (2008) The urge-to-cough: A respiratory sensation. In: Chung KF, Widdicombe JG (eds) Handbook of experimental pharmacology: Pharmacology and therapeutics of cough. Springer, Berlin, this volume
Dhuley J (1999) Antitussive effect of *Adhatoda vasica* extract on mechanical or chemical stimulation-induced coughing in animals. J Ethnopharm 67:361–365
Dicpinigaitis P (2007) Experimentally induced cough. Pulm Pharmacol Ther 20:319–324
Dicpinigaitis P (2008) Clinical cough II: Measuring the cough response in the laboratory.In: Chung KF, Widdicombe JG (eds) Handbook of experimental pharmacology: Pharmacology and therapeutics of cough. Springer, Berlin, this volume
Dierckx P, Leblanc G, Decoster A et al (1981) Double-blind study of glaucine in chronic cough. Int J Pharmacol Ther Toxicol 19:396–399
Douglas RM, Hemela H, D'Souza R et al (2004) Vitamin C for preventing and treating the common cold. Cocrane Database Syst Rev, Issue 4, Art. No.: CD000980
Duke JA, Ayenau ES (1985) Medicinal plants of China. Algonac. Reference Publications, Michigan, p 363
Eccles R (1994) Menthol and related cooling compounds. J Pharm Pharmacol 46:618–630
Eccles R (2002) The powerful placebo in cough studies. Pulm Pharmacol Ther 15:303–308
Eccles R (2006) Mechanisms of the placebo effect of sweet cough syrups. Respir Physiol Neurophysiol 152:340–348
Eccles R (2008) Central mechanisms IV: Conscious control of cough and the placebo effect. In: Chung KF, Widdicombe JG (eds) Handbook of experimental pharmacology: Pharmacology and therapeutics of cough, Springer, Berlin, this volume
Ernst E (1998) Complementary therapies for asthma: What patients use? J Asthma 35:667–671
Ernst E (2000) Breathing techniques–adjunctive treatment modalities for asthma? A systematic review. Eur Respir J 15:969–972
Ernst E, Pittler MH, Wider B (eds) (2006) The desktop guide to complementary and alternative medicine. Mosby, Elsevier
Ferley JP, Zmirou D, D'Adhemar D et al (1989) A controlled evaluation of a homeopathic preparation in the treatment of influenza-like symptoms. Br J Clin Pharmacol 27:329–335
Fetrow CW, Avila JR (2001) Professional's handbook of complementary and alternative medicines, 2nd edn. Springhouse Corporation, Springhouse, PA
Fontana GA, Widdicombe JG (2007) What is cough and what should be measured? Pulm Pharmacol Ther 20:307–312
Foster WM (2002) Mucociliary transport and cough in humans. Pulm Pharmacol Ther 15:277–282
Fuller RW (2003) Cough sensitivity and the use of provocation tests. In: Chung KF, Widdicombe JG, Boushey HA (eds) Cough: Causes, mechanisms and therapy. Blackwell, Oxford, pp 49–56
Gabrielian ES, Shukarian AK, Goukasova GI et al (2002) A double-blind, placebo-controlled study of *Andrographis paniculata* fixed combination Kan Jang in the treatment of acute upper respiratory tract infections including sinusitis. Phytomedicine 9:589–597
Geppetti P (2008) Cough sensors I: TRP membrane receptors on cough sensors. In: Chung KF, Widdicombe JG (eds) Handbook of experimental pharmacology: Pharmacology and therapeutics of cough. Springer, Berlin, this volume
Giles JT, Palat CT III, Chien SH et al (2000) Evaluation of Echinacea for treatment of the common cold. Pharmacotherapy 20:690–697
Haidl P, Kemper P, Butnarasu SJ et al (2001) Does the inhalation off L-menthol solution in the premedication for fiberoptic bronchoscopy affect coughing and the sensation of dyspnoea? Pneumologie 55:115–119

Haidvogl M, Riley DS, Heger M et al (2007) Homeopathic and conventional treatment for acute respiratory and ear complaints: A comparative study on outcome in the primary care setting. BMC Complement Altern Med 7:7

Hanacek J, Tatar M, Widdicombe JG (2006) Regulation of cough by secondary sensory inputs. Respir Physiol Neurobiol 152:282–297

Hasani A, Pavia D, Toms N et al (2003) Effect of aromatics on lung mucociliary clearance in patients with chronic airways obstruction. J Altern Complement Med 9:243–249

Higenbottam T (2002) Chronic cough and the cough reflex in common lung diseases. Pulm Pharmacol Ther 15:241–248

Hoffman T (2007) Ginger: An ancient remedy and modern miracle drug. Hawaii Med J 66: 326–327

Holloway E, Ram FS (2004) Breathing exercises for asthma. Cochrane Database Syst Rev Issue 1, Art No CD001277

Hosseinzadeh H, Ghenaati J (2006) Evaluation of the antitussive effect of stigma and petals of saffron (*Crocus sativus*) and its components, safranal and crocin in guinea pigs. Fitoterapia 77:446–448

Huntley A, White AR, Ernst E (2002) Relaxation therapies for asthma – a systematic review. Eur Respir J 57:127–131

Irwin RS, Baumann MH, Bolser DC et al (eds) (2006) Diagnosis and management of cough executive summary. Chest 129:1S–23S

Jiang Y, Li H, Li P et al (2005) Steroidal alkaloids from the bulbs of *Fritillaria puqiensis*. J Nat Prod 68:264–267

Kamei J, Nakamura R, Ichiki H et al (2003) Antitussive principles of Glycyrrhizae radix, a main component of preparations of Bakumondo-to (Mai-men-dong-tang). Eur J Pharmacol 469: 159–163

Korpas J, Tomori Z (1979) (eds) Cough and Other Respiratory Reflexes. Karger, Basel

Laude EA, Morice AH, Grattan TJ (1994) The antitussive effects of menthol, camphor and cineole in conscious guinea-pigs. Pulm Pharmacol 7:179–184

Lee L-Y, Gu Q (2008) Cough sensors III: Nicotinic membrane receptors on cough sensors. In: Chung KF, Widdicombe JG (eds) Handbook of experimental pharmacology: Pharmacology and therapeutics of cough. Springer, Berlin, this volume

Lin G, Li P, Chan SW (2001) Chromatographic analysis of Fritillaria isosteroidal alkaloids, the active ingredients of Beimu, the antitussive traditional Chinese medicinal drug. J Chromatogr A 935:321–338

Linde K, Barrett B, Wolkart K, Bauer R et al (2007) Echinacea for preventing and treating the common cold. Cochrane Database Syst Rev CD000530

Maa SH, Sun MF, Hsu KH et al (2003) Effect of acupuncture or acupressure on quality of life of patients with chronic obstructive asthma: A pilot study. J Altern Complement Med 9: 1396–1400

Mackensie AJ, Spoina D, Page CP (2004) Models used in the development of antitussive drugs. Drug Discov Today: Dis Models 1:297–302

Maher-Loughnan GP, Macdonald N, Mason AA et al (1962) Controlled trial of hypnosis in the symptomatic treatment of asthma. Br Med J 2:317–321

Manocha R, Marks GB, Kenchington P et al (2002) Sahaja yoga in the management of moderate to severe asthma: A randomized controlled trial. Thorax 57:110–115

May AJ, Widdicombe JG (1954) Depression of the cough reflex by pentobarbitone and some opioid derivatives. Br J Pharmacol 9:335–340

Mazzone S, Undem BJ (2008) Cough sensors V: Pharmacological modulation of cough sensors. In: Chung KF, Widdicombe JG (eds) Handbook of experimental pharmacology: Pharmacology and therapeutics of cough. Springer, Berlin, this volume

McElhaney JE (2006) Efficacy of COLD-fX in the prevention of respiratory symptoms in community-dwelling adults: A randomized, double-blinded, placebo-controlled trial. J Alter Complement Med 12:153–157

McGarvey LPA (2003) Clinical assessment of cough. In: Chung KF, Widdicombe JG, Boushey HA (eds) Cough: Causes, mechanisms and therapy. Blackwell, Oxford, pp 27–38

Medici TC, Shang H, Grosgurin P et al (1985) No demonstrable effect of sobrerol as an expectorant in patients with stable chronic bronchial diseases. Bull Exp Physiopathol Respir 21: 477–83
Melchart D, Linde K, Fischer P et al (1999) Echinacea for preventing and treating the common cold. Cochrane Database Syst Rev Issue 1, Art No. CD000530
Melchior J, Spasov AA, Ostrovskij OV et al (2000) Double-blind, placebo-controlled pilot and phase III study of activity of standardized *Andrographis paniculata* Herba Nees extract fixed combination (Kan jang) in the treatment of uncomplicated upper respiratory tract infection. Phytomedicine 7:341–350
Minamizawa K, Goto H, Shimada Y et al (2006) Effects of eppikahangeto, a Kampo formula, and Ephedra herba against laryngeal cough in guinea pigs. J Pharmacol Sci 101:118–125
Mitagoshi M, Amagaya S, Ogihara Y (1986) Antitussive effects of L-ephedrine, amygdalin, and makyokansekito (Ch medicine) using a cough model induced by sulphur dioxide in mice. Planta Med 4:275–278
Morice AH (2002) Epidemiology of cough. Pulm Pharmacol Therap 15:253–260
Morice AH (2003) Epidemiology of cough. In: Chung KF, Widdicombe JG, Boushey HA (eds) Cough: Causes mechanisms and therapy. Blackwell, Oxford, pp 1–18
Morice AH, Marshall AE, Higgins KS et al (1994) Effect of inhaled menthol on citric acid induced cough in normal subjects. Thorax 49:1024–1026
Morice AH, Fontana GA, Sovijarvi ARA et al (2004) The diagnosis and management of chronic cough. Eur Respir J 24:481–492
Morice AH, Merton MS, Mulrennan SA et al (2006) Opiate therapy in chronic cough. Am J Respir Crit Care Med 175:312–314
Morice AH, Fontana GA, Belvisi MG et al (2007) ERS guidelines on the assessment of cough. Eur Respir J 29:1–22
Nieman DC, Nehlson-Cannarella SI, Henson DA et al (1998) Immune response to exercise training and/or energy restriction in obese women. Med Sci Sports Exerc 30:679–686
Niimi A (2007) Geography and cough aetiology. Pulm Pharmacol Therap 20:383–387
Nishino T, Tagaito Y, Isono S (1996) Cough and other reflexes on irritation of airway mucosa in man. Pulm Pharmacol 9:285–292
Nishino T, Isono S, Tanaka A et al (2004). Laryngeal inputs in defensive airway reflexes in humans. Pulm Pharmacol Therap 17:377–382
Nosalova G, Strapkova A, Korpas J et al (1989) Objective assessment of cough suppressants under normal and experimental conditions. Drugs Exp Clin Res 15:77–81
Nosalova G, Kardosova A, Franova S (2000) Antitussive activity of a glucuronoxylan from *Rudbeckia fukgida* compared to the potency of two polysaccharide complexes from the same herb. Pharmazie 55:65–68
Nosalova G, Mokry J, Ather A et al (2007) Antitusive activity of the ethanolic extract of *Paederia foetida* (Rubiaceae family) in non-anaesthetized cats. Acta Vet Brno 76:27–33
Packman EW, London ST (1980) The utility of artificially induced cough as a clinical model for evaluating the antitussive effects of aromatics delivered by inunction. Eur Respir J 61:102–109
Papp R, Schuback G, Beck E et al (1998) Oscillacoccinum in patients with influenza-like syndromes: A placebo-controlled double-blind evaluation. Br Homeopath J 87:69–76
Paul IM, Beiler J, McMonagie et al (2007) Effect of honey, dextromethorphan, and no treatment on nocturnal cough and sleep quality for coughing children and their parents. Arch Pediatr Adolesc Med 161:1140–1146
Pavesi L, Subburaj S, Porter-Shaw K (2001) Application and validation of a computerized cough acquisition system for objective monitoring of acute cough–a meta-analysis. Chest 120:1121–1128
Pavord ID (2004) Cough and asthma. Pulm Pharmacol Ther 17:399–402
Poolsup N, Suthisisang C, Prathanturarug S et al (2004) Andrographis paniculata in the symptomatic treatment of uncomplicated upper respiratory tract infection: Systematic review of randomized controlled trials. J Clin Pharm Ther 29:37–45
Rabe A, Weiser M, Klein P (2004) Effectiveness and tolerability of a homeopathic remedy compared with conventional therapy for mild viral infections. Int J Clin Pract 58:827–832

Rubin BK (2003) Mucoactive agents in the treatment of cough. In: Chung KF, Widdicombe JG, Boushey HA (eds) Cough: Causes, mechanisms and therapy. Blackwell, Oxford, pp 269–283

Saab B-R, Pashayan N, El-Chemaly S et al (2006) Sesame oil use in ameliorating cough in children: A randomized controlled trial. Complement Ther Med 14:92–99

Saano V, Minker E, Joki S et al (1993) Influence of chinoin-170, a novel antitussive, on the mucociliary activity in respiratory airways of rats, rabbits, guinea-pigs and man. J Pharm Pharmacol 45:799–802

Sevelius H, McCoy JF, Colmore JP (1971) Dose response to codeine in patients with chronic cough. Clin Pharmacol Ther 12:449–455

Shah SA, Sander S, White CM et al (2007) Evaluation of Echinacea for the prevention and treatment of the common cold: A meta-analysis. Lancet 7:473–480

Shannon R, Baekey DM, Morris KF et al (2004) Production of reflex cough by brainstem respiratory networks. Pulm Pharmacol Ther 17:369–376

Skidmore-Roth L (2001) Mosby's handbook of herbs & natural supplements. Mosby, St. Louis, MO

Spasov AA, Ostrovskii OV, Chernikov MV et al (2004) Comparative controlled study of Andrographis paniculata fixed combination, Kan Jang and an Echinacea preparation as adjuvant, in the treatment of uncomplicated respiratory disease in children. Phytother Res 18:47–53

Steurer-Stey C, Russi W, Steurer J (2002) Complementary and alternative medicine in asthma – do they work? Swiss Med Wkly 132:338–344

Sutovska M, Nosalova G, Franova S et al (2007) The antitussive activity of polysaccharides from *Althaea officinalis* L., var. *Robusta, Arctium lappa* L., var *Herkules,* and *prunus persica* L., Batsch. Bratisl Lek Listy 108:93–99

Takahama K, Shirasaki T, Soeda F (2008) Central mechanisms III: Neuronal mechanisms of action of centrally-acting antitussives - electrophysiological and neurochemical study. In: Chung KF, Widdicombe JG (eds) Handbook of experimental physiology: Pharmacology and therapeutics of cough. Springer, Berlin, this volume

Tatar M, Hanacek J, Widdicombe JG (2008) The expiration reflex from the trachea and bronchi. Eur Respir J 31:385–390

Thomas M, Sheran J, Smith Net al (2007) AKL1, a botanical mixture for the treatment of asthma: A randomized, blind, placebo-controlled, cross-over study. BMC Pulm Med 7:4

Usmani OS, Belvisi MG, Patel HJ et al (2004) Theobromine inhibits sensory nerve activation and cough. FASEB J 19(2):231–233

Vedanthan PK, Kesavalu IN, Murphy KC et al (1998) Clinical study of yoga techniques in university students with asthma: A controlled study. Allergy Asthma Proc 19:3–9

Vickers AJ, Smith C (2004) Homeopathic Oscillococcinum for preventing and treating influenza and influenza-like syndromes. Cochrane Database Syst Rev: CD001957

Visweswaraiah NK, Telles S (2004) Randomized trial of yoga as a complementary therapy for pulmonary tuberculosis. Respirology 9:96–101

Widdicombe JG, Eccles R, Fontana G (2006) Supramedullary influences on cough. Respir Physiol Neurobiol 152:320–328

Yale SH, Liu K (2004) Echinacea purpurea therapy for the treatment of the common cold: A randomized double-blind, placebo-controlled clinical trial. Arch Intern Med 164:1237–1241

Ziment I (2001) Alternative therapies in asthma. In: Gershwin ME, Albertson TE (eds) Bronchial asthma: Principles of diagnosis and treatment, 4th edn. Humana, Towata, NJ, pp 255–278

Ziment I (2002) Herbal antitussives. Pulm Pharmacol Ther 15:327–324

Clinical Cough VI: The Need for New Therapies for Cough: Disease-Specific and Symptom-Related Antitussives

K.F. Chung

Contents

K.F. Chung
National Heart and Lung Institute, Imperial College London and Royal Brompton and Harefield NHS Trust, London SW3, UK
f.chung@imperial.ac.uk

K.F. Chung, J.G. Widdicombe (eds.), *Pharmacology and Therapeutics of Cough*, Handbook of Experimental Pharmacology 187,

Abstract Cough is a common symptom that can be self-limiting or persistent. Ideally, treatment of the underlying cause(s) of cough with specific treatments should eliminate cough. This approach may not be successful if no cause can be established or if the treatment of the cause fails. Suppression of cough may be disease-specific or symptom-related. There has been a long tradition in acute cough usually due to upper respiratory tract infections to use symptom-related antitussives. In chronic cough, suppression of cough may be achieved by disease-specific therapies, but in many patients it may be necessary to use symptomatic antitussives. The efficacy of some over-the-counter symptomatic antitussives is often no better than that of a placebo. Currently available cough suppressants include the centrally acting opioids such as morphine, codeine, pholcodeine, and dextromethorphan. Early studies reported success in reducing cough in patients with chronic bronchitis or chronic obstructive pulmonary disease (COPD); however, a carefully conducted blinded controlled study showed no effect of codeine on cough of COPD. Success with these cough suppressants may be achieved at high doses that are associated with side effects. A slow-release preparation of morphine has been shown to have some degree of efficacy, but this should be reserved for the most severe chronic cough patient, and for patients with terminal cancer who may also benefit from its analgesic effects. There are case reports of the success of centrally acting drugs such as amitriptyline, paroxetine, gabapentin, and carbamezepine in chronic cough. New agents derived from basic research such as new opioids such as nociceptin or antagonists of transient receptor potential vanniloid-1 may turn out to have antitussive effects. Efficacy of symptomatic cough suppressants must be tested in double-blind randomized trials using validated measures of cough in patients with chronic cough not responding to specific treatments. Patients with chronic cough need effective antitussives that could be used either on demand or on a long-term basis.

1 Introduction

Cough is a very common symptom, and is the most frequent reason for seeking an ambulatory health care visit in both adults and children (Chung and Pavord, 2008). Most commonly, a cough due to upper respiratory tract infection is transient, lasting for just over 1 week, but during which time it is common for sufferers to use antitussives, often using over-the-counter cough medicines. Consumers spend billions of dollars per year on over-the-counter medications for cough, and yet the evidence for their antitussive efficacy is not strong at all.

Chronic cough which persists for more than 3 weeks is also a common symptom in the general population and its prevalence in the USA and Europe is around 11–20% of the adult population (Barbee et al. 1991; Cullinan 1992; Lundback et al. 1991; Janson et al. 2001; Ford et al. 2006). Recent surveys have confirmed the link to cigarette smoking and also to environmental pollutants (Zemp et al. 1999; Cook and Strachan 1999; Pierse et al. 2006; Braun-Fahrlander et al. 1997; Gehring et al. 2002; Vedal et al. 1998). When patients with chronic cough present to a medical doctor, the task of the latter is to investigate for any causes of the cough. A chronic cough is most commonly associated with asthma, gastroesophageal reflux, and postnasal drip, and treatment of these specific conditions often leads to improvement in cough. These treatments may be referred to as disease-specific antitussives, in contrast to symptomatic antitussives, which are aimed directly at cough suppression. Often the underlying condition is not amenable to treatments or no underlying condition is diagnosed (idiopathic cough), and therefore symptomatic antitussives are needed to control cough since this is often associated with impairment in quality of life, depression, and physical symptoms.

While there are symptomatic antitussives that can be prescribed, the efficacy of these is only currently being accurately surmised using validated methods of cough measurement. The results of very few controlled clinical trials of antitussives have been published so far, and these are now addressing some of the inadequacies of currently available antitussives.

This chapter will discuss the treatment approaches to acute and chronic cough, highlight the currently available disease-specific and symptomatic cough suppressants, and look into the development of more efficacious antitussives, particularly for chronic cough. The treatment of chronic cough has two aims: specific treatments of causative factors (disease-specific), and symptomatic cough suppression.

2 Managing Chronic Cough with Disease-Specific Therapies

One example of the great progress made in the management of cough is the systematic diagnostic approach of finding the cause(s) of chronic cough, as advocated by all cough guidelines (Irwin 2006; Morice et al. 2004, 2006; Kohno et al. 2006). The management of chronic cough is dealt with in Morice and McGarvey (2008) and has been recently reviewed in detail (Pavord and Chung 2008). The physician needs to reach a diagnosis as to the cause(s) and apply specific treatment(s).

2.1 Eosinophil-Associated Cough

In conditions such as asthma or cough-variant asthma or eosinophilic bronchitis, inhaled corticosteroid therapy is used with great success in controlling cough. These airway conditions are all linked with an eosinophilic inflammation that is usually responsive to corticosteroid therapies for which there is very good evidence. In a study of patients with eosinophilic bronchitis, treatment with inhaled corticosteroids controlled cough, and was associated with an improvement in capsaicin cough sensitivity and a marked reduction in lung eosinophil counts (Brightling et al. 2000). Leukotriene receptor antagonists can also be effective in inhibiting asthma-associated cough (Dicpinigaitis and Dobkin 1999). Sometimes more specific treatments of eosinophilic inflammation may be used with success (Chung et al. 2006).

2.2 Gastroesophageal Reflux Disease

Gastroesophageal reflux disease (GORD) is commonly associated with chronic cough, and proton-pump inhibitors (PPI) or histamine H2-receptor antagonists are recommended. Early case reports have hailed up to 80–100% success rate of these treatments in controlling cough (Waring et al. 1995; Irwin et al. 1981), but a recent Cochrane meta-analysis of therapy in controlled blinded studies of chronic cough associated with GORD with PPIs concluded that there is "insufficient evidence to definitely conclude that GORD treatment with these agents is beneficial for cough associated with GORD in adults" (Chang et al. 2006). Nevertheless, it remains logical to give a trial of PPI therapy to patients in whom there is evidence of GORD, and empirically guidelines recommend using a trial of high doses for a period of at least 2–3 months (Morice et al. 2004), which remains to be tested in clinical trials.

The possibility of nonacid volume reflux (e.g., in patients treated with PPIs) or of abnormal upper esophageal motility problems as a cause of chronic cough has been evoked, but these associations have yet to be established as cause and effect. The use of promotility agents such as metoclopramide and domperidone, which are dopamine agonists, is recommended, but has not been studied properly. Success of surgical correction of GORD in patients whose cough has not responded to PPI therapy has been reported (Novitsky et al. 2002), and in this group there may have been patients with nonacid volume reflux who have benefited. However, a proper randomized controlled study is needed to determine whether surgery provides therapeutic benefit for cough caused by GORD. Increasingly, there is a recognition that reflux of gastric contents can also cause laryngeal abnormalities that in turn could cause cough, a condition labeled as laryngopharyngeal reflux by otolaryngologists (Ford 2005). Clearly, the treatment of cough associated with GORD remains unsatisfactory, and a lot of research is needed to establish the role and efficacy of specific treatments. This is very much linked to the issue as to whether gastroesophageal reflux or dismotility is a *cause* of chronic cough. Such patients may also benefit from nonspecific therapies in addition to specific treatments they receive related to GORD.

2.3 Postnasal Drip

Postnasal drip syndrome (PNDS), or nasal catarrh associated with rhinosinusitis, has been recognized as a cause of chronic cough mainly from clinical observations, although in the American College of Chest Physicians' guidelines this is now grouped under the heading of upper airway cough syndrome (Irwin 2006). This latter term is not ideal as it does not implicate the potential role of the larynx. Topical corticosteroids often with a nasal decongestant or antibiotics or antihistamines are used for treating rhinosinusitis, and first-generation antihistamines have been reported to be more effective than second-generation antihistamines in controlling cough, although this needs to be confirmed in controlled studies. Some otorhinolaryngologists do not agree that cough is a predominant symptom of patients with PNDS (O'Hara and Jones 2005), and it has been suggested that PNDS may be caused by laryngopharyngeal reflux, i.e., gastric reflux reaching the laryngopharyngeal area (Wise et al. 2006). There have been controlled studies that have provided evidence for the success of rhinosinusitis treatments on chronic cough.

There is some controversy as to whether cough in conditions usually associated with the production of mucus such as COPD or bronchiectasis should be suppressed as this could lead to mucus accumulation with increased risk of lung infections. In both COPD and bronchiectasis, an enhanced cough reflex has been demonstrated (Doherty et al. 2000; Torrego et al. 2006), and therefore part of the excessive cough which has been reported in COPD patients with the use of cough counters (Coyle et al. 2005) may be related to an enhanced cough reflex. In these conditions, it is difficult to reverse the previous condition with effective treatments.

2.4 Idiopathic Cough

More recent series, particularly those from the UK, have identified a significant proportion of patients labeled as having "idiopathic" cough, ranging from 7 to 46%, despite a thorough diagnostic workup (Chung and Pavord 2008). In these patients, all associated causes of cough had excluded and the cough had not responded to any specific treatments. Such patients tend to be middle-aged women and often give a history of cough onset around the menopause and may have organ-specific autoimmune disease, particularly autoimmune hypothyroidism (Mund et al. 2001, 2005). An alternative explanation is that the initiating cause of the cough may have disappeared, but its effect on enhancing the cough reflex may be more prolonged (Chung 2007). An example would be the transient appearance of an upper respiratory tract virus infection or an exposure to toxic fumes that results in prolonged damage of the airways' mucosa. The repetitive mechanical and physical effects of coughing bouts on airway cells could activate the release of various chemical mediators that could enhance chronic cough through inflammatory mechanisms (Niimi et al. 2005; Heino et al. 1990), providing a positive feed-forward system for cough persistence. It is quite possible that there is an induction of changes in the upper airways of inflammation and tissue remodeling induced by various causes associated with cough or by the act of coughing itself that could lead to an enhanced cough reflex, that in turn is responsible for maintaining cough. The cough becomes "idiopathic" when the primary inciting cause has resolved while cough is persistent.

3 Symptomatic Cough Suppressants

Cough suppressants have been traditionally divided into the central acting ones that include the opiates morphine, diamorphine, pholcodeine, and codeine, as being the most effective antitussives, and nonnarcotic synthetic derivatives of morphine, including dextromethorphan, which has no analgesic or sedative properties and is not usually addictive.

3.1 Currently Available Narcotic Antitussives

Opiates including morphine, diamorphine, and codeine are the most effective antitussive agents. At their effective doses they may cause physical dependence, respiratory depression, and gastrointestinal colic. Morphine and diamorphine should be used for severe distressing cough which cannot be relieved by other less potent antitussives, and are therefore usually confined to patients with terminal illness, such as bronchial carcinoma. These opioids also relieve anxiety and pain. They cause sedation, respiratory depression, and constipation. Opioids can exacerbate wheezing through the release of histamine, but this is rare. Diamorphine may be preferred to morphine because of its lower incidence of nausea and vomiting. Morphine may be given by mouth every 4 h, and also as a suppository. Diamorphine is preferably given by injection.

Codeine is the methylether of morphine and has long been considered the standard centrally acting antitussive drug against which the pharmacological and clinical effects of newer drugs have been measured. Codeine is probably the most commonly prescribed antitussive (with dextromethorphan) and is a frequent component of over-the-counter cough and cold medicines combined with decongestants and expectorants. Its antitussive activity when it is given orally has been reported in rather small studies dating back 30 years. Codeine has been reported to possess antitussive activity against pathological cough (Eddy et al. 1969b; Aylward et al. 1984) and against induced cough in normal volunteers (Empey et al. 1979).

Codeine should be used cautiously in patients with reduced hepatic function because hepatic glucuronidation inactivates it, although 10% of the oral dose may be demethylated to form morphine. It can be used without dose modification in patients with renal failure. Codeine toxicity manifests itself as respiratory depression and obtundation. Drowsiness may be an incapacitating side effect, together with nausea, vomiting, and constipation. Rarely, allergic cutaneous reactions such as erythema multiforme have been described. Codeine can cause physical dependence, but on a lesser scale than morphine. Adverse reactions in children have raised some concerns. Antitussive dosages of 3–5 mg/kg per day have caused somnolence, ataxia, vomiting, rash, and facial swelling and pruritis, while at a dosage above 5 mg/kg per day, respiratory depression requiring mechanical ventilation has been reported in 3% of children, with two deaths (von Muhlendahl et al. 1976).

Dihydrocodeine has no particular advantage over codeine and may cause more addiction than codeine. Pholcodeine is also reported to be as effective as codeine but has little or no analgesic effect. There seems little to choose between codeine and pholcodeine.

3.2 Currently Available Nonnarcotic Antitussives

Nonnarcotic antitussives include dextromethorphan, which is a synthetic derivative of morphine with no analgesic or sedative properties and which is usually included as a constituent of many compound cough preparations sold over the counter. It is reported to be as effective as codeine in suppressing acute and chronic cough when given orally (Eddy et al. 1969a; Aylward et al. 1984), with one study showing superiority over codeine (Matthys et al. 1983). Antitussive efficacy of a single 30-mg dose has been reported against cough associated with upper respiratory tract infections (Pavesi et al. 2001). Side effects are few with the usual dose, but at higher doses dizziness, nausea and vomiting, and headaches may occur. It should be avoided in patients with hepatic insufficiency as it undergoes metabolic degradation in the liver. Dextromethorphan should be used with caution also in patients receiving monoamine oxidase inhibitors as cases of central nervous system depression and deaths have occurred.

Other nonnarcotic preparations include noscapine and levopropoxyphene, although their antitussive efficacy has not been proven. Levodropropizine, a non-opioid antitussive with peripheral inhibition of sensory cough receptors, has a favorable benefit–risk profile compared with dextromethorphan (Catena and Daffonchio 1997); this is currently available in several European countries.

3.3 Effectiveness of Currently Available Symptomatic Antitussives Against Acute Cough

Antitussives have been tried in the control of acute cough due to the common cold (Freestone and Eccles 1997). A Cochrane review examined six trials of antitussives in which effects were compared with that of a placebo (Table 1) (Schroeder and Fahey 2004). Codeine was no more effective than a placebo in reducing cough symptoms (Eccles et al. 1992; Freestone and Eccles 1997). Two studies favored dextromethorphan over a placebo (Parvez et al. 1996; Pavesi et al. 2001), whereas a third did not show an effect (Lee et al. 2000). In the two studies with positive findings, ambulatory cough counts were found to be reduced by 19–36% using a 30-mg dose of dextromethorphan, but clinical cough scores were not recorded. Some studies have demonstrated an important effect of a placebo in inhibiting cough due to the common cold, and it is possible that there is an important placebo effect underlying the effects of antitussives (Eccles 2002).

The lack of effectiveness of dextromethorphan or the antihistamine diphenhydramine as compared with a placebo in children with upper respiratory tract infection is particularly important (Korppi et al. 1991; Taylor et al. 1993; Smith and Feldman, 1993; Schroeder and Fahey, 2004; West et al. 1975; Hutton et al. 1991; Clemens et al. 1997). This has been confirmed in a study of 100 children where dextormethorphan or diphenhydramine given as a single dose before bedtime was

Table 1 Symptomatic cough suppressants for acute cough in adults

Drug	Authors	*n*	Cough score	Cough count
Codeine (30 mg twice daily)	Eccles et al. (1992)	81	No effect	ND
Codeine (60 mg once)	Freestone and Eccles (1997)	80	No effect	No effect
Dextrometorphan (30mg once)	Parvez et al. (1996)	451	ND	Reduced by 19–36% ($p < 0.05$)
Dextrometorphan (30 mg once)	Lee et al. (2000)	44	Placebo = drug effect	Placebo = drug effect (> 50% effect)
Dextrometorphan (30 mg once)	Pavesi et al. (2001)	710	ND	Reduced by 12–17% ($p < 0.01$)
Moguisteine (600 mg per day)	Adams et al. (1993)	108	No effect (only in those with high scores)	ND

ND no data

found to be no better than a placebo in providing nocturnal symptom relief for children with cough and sleep difficulty from an upper respiratory tract infection (Paul et al. 2004). The American Academy of Pediatrics has highlighted the fact that the efficacy of antitussive preparations in children is lacking and that these medications may be potentially harmful (Committee on Drugs et al. 1997). Dosage guidelines for cough and cold mixtures have been extrapolated from adult data and clinical experience, and adverse effects and overdosage associated with these mixtures have been reported. Their recommendation is that cough due to acute viral airway infections is self-limiting and may be treated with fluids and humidity. The beneficial effects of buckwheat honey in the symptomatic relief of children's nocturnal cough and sleep difficulty with an upper respiratory tract infection may make this preferable to the use of antitussive mixtures (Paul et al. 2007).

3.4 Effectiveness of Currently Available Symptomatic Antitussives Against Chronic Cough

There have been very few studies of these antitussives in patients with chronic cough or chronic idiopathic cough. In fact, most of these studies have been performed in patients with chronic bronchitis and are at least 20 years old (Table 2). Codeine and dextromethorphan were previously shown to have an inhibitory effect on chronic cough of bronchitis or COPD (Matthys et al. 1983; Aylward et al. 1984; Sevelius and Colmore 1966; Sevelius et al. 1971). In studies involving larger groups of patients, both levodropropizine and moguisteine were reported to be effective in reducing cough frequency in chronic cough of bronchitis and COPD (Aversa et al. 1993; Allegra and Bossi 1988). However, in a recent well-conducted study of

Table 2 Studies of symptomatic cough suppressants in chronic bronchitis

Drug	Authors	*n*	Patient group	Effects
Codeine vs placebo	Aylward et al. (1984)	8	Chronic bronchitis	Decrease in coughs by 40% at 30-mg dose ($p < 0.05$)
Codeine vs placebo	Sevelius and Colmore (1966)	10	Chronic bronchitis	Decrease in coughs by 47% ($p < 0.01$)
Codeine vs placebo	Sevelius et al. (1971)	12	COPD	Decrease in coughs by 60% ($p < 0.001$)
Dextrometorphan vs placebo	Aylward et al. (1984)	8	Chronic bronchitis	Decrease in coughs by 50% at 630-mg dose ($p < 0.05$), and by 28% at 30-mg dose
20 mg dextrometorphan vs 20 mg codeine	Matthys et al. (1983)	16	Chronic bronchitis	Decrease in cough intensity dextrometorphan> codeine ($p < 0.0008$)
Levodropropizine vs placebo	Allegra and Bossi (1988)	194	Acute and chronic bronchitis	Decrease in cough severity, decrease in cough frequency by 72% ($p < 0.05$)
Moguisteine vs placebo	Aversa et al. (1993)	87	COPD, pulmonary cancer, fibrosis, unknown	Decrease in cough frequency by 42% ($p < 0.03$)
Moguisteine vs placebo	Adams et al. (1993)	108	Upper respiratory infections	.I' nighttime cough severity ($p < 0.05$)
Codeine vs placebo	Smith et al. (2006)	26	COPD	No effect on cough counts, capsaicin cough sensitivity and cough-specific quality of life

COPD chronic obstructive pulmonary disease

cough in COPD patients where cough scores, visual analogue scales, cough frequency using an ambulatory cough recorder, and capsaicin cough sensitivity were measured, a high dose of codeine (60 mg) was not effective in reducing cough (Smith et al. 2006). This casts some doubt on the results of some of the earlier smaller studies that reported beneficial effects on cough. More controlled trials of these antitussives in patients with chronic cough are needed.

Moguisteine, a peripherally acting antitussive, was not more effective than a placebo, apart from reduction of cough in a subgroup of participants with severer cough at night (Adams et al. 1993). It is to be noted that no efficacy of dextromethorphan or codeine on cough in children has been demonstrated. In a rare placebo-controlled study of children with acute nocturnal cough, neither dextromethorphan nor codeine was significantly more effective than the placebo in reducing the cough (Taylor et al. 1993).

The main issue with the opioids such as morphine, diamorphine, and codeine is that any potentially effective dose is usually associated with physical dependence, respiratory depression, and gastrointestinal colic. The use of morphine and diamorphine has been restricted to the severe distressing cough (usually associated with pain and distress) that often occurs in advanced lung cancer and with doses that cause sedation, respiratory depression, and constipation. In a randomized double-blind crossover study in 27 patients with severe chronic cough, slow-release morphine tablets (5 mg twice daily for 4 weeks) compared with a placebo (Morice et al. 2007) attenuated cough scores significantly by approximately 40% with an improvement in quality-of-life cough questionnaire scores, although citric acid cough response remained unchanged. Patients tolerated the treatment well despite 40% of them complaining of constipation and 25% of them complaining of drowsiness. This study dispels the notion that morphine would not be tolerable in chronic cough patients. This approach nevertheless should be used only in patients with the most distressing cough.

4 Symptomatic Cough Suppressants from the Literature

Several potential antitussives for chronic cough have been reported in the medical literature as case reports or open uncontrolled studies (Table 3).

4.1 Baclofen

Baclofen used as an antispasticity agent and also an agonist of the $GABA_B$ receptor was reported to have a small beneficial effect in two patients with chronic cough (Dicpinigaitis et al. 1998); the potential use of this compound is further supported by its ability to inhibit capsaicin-induced cough in normal subjects (Dicpinigaitis et al. 1998).

Table 3 Potentially new symptomatic cough suppressants

Baclofen ($GABA_B$ receptor agonists)
Amitriptyline
Gabapentin
Carbamezepine
New opioid receptor agonists
Nociceptin agonists
Tachykinin receptor antagonists
Bradykinin B2 receptor antagaonists
Antagonists of transient receptor potential vanniloid-1
Cannabinoids
Ion channel modulators
Local anesthetics

4.2 Centrally Acting Drugs

A range of centrally acting drugs have been reported to control cough in a group of patients in whom other measures have been unsuccessful. In five patients with cancer complaining of intractable cough not responding to codeine or morphine and of pruritius, paroxetine, a serotonin-reuptake inhibitor used for treatment of depression, was particularly effective (Zylicz and Krajnik 2004). There are positive reports of the effects of the tricyclic antidepressant amitriptyline in 12 patients and of gabapentin and carbamazepine, which are antiepileptic drugs, in two patients and one patient, respectively (Bastian et al. 2006; Mintz and Lee 2006; Jacome 1985). A randomized but open controlled study of amitriptyline (10 mg at nighttime) versus a combination of codeine/guaifenesin in patients with chronic cough "resulting from postviral vagal neuropathy," an entity that is recognized by otolaryngologists as triggered by an upper respiratory tract illness with injury to various branches of the vagus nerves, showed significant suppression of cough with amitriptyline (Jeyakumar et al. 2006). It is perhaps time that a double-blind controlled trial of amitriptyline in chronic cough be undertaken.

The need for further studies in the antitussive effects of these agents such as amitryptiline, gabapentin, and carbamezepine is supported by their use in the treatment of neuropathic pain, and there are similar processing peripheral and central mechanisms for chronic cough and chronic pain that could be controlled by these drugs (Gracely et al. 2007). In addition, it is also possible that similar processing mechanisms pertain to the sensation of dyspnoea, such that if cough and dyspnea were associated together as in COPD or in advanced lung cancer, a potential symptomatic antitussive might also improve dyspnoea (and pain). Capsaicin and voluntary coughs activate supramedullary pathways (Mazzone et al. 2007; Simonyan et al. 2007) and agents such as amitriptyline, gabapentin, and carbamezepine may have antitussive effects through inhibition of these pathways.

4.3 Expectorants and Mucolytics

The basis for using these agents as antitussives lies in the possibility that altering the volume of secretions or their composition will lead to suppression of the cough reflex. Mucolytic agents such as acetylcysteine, carbocisteine, bromhexine, and methylcysteine are often used to facilitate expectoration by reducing sputum viscosity in patients with chronic bronchitis. A small reduction in the exacerbation of bronchitis has been reported with orally administered acetylcysteine, accompanied by small improvement in cough, a decrease in volume of sputum, and some ease of expectoration (Aylward et al. 1980); however, other studies have failed to show beneficial effects on mucus clearance (Hautmeyers et al. 1999).

Aromatic agents such as eucalyptus and menthol have decongestant effects in the nose and can be useful in short-term relief of cough. Menthol inhibits capsaicin-induced cough in normal volunteers (Morice et al. 1994), and acts on a cold-sensitive nervous sensor. Demulcents also form an important component of many proprietary cough preparations and may be useful because the thick sugary preparation may act as a protective layer on the mucosal surface. The beneficial effects of honey in improving cough of the common cold in children (Paul et al. 2007) may result from such a demulcent effect.

5 Potential Novel Classes of Nonspecific Antitussives

Research into the neuroanatomical and neurophysiological mechanisms of sensory cough pathways has led to the identification of specific airway afferent nerve subtypes and of receptors and channels involved and therefore to potential cough targets. Importantly, this research is coupled with an understanding of the potential abnormalities and dysfunction of these pathways in the hypertussive state, so that these can be reversed or normalized. Most of the potential antitussives arising from this research have yet to be tried in cough patients. While this list of potential novel classes of nonspecific antitussives (Table 3) appears to focus on their effects at the periphery, it should be remembered that they may also have central effects, which remain more difficult to study. There have been recent reviews of the pharmacology of the sensory neural pathway of cough and of potential targets (Chung 2005; Bolser et al. 2006b; Dicpinigaitis 2006; Kollarik and Undem 2006).

5.1 Opioid Receptor Agonists

Opioid receptor agonists are classified by their activities at the opioid receptors μ, κ, and δ. The current compounds, morphine and codeine, are mostly μ-receptor agonists and may possess central as well as peripheral actions. BW443C, a μ-opioid receptor agonist and a pentapeptide polar agonist, as an aerosol acts on μ receptors on sensory receptors in the lung and has peripheral antitussive activity in the lungs (Adcock et al. 1988), but this has not been demonstrated in humans (Choudry et al. 1991).

Opioids may also act on κ-opioid receptors for their antitussive effects (Kamei et al. 1990). A δ-selective receptor agonist (SB221122) was shown to inhibit citric acid induced cough in the guinea pig (Kotzer et al. 2000), an effect prevented by a δ-receptor antagonist (SB244525). However, an orally selective δ-opioid receptor antagonist, TRK-851, with 100–250 times greater potency in inhibiting cough in rat and guinea pig than codeine, has been considered as a potential antitussive (Ueno et al. 2001).

Levodropropizine, a non-opioid antitussive and derivative of phenylpiperazinopropane, inhibits vagally conducted cough in the guinea pig by activating a reflex mediated by capsaicin-sensitive afferents, and not by a central mechanism of action (Lavezzo et al. 1992). It inhibits C-fiber activity induced by chemical stimuli

(Shams et al. 1996). It has been compared with dextromethorphan in patients with nonproductive cough and has been shown to have a more favorable benefit–risk profile (Catena and Daffonchio 1997).

5.1.1 Nociceptin

Nociceptin/orphanin is the endogenous peptide ligand for the orphan "opioid-like" NOP_1, which is a G-protein-coupled seven-transmembrane receptor. Nociceptin does not stimulate opioid receptors. NOP_1 receptors are widely distributed in the central nervous system and are also present in airway nerves in the guinea pig (Fischer et al. 1998), where nociceptin has been found to inhibit nonadrenergic, noncholinergic responses (Shah et al. 1998). Capsaicin-induced bronchoconstriction is attenuated by nociceptin (Corboz et al. 2000), an action possibly due to inhibition of tachykinin release from sensory C-fibers. Nociceptin administered intravenously or via the intracerebroventricular route suppresses capsaicin-induced and mechanically induced cough (McLeod et al. 2001; Bolser et al. 2001), effects blocked by an NOP_1 antagonist, J113397, but not by an opioid receptor antagonist. This NOP antagonist does not appear to penetrate the blood–brain barrier in the guinea pig when administered orally and remains effective as an inhibitor of capsaicin-induced cough, indicating that it also acts peripherally (McLeod et al. 2004).

Nociceptin inhibits the airway microvascular leakage induced in guinea-pig airways by intraesophageal hydrochloric acid infusion, acting probably at the prejunctional level by inhibiting tachykinin release (Rouget et al. 2004). Furthermore, the ability of nociceptin to block capsaicin-induced tachykinin release and bronchoconstriction has been traced to the activation of an inward-rectifier potassium channel (Jia et al. 2002). In the conscious guinea pig, nociceptin inhibited acid-induced cough; this effect was shown to result from a direct inhibitory effect of nociceptin on peripheral C-fibers caused by selective inhibition of acid-induced transient receptor potential vanniloid-1 (TRPV-1) activation (Lee et al. 2006). Many nociceptin agonists have been described (Chiou et al. 2007). A selective non-peptide nociceptin agonist is Ro-64–619, for which preclinical data have been summarized (Shoblock 2007). No data pertaining to humans have been published so far.

5.2 Tachykinin Receptor Antagonists

Tachykinins are present in capsaicin-sensitive primary afferent nerves and act through tachykinin receptor subtypes, NK1R, NK2R, NK3R (Geppetti et al. 1999). In rodents, capsaicin and other irritants can cause the release of tachykinins from peripheral nerve endings in the lungs via a local axon reflex. Tachykinins are potent bronchoconstrictors and increase microvascular permeability, and have various proinflammatory effects. These, together with a direct effect on myelinated Aδ-fibers, contribute to stimulation of cough. Tachykinins may enhance the responses of rapidly adapting receptors (RARs) and have also been implicated in the central "sensitization" of cough (Mutoh et al. 2000).

In the guinea pig, an NK2 receptor antagonist, SR48968, inhibited citric acid induced cough, while an NK1 receptor antagonist was ineffective (Advenier et al. 1992; Girard et al. 1995). A study in asthmatic subjects found no effect of CP-99,994 against bronchoconstriction and cough induced by hypertonic saline (Fahy et al. 1995). A non-peptide NK3 receptor antagonist (SB235375), with low penetrance into the central nervous system, inhibited citric acid induced cough and airways' hyperreactivity in the guinea pig (Hay et al. 2002; Daoui et al. 1998) but its development has been suspended.

5.3 Bradykinin B2 Receptor Antagonists

Bradykinin is a short peptide produced by the action of proteases, and can induce inflammation, and stimulate sensory nerve endings to induce the release of neuropeptides. Bradykinin activates RARs, and also causes bronchoconstriction (Bergren 1997). It also activates airway C-fibers and can cause coughing in the guinea pig and in patients with asthma. B2 receptor antagonist HOE 140 inhibited citric acid induced cough in the guinea pig (Featherstone et al. 1996). In the guinea pig, an inhibitor of angiotensin-converting enzyme caused sensitization of the cough reflex, and this was inhibited by a B2 receptor antagonist (HOE 140, icatibant) (Dicpinigaitis 2006). A number of B2 receptor antagonists have been developed (Bock and Longmore 2000), but these have not been tried for cough.

5.4 Antagonists of Transient Receptor Potential Vanilloid Receptor-1

Capsaicin, the pungent ingredient of chilli peppers, stimulates airway C- and Aδ-fibers, and also causes in guinea pigs and rats the release of neuropeptides in the airways, leading to airway smooth-muscle contraction and plasma extravasation. These lead to an increase in the activity of RARs. Capsaicin activates TRPV1 present on subpopulations of primary afferent neurones (Caterina et al. 1997). Capsazepine is a receptor antagonist of TRPV1, and blocks capsaicin-induced and citric acid induced cough (Lalloo et al. 1995). The TRPV1 receptor is localized to small-diameter afferent neurones in dorsal root and vagal sensory ganglia (Szallasi and Blumberg 1999) and is stimulated by protons, the endogenous receptor agonist of TRPV1 an eicosanoid, anandamide, and inflammatory mediators such as 12-hydroperoxyeicosatetraenoic acid and leukotriene B_4. TRPV1 channel activity is strongly modulated by the action of inflammatory mediators such as prostaglandins and bradykinin, through a protein kinase A or protein kinase C mediated receptor phosphorylation (Szallasi and Blumberg 1999; Premkumar and Ahern 2000).

An increase in expression of TRPV1 in epithelial airway nerves of the airways' mucosa of patients with nonasthmatic chronic cough has been reported. The increase was correlated with the degree of capsaicin cough responsiveness (Groneberg et al. 2004), supporting a role for TRPV1 in the enhanced cough reflex seen in chronic cough. Airway inflammation may modulate TRPV-1 sensitivity and an amplification of TRPV-1 may cause an increase in tussive sensitivity (McLeod et al. 2007a). Sensitization of TRPV-1 receptors can occur after ovalbumin sensitization of the guinea pig or after exposure to sulfur dioxide (McLeod et al. 2006, 2007b).

Novel TRPV1 antagonists such as iodoresiniferatoxin are very potent blockers, being 450-fold more potent that capsazepine. Iodoresiniferatoxin has been shown to be effective in inhibiting cough induced by citric acid and capsaicin in the guinea pig (Trevisani et al. 2004). Noncompetitive TRPV1 antagonists consisting of the trimers of *N*-alkylglycines are also effective in reducing capsaicin-induced neurogenic inflammation (Garcia-Martinez et al. 2002). McLeod et al. (2006) showed that *N*-(4-*tert*-butylphenyl)-4-(3-cholorphyridin-2-yl)-tetrahydropyrazine-1(2*H*)-carboxamide (BCTC) can inhibit allergen-induced cough in sensitized guinea pigs and a derivative of BCTC inhibits capsaicin-induced and citric acid induced cough (Leung et al. 2007). TRPV1 antagonists are currently undergoing trials in humans for migraine and pain (McLeod et al. 2007a).

5.5 Cannabinoids

Interest in the cannabinoids as modulators of cough was first shown by the demonstration that intravenous Δ^9-tetrahydrocannabinol reduced the amplitude of the cough response induced by electrical stimulation of the superior laryngeal nerve or by mechanical stimulation of the tracheal mucosa in anesthetized cats (Gordon et al. 1976). However, there are conflicting data regarding the receptor mediating these effects on cough. A selective CB2 agonist, JWH133, reduced citric acid induced cough in conscious guinea pigs, an action that was associated with inhibition of sensory nerve depolarization of the isolated vagus nerves induced by hypertonic saline, capsaicin, and prostaglandin E_2 (PGE_2) (Patel et al. 2003). This was suggestive of a peripheral mode of action of cannabinioids. However, an endogenous cannabinoid, anandamide, inhibited cough and bronchospasm induced by capsaicin in rodents through activation of CB1 cannabinoid receptors present on airway nerves (Calignano et al. 2000). In another study, CB1 or CB2 antagonists had no effect on anandamide-induced cough. In mice, the antitussive effects of a cannabinoid agonist against capsaicin-induced cough were shown to be via activation of CB1 receptors, probably acting centrally, as this was a μ2-opioid effect (Morita and Kamei 2003).

5.6 Ion Channel Modulators

5.6.1 Potassium Channels

Several potassium channels are located on vagal sensory neurones. A benzimidazalone compound, NS1619, is an opener of a large conductance Ca^{2+}-activated (BK_{Ca}) channel. It inhibits citric acid induced cough and the generation of action potentials in the guinea-pig tracheal Aδ- and C-fibers stimulated by hyperosmolality (Fox et al. 1997). These effects were prevented by iberiotoxin, a BK_{Ca} channel selective blocker. An ATP-sensitive K^+ channel opener, pinacidil, also inhibits cough induced by capsaicin in the guinea pig, an effect reversed by the ATP-sensitive K^+ channel blocker glibenclamide (Morita and Kamei 2000). Moguistine may work as an ATP-sensitive K^+ channel opener, underlying its antitussive activity (Morita and Kamei 2000).

5.6.2 Chloride Channels and Diuretics

Isotonic solutions of low chloride concentrations can stimulate action potential discharge of a subpopulation of Aδ-fibers and C-fibers in guinea pigs (Fox et al. 1995), and activate afferent fibers in the dog (Sant'Ambrogio et al. 1993). Low-chloride solutions induce cough in human, and the diuretic frusemide inhibits cough induced by low-chloride solutions but not by capsaicin (Ventresca et al. 1990). Frusemide inhibits to some extent airway afferent action potential discharge, and sensitizes slowly adapting receptors and desensitizes RARs in rat airways (Sudo et al. 2000). The mechanism by which frusemide works is unknown (Bolser et al. 2006b), but it may inhibit a neuronally expressed chloride transporter in the airways (Mazzone and McGovern 2006).

Frusemide-sensitive $Na^+/K^+/2Cl^-$ cotransporter (NKCC1) is expressed by the majority of neurones in the vagal sensory ganglia and by the peripheral terminals of low-threshold mechanosensors (cough receptors) in the guinea-pig trachea (Mazzone and McGovern 2006) The cotransporter allows the accumulation of intracellular chloride ions above the electrochemical equilibrium and opening of membrane chloride channels results in a depolarizing chloride current that contributes to the activation of sensory fibers. Lee et al. (Lee et al. 2005) have shown that bradykinin-induced depolarization of airway afferent nerves is inhibited by niflumic acid, a selective inhibitor of calcium-activated chloride channels, which reduces citric acid induced cough in anaesthetized guinea pigs (Mazzone and McGovern 2006).

5.7 γ-Aminobutyric Acid Receptors

γ-Aminobutyric acid (GABA) is an inhibitory neurotransmitter present in the central and peripheral nervous system, and the development of selective GABA receptor agonists and antagonists has led to the discovery of potential antitussive effects of these compounds. $GABA_B$ receptors modulate cholinergic and tachykininergic nerves (Chapman et al. 1993). Capsaicin-induced cough is inhibited in the conscious guinea pig by the $GABA_B$ receptor agonists, baclofen and 3-aminopropylphosphinic acid, an effect mediated through a central stimulation of $GABA_B$ receptors (Bolser et al. 1994). In human volunteers, there was an inhibitory effect of baclofen on capsaicin-induced cough (Dicpinigaitis et al. 1998), but only a small beneficial effect was shown in two patients with chronic cough (Dicpinigaitis and Dobkin 1997).

5.8 Local Anesthetics

Afferent nerves need voltage-gated sodium channels for action potential conduction from the nerve terminals to the central nervous system. Lignocaine and bupivicaine are local anesthetics by virtue of their sodium channel blocking activity, and block the cough response when delivered to the upper and lower airways by aerosol (Hansson et al. 1994). These agents also dampen upper-airway protective reflexes, and may occasionally induce bronchoconstriction, and therefore need to be used with care. Lignocaine inhalation inhibits cough at doses that do not affect reflex bronchoconstriction (Choudry et al. 1990). Lignocaine aerosol was also shown to be quickly effective in 62 patients with COPD (Chong et al. 2005); however, nebulized terbutaline was equally effective. Their duration of action is only of the order of 30 min or less. At present, these are usually reserved for the severest persistent coughers (Howard et al. 1977).

There has been some work in identifying specific sites of action of local anesthetics and of novel anesthetic-like molecules with antitussive activity (Carr 2006). Mexilitine, which has been shown to inhibit capsaicin cough response modestly in human (Fujimura et al. 2000), specifically inhibits action potential formation in guinea-pig tracheal mechanosensitive fibers which are cough receptors at concentrations that do not block action potential conduction along the sensory nerve axon (Carr 2006). This suggests that cough receptor nerve terminals express Na^+ channels that have properties that are different from those of their axons. In addition, there is evidence that myelinated cough afferents may have nonmyelinated terminals (Coleridge and Coleridge 1986), which may also explain these observations.

The quartenary ammonium compound RSD931, which inhibits spontaneous and histamine-evoked discharges from airway RARs but activates pulmonary C-fibers, reduces citric acid induced and capsaicin-induced cough in guinea pigs and rabbits (Adcock et al. 2003). However, this profile of effect was dissimilar to that of lignocaine, which was more generally inhibitory.

Other subtypes of Na^+ channels such as the tetrodotoxin (TTX)-resistant Na^+ channels are also expressed by airway afferent nerves. Capsaicin-sensitive neurones are sensitized by inflammatory mediators, such as PGE_2, an effect partly mediated by an increase in TTX-resistant Na^+ currents (Kwong and Lee 2005). These neurones are inactivated in the normal airways, but are recruited in inflamed airways to contribute to cough sensitization (Mazzone et al. 2005). Therefore, selective inhibitors of TTX-resistant Na^+ channels may be useful as cough suppressants in chronic cough.

5.9 5-Hydroxytryptamine

5-Hydroxytryptamine (5-HT), or serotonin, receptors are present in nodose ganglia and facilitate neural transmission in visceral C-fiber afferents. Activation of central 5-HT pathways mediates the antitussive activity of opiates in experiments performed in mice (Mazzone 2004). Infusion of 5-HT and of 5-hydroxytryptophan reduced cough responses to chloride-deficient solutions, but had no effect on capsaicin responses (Stone et al. 1993). In guinea pigs, an agonist at the 5-HT_{1A} receptor, 8-hydroxy-2-(di-*n*-propylamino)tetralin, showed both excitatory and inhibitory influences on the cough reflex to capsaicin, which may be initiated via postsynaptic stimulation of central serotoninergic neurones and/or coexisting peripheral sites (Undem and Carr 2001).

5.10 Other Potential Antitussives

Theobromine has been proposed as an antitussive after demonstration of an inhibitory effect on nerve conduction in the vagus nerves and on capsaicin cough responses in normal individuals (Usmani et al. 2005). No studies have been performed in cough patients. In a systematic review of methylxanthines in the treatment of cough in children, no conclusive evidence of an effect was found (Chang et al. 2005).

Because of the inflammatory response one observes in chronic cough and the plausibility that inflammatory factors may enhance the cough reflex, it is possible that certain specific anti-inflammatories may be good antitussives.

6 Evaluation and Clinical Use of Symptomatic Cough Suppressants

The recent American College of Chest Physicians' recommendations regarding the use of cough suppressants are summarized in Table 4 (Bolser et al. 2006a). The recommendations concentrate on patients with chronic and/or acute bronchitis with an upper respiratory tract infection. Cough suppressants are not recommended for cough resulting from upper respiratory tract infections, but there is a recommendation for peripheral and central cough suppressants in patients with acute or chronic bronchitis. Very little is mentioned regarding patients with chronic cough without bronchitis and the use of cough suppressants. Neither do the European guidelines on cough (Morice et al. 2004) specifically broach this issue, when clearly patients with "idiopathic" cough or patients who still cough despite primary treatment of the cause(s) and who suffer deterioration in quality of life would need relief from their cough. On the other hand, currently available cough suppressants are not very effective and have a low therapeutic ratio, so it may not be possible to make a clear recommendation.

Table 4 Recommendations of the American College of Chest Physicians' guidelines (2006) for symptomatic cough suppressants

1. In patients with chronic or acute bronchitis, peripheral cough suppressants, such as levodropropizine and moguisteine, are recommended for the short-term symptomatic relief of coughing.*Level of evidence, good; benefit, substantial; grade of recommendation, A (STRONG)*
2. In patients with cough due to URI, peripheral cough suppressants have limited efficacy and are not recommended for this use. *Level of evidence, good; benefit, none; grade of recommendation, D (NEGATIVE)*
3. In patients with chronic bronchitis, central cough suppressants, such as codeine and dextromethorphan, are recommended for the short-term symptomatic relief of coughing. *Level of evidence, fair; benefit, intermediate; grade of recommendation, B (MODERATE)*
4. In patients with cough due to URI, central cough suppressants have limited efficacy for symptomatic relief and are not recommended for this use. *Level of evidence, good; benefit, none; grade of recommendation, D (NEGATIVE)*

URI upper respiratory tract infection

With the advent of potential cough suppressants, randomized placebo-controlled, double-blind studies should be performed. So what about the type of cough patient for clinical trials of symptomatic antitussives? These should include patients with a persistent cough despite adequate treatment of any associated causes or without associated cause. Tools for the assessment of cough been developed and validated including cough-specific quality-of-life questionnaires, objective cough counts, and cough challenges that can be used in clinical trial environments (Yousaf and Birring 2008; Chung et al. 2006). Cough count is not the only aspect of cough that is important because the perceived severity by the patient is needed, and validated quality-of-life questionnaires are now available to assess this.

Although for many patients with chronic cough there may be adequate control with a careful diagnosis and specific treatments directed towards associated cause(s), for others, symptomatic cough suppressants may be needed. We need better treatments of the associated cause(s) and more efficacious symptomatic antitussives.

References

Adams R, Hosie J, James I, Khong T, Smith I et al (1993) Antitussive activity and tolerability of moguisteine in patients with acute cough: a randomized, double-blind, placebo-controlled study. Adv Ther 10:263–271

Adcock JJ, Schneider C, Smith TW (1988) Effects of codeine, morphine and a novel opioid pentapeptide BW443C, on cough, nociception and ventilation in the unanaesthetized guinea-pig. Br J Pharmacol 93:93–100

Adcock JJ, Douglas GJ, Garabette M, Gascoigne M, Beatch G, Walker M, Page CP (2003) RSD931, a novel anti-tussive agent acting on airway sensory nerves. Br J Pharmacol 138: 407–416

Advenier C, Giraud V, Naline E, Villain P, Emonds-Alt X (1992) Antitussive effect of SR 48968, a non-peptide tachykinin NK2 receptor antagonist. Eur J Pharmacol 250:169–173

Allegra L, Bossi R (1988) Clinical trials with the new antitussive levodropropizine in adult bronchitic patients. Arzneimittelforschung 38:1163–1166

Aversa C, Cazzola M, Clini , Dal Negro R, Maiorano V, Tana F, Allegra L (1993) Clinical trial of the efficacy and safety of moguisteine in patients with cough associated with chronic respiratory diseases. Drugs Exp Clin Res 19:273–279

Aylward M, Maddock J, Dewland P (1980) Clinical evaluation of acetylcysteine in the treatment of patients with chronic obstructive bronchitis: a balanced double-blind trial with placebo control. Eur J Respir Dis Suppl 111:81–89

Aylward M, Maddock J, Davies DE, Protheroe DA, Leideman T (1984) Dextromethorphan and codeine: comparison of plasma kinetics and antitussive effects. Eur J Respir Dis 65:283–291

Barbee RA, Halonen M, Kaltenborn WT, Burrows B (1991) A longitudinal study of respiratory symptoms in a community population sample. Correlations with smoking, allergen skin-test reactivity, and serum IgE. Chest 99:20–26

Bastian RW, Vaidya AM, Delsupehe KG (2006) Sensory neuropathic cough: a common and treatable cause of chronic cough. Otolaryngol Head Neck Surg 135:17–21

Bergren DR (1997) Sensory receptor activation by mediators of defense reflexes in guinea-pig lungs. Respir Physiol 108:195–204

Bock MG, Longmore J (2000) Bradykinin antagonists: new opportunities. Curr Opin Chem Biol 4:401–406

Bolser DC (2006a) Cough suppressant and pharmacologic protussive therapy: ACCP evidence-based clinical practice guidelines. Chest 129:238S–249S

Bolser DC (2006b) Current and future centrally acting antitussives. Respir Physiol Neurobiol 152:349–355

Bolser DC, Degennaro FC, O'Reilly S, Chapman RW, Kreutner W, Egan RW, Hey JA (1994) Peripheral and central sites of action of GABA-B agonists to inhibit the cough reflex in the cat and guinea pig. Br J Pharmacol 113:1344–1348

Bolser DC, McLeod RL, Tulshian DB, Hey JA (2001) Antitussive action of nociceptin in the cat. Eur J Pharmacol 430:107–111

Braun-Fahrlander C, Vuille JC, Sennhauser FH, Neu U, Kunzle T, Grize L, Gassner M, Minder C, Schindler C, Varonier HS, Wuthrich B (1997) Respiratory health and long-term exposure to air pollutants in Swiss schoolchildren. SCARPOL Team. Swiss Study on Childhood Allergy and Respiratory Symptoms with Respect to Air Pollution, Climate and Pollen. Am J Respir Crit Care Med 155:1042–1049

Brightling CE, Ward R, Wardlaw AJ, Pavord ID (2000) Airway inflammation, airway responsiveness and cough before and after inhaled budesonide in patients with eosinophilic bronchitis. Eur Respir J 15:682–686

Calignano A, Katona I, Desarnaud F, Giuffrida A, La Rana G, Mackie K, Freund TF, Piomelli D (2000) Bidirectional control of airway responsiveness by endogenous cannabinoids. Nature 408:96–101

Carr MJ (2006) Influence of mexiletine on action potential discharge and conduction in nodose Adelta afferent neurons innervating guinea pig isolated trachea. Pulm Pharmacol Ther 19: 258–263

Catena E, Daffonchio L (1997) Efficacy and tolerability of levodropropizine in adult patients with non-productive cough. Comparison with dextromethorphan. Pulm Pharmacol Ther 10:89–96

Caterina MJ, Schumacher MA, Tominaga M, Rosen TA, Levine JD, Julius D (1997) The capsaicin receptor: A heat-activated ion channel in the pain pathway. Nature 389:816–824

Chang AB, Halstead RA, Petsky HL (2005) Methylxanthines for prolonged non-specific cough in children. Cochrane. Database Syst Rev CD005310

Chang AB, Lasserson TJ, Kiljander TO, Connor FL, Gaffney JT, Garske LA (2006) Systematic review and meta-analysis of randomised controlled trials of gastro-oesophageal reflux interventions for chronic cough associated with gastro-oesophageal reflux. Br Med J 332:11–17

Chapman RW, Hey JA, Rizzo CA, Bolser DC (1993) GABAB receptors in the lung. Trends Pharmacol Sci 14:26–29
Chiou LC, Liao YY, Fan PC, Kuo PH, Wang CH, Riemer C, Prinssen EP (2007) Nociceptin/orphanin FQ peptide receptors: pharmacology and clinical implications. Curr Drug Targets 8:117–135
Chong CF, Chen CC, Ma HP, Wu YC, Chen YC, Wang TL (2005) Comparison of lidocaine and bronchodilator inhalation treatments for cough suppression in patients with chronic obstructive pulmonary disease. Emerg Med J 22:429–432
Choudry NB, Fuller RW, Anderson N, Karlsso JA (1990) Separation of cough and reflex bronchoconstriction by inhaled local anaesthetics. Eur Respir J 3:579–583
Choudry NB, Gray SJ, Posner J, Fuller RW (1991) The effect of 443C81, a mu opioid receptor agonist, on the response to inhaled capsaicin in healthy volunteers. Br J Clin Pharmacol 32: 633–636
Chung KF (2005) Drugs to suppress cough. Expert Opin Investig Drugs 14:19–27
Chung KF (2006) Measurement of cough. Respir Physiol Neurobiol 152:329–339
Chung KF (2007) Effective antitussives for the cough patient: An unmet need. Pulm.Pharmacol Ther 20:438–445
Chung KF, Pavord ID (2008) Prevalence, pathogenesis and causes of chronic cough. Lancet 371:1364–74
Chung KF, Hew M, Score J, Jones AV, Reiter A, Cross NC, Bain BJ (2006) Cough and hypereosinophilia due to FIP1L1-PDGFRA fusion gene with tyrosine kinase activity. Eur Respir J 27:230–232
Clemens CJ, Taylor JA, Almquist JR, Quinn HC, Mehta A, Naylor GS (1997) Is an antihistamine-decongestant combination effective in temporarily relieving symptoms of the common cold in preschool children? J Pediatr 130:463–466
Coleridge HM, Coleridge JCG (1986) Reflexes evoked from the respiratory tract and lungs. In: Cherniack NS, Widdicombe JG (eds) Handbook of physiology section 3: The Respiratory System, vol II, Control of Breathing Part 1. American Physiological Society, Bethesda, pp 395–429
Committee on Drugs (1997) Use of codeine- and dextromethorphan-containing cough remedies in children. Pediatrics 99:918–920
Cook DG, Strachan DP (1999) Health effects of passive smoking-10: Summary of effects of parental smoking on the respiratory health of children and implications for research. Thorax 54:357–366
Corboz MR, Rivelli MA, Egan RW, Tulshian D, Matasi J, Fawzi AB, Benbow L, Smith-Torhan A, Zhang H, Hey JA (2000) Nociceptin inhibits capsaicin-induced bronchoconstriction in isolated guinea pig lung. Eur J Pharmacol 402:171–179
Coyle MA, Keenan DB, Henderson LS, Watkins ML, Haumann BK, Mayleben DW, Wilson MG (2005) Evaluation of an ambulatory system for the quantification of cough frequency in patients with chronic obstructive pulmonary disease. Cough 1:3
Cullinan P (1992) Persistent cough and sputum: prevalence and clinical characteristics in south east England. Respir Med 86:143–149
Daoui S, Cognon C, Naline E, Emonds-Alt X, Advenier C (1998) Involvement of tachykinin NK3 receptors in citric acid-induced cough and bronchial responses in guinea pigs. Am J Respir Crit Care Med 158:42–48
Dicpinigaitis PV (2006) Current and future peripherally-acting antitussives. Respir Physiol Neurobiol 152:356–362
Dicpinigaitis PV, Dobkin JB (1997) Antitussive effect of the GABA-agonist baclofen. Chest 111:996–999
Dicpinigaitis PV, Dobkin JB (1999) Effect of zafirlukast on cough reflex sensitivity in asthmatics. J Asthma 36:265–270
Dicpinigaitis PV, Rauf K (1998) Treatment of chronic, refractory cough with baclofen. Respiration 65:86–88
Dicpinigaitis PV, Dobkin JB, Rauf K, Aldrich TK (1998) Inhibition of capsaicin-induced cough by the gamma-aminobutyric acid agonist baclofen. J Clin Pharmacol 38:364–367

Doherty MJ, Mister R, Pearson MG, Calverley PM (2000) Capsaicin responsiveness and cough in asthma and chronic obstructive pulmonary disease. Thorax 55:643–649

Eccles R (2002) The powerful placebo in cough studies? Pulm Pharmacol Ther 15:303–308

Eccles R, Morris S, Jawad M (1992) Lack of effect of codeine in the treatment of cough associated with acute upper respiratory tract infection. J Clin Pharm Ther 17:175–180

Eddy NB, Friebel H, Hahn KJ, Halbach H (1969a) Codeine and its alternates for pain and cough relief. 4. Potential alternates for cough relief. Bull World Health Organ 40:639–719

Eddy NB, Friebel H, Hahn KJ, Halbach H (1969b) Codeine and its alternates for pain and cough relief. 3. The antitussive action of codeine–mechanism, methodology and evaluation. Bull World Health Organ 40:425–454

Empey DW, Laitinen LA, Young GA, Bye CE, Hughes DT (1979) Comparison of the antitussive effects of codeine phosphate 20 mg, dextromethorphan 30 mg and noscapine 30 mg using citric acid-induced cough in normal subjects. Eur J Clin Pharmacol 16:393–397

Fahy JV, Wong HH, Geppetti P, Reis JM, Harris SC, Maclean DB, Nadel JA, Boushey HA (1995) Effect of an NK1 antagonist (CP-99,994) on hypertonis saline- induced bronchoconstriction and cough in male asthmatic subjects. Am J Resp Crit Care Med 152:879–884

Featherstone RL, Parry JE, Evans DM, Jones DM, Olsson H, Szelke M, Church MK (1996) Mechanism of irritant-induced cough: studies with a kinin antagonist and a kallikrein inhibitor. Lung 174:269–275

Fischer A, Forssmann WG, Undem BJ (1998) Nociceptin-induced inhibition of tachykinergic neurotransmission in guinea pig bronchus. J Pharmacol Exp Ther 285:902–907

Ford AC, Forman D, Moayyedi P, Morice AH (2006) Cough in the community: A cross sectional survey and the relationship to gastrointestinal symptoms. Thorax 61:975–979

Ford CN (2005) Evaluation and management of laryngopharyngeal reflux. J Am Med Assoc 294:1534–1540

Fox AJ, Urban L, Barnes PJ, Dray A (1995) Effects of casazepine against capsaicin- and proton-evoked excitation of single airway c-fibres and vagus nerve from the guinea-pig. Neurosci Lett 15:421–428

Fox AJ, Barnes PJ, Venkatesan P, Belvisi MG (1997) Activation of large conductance potassium channels inhibits the afferent and efferent function of airway sensory nerves in the guinea pig. J Clin Invest 99:513–519

Freestone C, Eccles R (1997) Assessment of the antitussive efficacy of codeine in cough associated with common cold. J Pharm Pharmacol 49:1045–1049

Fujimura M, Kamio Y, Myou S, Hashimoto T (2000) Effect of oral mexiletine on the cough response to capsaicin and tartaric acid. Thorax 55:126–128

Garcia-Martinez C, Humet M, Planells-Cases R, Gomis A, Caprini M, Viana F, De La PE, Sanchez-Baeza F, Carbonell T, De Felipe C, Perez-Paya E, Belmonte C, Messeguer A, Ferrer-Montiel A (2002) Attenuation of thermal nociception and hyperalgesia by VR1 blockers. Proc Natl Acad Sci USA 99:2374–2379

Gehring U, Cyrys J, Sedlmeir G, Brunekreef B, Bellander T, Fischer P, Bauer CP, Reinhardt D, Wichmann HE, Heinrich J (2002) Traffic-related air pollution and respiratory health during the first 2 yrs of life. Eur Respir J 19:690–698

Geppetti P, Tognetto M, Trevisani M, Amadesi S, Bertrand C (1999) Tachykinins and kinins in airway allergy. Expert Opin Investig Drugs 8:947–956

Girard V, Naline E, Vilain P, Emonds-Alt X, Advenier C (1995) Effect of the two tachykinin antagonists, SR 48968 and SR 140333, on cough induced by citric acid in the unanaesthetized guinea pig. Eur Respir J 8:1110–1114

Gordon R, Gordon RJ, Sofia D (1976) Antitussive activity of some naturally occurring cannabinoids in anesthetized cats. Eur J Pharmacol 35:309–313

Gracely RH, Undem BJ, Banzett RB (2007) Cough, pain and dyspnoea: Similarities and differences. Pulm Pharmacol Ther 20:433–437

Groneberg DA, Niimi A, Dinh QT, Cosio B, Hew M, Fischer A, Chung KF (2004) Increased expression of transient receptor potential Vanilloid-1 in airway nerves of chronic cough. Am J Respir Crit Care Med 170(12):1276–1280

Hansson L, Midgren B, Karlsson JA (1994) Effects of inhaled lignocaine and adrenaline on capsaicin-induced cough in humans. Thorax 49:1166–1168
Hautmeyers E, Gosselink R, Gayan-Ramirez G, Decramer M (1999) Effects of drugs on mucus clearance. Eur Respir J 14:452–467
Hay DW, Giardina GA, Griswold DE, Underwood DC, Kotzer CJ, Bush B, Potts W, Sandhu P, Lundberg D, Foley JJ, Schmidt DB, Martin LD, Kilian D, Legos JJ, Barone FC, Luttmann MA, Grugni M, Raveglia LF, Sarau HM (2002) Nonpeptide tachykinin receptor antagonists. III. SB 235375, a low central nervous system-penetrant, potent and selective neurokinin-3 receptor antagonist, inhibits citric acid-induced cough and airways hyper-reactivity in guinea pigs. J Pharmacol Exp Ther 300:314–323
Heino M, Juntunen-Backman K, Leijala M, Rapola J, Laitinen LA (1990) Bronchial epithelial inflammation in children with chronic cough after early lower respiratory tract illness. Am Rev Respir Dis 141:428–432
Howard P, Cayton RM, Brennan SR, Anderson PB (1977) Lignocaine aerosol and persistent cough. Br J Dis Chest 71:19–24
Hutton N, Wilson MH, Mellits ED, Baumgardner R, Wissow LS, Bonuccelli C, Holtzman NA, DeAngelis C (1991) Effectiveness of an antihistamine-decongestant combination for young children with the common cold: a randomized, controlled clinical trial. J Pediatr 118:125–130
Irwin RS (2006) Introduction to the diagnosis and management of cough: ACCP evidence-based clinical practice guidelines. Chest 129:25S-27S
Irwin RS, Carrao WM, Pratter MR (1981) Chronic persistent cough in the adult: The spectrum and frequency of causes and successful outcome of specific therapy. Am Rev Respir Dis 123: 413–417
Jacome DE (1985) La toux diabolique: neurogenic tussive crisis. Postgrad Med J 61:515–516
Janson C, Chinn S, Jarvis D, Burney P (2001) Determinants of cough in young adults participating in the European community respiratory health survey. Eur Respir J 18:647–654
Jeyakumar A, Brickman TM, Haben M (2006) Effectiveness of amitriptyline versus cough suppressants in the treatment of chronic cough resulting from postviral vagal neuropathy. Laryngoscope 116:2108–2112
Jia Y, Wang X, Aponte SI, Rivelli MA, Yang R, Rizzo CA, Corboz MR, Priestley T, Hey JA (2002) Nociceptin/orphanin FQ inhibits capsaicin-induced guinea-pig airway contraction through an inward-rectifier potassium channel. Br J Pharmacol 135:764–770
Kamei J, Tanihara H, Kasuya Y (1990) Antitussive effects of two specific k-opioid agonists, U-50, 448H and U-62, 066E, in rats. Eur J Pharmacol 199:281–286
Kohno S, Ishida T, Uchida Y, Kishimoto H, Sasaki H, Shioya T, Tokuyama K, Niimi A, Nishi K, Fujimura M, Matsuse H, Suzaki H (2006) The Japanese respiratory society guidelines for management of cough. Respirology 11(Suppl 4):S135–S186
Kollarik M, Undem BJ (2006) Sensory transduction in cough-associated nerves. Respir Physiol Neurobiol 152:243–254
Korppi M, Laurikainen K, Pietikainen M, Silvasti M (1991) Antitussives in the treatment of acute transient cough in children. Acta Paediatr Scand 80:969–971
Kotzer CJ, Hay DW, Dondio G, Giardina G, Petrillo P, Underwood DC (2000) The antitussive activity of delta-opioid receptor stimulation in guinea pigs. J Pharmacol Exp Ther 292: 803–809
Kwong K, Lee LY (2005) Prostaglandin E2 potentiates a TTX-resistant sodium current in rat capsaicin-sensitive vagal pulmonary sensory neurones. J Physiol 564:437–450
Lalloo UG, Fox AJ, Belvisi MG, Chung KF, Barnes PJ (1995) Inhibition by capsazepine of cough induced by capsaicin and citric acid, but not by hypertonic saline in awake guinea-pigs. J Appl Physiol 79:1082–1087
Lavezzo A, Melillo G, Clavenna G, Omini C (1992) Peripheral site of action of levodropropizine in experimentally-induced cough: role of sensory neuropeptides. Pulm Pharmacol 5:143–147
Lee MG, MacGlashan DW Jr, Undem BJ (2005) Role of chloride channels in bradykinin-induced guinea pig airway vagal C-fibre activation. J Physiol 566:205–212
Lee MG, Undem BJ, Brown C, Carr MJ (2006) Effect of nociceptin in acid-evoked cough and airway sensory nerve activation in guinea pigs. Am J Respir Crit Care Med 173:271–275

Lee PCL, Jawad MS, Eccles R (2000) Antitussive efficacy of dextromethorphan in cough associated with acute upper respiratory tract infection. J Pharm Pharmacol 52:1137–1142
Leung SY, Niimi A, Williams AS, Nath P, Blanc FX, Din QT, Chung KF (2007) Inhibition of citric acid- and capsaicin-induced cough by novel TRPV-1 antagonist, V112220, in guinea-pig. Cough 3:10
Lundback B, Nystrom L, Rosenhall L, Stjernberg N (1991) Obstructive lung disease in northern Sweden: respiratory symptoms assessed in a postal survey. Eur Respir J 4:257–266
Matthys H, Bleicher B, Bleicher U (1983) Dextromethorphan and codeine: objective assessment of antitussive activity in patients with chronic cough. J Int Med Res 11:92–100
Mazzone SB (2004) Sensory regulation of the cough reflex. Pulm Pharmacol Ther 17:361–368
Mazzone SB, McGovern AE (2006). Na + -K + -2Cl- cotransporters and Cl- channels regulate citric acid cough in guinea pigs. J Appl Physiol 101:635–643
Mazzone SB, Mori N, Canning BJ (2005) Synergistic interactions between airway afferent nerve subtypes regulating the cough reflex in guinea pigs. J Physiol 569:559–573
Mazzone SB, McLennan L, McGovern AE, Egan GF, Farrell MJ (2007) Representation of capsaicin-evoked urge-to-cough in the human brain using functional magnetic resonance imaging. Am J Respir Crit Care Med 176:327–332
McLeod RL, Parra LE, Mutter JC, Erickson CH, Carey GJ, Tulshian DB, Fawzi AB, Smith-Torhan A, Egan RW, Cuss FM, Hey JA (2001) Nociceptin inhibits cough in the guinea-pig by activation of ORL(1) receptors. Br J Pharmacol 132:1175–1178
McLeod RL, Jia Y, Fernandez X, Parra LE, Wang X, Tulshian DB, Kiselgof EJ, Tan Z, Fawzi AB, Smith-Torhan A, Zhang H, Hey JA (2004) Antitussive profile of the NOP agonist Ro-64–6198 in the guinea pig. Pharmacology 71:143–149
McLeod RL, Fernandez X, Correll CC, Phelps TP, Jia Y, Wang X, Hey JA (2006) TRPV1 antagonists attenuate antigen-provoked cough in ovalbumin sensitized guinea pigs. Cough 2:10
McLeod RL, Correll CC, Jia Y, Anthes JC (2007a) TRPV1 antagonists as potential antitussive agents. Lung 186(Suppl 1):S59–65
McLeod RL, Jia Y, McHugh NA, Fernandez X, Mingo GG, Wang X, Parra LE, Chen J, Brown D, Bolser DC, Kreutner W, Hey JA (2007b) Sulfur-dioxide exposure increases TRPV1-mediated responses in nodose ganglia cells and augments cough in guinea pigs. Pulm.Pharmacol Ther 20:750–757
Mintz S, Lee JK (2006) Gabapentin in the treatment of intractable idiopathic chronic cough: case reports. Am J Med 119:e13–e15
Morice AH, Marshall AE, Higgins KS, Grattan TJ (1994) Effect of inhaled menthol on citric acid induced cough in normal subjects. Thorax 49:1024–1026
Morice AH, Fontana GA, Sovijarvi AR, Pistolesi M, Chung KF, Widdicombe JG, O'Connell F, Geppetti P, Gronke L, De Jongste J, Belvisi M, Dicpinigaitis P, Fischer A, McGarvey L, Fokkens WJ, Kastelik J (2004) The diagnosis and management of chronic cough. Eur Respir J 24: 481–492
Morice AH, McGarvey L, Pavord I (2006) Recommendations for the management of cough in adults. Thorax 61(Suppl 1):i1–i24
Morice AH, Menon MS, Mulrennan SA, Everett CF, Wright C, Jackson J, Thompson R (2007) Opiate therapy in chronic cough. Am J Respir Crit Care Med 175:312–315
Morice AH, McGarvey L (2008) Clinical cough 1: Therapeutic treatments and management of chronic cough. In: Chung KF, Widdicombe JG (eds) Handbook of pharmacology. Pharmacology and therapeutics of cough. Springer, Berlin, this issue
Morita K, Kamei J (2000) Involvement of ATP-sensitive K(+) channels in the anti-tussive effect of moguisteine. Eur J Pharmacol 395:161–164
Morita K, Kamei J (2003) Antitussive effect of WIN 55212–2, a cannabinoid receptor agonist. Eur J Pharmacol 474:269–272
Mund E, Christensson B, Larsson K, Gronneberg R (2001) Sex dependent differences in physiological ageing in the immune system of lower airways in healthy non-smoking volunteers: study of lymphocyte subsets in bronchoalveolar lavage fluid and blood. Thorax 56:450–455

Mund E, Christensson B, Gronneberg R, Larsson K (2005) Noneosinophilic CD4 lymphocytic airway inflammation in menopausal women with chronic dry cough. Chest 127:1714–1721

Mutoh T, Bonham AC, Joad JP (2000) Substance P in the nucleus of the solitary tract augments bronchopulmonary C fiber reflex output. Am J Physiol Regul Integr Comp Physiol 279:R1215–R1223

Niimi A, Torrego A, Nicholson AG, Cosio BG, Oates TB, Chung KF (2005) Nature of airway inflammation and remodeling in chronic cough. J Allergy Clin Immunol 116:565–570

Novitsky YW, Zawacki JK, Irwin RS, French CT, Hussey VM, Callery MP (2002) Chronic cough due to gastroesophageal reflux disease: efficacy of antireflux surgery. Surg Endosc 16:567–571

O'Hara J, Jones NS (2005) The aetiology of chronic cough: a review of current theories for the otorhinolaryngologist. J Laryngol Otol 119:507–514

Parvez L, Vaidya M, Sakhardande A, Subburaj S, Rajagopalan TG (1996) Evaluation of antitussive agents in man. Pulm Pharmacol 9:299–308

Patel HJ, Birrell MA, Crispino N, Hele DJ, Venkatesan P, Barnes PJ, Yacoub MH, Belvisi MG (2003) Inhibition of guinea-pig and human sensory nerve activity and the cough reflex in guinea-pigs by cannabinoid (CB2) receptor activation. Br J Pharmacol 140:261–268

Paul IM, Yoder KE, Crowell KR, Shaffer ML, McMillan HS, Carlson LC, Dilworth DA, Berlin CM Jr (2004) Effect of dextromethorphan, diphenhydramine, and placebo on nocturnal cough and sleep quality for coughing children and their parents. Pediatrics 114:e85–e90

Paul IM, Beiler J, McMonagle A, Shaffer ML, Duda L, Berlin CM Jr (2007) Effect of honey, dextromethorphan, and no treatment on nocturnal cough and sleep quality for coughing children and their parents. Arch Pediatr Adolesc Med 161:1140–1146

Pavesi L, Subburaj S, Porter-Shaw K (2001) Application and validation of a computerized cough acquisition system for objective monitoring of acute cough: a meta-analysis. Chest 120:1121–1128

Pavord IM, Chung KF (2008) Chronic cough: treatment and management. Lancet

Pierse N, Rushton L, Harris RS, Kuehni CE, Silverman M, Grigg J (2006) Locally generated particulate pollution and respiratory symptoms in young children. Thorax 61:216–220

Premkumar LS, Ahern GP (2000) Induction of vanilloid receptor channel activity by protein kinase C. Nature 408:985–990

Rouget C, Cui YY, D'Agostino B, Faisy C, Naline E, Bardou M, Advenier C (2004) Nociceptin inhibits airway microvascular leakage induced by HCl intra-oesophageal instillation. Br J Pharmacol 141:1077–1083

Sant'Ambrogio FB, Sant'Ambrogio G, Anderson JW (1993) Effect of furosemide on the response of laryngeal receptors to low-chloride solutions. Eur Respir J 6:1151–1155

Schroeder K, Fahey T (2004) Over-the-counter medications for acute cough in children and adults in ambulatory settings. Cochrane Database Syst Rev CD001831

Sevelius H, Colmore JP (1966) Objective assessment of antitussive agents in patients with chronic cough. J New Drugs 6:216–233

Sevelius H, McCoy JF, Colmore JP (1971) Dose response to codeine in patients with chronic cough. Clin Pharmacol Ther 12:449–455

Shah S, Page CP, Spina D (1998) Nociceptin inhibits non-adrenergic non-cholinergic contraction in guinea-pig airway. Br J Pharmacol 125:510–516

Shams H, Daffonchio L, Scheid P (1996) Effects of levodropropizine on vagal afferent C-fibres in the cat. Br J Pharmacol 117:853–858

Shoblock JR (2007) The pharmacology of Ro 64–6198, a systemically active, nonpeptide NOP receptor (opiate receptor-like 1, ORL-1) agonist with diverse preclinical therapeutic activity. CNS Drug Rev 13:107–136

Simonyan K, Saad ZS, Loucks TM, Poletto CJ, Ludlow CL (2007) Functional neuroanatomy of human voluntary cough and sniff production. Neuroimage 37:401–409

Smith J, Owen E, Earis J, Woodcock A (2006) Effect of codeine on objective measurement of cough in chronic obstructive pulmonary disease. J Allergy Clin Immunol 117:831–835

Smith MB, Feldman W (1993) Over-the-counter cold medications. A critical review of clinical trials between 1950 and 1991. J Am Med Assoc 269:2258–2263

Stone RA, Worsdell YM, Fuller RW, Barnes PJ (1993) Effects of 5-hydroxytryptamine and 5-hydroxytryptophan infusion on the human cough reflex. J Appl Physiol 74:396–401

Sudo T, Hayashi F, Nishino T (2000) Responses of tracheobronchial receptors to inhaled furosemide in anesthetized rats. Am J Respir Crit Care Med 162:971–975

Szallasi A, Blumberg PM (1999) Vanilloid (Capsaicin) receptors and mechanisms. Pharmacol Rev 51:159–212

Taylor JA, Novack AH, Almquist JR, Rogers JE (1993) Efficacy of cough suppressants in children. J Pediatr 122:799–802

Torrego A, Haque RA, Nguyen LT, Hew M, Carr DH, Wilson R, Chung KF (2006) Capsaicin cough sensitivity in bronchiectasis. Thorax 61:706–709

Trevisani M, Milan A, Gatti R, Zanasi A, Harrison S, Fontana G, Morice AH, Geppetti P (2004) Antitussive activity of iodo-resiniferatoxin in guinea pigs. Thorax 59:769–772

Ueno S, Saitoh A, Kawai K, Natsume K, Sakami S, Maeda M, Ito H, Ohtaki M, Morita K, Kamei J, Endoh T, Nagase H (2001) The rational drug design and antitussive effects of a novel opioid delta receptor antagonist TRK-851. Jpn J Pharmacol 85:S32–S34

Undem BJ, Carr MJ (2001) Pharmacology of airway afferent nerve activity. Respir Res 2:234–244

Usmani OS, Belvisi MG, Patel HJ, Crispino N, Birrell MA, Korbonits M, Korbonits D, Barnes PJ (2005) Theobromine inhibits sensory nerve activation and cough. FASEB J 19:231–233

Vedal S, Petkau J, White R, Blair J (1998) Acute effects of ambient inhalable particles in asthmatic and nonasthmatic children. Am J Respir Crit Care Med 157:1034–1043

Ventresca P, Nichol GM, Barnes PJ, Chung KF (1990) Inhaled furosemide inhibits cough induced by low chloride content solutions and not by capsaicin. Am Rev Respir Dis 142:143–146

von Muhlendahl KE, Scherf-Rahne B, Krienke EG, Baukloh G (1976) Codeine intoxication in childhood. Lancet 2:303–305

Waring JP, Lacayo L, Hunter J, Katz E, Suwak B (1995) Chronic cough and hoarseness in patients with severe gastroesophageal reflux disease. Diagnosis and response to therapy. Dig Dis Sci 40:1093–1097

West S, Brandon B, Stolley P, Rumrill R (1975) A review of antihistamines and the common cold. Pediatrics 56:100–107

Wise SK, Wise JC, DelGaudio JM (2006) Association of nasopharyngeal and laryngopharyngeal reflux with postnasal drip symptomatology in patients with and without rhinosinusitis. Am J Rhinol 20:283–289

Yousaf N, Birring SS (2008) Clinical cough III: measuring cough in the patient. In: Chung KF, Widdicombe JG (eds) Handbook of pharmacology: Pharmacology and therapeutics of cough. Springer, Berlin, this issue

Zemp E, Elsasser S, Schindler C, Kunzli N, Perruchoud AP, Domenighetti G, Medici T, Ackermann-Liebrich U, Leuenberger P, Monn C, Bolognini G, Bongard JP, Brandli O, Karrer W, Keller R, Schoni MH, Tschopp JM, Villiger B, Zellweger JP (1999) Long-term ambient air pollution and respiratory symptoms in adults (SAPALDIA study). The SAPALDIA Team. Am J Respir Crit Care Med 159:1257–1266

Zylicz Z, Krajnik M (2004) What has dry cough in common with pruritus? Treatment of dry cough with paroxetine. J Pain Symptom Manage 27:180–184

Index

Printing: Krips bv, Meppel, The Netherlands
Binding: Stürtz, Würzburg, Germany